Engineering Principles in Biotechnology

Engineering Principles in Biotechnology

Wei-Shou Hu

Department of Chemical Engineering and Materials Science
University of Minnesota, USA

This edition first published 2018
© 2018 by John Wiley & Sons Ltd

The right of Wei-Shou Hu to be identified as the author of this work has been asserted in accordance with law.

Registered Office(s)
John Wiley & Sons, Inc., 111 River Street, Hoboken, NJ 07030, USA
John Wiley & Sons Ltd, The Atrium, Southern Gate, Chichester, West Sussex, PO19 8SQ, UK

Editorial Office
The Atrium, Southern Gate, Chichester, West Sussex, PO19 8SQ, UK

For details of our global editorial offices, customer services, and more information about Wiley products visit us at www.wiley.com.

Wiley also publishes its books in a variety of electronic formats and by print-on-demand. Some content that appears in standard print versions of this book may not be available in other formats.

Library of Congress Cataloging-in-Publication Data

Names: Hu, Wei-Shou, 1951- author.
Title: Engineering fundamentals of biotechnology / by Wei-Shou Hu.
Description: First edition. | Hoboken, NJ, USA : John Wiley & Sons, Inc.,
    2018. | Includes bibliographical references and index. |
Identifiers: LCCN 2017016663 (print) | LCCN 2017018764 (ebook) | ISBN
    9781119159032 (pdf) | ISBN 9781119159049 (epub) | ISBN 9781119159025
    (cloth)
Subjects: | MESH: Bioengineering | Biotechnology
Classification: LCC R855.3 (ebook) | LCC R855.3 (print) | NLM QT 36 | DDC
    610.285–dc23
LC record available at https://lccn.loc.gov/2017016663

Cover design by Wiley
Cover images: (Background) © STILLFX/Gettyimages; (Illustrations) Courtesy of Wei-Shou Hu

Set in 10/12pt WarnockPro by SPi Global, Chennai, India

Printed in the UK

This book is dedicated to Jenny, Kenny, and Sheau-Ping.

Contents

Preface

Bioprocesses use microbial, plant, or animal cells and the materials derived from them, such as enzymes or DNA, to produce industrial biochemicals and pharmaceuticals. In the past two decades, the economic output from bioprocesses has increased drastically. This economic growth was the result of the translation of numerous discoveries to innovative technologies and manufactured products. The success has brought together numerous scientists and engineers of different disciplines to work together to break new ground. The task of taking biotechnological discoveries to a successful product or process requires a multidisciplinary team consisting of engineers and chemical and biological scientists to work synergistically. The success of a project, a team, or even a company in biotechnology often hinges on the ability of scientists and engineers of different specialties to work effectively together. This book has been written with this important characteristic of the bioprocess industry in mind. A major goal of the book is to give students the necessary vocabulary and critical engineering knowledge to excel in bioprocess technology.

This textbook is based on a biochemical engineering course that has been offered at the University of Minnesota for a number of years. The contents are intended for a semester course of about 14 weeks of three lecture-hours a week. Although the majority of the students taking this course are senior undergraduate and graduate students from chemical engineering and bioengineering, nearly one-third are graduate students from a life science background. An emphasis of the content and writing of the book is thus the fundamental engineering principles, the quantitative practice, and the accessibility of analysis for students of different backgrounds. The target audience of the book is not only students taking the biochemical engineering or bioprocess engineering courses given in chemical engineering or bioengineering programs but also students in biotechnology programs that are outside of the chemical engineering disciplines, especially in countries outside North America.

In writing this book, I assumed that the students have had at least one biology course, and have fundamental knowledge of carbohydrates, DNA, RNA, proteins, and other biomolecules, as is the case for most engineering students nowadays. Nevertheless, students from both engineering and life science backgrounds will encounter new vocabulary and new concepts that will help them in cross-disciplinary communication once they join the biotechnology workforce.

Chapters 1 and 2 give an overview of organisms, cells and their components, how they become the product, and what the bioprocesses that produce them look like. Chapters 3 and 4 use basic biochemical reactions, especially the energy metabolism pathways, to familiarize engineering students with analysis of biochemical systems and to introduce the concepts of material balance and reaction kinetics to students with a life science background. For all students, these chapters introduce them to kinetic analysis of binding reactions, gene expression, and cellular membrane transport.

Chapters 5 and 6 cover the quantitative description of cell growth and the steady-state behavior in a continuous bioreactor. This paves the way for dealing with different types of bioreactors. Chapters 7, 8, and 9 are the core of bioreactor engineering, dealing with subjects important to process development. These chapters draw upon extensive practical interactions with industry to make them more relevant to bioprocess technology.

The next three chapters – 10, 11, and 12 – discuss three segments of bioprocesses. Cell culture processes, the subject of Chapter 10, currently produce goods valued over US$100 billion per annum. After introducing cell culture processes, the evolution of biomanufacturing and its life cycle is discussed. Chapters 11 and 12 look to the future on cell-based therapy and on the technologies arising from synthetic biology. In dealing with stem cells, the kinetic description of cellular differentiation is also introduced, and in discussing synthetic pathways the importance of using a stoichiometric relationship to determine the maximum conversion yield is reiterated. The last two chapters, 13 and 14, highlight the bioseparation processes. The overall strategy and the key concepts of various unit operations in bioseparation are covered briefly in Chapter 13. Chapter 14 focuses on the basic quantitative understanding of chromatography.

Writing this book has been a long undertaking. Many of my former and current graduate students have helped in formulating the problem sets and the examples. In preparing the book, I also took ideas from many textbooks on biochemical engineering, especially *Bioprocess Engineering: Basic Concepts* by Shuler and Kargi; *Biochemical Engineering Fundamentals* by Bailey and Ollis; *Fermentation and Enzyme Technology* by Wang, Cooney, Demain, Dunnill, Humphrey, and Lilly; and *Biochemical Engineering* by Aiba, Humphrey, and Millis. I extend my gratitude to my colleagues at the University of Minnesota, especially Arnold G. Fredrickson, Friedrich Srienc, Edward Cussler, Ben Hackel, Kechun Zhang, Samira Azarin, Efie Kokkoli, Yiannis Kaznessis, and Prodromos Daoutidis, for their stimulating discussion that helped shape the book. Finally, I thank Kimberly Durand for her editorial devotion to this book.

About the Companion Website

Don't forget to visit the companion website for this book:

www.wiley.com/go/hu/engineering_fundamentals_of_biotechnology

There you will find valuable material designed to enhance your learning, including:

- Panels in PowerPoint

Scan this QR code to visit the companion website:

1

An Overview of Bioprocess Technology and Biochemical Engineering

1.1 A Brief History of Biotechnology and Biochemical Engineering

For thousands of years, humans have harnessed the metabolic activities of microbes. Microbes are important contributors to the generation of many foods, including bread, cheese, and pickled vegetables. Microbes are our unwitting life partners, but humans were not even aware of their existence until Antony van Leeuwenhoek discovered microorganisms. In the 1860s, Louis Pasteur discovered that microbes are responsible for lactic acid and the ethanol fermentation of sugar. He directly linked microbial metabolism to the synthesis of products.

Before the turn of the twentieth century, both researchers and food producers began to use microbes more purposefully. They were increasingly used to ferment milk and wine. This period is now considered to be the dawn of microbiology, or applied microbiology. In the early years, microbiology as a scientific field was closely linked to food microbiology, due to the positive roles of microbes in food fermentation as well as food spoilage and resulting illness.

1.1.1 Classical Biotechnology

In the beginning of the twentieth century, the use of microorganisms was extended beyond fermenting foods to the production of chemical compounds. Lactic acid was produced by fermentation using *Lactobacillus* spp. This marked the start of genuine industrial microbial fermentation. One of the first workhorses was the anaerobic bacterium *Clostridium acetobutylicum*, which was used to ferment sugar to acetone, ethanol, and butanol. Citric acid production using the mold *Aspergillus niger* also came about in the 1920s.

Penicillin production, the predecessor of modern fermentation, did not start until the 1940s. Alexander Fleming discovered penicillin after observing the inhibition of bacterial growth by a compound produced by a green mold. This observation led to the development of the bioprocess we know today. In nature, the compound was produced only at low concentrations. Thus, a large volume of culture was needed to generate the amount needed for clinical trials.

The demand for large quantities of penicillin led to the development of submerged culture in liquid medium, as opposed to the traditionally used agar-surface culture.

Engineering Principles in Biotechnology, First Edition. Wei-Shou Hu.
© 2018 John Wiley & Sons Ltd. Published 2018 by John Wiley & Sons Ltd.
Companion Website: www.wiley.com/go/hu/engineering_fundamentals_of_biotechnology

Along with the use of liquid-submerged culturing techniques came the search for a better medium composition and the utilization of corn steep liquor, a practice that continued for more than five decades.

The successful use of microbial fermentation to produce chemicals and natural products began a long and prosperous period. During this time, research laboratories and pharmaceutical and food companies isolated microorganisms from various sources (e.g., soil, gardens, and forests) to look for microbial species that produce various useful compounds. In addition to penicillin, many other microbial natural products with antibiotic activities were discovered. In the 1950s, *Corynebacterium glutamicum* began to be used to produce glutamic acid, which was used in a common food seasoning, monosodium glutamate. This led the way to a large amino acid industry. This period is considered to be the "classical period" of biotechnology (Figure 1.1).

Unlike the earlier solvent-producing bacteria, the molds and bacteria used in the production of antibiotics, amino acids, and other biochemicals are aerobic microorganisms. In order to produce a large quantity of a product, the volume of the culture and cell concentration must be increased, which in turn increases the demand for oxygen by microbes in culture. The demand for oxygen during scaling up led to the use of stirred-tank bioreactors with continuous sparging of air. A scaled-up process also produces more heat from the increased cellular metabolism and mechanical agitation.

The technical challenges associated with process scale-up spurred a golden period of bioprocess research. Many technical advances were made in bioreactor design to enhance oxygen transfer, sterility control, and process performance during the 1950s and 1960s.

In the second half of the twentieth century, microbiology became the core of industrial biotechnology. Many more antibiotics were discovered. The spectra of new secondary metabolites expanded from antibacterial (e.g., streptomycin, actinomycin, and cepharosporin) products to antifungal (e.g., nystatin and fungicidin), anticancer (e.g., mitomycin), and immunomodulating (e.g., cyclosporin) products. Many other microbial metabolites found their applications in the food, chemical, agricultural, and pharmaceutical industries and were successfully commercialized. Nucleosides produced by microorganisms were used to season food. Fermentation-derived lysine became an important additive to animal feed, and supported a vast industry of farm animals. Citric acid fermentation became a bioproduced commodity chemical.

In addition to metabolites, enzymes produced by microorganisms or isolated from plant and animal tissues have been utilized in food processing. For instance, amylases are used in starch processing, renin is used in cheese fermentation, and various proteases are used for protein hydrolysis. Many enzymes are also used in biocatalytic processes to produce new products. Glucose isomerase catalyzes the isomerization of the glucose molecules derived from cornstarch into high-fructose corn syrup, a staple ingredient in many processed foods. Penicillin isomerases are used to convert the side chain in penicillin from a phenylacetyl group to different acyl groups. Unlike the original penicillin G discovered by Fleming, these new penicillin-derived antibiotics are not sensitive to acid hydrolysis within the human digestive system. They also have expanded antimicrobial spectra.

During the decades following World War II, there was an unprecedented expansion of government-funded research in universities and other research institutions. The advances in fundamental biochemistry, biophysics, and molecular biology enhanced

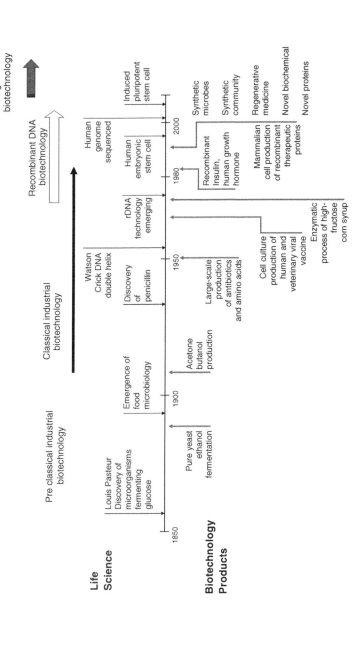

Figure 1.1 Milestones in biotechnology and historical advances in biochemical engineering.

our understanding of the nature of life and the universe. These fundamentals enabled mechanism-based discoveries that paved the way for the modern field of biotechnology, touched upon many aspects of human life, and spurred economic growth. The statins, which inhibit the rate-controlling enzyme in cholesterol biosynthesis, became star drugs for controlling cholesterol metabolism. Another example is mitomycin, an anticancer drug that is toxic because it crosslinks DNA molecules.

In the second half of the twentieth century, researchers gained a basic understanding of the structure and biochemistry of DNA and its role in genetics. These advances led to contemporary molecular biology. New understanding of the regulation of gene expression transformed industrial biotechnology and many other economic sectors. The invention of recombinant DNA technology then allowed the never-before-imagined insertion of engineered DNA sequences into a host organism for expression.

1.1.2 Recombinant DNA

Recombinant DNA (rDNA) technology enabled a new generation of products. It enabled a human protein to be produced in a host cell, be it a bacterium or a cultured human or hamster cell (Figure 1.2). It allowed us to modify the metabolic pathway of an organism by amplifying, deleting, or changing an enzyme in the pathway. Importantly, this enabling technology also spurred many venture-capital-financed startups like Genentech, Cetus, and Biogen. Thus began a new era of entrepreneur-driven innovation, forming the early stages of the next rapid expansion phase of industrial biotechnology.

Using rDNA technology, human insulin was produced in *Escherichia coli* by Genentech and licensed by Eli Lilly Company in 1981. Prior to that, type I diabetes patients had to use insulin isolated from pig pancreas, which has one amino acid different from the human form. After 1981, the insulin being used by patients was identical to the insulin that is produced in humans. A new era of producing human proteins for disease treatment then followed. Although difficult to produce before the age of recombinant technology, many of these proteins could now be cloned into a host cell and produced in sufficient quantities for treating patients. The benefit of producing these therapeutic proteins in a host cell lies not only in their increased availability; they also have increased purity. Proteins isolated from pooled donor blood might harbor harmful contaminants such as bloodborne viruses. By using rDNA technology for protein production, such danger is eliminated.

The list of heterologous proteins (or proteins produced using a different species of host cell) expressed in *E. coli* and yeast cells includes: interferon, human growth hormone, and cytokines for therapeutic uses; hepatitis B surface antigen as a vaccine; and bovine growth hormone for use in cows. However, many human proteins need to be modified after they are translated to produce the final biologically active form (Figure 1.3). Such posttranscriptional modifications, including glycosylation, complex disulfide bond formation, phosphorylation, and γ-carboxylation, are not carried out in microbes in the same way as they are in humans. To ensure their faithful expression with the requisite posttranslational modifications, mammalian cells are often employed.

Prior to the rDNA era, mammalian cells were used primarily for research and to produce viruses for use as vaccines. The use of Chinese hamster ovary (CHO) cells to produce recombinant therapeutic proteins (e.g., coagulation factor VIII for hemophiliac patients, and erythropoietin, a stimulator for red blood cell formation, for severe

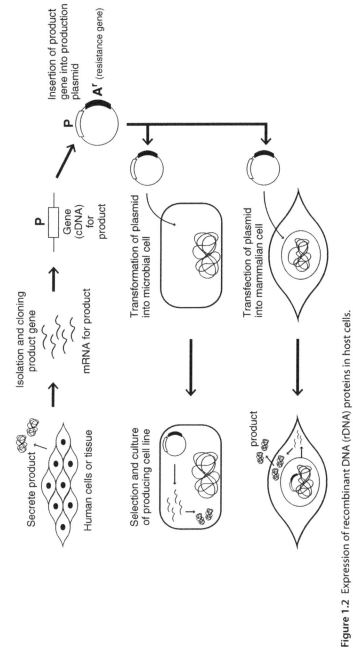

Figure 1.2 Expression of recombinant DNA (rDNA) proteins in host cells.

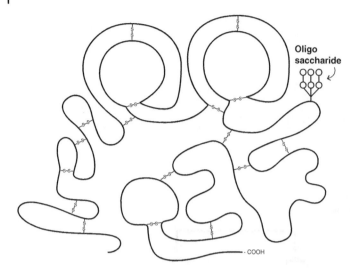

Figure 1.3 Examples of posttranslational modifications that necessitate the expression of heterologous therapeutic proteins in mammalian cells. The posttranslational modifications frequently encountered include disulfide bond formation and glycosylation.

anemia patients) led to an expansion of the use of these factors in the clinic and their commercial success. In the past quarter century, the commercial value of these so-called "biologics" grew to over US$80 billion per annum worldwide by 2015. Most notably, the success of antibody-based proteins and, more recently, fusion proteins containing the constant region of immunoglobulin molecules (Fc proteins) has primarily contributed to this expansion.

1.1.3 A Typical Bioprocess

Most biotechnological processes using microbes or mammalian cells for producing products follow a similar *modus operandi* (Panel 1.1). At the heart of the process are the "producing cells." The producing cells, once established and characterized, are replicated in large quantities and aliquoted into smaller quantities. Each aliquot is used to start a separate production process, sometimes called a "production train" or "production run." Typically, microbial-producing cells are lyophilized (i.e., freeze-dried), and mammalian cells are cryopreserved using liquid nitrogen.

Panel 1.1 Unit Operations of a Typical Bioprocess

– Sterilization
– Raw materials preparation
– Innoculum preparation (cell expansion)
– Production in bioreactors
– Recovery
– Derivatization
– Formulation
– Packaging

To start a production process, cells in a vial are reactivated after being placed into a test tube or flask of media (Figure 1.4). This starter culture is serially transferred into larger culture volumes, until the necessary number of cells needed for inoculation into production fermenters is reached. During serial expansion, the volume ratio for consecutive cultures ranges from 1:20 to 1:50 for microbial cells and from 1:5 to 1:10 for mammalian cells (Figure 1.5). During the cell expansion phase, cells are maintained in a state of rapid growth.

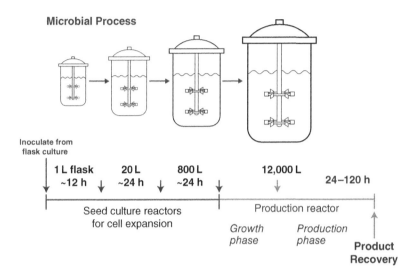

Figure 1.4 A typical manufacturing process of microbial product.

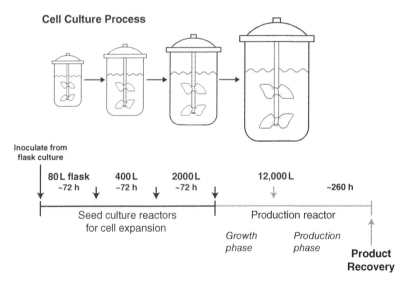

Figure 1.5 A typical manufacturing process of therapeutic proteins by mammalian cells. The duration of each expansion stage is longer and the size ratio of consecutive bioreactors is smaller than those in microbial fermentation.

In the production reactor, cells are further expanded until the cell density reaches a high level. In many cases, the product is formed when cells are in the expansion stage. In other cases, the majority of product is formed in the production stage, after the cell expansion stage is over. The production phase may be triggered by adding an inducer for the desired product to induce gene expression, or otherwise changing culture conditions to favor production formation. To enhance productivity, researchers often prolong the duration of the viability and activity of cells in the production phase.

The running time of the production reactor is typically longer than it was in each seed culture reactor. Upon the completion of fermentation, culture broths are processed to recover product. In many cases, the fermentation process produces an intermediate product. After recovery, the intermediate product has to be further modified to become the final product.

1.1.4 Biochemical Engineering and Bioprocess Technology

In the 1940s, the emergence of submerged cultures for microorganisms led to the adoption of stirred-tank reactors. The subsequent research on bioreactor development coincided with a period of new trends for reactor analysis in chemical engineering. For example, the need to sterilize the media and the reactor before fermentation led to research on the kinetics of thermal cell death. The search for better media supported high-productivity stimulated studies on microbial growth requirements and inspired the application of stoichiometric principles to microbial fermentation.

Like the chemical reaction engineering research, kinetic models of cell growth and product formation for microbial culture also began to take shape (Figure 1.6). A mathematical formula describing the response of growth rate of *E. coli* to glucose concentration, often called the Monod model, began to be adopted in the 1950s.

Driven by the wide application of continuous stirred-tank reactors in chemical reaction engineering research, continuous culture became a powerful tool to explore the dynamic behavior of microorganisms grown either as a single-population pure culture or as a mixed culture of multiple microorganisms. These studies set the foundation for quantitative analysis of cell cultivation for the next few decades.

Sustaining a sufficient oxygen level is pivotal when scaling up a microbial culture. Studies in gas–liquid interfacial mass transfer led to the development of principles for scaling up. As they grow, many molds and mycelial bacteria generate a culture broth that is essentially a non-Newtonian fluid. Alternatively, they may grow as an aggregated cell mass (i.e., a pellet), with some pellets being relatively large in size. The diffusion of oxygen inside a microbial pellet is more limited than diffusion through culture medium. For this reason, intensified studies have been devoted to transport phenomena in fermentation.

In some processes, cells or enzymes are immobilized in porous solids so that they can be retained in a continuous biocatalytic reactor and reused. The transfer of substrate and product in the solid phase may restrict the production rate. This interplay between reaction and diffusion is another area of study for biochemical engineers.

Following the development of models for describing the kinetics of microbial growth came the need to better control bioprocesses for maximal productivity. In the late 1960s, much effort was devoted to online instrumentation, leading to the use of computer-controlled bioreactors in the 1970s. Online nutrient feeding based

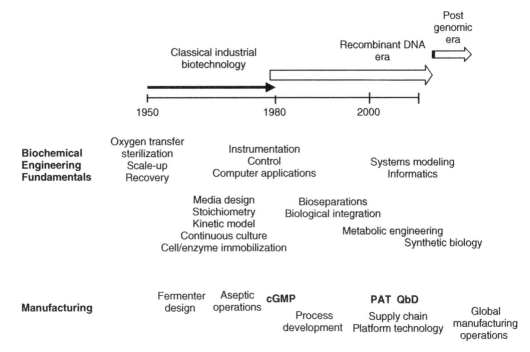

Figure 1.6 Subjects of studies in biochemical engineering as bioprocessing advances. In the manufacturing of industrial products, technological development evolved from reactor design to aspect control (to reduce the contamination rate in the 1950–1970s) to implementing current Good Manufacturing Practice (cGMP) and establishing process development infrastructure in the 1980s–1990s. More recently, emphasis has been on developing Process Analytical Technology (PAT), Quality by Design (QbD), and platform technology that will be suitable for different products. Current interests include the search for transformative technology that may further cut the process development timeline and reduce the cost of goods.

on computer-administered algorithms enabled the exploration of controlling a cell's physiological state. Such efforts extended into the 1980s. Increased computational power and the drive to further enhance productivity also facilitated the development of mathematical models, online data acquisition, signal processing, and control strategies.

The beginning of recombinant protein research triggered major changes throughout the field of biochemical engineering. The demand for the new products posed two challenges: first, the purification of product proteins from recombinant *E. coli* cells; and second, the cultivation of mammalian cells in large-scale bioreactors. Many new conceptual frameworks for bioreactor and product isolation were explored in the 1980s and the first half of the 1990s. These increased research efforts and expanded corps of biochemical engineers contributed significantly to the rapid expansion of the rDNA protein-based industry.

Genome sequencing in the past two decades has had huge impact on biological sciences and health care, as well as many technological areas. Importantly, genomic information has facilitated the rapid discovery of proteins and enzymes from remote species. Some of these have proven useful for modifying a pathway or even synthesizing a new pathway, and they have led to advancements in metabolic engineering and synthetic biology. Advances in desktop computing have also made mathematical analysis readily accessible to engineers. This has led to the permeation of systems analysis into biochemical engineering research.

Along with the development of biotechnology and biochemical engineering, manufacturing bioprocess technology has also evolved over the years. The field's early focus was on bioreactor design and operation. The subsequent need for generating more sophisticated protein molecules (e.g., for injection into patients) has since facilitated the development of cGMP (current Good Manufacturing Practice) in biomanufacturing. This ensures a consistently high-quality product. cGMP practices have led to the emphasis on PAT (Process Analytical Technology) and QbD (Quality by Design) in product development, to better monitor process variables in real time and to implement control strategies to ensure the quality of final products.

Biochemical engineers in industrial environments have also been increasingly engaged in ensuring a steady and consistent global supply of raw materials and product inventory. Market globalization has resulted in a more distributed manufacturing process, meaning that many products are produced in multiple sites over the globe, but that all sites must produce products of the same quality. For this reason, contract manufacturing organizations (CMOs), which specialize in making products for the "innovator" or the originator of products, have become more common. This allows the originator to focus on the function it performs better, such as the research and development of new technologies and products.

To speed up product development, many firms have adopted the strategy of platform technology. In this scenario, the same platform is used for different products. For example, a host cell that has already been optimized to secrete enzyme products and is being actively used in product development may also be used to produce all future industrial enzymes, even if another host cell may be theoretically better. This is because the time and effort required to optimize a new host cell can be very high and costly.

1.2   Industrial Organisms

The workhorses of biotechnology encompass a wide array of organisms, from prokaryotic and eukaryotic cells to multicellular organisms, like plants or animals. The vast majority of industrial cultures carried out in bioreactors employ unicellular organisms (like bacteria, yeast, and mold) or cells derived from tissues of animals and plants. In some cases, somatic embryos or shoots of plants may be propagated in a bioreactor. In a few rare cases, an entire organism (such as insect larvae) may be grown in a reactor environment to produce vaccines or bio-insecticides.

Until the introduction of rDNA, virtually all industrial products were derived from their native producers. However, few of these natural organisms are able to produce their product at a level that is suitable for economical industrial production. To overcome this limitation, the organisms are mutated or otherwise improved to enhance their productivity.

In the rDNA era, this changed. Now, many proteins can be easily expressed in host cells for which genetic tools are widely available, such as *E. coli*, the yeast *Saccharomyces cerevisiae*, and even some mammalian cells. An entire metabolic pathway can be introduced into a microorganism to make it produce a new and nonnative compound at a very high level.

Industrial microorganisms can be classified according to the class of products they produce or by their taxonomy (Panels 1.2 and 1.3). Some genera have a certain propensity for producing certain classes of compounds. For example, members of *Streptomyces* are known for being producers of antibiotics, while species in *Corynebacterium* are most amenable for producing amino acids.

Panel 1.2 Industrial Organisms

Prokaryote
 Eubacteria
  Unicellular
  Mycelial – Actinomycetes
   (Streptomyces, Nocardia)
Archaea (Extremophiles)
Eukaryotic Cells
 Fungi
  Yeast
  Mold
 Algae
 Protozoa
 Insect cells
 Plant cells, tissues, and organs
 Animal cells, tissues, and organs

Transgenic Animals and Plants

Panel 1.3 Biotechnological Products

Metabolic Process
 Food
 Waste treatment
 Mineral leaching
 Bioremediation
 Degradation of toxic compounds
Metabolites
 Primary metabolites
 Secondary metabolites
Secreted Biopolymers
 Extracellular polysaccharides
 Extracellular proteins
Cells, Tissues, and Their Components
 Cells
 Enzymes
 Polysaccharides
 Viruses

The boundaries among those product-based classes became blurred with the introduction of rDNA. A few microorganisms (such as *E. coli*, a few species of yeast, as well as CHO cells) have become the favorite hosts for industrial production for a wide range of products. However, the overall vast majority of products are still produced by their native producers after undergoing some strain improvement work in the laboratories.

1.2.1 Prokaryotes

1.2.1.1 Eubacteria and Archaea

Prokaryotic microorganisms constitute the majority of industrial organisms. Industrially important unicellular microorganisms include eubacteria ("true bacteria") and archaea. Microorganisms in both orders lack a nucleus, and their cytosol and its chromosome are enwrapped in a cytoplasmic membrane, which is encased in a cell wall.

The industrially important bacteria are very diverse. A common classification of bacteria is based on Gram staining of their cell wall. They are generally classified as Gram-positive (e.g., *Bacillus*) or Gram-negative (e.g., *E. coli*) bacteria, both of which are represented among industrial bacteria. They may also be aerobic and require oxygen to grow (e.g., *Corynebacterium*), or be anaerobic (e.g., *Clostridium*). For example, *Clostridium acetobutylicum* was used to produce acetone and butanol over a hundred years ago, while *Corynebacterium glutamicum* has been used in the production of glutamic acid since the 1950s.

Some bacteria are capable of differentiation; for instance, *Bacillus* can differentiate to form spores. One family of microorganisms, the Actinomycetaceae, includes many species in the *Streptomyces* genus and produces many important antibiotics. The Actinomycetaceae bacteria grow as mycelia. The mycelia protrude into the air, where they become aerial mycelia, and on a solid agar medium, mycelia protrude into the agar, where they form substrate mycelia. The aerial mycelia can fragment and form spores, which help the bacteria to counter adverse environmental conditions.

Microorganisms in the Archaea domain resemble eubacteria in most respects. They are morphologically similar and can be Gram-positive or negative. Members of the Archaea domain have rather unusual metabolic characteristics. Many are extremophiles, meaning they thrive in very extreme and hostile environments. They can be found living in conditions with high salt concentrations (extreme halophile) or high temperatures (extreme thermophile).

The methanogenic archaea, which grow under strict anaerobic conditions, use CO_2 and H_2 to produce CH_4 and derive energy. Other than photosynthetic organisms, few living systems are capable of fixing CO_2; these archaea use H_2 as an energy source instead of light for this process. Microorganisms in this order produce many useful enzymes that are stable in extreme conditions, such as high-temperature stable DNA polymerases, used in polymerase chain reactions (PCRs), and other high-temperature industrial enzymes. However, their preference of growing under extreme conditions also poses challenges for their cultivation.

1.2.2 Eukaryotic Microorganisms

Eukaryotes all share a common characteristic: a true nucleus. Eukaryotic organisms encompass microorganisms like yeasts, molds, and algae. Cells of plants and animals are also eukaryotic cells. In eukaryotic cells, lipid bilayer membranes are present not only on the cell surface, but also surrounding the various organelles. These lipid bilayers segregate the cell into different compartments, and allow each compartment to have different kinds of chemical environments for carrying out specialized functions. For example, the highly oxidative environment of the lumen of mitochondria, from which most cellular energy [adenosine triphosphate (ATP)] is derived, is segregated from other parts of the cell. Protein molecules that are destined for secretion are translocated into the endoplasmic reticulum (ER) while they are being translated, so that protein folding can proceed with the help of the various chaperone proteins in the ER. Such compartmentalization gives eukaryotes functional complexity and sets them apart from prokaryotes.

1.2.2.1 Fungi

Both yeasts and filamentous molds are fungi. The best-known industrial yeast is the ethanol-fermenting *S. cerevisiae*. Some yeast strains are used to produce citric acid, enzymes, and specialized lipids. Others are capable of growing on methanol or hydrocarbons.

Molds, as commonly seen in spoiled foods, are widely used in the food industry. Many industrial enzymes for hydrolyzing protein (proteases), starch (amylases), or cellulose (cellulases) are produced by molds. Molds produce many important antibiotics, including penicillin and cephalosporin. Recently, yeast cells have also been used for producing heterologous recombinant proteins. A notable one is the hepatitis B surface antigen, which is used as vaccine against hepatitis B virus.

1.2.2.2 Algae

Algae are used for the industrial production of specialty compounds, such as carotenoids. They are more frequently used as a dried cell mass for health food supplements and animal feed, without extracting the product out of cells. Different from traditional bioprocesses, algae are also used in environmental processes, often in an open environment (such as open ponds or circulating pipes). In some cases, the cultivation is carried out as a mixed culture of algae and other microorganisms. These cultures have the capability of performing photosynthesis and fixing CO_2, and they are being explored as a vehicle for biodiesel production.

1.2.3 Multicellular Organisms and Their Cells

1.2.3.1 Insect Cells

Insect cells can be used for the production of an insect virus, which is used as an insecticide. Some insect cells are also good vehicles for expressing different proteins that can subsequently be used as vaccines, especially in veterinary applications. The use of insect cells for producing human protein pharmaceuticals, however, has not been successful. Most human therapeutic proteins are glycosylated, but the glycosylation carried out in insect cells gives rise to glycans of different structures than those synthesized in mammals. Thus, proteins made in insect cells are unsuitable for administration to patients.

1.2.3.2 Plant Cells, Tissues, and Organs

Plant tissues have long been the source of many medicinal substances, pigments, fragrances, and enzymes. It is thus natural to try to isolate and cultivate plant cells, or even tissue, for the production of those compounds. Isolated plant cells may be grown as relatively disperse cells, or as cell clusters, called "callus." Some plant metabolites are produced in differentiated tissues and transported to another tissue for maturation. The complexity of biosynthesis in plant tissues is greater than it is in microbial metabolites.

Although much progress has been made over the past couple of decades, the concentrations of product that can be achieved using cultured plant cells are still generally low. However, the successful transformation of plant metabolite production to an industrial bioreactor operation will ensure steady supplies and minimize the susceptibility to supply chain interruption due to natural or social-political incidents.

Plant tissues or organs are cultured for propagation purposes. Bioprocess methods may be applied to grow somatic embryos or shoots (stems to allow for generating cultivars), a process often called "micropropagation." By using somatic embryos, it is possible to generate clones of progeny of the same somatic cell, thus greatly enhancing the consistency of the product.

1.2.3.3 Animal Cells, Tissues, and Organs

Animal tissues have long been used to produce many enzymes for use in medicine. Insulin was first isolated from pig pancreas before its recombinant product became available. The pituitary gland of cadavers was the source of human growth hormone to treat dwarf syndrome. Factor VIII isolated from human blood was used to treat hemophilic patients. Animal cells isolated from tissues and cultivated *in vitro* are now used to produce virus particles for vaccines. Prior to the wide application of cultured animal cells, virus production was often carried out using animals. The shift of viral vaccine production to cultured animal cells has made the production process more robust and the product more amenable to quality control.

Many human proteins of therapeutic value are now produced as heterologous proteins in host cells. Some are produced in *E. coli* (such as insulin and human growth hormone) or yeast (such as cytokines and granulocyte stimulating factor). However, many human proteins require glycosylation and other posttranslational modifications to be clinically effective. The production of those proteins typically requires the use of mammalian cells as the production vehicle.

Unlike traditional biochemical products, which are produced by a large array of microorganisms, the host mammalian cells used for rDNA production are limited to a handful of cell lines. In fact, CHO cells are used for the production of over 80% of mammalian cell-based rDNA products. Other cells used include mouse myeloma-derived cells and Syrian hamster kidney cells.

1.2.4 Transgenic Plants and Animals

Genetically modified organisms (GMOs), especially transgenic plants, have been used in the field since the 1990s. An engineered resistance to insects and herbicide, carried in transgenic plants like corn, soybeans, and cotton, has greatly increased the yield of those products. The resistance to insecticide can be accomplished by cloning a protein, often derived from *Bacillus thuringiensis*, into the plant genome. The protein is lethal to many insects once ingested. Herbicide tolerance allows for selective removal of weeds and preservation of the engineered plant (e.g., soybean) upon herbicide application.

For corn and cotton, GMOs are planted in more acreage than nonengineered crops in the United States.

Transgenic farm animals, on the other hand, have faced more resistance in customer acceptance and regulatory approval. Many cows are given bovine growth hormone (BGH) to increase their efficiency of feed consumption and milk production. Transgenic cows that endogenously produce a higher level of BGH similarly produce more milk. Farmed salmon have been engineered to express higher levels of growth hormone, to extend their growth season and speed their growth. Pigs have also been engineered to produce less polluting excretes. These modifications, however, are not acceptable to most consumers, so the use of GMO animals in the consumer market is still being debated.

In the past decade, transgenic organisms, including plants and animals, have been explored as a production vehicle for biopharmaceuticals. Antitrypsin and protein C (a clotting factor) have been produced in the milk of goats and pigs, respectively. Recombinant antibody has been produced in corn. It is anticipated that the costs of goods for products produced in transgenic organisms will be substantially lower than those produced by cultured cells in bioreactors.

Transgenic technology, while widely used in crops, is still in the exploratory stage for animals. Environmental concerns are a lingering issue. This is even more applicable to plants, as they are cultivated in open fields and it is unavoidable that escaped pollen will eventually cross-contaminate native plant species.

1.3 Biotechnological Products

Biotechnological products can be classified by their chemical nature (e.g., carbohydrates or proteins), application (e.g., agricultural or medical), or source (e.g., animal or microbial products) (Panel 1.3). Tables 1.1 to 1.6 list examples of biotechnological products by their areas of application. In this section, we will group and discuss products by their source or chemical nature.

1.3.1 Metabolic Process

Humans have been cohabitating with microbes since the dawn of existence. For thousands of years, humans have unwittingly taken advantage of their metabolic processes while making wine, bread, and cheese. In more recent decades, with specific knowledge of the microorganisms involved, we intentionally select specific variants of microbes to better control those processes. For example, by using a yeast strain that has a particular pattern of carbon dioxide release, one can better control the texture of bread. In such applications, no particular compound is isolated as a product; rather, it is the process of metabolism that is being exploited.

Metabolic process is the centerpiece of a number of biotechnological applications. In wastewater treatment, a mixture of microbes grows and metabolizes the organic materials in a waste stream to reduce the level of organics below a threshold. Likewise, in some detoxification processes, microbes make certain compounds less toxic, such as chlorinated organics, pesticides, and herbicides. In other cases, microorganisms reduce the metallic oxide in mineral ore to facilitate the extraction of metals from the ore.

Table 1.1 Microorganisms used in food processing.

Food and beverages	
Yeast	
Saccharomyces cerevisiae	Baker's yeast, wine, ale, sake
Bacteria	
Lactobacillus sanfranciscensis	Sour French bread
Streptococcus thermophilus	Yogurt
Lactobacillus bulgaricus	Yogurt
Gluconobacter suboxidans	Vinegar
Mold	
Penicillium roquefortii	Blue-veined cheeses
Penicillium camembertii	Camembert and brie cheeses
Aspergillus oryze	Sake (rice starch hydrolysis)
Agriculture	
Entomopathogenic bacteria	
Bacillus thuringiensis	Bio-insecticides

Note: Their metabolic process contributes to the product. Microbial cells are also used directly as product, as in insecticide and baker's yeast.

Table 1.2 Some primary metabolites and their producers.

Industrial chemicals	
Yeast	
Saccharomyces cerevisiae	Ethanol
Bacteria	
Clostridium acetobylicum	Acetone and butanol
Xantomonas campestris	Polysaccharides
Mold	
Aspergillus niger	Citric acid
Bacteria	
Corynebacterium glutamicum	L-lysine, monosodium glutamate

Note: Many are industrial chemicals or bulk biochemicals.

Table 1.3 Examples of additional metabolites as industrial chemicals.

Other industrial chemicals and producing organisms	
Polysaccharides	
Bacteria	
Leuconostoc mesenteroides	Dextran
Xanthomonas campetris	Xanthan gum
Vitamins	
Yeast	
Eremothccium ashbyi	Riboflavin
Bacteria	
Pseudomonas denitrificans	Vitamin B_{12}
Propionibacterium	Vitamin B_{12}

Table 1.4 Examples of industrial proteins produced in native producers and in recombinant hosts.

Secreted protein products and producing organisms	
Industrial proteins	
Enzymes	
Mold	
Aspergillus oryzae	Amylases
Aspergillus niger	Glucoamylase
Trichoderma reesii	Cellulase
Saccharomycopsis lipolytica	Lipase
Aspergillus	Pectinases and proteases
Endothia parasitica	Microbial rennet
Bacteria	
Bacillus	Proteases
Heterologous industrial proteins	
Aspergillus niger	Albumin (for industrial use)
E. coli	Bovine growth hormone

1.3.2 Metabolites

Microbial metabolites constitute the bulk of bioproducts. In the first half of the twentieth century, microbiologists learned how to harness microbes to make metabolites. Natural microbes make metabolites for their own needs, usually only at the time and in the exact quantity that is needed. In the laboratory, microbiologists first identify and isolate microbes that produce those useful metabolites, then introduce mutations in them to make them produce those metabolites, typically at levels thousands of times above what is natural in order for the production to be economical. For almost every common metabolite that is useful to humans (e.g., acetic acid, ethanol, citric acid, and lysine), scientists have isolated organisms to produce it.

Metabolites are largely divided into primary metabolites and secondary metabolites. "Primary metabolites" are terminal products or intermediates of the cell's catabolism and anabolism. These are generally the intermediates of energy metabolism (such as ethanol, citric acid, or succinic acid) or products of the biosynthetic pathways of cellular building blocks (such as glutamic acid and lysine) (Table 1.2). Primary metabolites are essential for the growth of the producers.

In contrast, many natural products produced by microorganisms and plants are seemingly nonessential, meaning that if cells were not able to synthesize the product, their growth would not be affected. These metabolites are called "secondary metabolites" (Table 1.5). Many pharmaceutical drugs, such as antibiotics, fall into this category. These secondary metabolites are usually produced after the period of rapid growth is over. In nature, antibiotic biosynthesis is possibly a mechanism of the microbe's self-defense. Under adverse conditions, when growth ceases, antibiotics are produced to retard the invasion of other microorganisms into the producer's habitat.

Following the discovery of penicillin, scientists began to search for microbes in soils, lakes, oceans, and plants that produce metabolites with antimicrobial activities. These products are often called natural products. These activities have greatly expanded the

Table 1.5 Some secondary metabolite producing organisms and their products as drugs.

Pharmaceuticals	
Drugs	
Mold	
Penicillium chrysogenum	Penicillins
Cephalogporium acremonium	Cephalosporins
Rhizopus nigricans	Steroid transformation
Bacteria	
Streptomyces	Amphotericin B, kanamycins, neomycins, streptomycin, tetracyclines, and others
Bacillus subtilis	Bacitracin
Bacillus polymyxz	Polymyxin B
Mycobacterium	Steroid transformation
Norcadia autotrophica	Pravastatin

Table 1.6 Examples of pharmaceutically important biologics, including viral vaccines and recombinant proteins.

Pharmaceuticals	
Biologic products and their sources	
Mammalian cells for virus vaccines	
Human lung fibroblast MRC-5	Attenuated hepatitis C virus
Monkey kidney epithelial (Vero) cell	Inactivated polio virus
Heterologous expression of human proteins	
Mammalian cells as hosts	
Chinese hamster ovary (CHO) cells	Antibodies Interferon, tissue plasminogen (tPA) activator, erythropoietin (EPO)
Human kidney 293 cell lines	Adenovirus for gene therapy
Bacteria as hosts	
Escherichia coli	Insulin, human growth hormone, bovine growth hormone
Yeast as hosts	
Saccharomyces cerevisiae	Hepatitis B virus surface antigen (vaccine against hepatitis B)
Saccharomyces cerevisiae, Pichia pastoris	Serum albumin

repertoire of bioactive metabolites. Many have found medical and industrial applications, including tetracyclines as antibiotics, mitomycin and taxol as anticancer drugs, cyclosporin as an immune suppressor, and various statins for suppressing cholesterol biosynthesis.

1.3.3 Cells, Tissues, and Their Components

A large number of biotechnological products are the biomass of the organism. Baker's yeasts are notable industrial products (Table 1.1). *B. thuringiensis* (BT) is used as a bio-insecticide (Table 1.1). Plant tissues are used widely for the extraction of secondary metabolites for medical use, such as the malaria drug quinine. Plant shoots are used as cultivars. Human cartilage procured from cadavers is used for joint repair.

The potential of producing human tissues through *in vitro* cultivation or tissue engineering has been a growing field. The US Food and Drug Administration (FDA) has approved the culture of a human skin equivalent for patients with severe ulcers or burns. *In vitro* expansion of somatic plant embryos, and their application as artificial seed, has been a reality for nearly three decades.

1.3.3.1 Viruses

Viral vaccines are important to human and animal health care. Most vaccines are produced by infecting cultured animal cells and harvesting the virus released into the culture fluid (Table 1.6). Two major kinds of viruses are used to produce vaccine: attenuated or inactivated. Attenuated viruses are unable to elicit a pathological response, but can still replicate and induce an immunogenic response.

Inactivated viruses are viruses that have been rendered harmless by chemical treatments, such as formalin. Attenuated viruses are live, but have been adapted to become harmless. They are more potent and are required only in very small doses to elicit an immunogenic response. However, with attenuated viruses, there is always a small risk of genetic reversion to render the virus pathogenic. Another disadvantage of attenuated viruses is that they need to be shipped at low temperature to preserve biological activity.

1.3.4 Secreted Enzymes and Other Biopolymers

Many microorganisms secrete proteins, especially enzymes that convert the resources available in the environment to a form that can be used as nutrients. For example, *Bacillus* and *Aspergillus* produce proteases outside of the cell to hydrolyze proteins into amino acids, which are then used as nutrients by those organisms. A mold, *Trichoderma reesei*, produces cellulase to hydrolyze cellulose and use the hydrolysis product for growth (Table 1.4).

In addition to enzymes, microbes also produce other polymers and secrete them into the extracellular space. For instance, *Xanthomonas campestris* produces a polysaccharide called xanthan gum, which makes solutions very viscous and is used as a thickening agent in ketchup and some processed foods (Table 1.2). Many microorganisms that are capable of using hydrocarbon as a carbon source also produce emulsifiers (often lipid-conjugated polysaccharides) to help dispense hydrocarbon droplets and to increase the surface area of this carbon source for their consumption.

1.3.5 Recombinant DNA Products

rDNA technology opened a new era in biotechnology (Table 1.6). While the transformation has been visible in medicine, agriculture, and chemistry, its impact was most immediate in medicine. Many therapeutic proteins, which were at one point in short supply, became more widely available through rDNA technology. Success in the agriculture sector was also apparent through the cloning and engineering of the gene responsible for a desired trait, such as insect resistance, into a plant.

rDNA technology also enables one to alter the metabolic pathway of an organism to enhance the production of a metabolite, or even to introduce a new pathway for a novel product. Such metabolic engineering is now common practice in both microorganisms and plants.

1.3.5.1 Heterologous rDNA Proteins

The deficiency of some biologically active proteins has been at the root of many congenic diseases and has been known to be the cause of many pathological conditions. Before the rDNA era, many therapeutic proteins, including Factor VIII, insulin, and human growth hormone, were isolated from human blood, cadavers, or animal tissues. Production relied on pooling a large number of procured human or animal

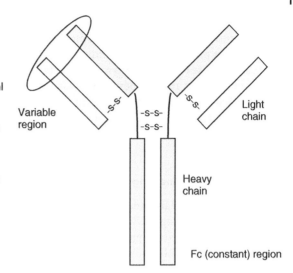

Figure 1.7 Structure of an immunoglobulin-G (IgG) molecule. A major class of therapeutic proteins that has emerged in the past 15 years is antibodies, in particular IgG. For example, anti-VEGF (vascular endothelial growth factor) suppresses blood vessel formation in some tumors. Each IgG molecule has two heavy chains and two light chains that are linked by disulfide bonds and segregated into a constant region and variable region. The variable region recognizes antigen. Each IgG has two antigen-binding sites.

tissues. The detection and exclusion of contaminated or contagious tissues was not always easy. Compounding this difficulty were previously unknown or undetectable infectious agents, such as HIV in the 1980s. rDNA technology allows these proteins to be produced under well-controlled conditions in large quantities, free of human virus contamination.

The use of a recombinant protein as a therapeutic agent quickly expanded beyond supplementing a deficiency in patients. Antibodies and other binding proteins are now also used to block disease-causing processes (Figure 1.7). For example, trastuzumab (trade name, Herceptin), an immunoglobulin G (IgG) against epidermal growth factor receptor that is overexpressed in some breast cancers, is used to treat HER2-positive breast cancer patients.

Many of those proteins are not native to the human body, but are designer proteins made of domains of human proteins (such as the Fc region of the IgG molecule) and an activity domain for the induction of a biological response. An example of such a fusion protein is etanercept (trade name, Enbrel), which is the fusion product of tumor necrosis factor receptor 2 (TNFR2) and the Fc fragment of IgG. It binds to tumor necrosis factor α (TNFα), the master regulator of the inflammatory response, and is used to treat rheumatoid arthritis.

These protein medicines, along with viruses and cells, are classified as "biologics" by drug regulatory agencies. This is to distinguish them from the traditional chemical drugs, which are called "drugs." Traditional drugs can be very well characterized in terms of their chemical composition, structure, purity, and contaminants. Once characterized, the quality of the product is not in doubt. Biologics, on the other hand, cannot be completely defined by their chemical composition, and sometimes a complete chemical characterization is not even possible. For example, the chemical composition of cells or viruses cannot be entirely defined due to their enormous complexity. Furthermore, their chemical composition may differ in different growth conditions. For instance, a therapeutic protein may have the same primary amino acid sequence, but may have a different glycosylation profile depending on how it was produced. Biologics are thus much more prone to variations caused by process changes. The regulation on their commercial availability is thus much more stringent than it is for drugs.

The use of rDNA for protein production goes beyond human applications. It has also become a powerful vehicle for the production of industrial proteins and enzymes. Many microbes that produce attractive industrial proteins are not easily cultured at an industrial scale because of any one of a number of factors, such as slow growth, low productivity, special nutrient requirements, unfavorable temperature, or possible pathogenicity. Those proteins are instead produced in a number of industrial hosts, including *E. coli*, yeasts, and fungi. Notable products include enzymes for food processing, proteases used in detergents, BGH for the cattle industry, and human collagen and serum albumin for medical use.

1.3.6 Metabolic Engineering and Synthetic Pathways

The repertoire of metabolite products grew dramatically during the classical period of industrial biotechnology. The capability to produce a metabolite is largely dependent on the producing microorganism. Often, a particular species or strain is used for a particular compound because of the characteristics of the organism. For example, *C. glutamicum* is used to produce glutamic acid and *A. niger* is used to produce citric acid.

Once the organism producing a compound is isolated, extensive strain improvement is performed to increase its production by orders of magnitude higher than what the microorganism would normally produce. This is often accomplished by a series of treatments by chemicals (mutagens) or irradiation, to increase the mutation frequency so that mutants with the desired properties can be found and isolated (Figure 1.8).

rDNA technology changed the landscape of metabolite production. With rDNA technology, one or several genes can be introduced to change a cell's metabolic reactions in a specific and planned manner. This approach is often referred to as metabolic engineering.

One of the first examples of metabolic engineering was the introduction of an enzyme into a methanol-utilizing bacterium to enable it to use a more energy-efficient pathway to incorporate ammonium into organic nitrogen. Similarly, rDNA was used to enhance amino acid productivity in *Corynebacterium*, by replacing its enzymes with ones that are less prone to feedback inhibition and more energetically favorable, and by changing its transporters for better substrate uptake and product secretion.

The reach of rDNA goes beyond enhancing the productivity in traditional producers like *Corynebacterium*. It also enabled a number of amino acids, including phenylalanine and lysine, to be produced in *E. coli*. Amino acid biosynthesis in *E. coli* is tightly regulated to minimize wasteful overproduction. To engineer *E. coli* to overproduce and excrete an amino acid to a level sufficient for industrial production, biotechnologists replace *E. coli*'s rate-controlling enzymes with ones that are not subject to feedback inhibition or repression.

Through metabolic engineering, microbes can also be made to shunt their metabolite to a new branch to generate a new metabolite. For example, *E. coli* has been engineered to produce ethanol at an efficiency only previously seen from the yeast *S. cerevisiae*.

Metabolic engineering has also been applied to create a synthetic pathway to produce novel compounds. For instance, *E. coli* has been engineered to synthesize the antimalaria drug artemisinin, which is normally produced only in the plant *Artemisia*.

Metabolite Producing Strain Improvement

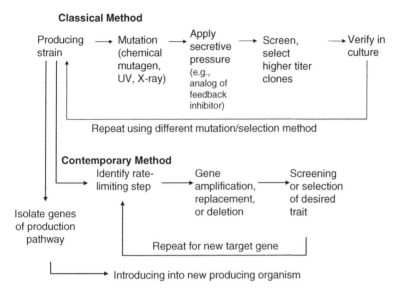

Figure 1.8 Classical and contemporary scheme of strain improvement for the producing microorganisms of industrial chemicals. The classical way is to introduce mutations in the producing organism through mutagenesis to increase the frequency of mutations on the genes that may affect the synthesis of the target metabolite. Sometimes, a selective strategy is available to enrich the mutated higher producing cells. For example, an analog (a compound that has a similar structure to a native compound, has some similar chemical properties such as exerting inhibition on an enzyme, but cannot be metabolized) may be used to cause inhibition of biosynthetic enzyme. Consequently, its biosynthesis in the wild-type microorganism is suppressed, causing cell growth to be retarded. Any mutant that is no longer inhibited by the analog will survive. Such an analog can thus selectively enrich feedback inhibition deregulated mutants. In any case, extensive screening of higher producers after mutagenesis is necessary. The modern method uses direct cloning to introduce, enhance, or eliminate a gene (or genes) to enhance the high-productivity trait. It can speed up the process of obtaining a high producer dramatically compared to the classical method.

1.4 Technology Life Cycle, and Genomics- and Stem Cell-Based New Biotechnology

1.4.1 The Story of Penicillin and the Life Cycle of Technology

Discoveries in microbiology (from the late nineteenth century through the first half of the twentieth century) paved the way for classical biotechnology. Penicillin has proven to be one of the most influential discoveries of this period. The campaign that translated this discovery into a groundbreaking therapeutic medicine also laid the foundation for many other products and new technologies.

Alexander Fleming's discovery in the laboratory was only one piece of this impressive story. Chemists and biochemists pushed forward to isolate and characterize the compound. Medical researchers and clinicians crusaded to test its effectiveness in animal models and patients. These visionaries had to lead the way and gather the necessary resources to convince manufacturers to make sufficient quantities for use in clinical trials. Only through this combined effort did penicillin become available for more widespread use throughout medicine.

At the time of its discovery, the amount of penicillin attainable in culture was extremely low, making the cost of goods extremely high (Figure 1.9). This initial limited availability made penicillin a high-profit-making drug. Subsequent efforts in strain improvement and process development resulted in a rapid increase in productivity and expanded the capacity of production. Meanwhile, the cost of the drug decreased substantially, but generally remained high and profitable for a couple of decades.

Over time, all products enter a second phase with the end of a product's patent and the entry of generic versions of the drug into the market. At this point, competition begins to drive down the price and profit margin for drugs. In the case of penicillin, no patent was ever applied for. Nevertheless, the maturation of technology allowed the cost of goods to decrease substantially. The lower price allowed more people in more countries to have access to it, and the market expanded quickly. Usually, the total profit from making and marketing a drug is still sufficiently attractive to keep the original lead makers in the market. In this period, increasing competition spurs process intensification and enhances productivity and robustness.

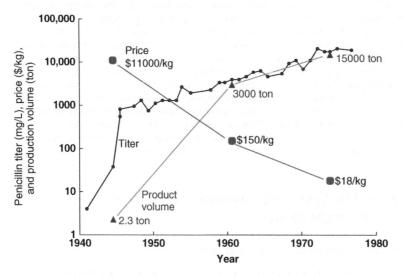

Figure 1.9 The evolvement of penicillin as a technological product. Initial low productivity and production level were quickly enhanced by orders of magnitude due to the high demand created by the revolutionary nature of the product. This is followed by a slower but steady growth driven by the reduced price and wider affordability in different regions of the world. The latter slower growth period entails a much larger total volume of goods and sales. Eventually, the commercial value of penicillin is too low to be profitable for major pharmaceutical companies. The history of penicillin is a general reflection of natural products in classical biotechnology. The decline in classical biotechnological products is supplanted by recombinant DNA (rDNA) technology. The same life cycle will be repeated for rDNA technological products in the years to come.

As a class of drugs, antibiotics (and natural products in general) inspired many discoveries. The number of drugs in this class increased drastically in the 1960s and 1970s. Initially being profitable, the landscape of competitions subsequently drove down the price of antibiotics, and the technology used to produce them became widely practiced. In the 1990s, traditional Western pharmaceutical companies found it more profitable to move on to the new class of enabling biopharmaceutical products. This life cycle pattern is generally applicable for a successful product or class of products and will likely be replayed in the rDNA era.

After the emergence of classical biotechnology, one witnessed an infusion of research funds in life and biochemical sciences from governments and commercial firms that drove the push for new technologies. Rather than traditional pharmaceutical companies, alliances of venture capitals and scientists propelled rDNA into a transformative technology. In the 1980s, the growth opportunity for startup companies across North American campuses was unprecedented. These companies are now global giants like Genentech, Amgen, and Biogen.

Two decades after the introduction of the first wave of rDNA-based biologics, we are seeing the emergence of "biosimilars." This term is used to describe the copy version of therapeutic proteins. Because of the complexity of biologic drugs, regulatory agencies do not permit the same substance produced by companies other than the original producer to enter the market without at least some clinical trials after the original patent expires. These "follow-on" biologics are thus called biosimilars. This distinguishes them from generic small-molecule drugs, which can generally be introduced without conducting clinical trials.

Because of the complexity of the product proteins and the regulatory requirements for product entry, the impact of biosimilars on the landscape of biologics will be more gradual than that of traditional chemical drugs. However, the transition of rDNA biotherapeutics from the exponential expansion phase to a slower growing and momentous stage is inevitable. This transition represents technological maturation in advanced countries but bestows arising opportunities and challenges in other regions of the world.

1.4.2 Genomics, Stem Cells, and Transformative Technologies

While rDNA technology is propelling technological advancement, huge strides have also been made in new areas of science. Mouse embryonic stem cells were isolated in the beginning of the 1980s. Human embryonic stem cells were not established until 1997. These embryonic stem cells are able to differentiate into cells constituting all adult tissues. This ability is called pluripotency.

A decade later, scientists discovered that they could reprogram somatic cells to become pluripotent cells [called induced pluripotent cells (iPSCs)]. This made it possible to generate pluripotent cells from any individual's somatic cells, like skin or liver cells, for possible tissue regeneration. Stem cell technology will likely transform the face of medicine in the future.

Another significant advancement is in the field of genomics. The human genome was sequenced at the dawn of the twenty-first century. Prior to that, a number of yeasts, fungi, and bacteria had their genomes sequenced. Simultaneously, methods were being developed to engineer genomes. While rDNA accomplished the insertion and deletion of genes, altered segments constitute only a tiny fraction of the genome, even for the smallest bacterial genomes. We are now on the verge of being able to reorganize the cellular genome globally, rather than locally, and to understand how the genome is globally regulated and controlled.

In the twenty-first century, biotechnologists and biochemical engineers will need to learn and relearn systems analysis of biochemical sciences, as new technology is likely to involve regulations at both genomic and biochemical levels in everything from microbes to plant and stem cells to genome-engineered synthetic organisms and communities.

Further Reading

Aiba, S, Humphrey, AE & Millis, NF 1973, *Biochemical Engineering*, 2nd edn, University of Tokyo Press, Tokyo.

American Chemical Society 1999, *The Discovery and Development of Penicillin 1928–1945*, The Alexander Fleming Laboratory Museum. Available from: https://www.acs.org/content/dam/acsorg/education/whatischemistry/landmarks/flemingpenicillin/the-discovery-and-development-of-penicillin-commemorative-booklet.pdf. [19 July 2016].

Bailey, JE & Ollis, DF 1986, *Biochemical Engineering Fundamentals*, 2nd edn, McGraw-Hill, New York.

Baltz, RH, Davies, JE & Demain, AL 2010, *American Society for Microbiology*, 3rd edn, ASM Press, Washington, DC.

Problems

A1 List three products for each of the following categories:
- rDNA protein by microbiological process
- Industrial specialty chemicals
- Industrial commodity chemicals

A2 Give a short answer to each of the following:
- Explain the reasons for using mammalian cells to produce therapeutic proteins.
- Describe how animal cells are engineered to synthesize a heterologous rDNA protein.
- Give the names of two primary metabolite products and two secondary metabolite products. Explain what primary and secondary metabolites are.

A3 Do most secondary metabolites exhibit growth-associated or non-growth-associated production kinetics?

A4 Match each term in the first column in Table P.1.1 to a term in the second column. Each term in the second column can be used only once.

A5 Many IgG antibody molecules have important pharmaceutical applications. Are they produced in animal cells or in bacteria? What structural characteristics of IgG dictate the selection of producing cells?

Table P.1.1 Matching terms with close relationships.

A. Penicillin	a. *Saccharomyces cerevisiae*
B. Lysine	b. Primary metabolism
C. Chinese hamster ovary cell	c. Antibody
D. Ethanol	d. Secondary metabolism
E. Glucose isomerase	e. *Clostridium acetobutylicum*
	f. Tissue-type plasminogen activator
	g. High-fructose corn syrup

A6   List two examples of classical biotechnology products produced using each category of microorganism: bacteria, yeast, and mold. Write down the name of the
genus and species.

A7   Microbial processes are used to produce product G from substrate A, as shown in
Figure P.1.1. Part of the pathway is also used to produce product E. Both E and G
are required for cell growth. The first enzyme is inhibited by G and E individually,
and repressed by E.
  • Discuss briefly how microbiologists use traditional mutation, selection, and
    screening approaches to direct more substrate A to product G.
  • Describe how to use modern metabolic engineering approaches (on genes or
    proteins) to enhance the production of G.

Figure P.1.1 A metabolic pathway producing *G* from
A with feedback inhibition (solid curve) and
feedback repression (dashed curve).

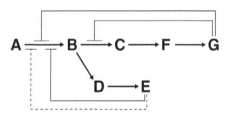

2

An Introduction to Industrial Microbiology and Cell Biotechnology

2.1 Universal Features of Cells

A cell is the basic unit of a living system. What are its critical features? First, there must be a barrier partitioning the internal environment of the cell from the external environment (Panel 2.1). There must also be transport mechanisms in place to allow desired materials (e.g., food and water) to cross the barrier and enter the cell. Using a process called anabolism, cells must be able to utilize these desired materials to make essential cellular components and to grow. Cells must also be able to derive chemical potential energy from the materials they acquire from the environment (using a process called catabolism), or from light. The chemical potential energy derived from catabolism is used in anabolism to make more cell mass and to maintain cells as a separate entity from the environment, as it takes energy to keep intracellular and extracellular environments apart.

Panel 2.1

A living organism has:

1) A barrier (i.e., lipid bilayer membrane)
2) Energy conversion (i.e., electron transfer and energy capture mechanism)
3) Catabolism and anabolism of organic materials
4) A transport system
5) Reproduction information

Lastly, cells must have the capability to reproduce and have a system in place to pass on genetic information. This genetic material must be replicated with a very high degree of fidelity, but also allow for some infidelity or "flexibility," so that small variations can arise, leading to some degree of diversity in the population. Such diversity allows some of the population that are more fit to survive in adverse environments. In this way, through the course of evolution, diverse species emerge, thrive, subside, and even become extinct.

Interestingly, despite a long history of evolution and vast diversity, the cells of all living organisms continue to share common features. They all use lipid bilayer membranes as a cellular barrier, the same set of materials to build cell mass, the same core pathways for catabolism and for using chemical potential energy, and the same genetic information system.

Engineering Principles in Biotechnology, First Edition. Wei-Shou Hu.
© 2018 John Wiley & Sons Ltd. Published 2018 by John Wiley & Sons Ltd.
Companion Website: www.wiley.com/go/hu/engineering_fundamentals_of_biotechnology

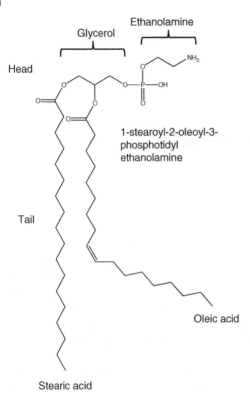

Head

Glycerol

Ethanolamine

NH₂

1-stearoyl-2-oleoyl-3-phosphotidyl ethanolamine

Tail

Oleic acid

Stearic acid

Figure 2.1 A phospholipid molecule. A hydroxyl group of the glycerol backbone is linked to a phosphate. A saturated fatty acid and an unsaturated fatty acid are attached to the other two hydroxyl groups.

2.2 Cell Membranes, Barriers, and Transporters

It is generally believed that life on earth originated in an aqueous environment. Even today, all cells in living systems reside in an aqueous environment. Their intracellular contents are separated from the external environment by a membrane consisting of a double lipid layer (i.e., a lipid bilayer). The lipids are amphipathic, meaning that they have a hydrophilic (water-loving) head group, with a negatively charged phosphate, and a hydrophobic (water-averting) tail group made of two fatty acid molecules (Figure 2.1). When suspended in an aqueous solution, amphipathic molecules, like soap, can form spherical-shaped micelles. Within micelles, the hydrophilic head group faces outside, in contact with its aqueous surroundings, while the hydrophobic tails are shielded within the spherical interior (Figure 2.2). Micelles may also encapsulate other hydrophobic molecules, as they maintain a non-aqueous environment within their core. In contrast, since a cell is enclosed by a lipid bilayer membrane, it can have an aqueous environment both inside and outside, which allows the interior of a living system to be in an aqueous phase.

Lipid bilayer membranes have a very high barrier property, due to the fatty acid tails of the structure. These tails organize into a highly ordered liquid crystal structure at very mild temperatures (Figure 2.3). They are fluidic with a high lateral diffusivity, but only a few very small molecules (like oxygen and ethanol) can freely diffuse across the bilayer. Except for O_2 and H_2O, the influx of nutrients and excretion of metabolites are mediated by specialized transport mechanisms. Cells therefore have a large number of transporters to allow molecules to cross their membrane in both directions.

Figure 2.2 Lipids forming micelles in aqueous phase. A lipid bilayer allows both the interior and exterior of the lipid droplet to be aqueous phase.

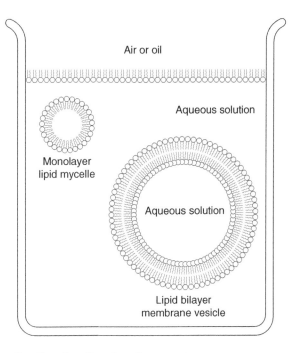

Figure 2.3 Lipid bilayer membrane in a fluid state.

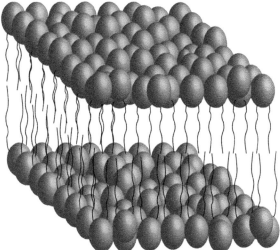

2.3 Energy Sources for Cells

Cells must acquire energy to perform the synthetic work required to make new cellular materials. They also need energy to maintain their intracellular environment at optimal conditions. Unlike the cells' external environment, which may undergo major perturbations, intracellular environments are relatively constant. The interior and exterior of cells differ with respect to the concentration of salts, metabolites, and macromolecules (like proteins and nucleic acids). This difference in solute concentrations causes an osmotic pressure difference between the intracellular and extracellular environments.

Even though the diffusion rate of a solute across the membrane is very small, solute concentrations on either side of the membrane will equalize, given enough time. To maintain their internal environment, cells use energy to selectively move solutes: some are actively brought in, others are barricaded inside, and still others are pumped out. Cells may need to physically move around and will use energy to perform mobility or other mechanical work. An example of mechanical work is when a cell undergoes division: cellular components, such as organelles, have to be moved around intracellularly to distribute to two daughter cells.

Energy exists in many forms: electrical energy, kinetic energy, thermal energy, and so on. A living system can only use photo energy and/or chemical potential energy as an energy source. Ultimately, the most fundamental source of energy for all living systems is photo (or light) energy. Plants and other photosynthetic microorganisms can "harvest" photo energy and convert it to chemical potential energy. This chemical potential energy, in the form of organic materials in plants and photosynthetic microorganisms, sustains the life of all non-photosynthetic organisms.

2.3.1 Classification of Microorganisms According to Their Energy Source

The most fundamental way to classify organisms is according to their energy source. Organisms can be categorized by those that derive their energy from light (phototrophs) versus those that derive their energy from chemical compounds (chemotrophs) (Figure 2.4). Among those that use photo energy, some use only inorganic compounds (e.g., CO_2) as their source of carbon. These are called photolithotrophs and are capable of "fixing" CO_2 into organic substances using the energy that they derive from light. Some photosynthetic microorganisms can use light as an energy source

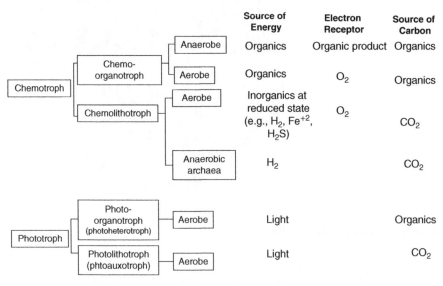

Auxotroph: Uses inorganic compounds as carbon sources.
Heterotroph: Uses organic compounds as carbon sources.

Figure 2.4 Classification of microorganisms according to their energy source, electron receptor, and carbon.

but cannot fix CO_2. They must therefore use additional organic materials as carbon sources (photo-organotrophs and photoheterotrophs). Those organisms that do not perform photosynthesis are called chemotrophs. They derive chemical potential energy from either organic compounds (chemolithotrophs) or inorganic compounds (chemo-organotrophs).

Most organisms use respiration to derive energy from compounds with a high chemical potential energy (i.e., at a more reduced state). Respiration occurs in either the presence (aerobic) or absence (anaerobic) of oxygen. During respiration, electrons flow from the energy source (an electron donor) to a terminal electron acceptor. In an aerobic process, the ultimate electron acceptor is oxygen. In anaerobic respiration, electrons eventually pass on to an acceptor, such as nitrate (NO_3^-) or fumarate ($^-OOC\ CH=CH\ COO^-$). In this process, the energy source (i.e., glucose) is oxidized while the electron acceptor is reduced and, thus, receives an electron. For instance, oxygen is reduced to water, nitrate to nitrite or nitric oxide, and fumarate to succinate. The reduced product is then excreted out of the cell.

Another process of deriving energy from an organic compound is by fermentation. Fermentation is also an anaerobic process, but it does not involve respiration or transferring electrons along the membranes. Rather, the electrons removed from glucose are stored in NADH and adenosine triphosphate (ATP) molecules. Glucose is first oxidized to an intermediate compound, pyruvate. The energy release from the oxidation is used to convert NAD^+ to a reduced (thus at a higher energetic state) NADH, and to convert adenosine diphosphate (ADP) to ATP. NADH and ATP are energy carriers, meaning they transit between a high-energy form (NADH and ATP) and a low-energy form (NAD^+ and ADP). NADH is then re-oxidized, reverting to NAD^+ in the second segment of fermentation that reduces the intermediate compound, pyruvate, to the fermentation product, such as ethanol and lactate. ATP is used for cellular work and is converted to ADP. Both NAD and ADP are recycled to ferment more glucose (Panel 2.2).

Panel 2.2 Anaerobic Metabolism of Glucose

Oxidation of glucose to pyruvate:

$$\text{Glucose} + 2\ \text{ADP} + 2\ \text{P}_i + 2\ \text{NAD}^+ \rightarrow 2\ \text{Pyruvate} + 2\ \text{ATP} + 2\ \text{NADH} \qquad (2.1)$$

Reduction of pyruvate (regeneration of NAD):

$$2\ \text{Pyruvate} + 2\ \text{NADH} \rightarrow 2\ \text{Lactate} + 2\ \text{NAD}^+ \qquad (2.2)$$

Net:

$$\text{Glucose} + 2\ \text{ADP} + 2\ \text{P}_i \rightarrow 2\ \text{Lactate} + 2\ \text{ATP} \qquad (2.3)$$

The process of breaking down organic molecules to generate a chemical energy reservoir of ATP is called catabolism. In breaking down the carbon source, whether aerobically or anaerobically, various reaction intermediates (or catabolites) are also produced. These catabolites are then used in anabolism to make cellular materials using the chemical potential energy derived from catabolism.

2.4 Material and Informational Foundation of Living Systems

2.4.1 All Cells Use the Same Molecular Building Blocks

All living systems are built from the same basic types of compounds: monosaccharides, amino acids, nucleotides, and lipids. These basic sets of compounds are called "building blocks." They are used to synthesize the major cellular materials, polysaccharides, proteins, nucleic acids, and cell membranes (Panel 2.3). While proteins, polysaccharides, and nucleic acids are polymers of their corresponding building blocks, the bilayer membrane is an assembly of the individual lipid molecules, but not a polymer. In addition to using the same building blocks, all organisms use the same basic types of molecules (ATP, NADH, NADPH, etc.) to store and transfer chemical potential energy. They all use the same mechanism of gene expression to pass their genetic information to their offspring. The chemical nature of living systems is thus very invariable, or conserved. When a property remains very similar among organisms that have diverged in evolution, the property is often referred to as being conserved. From the simplest microorganisms to the more sophisticated flowering plants and animals, the same sets of building blocks, biopolymers, and energy carriers and the same mechanisms of information transfer are used.

Panel 2.3 Building Blocks

- Carbohydrates (monosaccharides)
- L-amino acids
- Nucleotides
- Fatty acids, lipids

   This set of basic units is used to make biopolymers of oligo- and poly-saccharides, proteins, nucleic acids, and membranes.

2.4.2 Genes

The information that an individual organism uses to reproduce itself is stored in its DNA. Genetic information is realized in the cell through its transcription into RNA and, furthermore, the translation of RNA into protein (Figures 2.5 and 2.6). To ensure that genetic information is faithfully reproduced, living systems replicate DNA with a very high degree of fidelity. The processes of processing genetic information into RNAs and proteins are also highly regulated and checked for errors.

   The basic unit of heredity is a gene. The old notion (central dogma) was that one gene encodes one protein, and that one gene is a single segment of a DNA molecule. We now know that many different proteins may be made from the same DNA segment. Also, a single RNA transcript can be spliced differently to make many different messenger RNAs (mRNAs).

   We have also learned other surprising facts about genetic regulation. For instance, in eukaryotic cells only some regions in the transcript (exons), but not others (introns), are translated into a protein (Figure 2.6). Also, the same transcript can be spliced differently (alternative splicing), in different stages of cell growth or in different tissues in the same

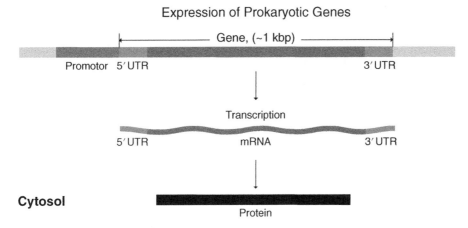

Figure 2.5 Gene expression from DNA to protein in prokaryotic cells.

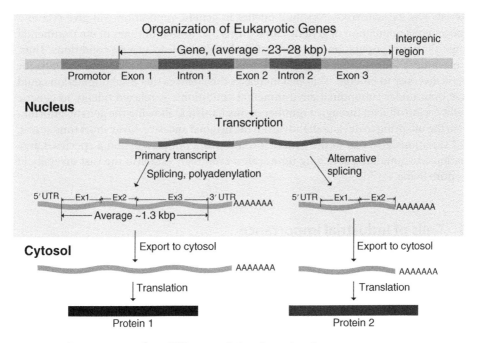

Figure 2.6 Gene expression from DNA to protein in eukaryotic cells.

organism, to give rise to different protein molecules. In some cases, a segment of DNA may be transcribed into an RNA species (noncoding RNA) that influences the expression of other genes without being translated into a protein. A gene is thus a segment of DNA that encodes a protein or a functional unit of RNA, such as tRNA (transfer RNA), rRNA (ribosomal RNA), or other noncoding RNA. Furthermore, a segment of DNA may be the locus of multiple genes.

2.4.3   Genetic Information Processing

The fidelity of the information-processing cell machinery is critical to the well-being of an organism. An adult human body consists of over roughly one trillion (10^{12}) cells. From fertilization to adulthood, the original fertilized egg cell goes through more than 40 cell divisions and, thus, an equal number of genome replication cycles. The DNA replication must be faithful to ensure all trillion cells have identical genetic information. Furthermore, the DNA in each cell must be faithfully decoded to produce the proteins that make up our form and carry out our bodily functions. A high-fidelity processing of the information ensures that the genetic blueprint accurately delivers the intended final product. The high fidelity in DNA replication and genetic information processing is crucial for preserving identity in offspring and the survival of the species. For example, the offspring of an *Escherichia coli* variant that has survived a harsh environment must retain the same advantageous trait to continue to survive.

However, DNA replication is not completely error-free. Errors have to be identified and corrected through additional error-checking (or "proofreading") and repair processes. A very high fidelity in DNA replication thus carries an energetic cost due to enhanced error detection and repair. Worse, an overly high fidelity in replicating genetic materials has its own risks. Absolute fidelity in genetic replication will give rise to a homogeneous population of genetic clones. This means that in times of environmental change, the entire population could be uniformly vulnerable to adverse conditions. Thus, the machinery for reproducing genetic information must also be flexible, to allow for genetic diversity to emerge. This also increases the chances that a subpopulation could thrive even under suboptimal environmental conditions. A relaxed fidelity in genetic heredity is introduced through a number of mechanisms, allowing the genetic information in the offspring to deviate slightly from the original ancestor. Over short time scales, small variations give rise to the genetic diversity of a population and a species. Large variations, accumulated over long time scales, eventually establish the vast diversity of the entire living world.

2.5   Cells of Industrial Importance

The cells used in industrial production include both microorganisms and singly grown cells derived from tissues of multicellular organisms, such as insects, fish, or mammals (Figure 2.7). All microorganisms are unicellular, consisting of a single cell. Microorganisms, from bacteria to fungi, can be grossly divided into two categories, prokaryotes and eukaryotes. Their category reflects the presence or absence of a nucleus and other membrane-bound organelles, such as mitochondria and chloroplasts. The prokaryotes include bacteria and archaea (Figure 2.8). Eukaryotic microorganisms include fungi (which encompass yeasts and molds) and protozoans.

In eukaryotic cells, the intracellular space is compartmentalized into different organelles by lipid bilayer membranes. These membranes act as a barrier to create different chemical environments and to allow specialized cellular functions to be carried out in different organelles.

Another way that cells have increased their functional complexity is by acquiring the ability to differentiate. Many bacteria and fungi are capable of differentiation and enter

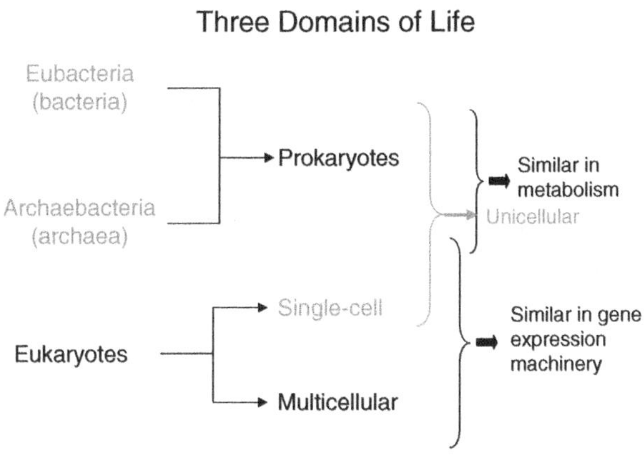

Three Domains of Life

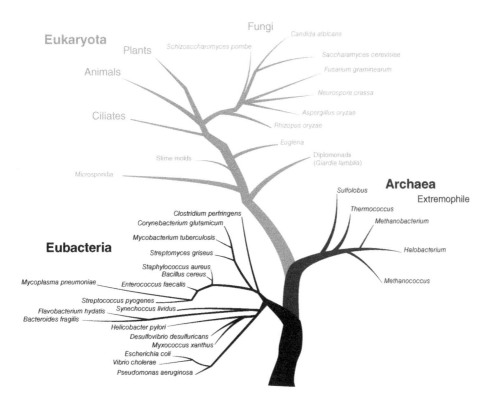

Figure 2.7 Organisms in the three domains of life.

Figure 2.8 Positions of representative industrial organisms in the tree of life.

different growth cycles in different times of their life. For example, many bacteria and molds grow as vegetative cells when conditions are favorable, and form spores when encountering an adverse growth environment. Spores give them a higher degree of tolerance to harsh environments. Once the conditions again become favorable, spores germinate and return to vegetative growth. The differentiation in unicellular organisms is thus "reversible"; cells can cycle through different differentiation states. Many natural products, such as antibiotics, are produced when microorganisms differentiate.

All multicellular organisms are eukaryotes. These cells are capable of differentiating into different tissues, in which each has a unique functional role, just like leaves and flowers do for plants. For industrial applications, differentiated cells or even tissues may be isolated from plant or animal tissues and expanded in culture to produce products. For example, plant shoots and roots may be grown in culture to produce secondary metabolites; prior to the recombinant DNA era, cells isolated from kidney tissue were explored for the production of urokinase.

2.5.1 Prokaryotes

Prokaryotes encompass two taxonomic groups: the subkingdom Eubacterium and the domain Archaea. Eubacterium is at the root of the tree of life (i.e., this kingdom emerged earliest on this planet). Archaea falls between Eubacterium and the eukaryotes. Cells of the Archaea domain are similar to bacteria in size and morphology. The most important distinction between archaea and eukaryotes is the absence of a nucleus. Both bacteria and archaea are, in general, physically smaller than eukaryotes, with a typical length or diameter of only 1 to a few microns. Their genomes are also smaller, typically on the order of a few million base pairs (bp) that are often organized into a single chromosome (although some larger bacterial genomes have more than one).

A prokaryote's average gene size, in terms of the length of DNA coding for a protein (i.e., a protein-coding sequence), is about 1000 base pairs (1 kbp). Without a nucleus, mRNA is transcribed directly in the cytoplasm and is also translated into proteins there. A typical bacterium has 3000–4500 genes. The parasitic bacteria *Microplasma* have very small genomes (only slightly more than 300 genes). The secondary metabolite–producing *Streptomyces* have upward of 7000 genes. Prokaryotes have evolved to reside in all sorts of different conditions on earth. Some species in the Archaea domain are hyperthermophiles that can grow at temperatures of 80–108 °C, a condition under which no eukaryote has been found to live.

2.5.2 Eubacteria

Bacteria are generally classified into two major groups, according to how their cell wall appears after staining with crystal violet and iodine (Gram stain). After staining and washing with ethanol, and then counterstaining with safarin to increase the contrast between stained and unstained cells, those cells whose walls are colored purple are called "Gram-positive." Those not stained are called "Gram-negative."

The cell walls of Gram-positive and negative bacteria have very different thicknesses and chemical compositions, which cause differences in their retention of the stain. Despite its long use in the characterization of bacteria, the Gram stain is not the most important criterion when comparing the evolutionary closeness of microorganisms. Nevertheless, this classification is still among the most basic methods for bacterial identification.

2.5.2.1 Cell Wall and Cell Membrane

The thickness of the cell wall layer differentiates Gram-positive from Gram-negative bacteria. The wall is composed of peptidoglycan, a polymer of glycan (modified

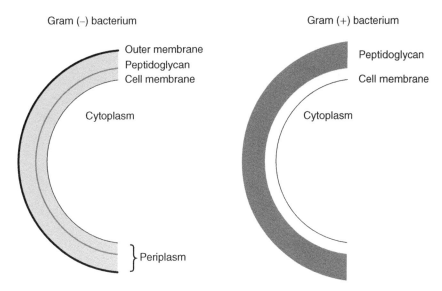

Figure 2.9 Cell wall organization in bacteria.

carbohydrate) that is cross-linked by peptides. It gives the bacteria their mechanical sturdiness and rigid structure (Figure 2.9).

In Gram-positive bacteria, a cell wall sits outside of the lipid bilayer membrane. The lipid bilayer controls the material exchange between the cell's interior and the environment. The cell wall is 15–30 nm thick and composed primarily of peptidoglycan. It is mixed with neutral or acidic polysaccharides, glycolipids, lipids, and other compounds. Gram-negative bacteria have two membranes. The inner membrane is a lipid bilayer that resides underneath a peptidoglycan layer. The outer membrane encircles the inner layer of peptidoglycan and the inner membrane, and it has an outer cover of lipopolysaccharide. It consists of a lipid core, facing inward, and oligosaccharide (O-antigen), facing outward.

For some pathogens, the O-antigen plays a critical role in their virulence. The outer membrane consists of saturated fatty acids, has a low permeability for many compounds, and provides barriers to hydrophobic compounds. Many proteins sit embedded within the outer membrane, and some serve as sensors or participate in transport.

In Gram-negative bacteria, there is a space in between the outer membrane and the cell membrane (cytoplasmic membrane) called the periplasmic space (periplasm) (Figure 2.10). The outer part of the periplasm is linked to the outer membrane by peptidoglycans. The periplasm is not just "space"; it is chemically complex. In fact, some of the more abundant cellular proteins are located in the periplasm, possibly involved in solute transport, hydrolysis of substrate, detoxification, or solute binding.

In the industrial production of recombinant proteins in *E. coli* cells, the periplasm is a location often used for depositing those products. By placing the protein product in the periplasm, rather than inside the cell, one does not need to break up the entire cell in order to release the product. This also allows the product protein to be folded in the more highly oxidative, extracellular environment. In some cases, placing a protein product in the periplasm also facilitates disulfide bond formation. This is in contrast to the expression of those heterologous proteins in the cytosol. Under the conditions

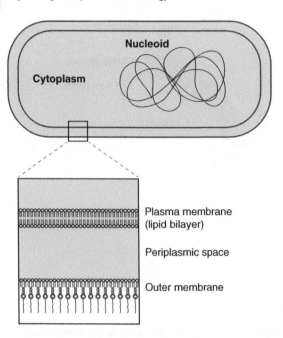

Figure 2.10 Periplasmic and cytoplasmic overexpression of proteins in *E. coli.*

Plasma membrane (lipid bilayer)

Periplasmic space

Outer membrane

of industrial production (i.e., a very high level of expression), the translated protein molecules typically fail to fold correctly and aggregate to each other through interactions of their hydrophobic regions. These aggregates form particles (called "inclusion bodies") that are visible under a transmission electron microscope. In industrial production, these inclusion bodies are extracted from cells after breaking up of the cell wall, and they need to be solubilized and refolded again to become active molecules.

The lipid bilayer that forms the membrane consists of phospholipids and encloses the cytoplasm. The soluble portion of the cytoplasm and cytosol has a very high concentration of protein (100–300 mg/mL). In addition to the cytosol, the nucleoid and ribosomes are distinctive in cytoplasm. The nucleoid, located approximately in the center of the cell, is where DNA and RNA synthesis occurs. A bacterial chromosome is a large DNA molecule sitting adjacent to the nucleoid. The DNA is tightly coiled and bound by many proteins. If completely extended, an *E. coli* (about 2 μm in size) DNA molecule is nearly 1 mm long. The ribosome is the machinery for making proteins. It is a complex particle, consisting of many ribosomal proteins and ribosomal RNA of different types (5S, 16S, and 23S). Each cell contains thousands of ribosomes. Bacterial ribosomes are complex [with a size of 22×30 nm, or 2.7×10^6 Dalton (a Dalton is a measurement of molecular weight)], but they are not the largest complexes in the cytoplasm. Although most enzymes in the cells are either "soluble" or reside on the membrane, some enzymes also form large multi-unit complexes. The pyruvate dehydrogenase complex, for example, is a complex of multiple copies of three enzymes that convert pyruvate to acetyl CoA and CO_2. Its total size is 4.6×10^6 Daltons.

2.5.2.2 Membrane and Energy Transformation
In bacteria and archaea, the cell membrane is the site of energy generation. Some membrane proteins act as a proton pump, to keep the two sides of the membrane at slightly different pH levels. The chemical potential energy derived from the catabolism of carbon sources [e.g., from the catabolism of glucose through glycolysis and the tricarboxylic acid

Figure 2.11 Electron transport across bacterial cytoplasmic membrane and generation of ATP.

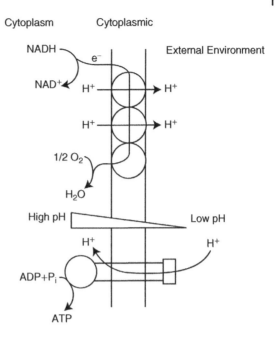

(TCA) cycle] is stored as reducing electrons in NADH. High-energy electrons in NADH are passed down to lower energy states through a series of membrane proteins, which act as electron relays. Electrons are eventually delivered to oxygen, which reduces it to H_2O (Figure 2.11). Oxygen is thus the ultimate electron acceptor. As an electron travels down its energy ladder, proton pumps drive protons across the membrane and against the pH gradient.

The typical growth environment of bacteria has a relatively neutral pH, whereas the intracellular pH is somewhat alkaline. *E. coli* cells have an intracellular pH of around 7.8. The pumping of protons, in addition to generating the proton concentration gradient, also creates a charge difference between the two sides of the membrane (with the outside being more negative). The resulting electromotive force attracts protons to flow back over the membrane and into the cytosol. Protons are transported through a membrane protein, ATP synthase (which actually consists of many protein molecules), along the pH gradient. As protons flow back across the membrane through ATP synthase, ATP is generated.

Some bacteria have adapted to grow in a very high-alkaline environment, which means their extracellular membrane has a higher pH than the interior of their cytoplasmic membrane. They have developed a mechanism for using a sodium pump and sodium gradient for maintaining an electromotive force to drive ATP generation.

In phototrophs, the intracytoplasmic membrane is developed as a photosynthetic apparatus. It exists in different forms, such as a flat sheet, a vesicle, or sacs. A similar membranous cytoplasmic structure is also seen in methanotrophs and in bacteria that are capable of nitrogen fixation (i.e., converting N_2 to NH_3).

2.5.2.3 Differentiation

As the microbial world evolves to become more complex, structural and functional specialization not only evolves in the internal (intracellular) space but also extends over a

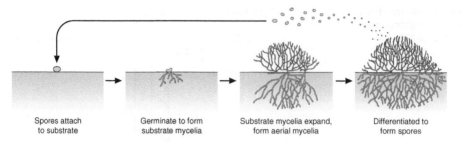

| Spores attach to substrate | Germinate to form substrate mycelia | Substrate mycelia expand, form aerial mycelia | Differentiated to form spores |

Figure 2.12 Differentiation of *Streptomycetes* from vegetative cells to spore.

population, over space, and over time. Different individual organisms of a population develop to serve different functions or morph into different shapes (morphology), typically under different conditions. These microorganisms also remain capable of returning to their fast-growing and undifferentiated stage (Figure 2.12).

The most notable differentiation in bacteria is the sporulation that happens in unfavorable environments. In *Bacillus*, the spores are smaller than vegetative cells, but otherwise reasonably similar in shape. In mycelial bacteria *Actinomycetes*, the vegetative cells form mycelia. First, the mycelia grow into the solid that the bacterium rests on to develop substrate mycelia. Later, mycelia grow into air to form aerial mycelia. It is the aerial mycelia that form spores. As spores rise from the surface of the soil, they have a better chance of being carried afar to a new environment, thus ensuring the survival of the species. Similar differentiation events, from vegetative growth to sporulation, are also seen in molds. In the course of differentiating to spores, many microorganisms also produce secondary metabolites. Numerous secondary metabolites have become antimicrobial, anticancer, or immunosuppressive drugs.

2.5.3 Archaea

In most respects, microorganisms in the Archaea domain resemble the eubacteria. They are morphologically similar to bacteria and may stain positive or negative by the classical Gram stain. Until about three decades ago, they were considered to be bacteria. They are no longer classified that way, because their cell walls do not contain peptidoglycan.

Structurally, they have many characteristics that are similar to eukaryotes (Panel 2.4). They have histone proteins that allow DNA to be tightly packaged into chromosomes (histone is not seen in bacteria). Their RNA polymerase, with eight to ten subunits, resembles that of eukaryotes. The sensitivity of their ribosome to translation-inhibitory antibiotics is also more similar to eukaryotes than to bacteria.

Panel 2.4 Similarity of Archaea to Eukaryotes

1) Membrane lipid has ether linkage, not ester linkage, to glycerol.
2) Cell wall has no peptidoglycan (only in bacteria).
3) Contain histones similar to eukaryotes (only in eukaryote).
4) RNA polymerase (has 8–0 subunits) similar to eukaryote.
5) Ribosomal sensitivity to antibiotic differs from bacteria.
   More similar to cytosolic ribosome in eukaryotes.

Most species in Archaea have rather unique physiological characteristics. Many are extremophiles, meaning they can grow in very extreme and hostile environments. Such conditions might include high salt levels (extreme halophile), the very low pH in sulfidic ore (extreme acidophile), the very high pH of carbonate-rich springs (extreme alkaliphile), and the high temperature of deep-sea volcanic vents (extreme thermophile) (Panel 2.5). "Methanogenic archaea," while growing under strict anaerobic conditions, use CO_2 and H_2 to produce CH_4 and derive energy. Other than photosynthetic organisms, few microorganisms are capable of fixing CO_2. These archaea use H_2 as an energy source instead of light for this process.

Panel 2.5 Archaea

- **Methanogenic**
  - Obligate anaerobic, reduces CO_2 to CH_4
- **Extremely halophilic**
  - Grows in 3–5 M NaCl
  - Has bacteiorhodopsin (harvests light energy)
  - Has halorhodoposin (chloride pump)
- **Extremely thermophilic**
  - Grows in 55–100 °C

Some industrial enzyme-catalyzed processes occur at an elevated temperature. During the course of polymerase chain reaction (PCR), for instance, the enzyme (DNA polymerase) in the reaction mixture undergoes numerous high-temperature cycles and is prone to heat inactivation. For those applications involving extremely harsh conditions, enzymes isolated from extremophiles of the Archaea domain have become the workhorses.

2.5.4 Eukaryotes

Eukaryotic cells increase their complexity by developing membranous structures inside of their cells. Their intracellular space is segregated into specialized environments for different functions. This elaborate compartmentalization of cellular spaces is accompanied with an increase in cell size. The eukaryotic genomes are larger as well. A eukaryotic gene consists of not only the sequences that will become the mature mRNA (called exons) but also segments that disperse in the gene (called introns). Introns, while being initially transcribed into pro-mRNA, are later removed (or "spliced") to give rise to mature mRNA that includes only exons. An average gene in mammals is about 2–28 kbp, even though the average protein-coding regions only correspond to 1.3 kbp.

In general, the genome of eukaryotes encodes a larger number of genes. For example, fruit flies have about 17,000 genes, while mammals have over 30,000 genes, as opposed to the 3000–5000 range seen in bacteria. Some simpler eukaryotes, like the budding yeast *Saccharomyces cerevisiae,* have only about 6000 genes. This is less than the antibiotic-producing *Streptomyces coelicolor*, which has more than 7000 genes. However, its genome size is about 12 Mbp, which is substantially larger than the 7 Mbp genome size of *S. coelicolor.*

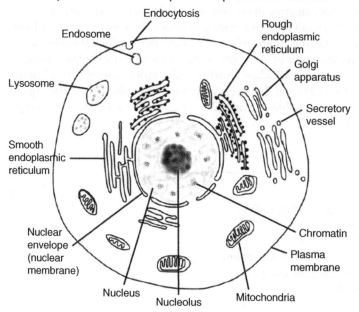

Organelles in an animal cell

Figure 2.13 Animal (eukaryotic) cells with organelles.

In addition to the larger size and generally larger number of their genes, eukaryotes also have repetitive sequences in their genome. These segments are sequentially similar, can vary in size, and are dispersed throughout the genome. These repetitive sequences and many other regions of the genome do not code for proteins. But they are not "junk," as they were once thought to be. They may help regulate gene expression, for example, by allowing transcriptional regulators to bind, or by being transcribed into noncoding RNAs and micro-RNAs, elements that are increasingly being found to have vast impact on global gene expression.

The compartmentalized cellular space of eukaryotes enables the distribution of functions to specialized environments (Figure 2.13). The presence of organelles also allows different chemical environments, or even separate phases (such as fatty acids or other hydrocarbons that are immiscible), to be kept spatially isolated. Lysosome contains an acidic environment, allowing for the hydrolysis of cellular waste. Many fungi also have a high capacity for protein secretion using specialized secretory pathways in the endoplasmic reticulum and Golgi apparatus.

2.5.4.1 The Nucleus

In eukaryotes, DNA molecules are enclosed inside a double-membrane-encapsulated nucleus. The genome size of eukaryotes is larger than that of bacteria [*E. coli* ($\sim 4.0 \times 10^6$ bp)], ranging from moderately larger in yeast ($\sim 1.2 \times 10^7$ bp) to three orders of magnitude larger in mammals ($\sim 2–3 \times 10^9$ bp). Unlike bacterial DNA, which typically forms a single linear or circular molecule, eukaryotic cells organize their DNA into separately packed structures, called chromosomes. By doing so, eukaryotes are able to contain a large amount of DNA in a small space.

The compaction of chromosomes is achieved by forming DNA–protein (with the protein primarily being histones) complexes. The nuclear compartment segregates DNA–RNA synthesis and ribosome assembly from metabolic processes and protein synthesis that occurs in the cytoplasm. Ribosomes are assembled in the nucleus (specifically, in the nucleoli) and are exported into the cytoplasm for protein synthesis.

The complex tasks of sorting out which segments of DNA or which genes are to be transcribed into RNA at any given moment occur in the nucleus. A large array of transcription factors and other transcription regulators exert their control by acting on DNA in the nucleus. The transcribed RNA (call pro-mRNA) is first spliced, to remove introns. The resulting mature mRNA consists of a 5′-UTR (untranslated region), the protein-coding regions (exons), and a 3′-UTR. Additionally, it is polyadenylated at its 3′ end with up to a few hundred bases of adenosine. Polyadenylation is a signal for the mature mRNAs to then be exported to the cytosol for translation.

2.5.4.2 Mitochondrion

Mitochondria have a size similar to bacteria. They are thought to have originated from a bacterial-like structure that was acquired by primitive eukaryotes. Mitochondria are the power plants of the cell. Cells increase or decrease the number of these power plants according to the demand. A high-energy-demanding animal cell can have as many as 3000 mitochondria.

Mitochondria have double membranes. Only the inner membrane is a true lipid bilayer with true barrier properties that prevent free exchange of solutes between mitochondrial lumen (cell matrix) and intermembrane space (Figure 2.14). The main

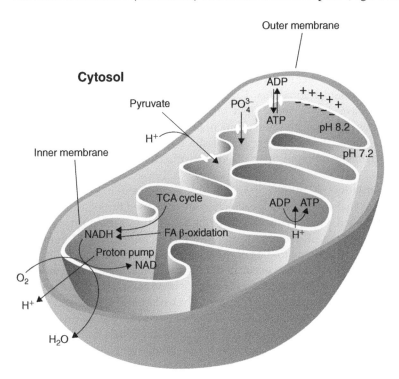

Figure 2.14 A mitochondrion and major energy generation activities taking place in the inner membrane.

ATP generation process happens via electron transfer and occurs across the inner mitochondrion membrane. Combined, the mitochondrial membrane surface area is greater than that of the cytoplasmic membrane. Mitochondria, being the site of reactions between electrons and oxygen, are rich in potentially damaging free radical species. By confining those reactions within mitochondria, cells can potentially reduce cellular damage.

Metabolically, mitochondria are the sites where the Kreb cycle and fatty acid oxidation take place. Pyruvate generated in glycolysis is transported into the mitochondria and further oxidized. The chemical potential energy of the oxidation reaction is preserved in NADH and FADH, while the carbon skeleton is converted to CO_2. In the electron transport chain, electrons from NADH and FADH are transferred to the final electron acceptor, O_2. As the electron is passed through the protein complexes of the electron transport chain, a proton is pumped out of the mitochondrial inner membrane against the proton concentration outside (which is ten times higher). The lower pH (by 1.0-unit pH) outside the mitochondria then drives the proton to return to the lumen of the mitochondria through ATP synthase. As the proton passes through ATP synthase, ATP is generated from ADP.

The mitochondrion resembles a bacterium in a few respects. First, its size. It also has its own genome, which is a circular DNA molecule. Mitochondrial DNA is regulated and replicated separately from genomic DNA. The biogenesis (increase in number) of mitochondria is also independent of cell division.

In addition to mitochondria, photosynthetic eukaryotic organisms require chloroplasts. The chloroplast, also thought to have originated from photosynthetic bacteria, is a major energy-processing center in the cells. It converts photo energy to chemical potential energy.

2.5.4.3 Endoplasmic Reticulum and Golgi Apparatus

In addition to mitochondria, other major eukaryotic organelles include the endoplasmic reticulum (ER) and the Golgi apparatus. Except for the nucleus and mitochondria, all other organelles have only a single lipid bilayer membrane. The ER and Golgi apparatus are the sites of folding and processing of proteins. These proteins ultimately either are excreted out of the cell or are intended for special locations such as the lysosome or cytoplasmic membrane. Specialized biosynthesis and metabolism are also performed here.

Through the translation of mRNA, proteins are synthesized in the cytoplasm (Figure 2.15). Proteins that are destined for secretion are translated through an mRNA–ribosome complex located on the cytosolic side of the ER. As they are being translated, they translocate through the ER membrane and into the ER lumen. The ER lumen is rich in other proteins that facilitate folding of the synthesized protein.

Posttranslational modifications also occur in the ER and Golgi apparatus. Examples of such modifications include the formation of disulfide bonds and the addition of carbohydrates to the protein molecule on specific sites (glycosylation).

Many specialized chemical reactions also happen here. For example, parts of cholesterol biosynthetic enzymes reside in the ER. Many detoxification reactions of toxins and other water-insoluble molecules occur in the ER of liver cells (hepatocytes). Hepatocytes also secrete many proteins, including albumin and coagulation factors that constitute many of the protein components in blood. Liver hepatocytes, which are

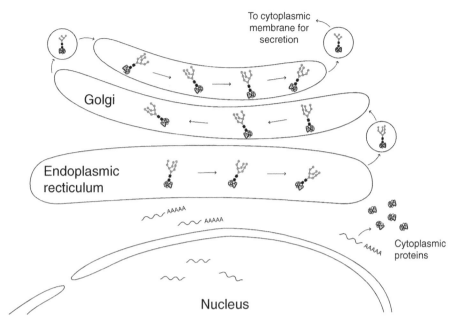

Figure 2.15 Expression and secretion of an extracellular protein in a eukaryotic cell.

specialized for protein secretion, have an abundance of "rough ER," so named because of the accumulation of a large number of mRNA–ribosome complexes on the outside of the ER membrane. Conversely, those more specialized in oxidative detoxification and xenobiotic metabolism are rich in "smooth ER."

2.5.4.4 Other Organelles

The lysosome is a spherical organelle with a low interior pH. When cellular materials exceed their natural lifespan and are no longer needed by the cell, they are degraded here. Every catalytic reaction that relies on an enzyme has a certain probability of going awry. For example, a reactant may fail to convert and "seize up" the enzyme, or an amino acid in the enzyme may accidentally become modified. Even in the cellular environment, some amino acids in the protein may get improperly oxidized. The accumulation of such modified amino acids may render a protein nonfunctional. Most proteins, thus, have finite lifespans and are "turned over" or "recycled" to prevent them from functioning improperly and damaging the cell. Proteins that need to be turned over are tagged by ubiquitin and sent to the proteasome for degradation.

The lysosome contains a large number of digestive enzymes. It has a proton pump on its membrane to maintain its low internal pH (pH 5.0). Some cells in higher organisms are specialized "scavengers," which engulf foreign particles or dead cells. The lysosome is the site in these specialized cells where engulfed particles are degraded.

In some fat cells, the metabolism of lipid occurs in the peroxisome. These reactions generate large amounts of reactive oxygen and have to be contained in these specialized organelles.

Eukaryotic cells are also able to endocytose external particles, by enclosing the particles in a cellular membrane and internalizing them as vesicles. This capability is not seen in prokaryotes. In their case, nutrient uptake occurs solely via absorption through the membrane, and is usually mediated by various transport proteins.

2.5.4.5 Cytosol

The cytosol is where most of the metabolic reactions and protein synthesis occur. The majority of cellular energy is generated in mitochondria as ATP. It is then exported into the cytosol and used for biosynthesis and other work. Eukaryotic cells, being larger in size, also exhibit cytoplasm streaming and complex intervesicle transport. They have cytoskeletons of intracellular mechanical fibrous structure. This structure includes microtubules, actin filaments, and intermediate filaments, which provide mechanical support for sustaining the large cell's shape and movement. The cytoskeleton also functions to translocate cytoplasmic organelles and provides the mechanical force for mitosis and cell division.

Example 2.1   Comparison of Oxygen Flux in *E. coli* and Human Cells

We will perform an order-of-magnitude estimation of the material flow rate across the surface of an actively growing *E. coli* cell. An *E. coli* cell can be approximated as a cylinder of 1 μm in diameter and 2 μm in length. When an *E. coli* cell is growing in a medium containing glucose and ammonium, the yield coefficient based on glucose is 0.3 g/g. In other words, 0.3 g of dry cell biomass is formed for every gram of glucose consumed. For each mole of glucose consumed, the *E. coli* cell also consumes 5.5 moles of oxygen. It takes 1 h for an *E. coli* cell to reproduce itself and generate twice the biomass.

A human cell (20 μm in diameter) consumes 5×10^{-10} mmol/cell-h of oxygen.

We will calculate the flux of glucose and oxygen across the cellular membrane of an *E. coli* cell, then compare it to a human cell.

Solution

Neglecting the area of the two ends of the cylinder, the surface area of an *E. coli* cell is the perimeter multiplied by the length of the cylinder. The surface area (a) and volume (V) are:

$$a = \pi \times 1\mu m \times 2\mu m = 6.28 \times 10^{-8} cm^2$$

$$V = \pi \times \left(\frac{1}{2}\mu m\right)^2 \times 2\mu m = 1.57 \times 10^{-12} cm^3$$

The specific density of the cell is about 1.02. For the purpose of an order-of-magnitude analysis, we can consider it to be 1.0. The mass of the content of a single cell is $\sim 1.6 \times 10^{-12}$ g.

The cell volume consists of 75% water and 25% dry biomass. The amount of dry biomass in a cell is:

$$m_{dry\ cell} = 4 \times 10^{-13} g$$

The yield of glucose is 0.3 g/g:

$$\frac{m_{glc\ consumed}}{m_{dry\ cell}} = \frac{1g}{0.3g} \quad m_{glc} = 1.3 \times 10^{-12} g$$

It takes 1.3×10^{-12} g of glucose to produce the biomass of a single *E. coli* cell in 1 h. On a molar basis, the specific glucose consumption rate is:

$$q_{glc} = \frac{1.3 \times 10^{-12} g/h}{180 g/mol} = 7.22 \times 10^{-15} mol/h$$

The amount of oxygen consumed can then be calculated:

$$q_{O2} = 5.5 \times q_{glc} = 3.97 \times 10^{-14} mol/h$$

The oxygen flux for an *E. coli* cell is:

$$3.97 \times 10^{-14} mol/h \div 6.28 \times 10^{-8} cm^2 = 6.32 \times 10^{-7} mol/h \cdot cm^2$$

We then proceed to calculate the oxygen flux for a human cell. The volume and surface area of a human cell are:

$$a = 4\pi r^2 = 4 \cdot \pi \cdot \left(\frac{20 \times 10^{-4} cm}{2} \right)^2 = 1.26 \times 10^{-5} cm^2$$

$$V = \frac{4}{3}\pi r^3 = 4.19 \times 10^{-9} cm^3$$

$$5 \times 10^{-10} \frac{mmol}{cell} \cdot h \div 1.26 \times 10^{-5} \frac{cm^2}{cell}$$

$$= 3.97 \times 10^{-5} \frac{mmol}{cm^2} \cdot h = 3.97 \times 10^{-8} mol/h \cdot cm^2$$

The oxygen flux is about 10 times higher in an *E. coli* cell than in human cells.

One can see that the volume of a human cell is three orders of magnitude larger than an *E. coli* cell, but its oxygen consumption rate is only about 100 times higher.

This example also illustrates a typical engineering analysis of biological systems. In the process of analysis, we often make necessary assumptions to get an estimate of various values.

2.6 Cells Derived from Multicellular Organisms

Cells from plants and animals can be isolated from different tissues and grown in culture. While most are used mainly in research, many are industrially and economically important. For example, Chinese hamster ovary (CHO) cells (isolated from the Chinese hamster ovary) are used as host cells for the production of many therapeutic proteins. A number of human cell lines are used for the production of viral vaccines. In general, those cells were isolated from dissociated tissues and then placed under culture conditions that are permissive for their proliferation. Depending on the tissue and species of origin, the derived cells have rather different characteristics.

While cultured cells retain the basic structure and characteristics of eukaryotic cells, they also bear one fundamental difference from unicellular organisms. In unicellular organisms, each individual cell thrives and survives separately. In multicellular organisms, the fight for survival and competition is among the organisms, but not among individual cells.

The growth and differentiation of individual cells are regulated and coordinated by developmental events, but are not based on nutrient abundance. The barrier with the environment is also at the level of the organism, and not at the cellular level. When encountering adverse conditions, the organism uses its barrier like a skin, to protect the cells in the body and to maintain "optimal" conditions. For this reason, cells derived from multicellular organisms often have a narrower range of tolerable growth conditions than unicellular organisms. For instance, *E. coli* cells can grow under a wide range of temperature (20–39 °C), pH levels (6–8), and osmotic pressure. Cells isolated from a chicken, mouse, or human would have a much narrower range, beyond which growth ceases.

2.7 Concluding Remarks

In the past century, humans have successfully used the metabolic activities of a large number of microorganisms to prepare better foods, produce industrial biochemicals, and create new pharmaceuticals. Those microorganisms reside naturally in vastly diverse environments and have adapted to tremendously different habitats by possessing various metabolic and synthetic capabilities.

In the past two decades, we have also begun to exploit the biological capacity of cells isolated from multicellular organisms, particularly to produce pharmaceutical biologics. It is humbling to note that the number of organisms that humans have isolated, cultured, and identified is only a small fraction of the whole. More recently, we have begun to explore the genetic capacity of organisms. Previously, this was beyond our reach. Now, even genes from exotic organisms are increasingly being used to enrich the synthetic capability of host cells. With advances in genome technology and genome-engineering tools, we may begin to see an expanding repertoire of host organisms and their genome-engineered counterparts for the biomanufacturing of new classes of compounds.

Further Reading

Alberts, B, Bray, D, Hopkin, K, Johnson, AD, Lewis, J, Raff, M, Roberts, K, Walter, P 2013, '1 Cells: The Fundamental Units of Life' in *Essential Cell Biology*. Garland Science, New York.

White, D, Drummond, J & Fuqua, C 2012, 'Chapter 1, Structure and Function' in *Physiology and Biochemistry of Prokaryotes*, 4th edn. Oxford University Press, New York.

Problems

A1 **True or False**: Microorganisms in Archaea are phylogenetically grouped between eubacteria and eukaryotes because they have organelles (like eukaryotes) but also have cell walls (unlike eukaryotes).

A2 **True or False**: All higher organisms that undergo sexual reproduction are diploid at some stage in their life cycle and haploid in another; and in between the diploid and haploid states, their chromosomes are re-sorted, a source of genetic diversity.

A3 **True or False**: An extremophilic bacterium isolated from a hot spring can reduce inorganic sulfur (SO_4^{2-}) to derive energy.

A4 **True or False**: A hydrophilic amino acid is more likely to be located on the exterior surface of a folded protein rather than on its interior.

A5 **True or False**: DNA is connected by peptide bonds.

A6 Which of the following is a correct statement about hydrophobic interactions?
   a) They are driven by the entropy change in structured water.
   b) They are important for protein folding.
   c) They bring two hydrophobic species together.

A7 Which is the strongest type of chemical bond?
   a) van der Waals interactions
   b) Ionic interactions
   c) Covalent bonds
   d) Hydrogen bonds

A8 Which of the following is NOT a common element for biomolecules?
   a) Phosphorus
   b) Hydrogen
   c) Silicon
   d) Sulfur

A9 How many bonds does phosphorus normally form?
   a) 1
   b) 2
   c) 3
   d) 4
   e) 5

A10 Prokaryotes possess which of the following structures:
   a) Chromatin
   b) Nucleoid
   c) Lipid bilayer
   d) Ribosome

A11 Which of the following types of microorganisms is not frequently seen in archaea?
   a) Methanogen
   b) Extreme halophile
   c) Ethanogen
   d) Extreme thermophile

A12 Which of the following is a property of archaea?
   a) Morphologically similar to eubacteria
   b) Uses the same DNA-compacting agents as eubacteria
   c) DNA replication happens in the nucleus.
   d) All of the above
   e) None of the above

A13 List three commonly seen ionic interactions in a cell.

A14 A thermophilic single-cell microorganism isolated from a hot spring in Yellow-stone National Park can use inorganic elements as an energy source. Which of the following statements is true about this microorganism?
a) It is more likely to belong to archaea than eubacteria.
b) It may use CO_2 as a carbon source.
c) Fe^{+3} cannot be an energy source for this microbe.
d) It has a cell wall.

A15 A chemolithotrophic archaea can grow anaerobically on H_2 and CO_2. It oxidizes H_2 to H_2O using the oxygen atom from CO_2. Which compound is the energy source? Which is the electron acceptor? Which is the carbon source?

A16 What are the building blocks of biological systems, and what are the polymers or assembly that are built from those building blocks?

A17 In a few sentences, describe each of the following items:
a) Halophile
b) Haploid
c) Lysosome
d) Endoplasmic reticulum
e) Alternative splicing

A18 Which of the following are among the top five most abundant elements in the cell?
a) C
b) H
c) O
d) N
e) P
f) S
g) Fe
h) Mg

A19 **Scale dimensions:** The dimensions important to engineering analysis are length, mass, and time. Plot the scales of length and mass relevant to living systems, from subcellular components, cells, and organisms (from viruses to multicellular organisms). Construct a timescale to indicate: (1) the characteristic time of an enzymatic reaction (e.g., how long it takes for a peptide bond to be hydrolyzed by chymotrypsin); (2) the doubling time of *E. coli*, yeast, and a human cell; and (3) the lifespans of a fruit fly, a *Caenorhabditis elegans* organism, a mouse, and a human.

3

Stoichiometry of Biochemical Reactions and Cell Growth

3.1 Stoichiometry of Biochemical Reactions

Cell growth is a process of chemical conversion. External nutrients are taken up by cells and, through biochemical reactions, a portion of these nutrients are used to generate chemical potential energy. This energy is then used to transform the other portion of nutrients into cellular materials. In this process of chemical conversion, some metabolites may be excreted as products. As the cell gains cellular materials and increases in size, cell replication can occur. In this chapter, we will focus on the material aspects of cell growth and product formation.

Like other chemical reactions, a biochemical reaction can be represented by a stoichiometric equation consisting of the molecular formulas of the reactants and products and their corresponding stoichiometric coefficients. As an example, we will begin with glycolysis reactions. The stoichiometric equations representing glycolysis are shown in Panels 3.1, 3.2, and 3.3. A metabolic pathway can also be depicted with the flow of carbon skeletal compounds. In this format, the "main" compounds (whose structure is being transformed) are lined up in the main path, and the other co-substrates, cofactors, and reaction participants [such as water (H_2O) and carbon dioxide (CO_2)] are positioned to the side.

The glycolysis pathway is described in Figure 3.1. In this case, since the molecular transformation centers on glucose ($C_6H_{12}O_6$), only glucose and the molecules derived from it are listed in the central flow. Considering the flow of carbon as the main artery, the pathway from glucose to pyruvate ($C_3H_4O_3$) is linear, without any branching point.

The pyruvate generated from glycolysis may then continue on to the tricarboxylic acid (TCA) cycle (Figure 3.2), where it can be decarboxylated to form acetyl coenzyme A (acetyl CoA) or can be reduced to ethanol (C_2H_6O) in yeast, or lactate ($C_3H_6O_3$) in mammals (Panel 3.4). At the node of pyruvate, the material flux thus diverges into two branches. Thus, metabolic pathways may be linear or branched. At the node of glucose-6-phosphate (G6P), another branch is formed that leads to the pentose phosphate pathway (PPP) (Figure 3.5). In the PPP, a portion of G6P is diverted to generate five-carbon sugars for nucleotide synthesis. Even closer examination will reveal another small side stream, which diverts pyruvate into alanine. Furthermore, dihydroxy acetone-3-phosphate may be used to form glycerol-3-phosphate (G3P).

Engineering Principles in Biotechnology, First Edition. Wei-Shou Hu.
© 2018 John Wiley & Sons Ltd. Published 2018 by John Wiley & Sons Ltd.
Companion Website: www.wiley.com/go/hu/engineering_fundamentals_of_biotechnology

Panel 3.1 Glycolysis Reactions

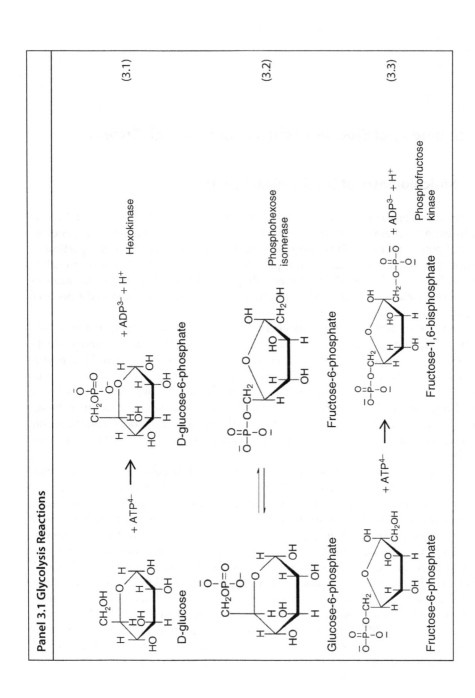

Panel 3.2

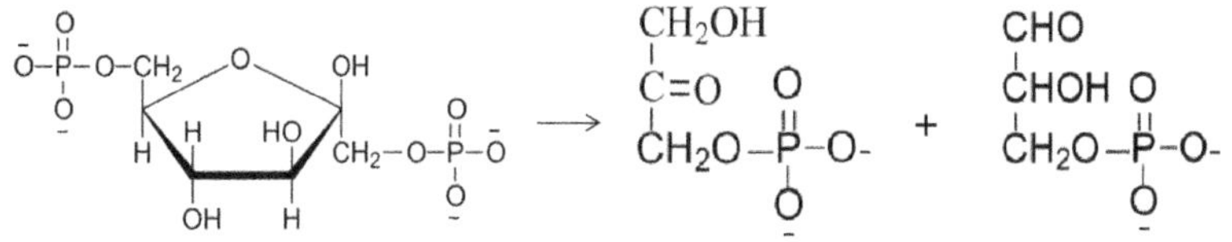

Fructose-1,6-bisphosphate → Dihydroxyacetone phosphate + Glyceraldehyde 3-phosphate   Aldolase   (3.4)

Dihydroxyacetone phosphate ⇌ Glyceraldehyde 3-phosphate   Triose phosphate isomerase   (3.5)

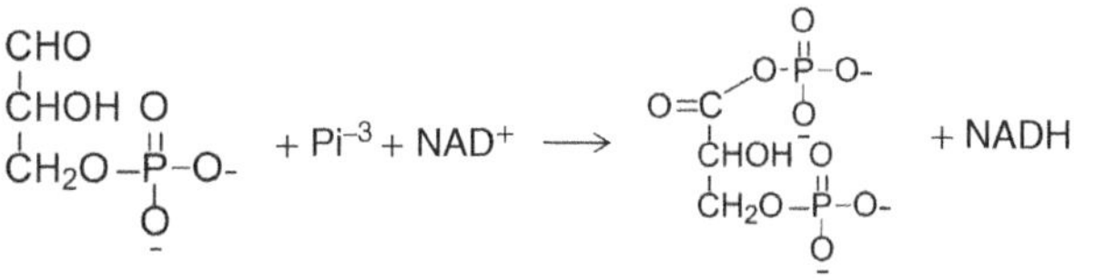

Glyceraldehyde 3-phosphate + Pi^{-3} + NAD$^+$ → 1,3 Bisphosphoglycerate + NADH   Glyceraldehyde-3 phosphate dehydrogenase   (3.6)

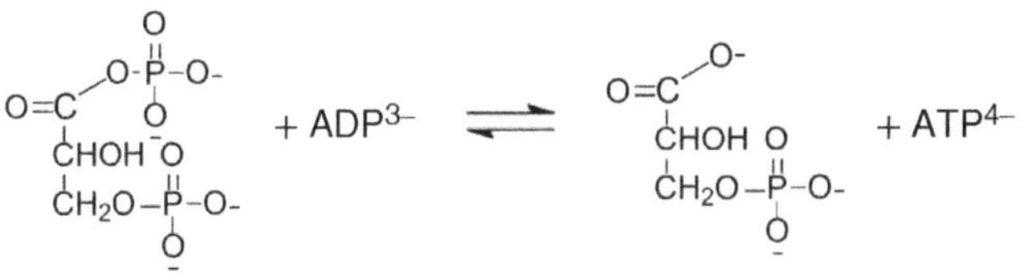

1,3 Bisphosphoglycerate + ADP^{3-} ⇌ 3-Phosphoglycerate + ATP^{4-} Phosphoglyceate kinase   (3.7)

Panel 3.3

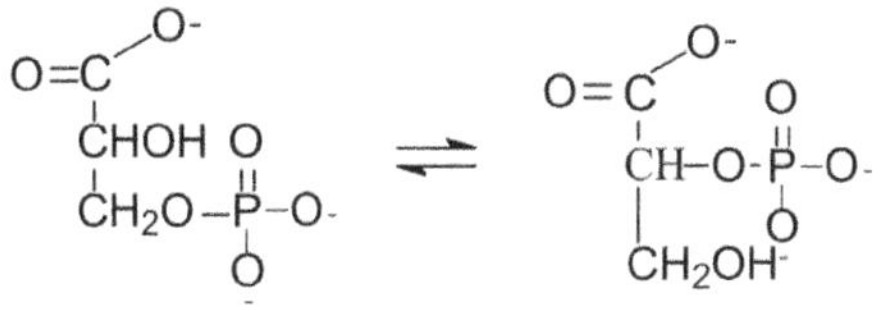

3-Phosphoglycerate ⇌ 2-Phosphoglycerate   Phosphoglycerate mutase   (3.8)

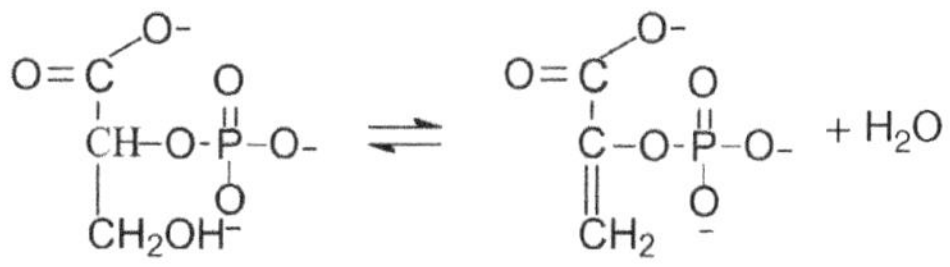

2-Phosphoglycerate ⇌ Phosphoenol pyruvate + H$_2$O   Enolase   (3.9)

(Continued)

Panel 3.3 (Continued)

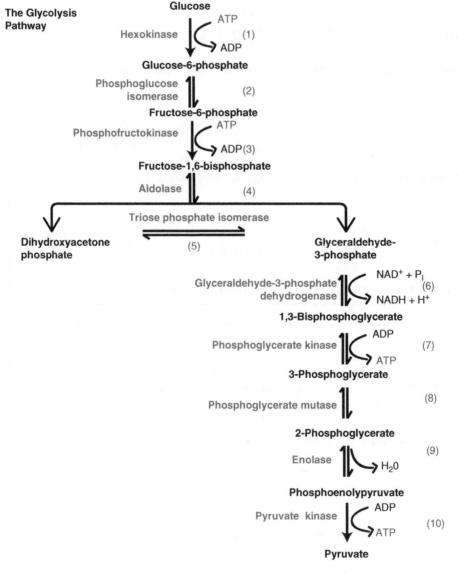

Phosphoenol pyruvate → Pyruvate + ATP⁴⁻, Pyruvate kinase (3.10)

Figure 3.1 Glycolysis pathway. The stoichiometric reaction equations are shown in Panels 3.1–3.3. The corresponding reaction number is shown beside each reaction step.

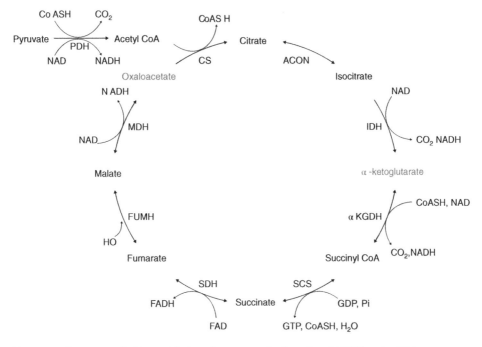

Figure 3.2 Energy metabolism: oxidation of pyruvate, tricarboxylic acid (TCA) cycle. PDH: pyruvate dehydrogenase; CS: citrate synthase; ACON: aconitase; IDH: isocitrate dehydrogenase; αKGDH: α-ketoglutarate dehydrogenase; SCS: succinyl CoA synthetase; SDH: succinate dehydrogenase; FHMU: fumarase; MDH: malate dehydrogenase.

Panel 3.4 Reduction of Pyruvate and Regeneration of NAD		
Anaerobic growth of yeast: ethanol fermentation		
$CH_3COCOO^- \longrightarrow CH_3CHO + CO_2$ Pyruvate $\qquad$ Acetaldehyde	Pyruvate decarboxylase	(3.11)
$CH_3CHO + NADH + H^+ \longrightarrow CH_3CH_2OH + NAD^+$ Acetaldehyde $\qquad\qquad$ Ethanol	Acetaldehyde dehydrogenase	(3.12)
Hypoxic exercising muscle: lactate formation		
$CH_3COCOO^- + NADH + H^+ \longrightarrow CH_3 \cdot CHOH \cdot COO^- + NAD^+$ Pyruvate $\qquad\qquad\qquad$ Lactate	Lactate dehydrogenase	(3.13)

Material balance is performed on a set of biochemical reactions to determine how much of the input materials are channeled to different branches, to evaluate the efficiency of resource utilization, and to calculate the theoretical conversion from a raw material to a product. When designing a new synthetic pathway, material balance is used to help assess the efficiency of alternative routes for maximizing process performance.

In general, material balance needs to be performed only at the branching points of pathways, because the molar fluxes of reactions in a linear pathway are related by their stoichiometric coefficients. For example, considering glycolysis as a linear pathway, the ratio of the molar flux of the hexokinase reaction to the pyruvate kinase reaction is 1:2.

The number of branches a pathway has depends on how detailed the pathway is. Depending on the desired accuracy, some branches with a small flux may be neglected. At the branching of pyruvate to lactate or the TCA cycle, all of the glucose carbons coming downstream go to one of the two branches. If the objective is to determine the magnitude of flux of glycolysis going to each branch in an exercising muscle, the PPP may be ignored since it typically diverts only a low percentage of the total glucose carbons from glycolysis. The amount of carbons that are diverted to alanine and glycerol-3-phosphate is even smaller than the PPP. Therefore, when detailing glycolysis flux in an exercising muscle, lactate and the TCA cycle are included, whereas the PPP and other smaller branches may be neglected, unless greater detail is needed.

The stoichiometric equation of the overall reaction can be obtained by summing up all of the component reactions. The overall reaction of glycolysis from glucose to pyruvate is shown in Panel 3.5. It is important to remember that a typical flow presentation does not show the stoichiometric coefficients. Writing down the stoichiometric equation of the overall reaction helps to ensure that the stoichiometric coefficient is accurate. We will employ two pathways, the glycolysis pathway and the PPP, to illustrate the systematic practice of material balance on biochemical pathways.

Panel 3.5 Overall Reaction of Glycosis

Overall reaction of glycolysis:

$$
\begin{array}{c}
\text{CHO} \\
|\\
\text{HCOH} \\
|\\
\text{HOCOH} \\
|\\
\text{HCOH} \\
|\\
\text{HCOH} \\
|\\
\text{CH}_2\text{OH}
\end{array}
+ 2\,ADP^{3-} + 2\,NAD^+ + 2\,P_i^{3-} \longrightarrow 2
\begin{array}{c}
O = C \diagdown O^- \\
|\\
C = O \\
|\\
CH_3
\end{array}
+ 2\,ATP^{4-} + 2\,NADH + 2\,H_2O
$$

$$(3.14)$$

Overall reaction of ethanol formation from pyruvate:

$$CH_3COCOO^- + NADH + H^+ \longrightarrow C_2H_5OH + CO_2 + NAD \qquad (3.15)$$

Overall reaction of lactate formation from pyruvate:

$$CH_3COCOO^- + NADH + H^+ \longrightarrow CH_3 \cdot CHOH \cdot COO^- + NAD^+ \qquad (3.13)$$

3.1.1 Metabolic Flux at Steady State

In energy metabolism, pyruvate generated in glycolysis is channeled into two streams. One goes into a reductive branch to become lactate or ethanol; the other enters the oxidative branch of the TCA cycle. The reductive branch is active in an exercising and lactate-excreting muscle and in ethanol-fermenting yeast. The overall reactions for glycolysis and the formation of ethanol or lactate are shown in Panel 3.4. Writing up the overall reaction as in Panel 3.4 is materially more revealing than the flowchart pathway shown in Figure 3.2. From the overall reaction of glycolysis, it is clear that the reactants glucose, adenosine diphosphate (ADP), phosphate, and NAD are at a stoichiometric proportion of 1:2:2:2. The products pyruvate, adenosine triphosphate (ATP), and NADH are at stoichiometric proportions of 2:2:2. NAD^+ has one positive charge. We will retain its "+" in formula equations, but use a simplified notation of "NAD" when referring to it in text.

3.1.1.1 NAD/NADH Balance in Glycolysis

Let us consider an *in vitro* reaction system in which we have added 10 glycolysis enzymes that will enable the conversion of glucose to pyruvate. We add the reactants in a stoichiometric amount, according to the overall reaction. We will also need to add a small amount of ATP in order to initiate the front portion of the reaction. The reactions take place in series, converting glucose to glycolytic intermediates, then to pyruvate. When the reaction is complete and all of the reactants are consumed, one can expect a stoichiometric amount of products, according to the overall stoichiometric equation. For instance, if the reaction system consumes 1 mole of glucose and 2 moles each of ADP, phosphate, and NAD, at the end of the reaction there should be a net of 2 moles each of pyruvate, ATP, and NADH.

Now, let us assume that we will use the system to mimic glycolysis in the cell. With a continuous supply of glucose from the surrounding environment, cells in our body will continue to consume glucose without the addition of other reactants, like ATP or NAD. Therefore, we need to continuously add glucose in our reaction system, but we only need to add a fixed amount of ADP, phosphate, and NAD at the beginning. After ADP, phosphate, and NAD are all consumed (and pyruvate, ATP, and NADH are generated), the addition of more glucose will not advance glycolysis because the other reactants are exhausted.

What is the difference, then, between a cellular system and the *in vitro* system described above? Inside cells, ATP is continuously used to synthesize various cellular materials, to maintain the concentration gradients of many cellular materials, as well as to perform other cellular "work." In these reactions, ATP is hydrolyzed to ADP and phosphate. ADP and phosphate are thus continuously recycled to sustain glycolysis.

Like ATP, NADH is a high-energy carrier in cellular reactions; it cycles between NADH and NAD^+ to deliver reductive chemical potential energy. Unlike ATP, NADH does not participate directly in synthetic reactions or other cellular work. In aerobic organisms, it gets regenerated back to its oxidized state, NAD, through an electron transport chain to react with oxygen and form water. During anaerobic growth in yeast, NADH cannot be regenerated through a reaction with oxygen. Rather, it is reverted to NAD through the reduction of pyruvate to form ethanol. In exercising muscle cells, oxygen cannot be supplied fast enough to oxidize NADH through an electron transport chain. To sustain glycolysis and generate ATP, pyruvate derived from glycolysis is reduced to lactate and simultaneously oxidizes NADH to NAD. In glycolysis, 1 mole of glucose is converted to 2 moles of pyruvate. By converting 2 moles of pyruvate to 2 moles of lactate or ethanol, 2 moles of NAD are then regenerated. The net overall reaction equations (Panel 3.6) result in the conversion of 1 mole of glucose and 2 moles of ADP to 2 moles of lactate or 2 moles of ethanol, with a net synthesis of 2 ATP. In a cellular system, since ADP is regenerated through various cellular processes, reactants can be continuously supplied to allow glycolysis to proceed with ongoing and continued production of ethanol or lactate.

Panel 3.6

Overall reaction of ethanol fermentation:

$$C_6H_{12}O_6 + 2\ ADP + 2\ P_i \longrightarrow 2\ C_2H_5OH + 2\ CO_2 + 2\ ATP \tag{3.16}$$

Overall reaction of lactate fermentation:

$$C_6H_{12}O_6 + 2\ ADP + 2\ P_i \longrightarrow 2\ CH_3 \cdot CHOH \cdot COOH + 2\ ATP \tag{3.17}$$

In an *in vitro* system, neither NADH nor ADP is being regenerated to sustain the consumption of glucose through glycolysis. To allow for continuous consumption of glucose, one may include the enzyme lactate dehydrogenase, which catalyzes the conversion of pyruvate and NADH to lactate and NAD, plus a mechanism of ADP regeneration.

3.1.1.2 Oxidative Metabolism and NADH

The completion of aerobic metabolism of glucose (i.e., the breaking down of pyruvate to 3 CO_2) requires an active TCA cycle. Macroscopically, the complete oxidation of glucose through enzyme-catalyzed reactions and through chemical combustion shares the same stoichiometric reaction equation. One mole of glucose is broken down to 6 moles each of CO_2 and H_2O, with the consumption of 6 moles of O_2. In fine detail, the former also entails the conversion of cellular energy carriers from their low energetic state (NAD, ADP, and FAD) to high energetic state (NADH, ATP, and $FADH_2$). In the latter, the change in chemical potential energy from the combustion reaction is released as heat.

The breaking down of pyruvate to CO_2 occurs through the TCA cycle, but no TCA cycle reactions involve O_2. First, pyruvate (generated in glycolysis) is decarboxylated (i.e., it loses 1 CO_2) to form acetyl CoA before entering the TCA cycle (Panel 3.7). Once in the TCA cycle, acetyl CoA is further oxidized to 2 CO_2 through a total of five reactions (Eqs. 3.19 to 3.23 in Panels 3.7 and 3.8). After removing the two carbon atoms as CO_2, the remaining three reactions in the TCA cycle recover the chemical potential energy and regenerate oxaloacetate (Eqs. 3.24 to 3.26 in Panel 3.9).

The TCA cycle is typically depicted in a circular manner, which gives a clearer view of the overall cycle. However, the stoichiometric aspects of the cycle are better depicted through individual reactions. As an overall reaction, the TCA cycle can be written as Eq. 3.27. The oxidation of pyruvate generates 3 CO_2, 4 NADH, and 1 $FADH_2$. Under aerobic metabolism, each glucose molecule yields 2 NADH during glycolysis and 6 CO_2, 8 NADH, and 2 $FADH_2$ after glycolysis (including the pyruvate dehydrogenase reaction, which converts pyruvate to acetyl CoA, and TCA cycle reactions). The overall reaction of glycolysis through the TCA cycle is shown in Eq. 3.28 (Panel 3.10).

It is interesting to note that until this stage, there is no O_2 participation. The 10 NADH and 2 $FADH_2$ are then used to generate ATP through oxidative phosphorylation. Through an electron transport chain, NADH and $FADH_2$ pass their high-energy electrons to a lower energy state, which allows H to react with O_2 and form H_2O while also generating a H^+ gradient across the eukaryotic mitochondrial membrane or the prokaryotic plasma membrane (Panel 3.11). The H^+ gradient is used to generate ATP. The oxidation of NADH and $FADH_2$ through electron transfer regenerates NAD and

FAD, which sustains TCA cycle activity. In the absence of oxygen, or in the presence of an electron transport chain inhibitor like cyanide, NAD and FAD will be depleted and the TCA cycle will be stalled.

Panel 3.7 Oxidation of Pyruvate via the TCA Cycle

Enzyme

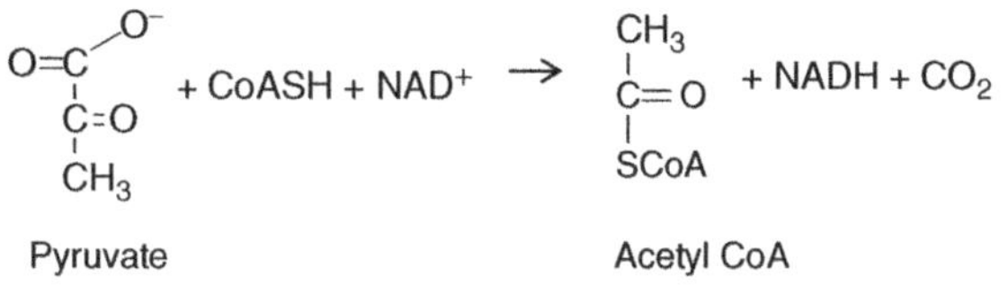

$$O=C{\overset{O^-}{\diagup}}\ \underset{CH_3}{\overset{|}{\underset{|}{C=O}}} + CoASH + NAD^+ \rightarrow \underset{SCoA}{\overset{CH_3}{\underset{|}{\overset{|}{C=O}}}} + NADH + CO_2$$

Pyruvate → Acetyl CoA

Pyruvate dehydrogenase   (3.18)

Tricarboxylic acid (Krebs) cycle reactions:

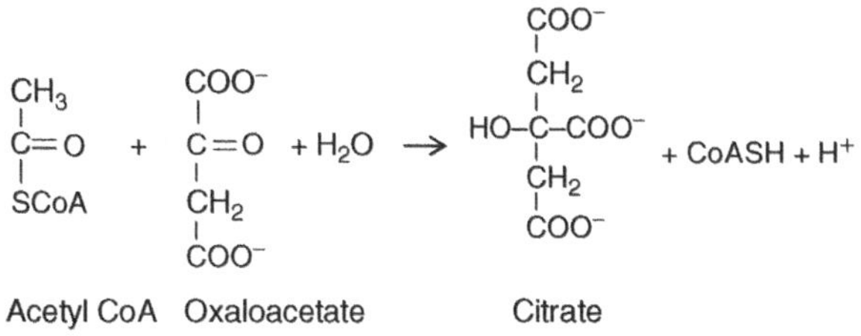

$$\underset{SCoA}{\overset{CH_3}{\underset{|}{\overset{|}{C=O}}}} + \underset{\underset{COO^-}{|}}{\overset{COO^-}{\underset{|}{\overset{|}{C=O}}}} + H_2O \rightarrow \overset{COO^-}{\underset{COO^-}{\underset{|}{\overset{CH_2}{\underset{|}{HO-C-COO^-}}}}} + CoASH + H^+$$

Acetyl CoA   Oxaloacetate → Citrate

Citrate synthase   (3.19)

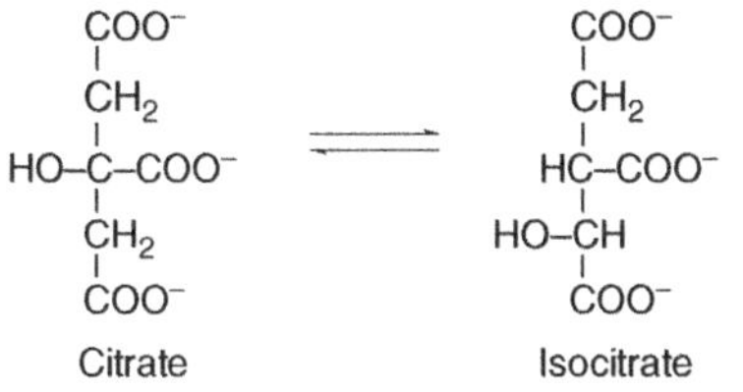

$$\overset{COO^-}{\underset{COO^-}{\underset{|}{\overset{CH_2}{\underset{|}{HO-C-COO^-}}}}} \rightleftharpoons \overset{COO^-}{\underset{COO^-}{\underset{|}{\overset{CH_2}{\underset{|}{HC-COO^-}}}}}$$

Citrate → Isocitrate

Aconitase   (3.20)

Panel 3.8

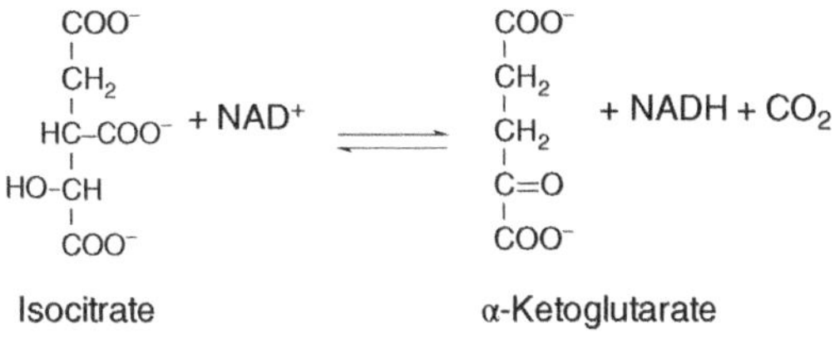

$$\overset{COO^-}{\underset{COO^-}{\underset{|}{\overset{CH_2}{\underset{|}{HC-COO^-}}}}} + NAD^+ \rightleftharpoons \overset{COO^-}{\underset{COO^-}{\underset{|}{\overset{CH_2}{\underset{|}{\overset{CH_2}{\underset{|}{C=O}}}}}}} + NADH + CO_2$$

Isocitrate → α-Ketoglutarate

Isocitrate dehydrogenase   (3.21)

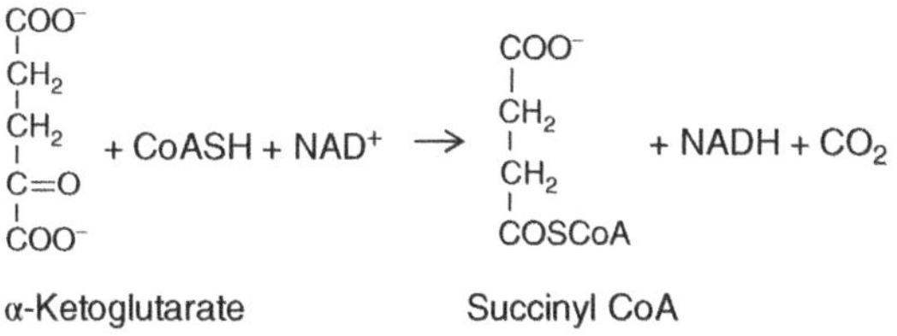

$$\overset{COO^-}{\underset{COO^-}{\underset{|}{\overset{CH_2}{\underset{|}{\overset{CH_2}{\underset{|}{C=O}}}}}}} + CoASH + NAD^+ \rightarrow \overset{COO^-}{\underset{COSCoA}{\underset{|}{\overset{CH_2}{\underset{|}{CH_2}}}}} + NADH + CO_2$$

α-Ketoglutarate → Succinyl CoA

α-Ketoglytarate dehydrogenase complex   (3.22)

(Continued)

Panel 3.8 (Continued)

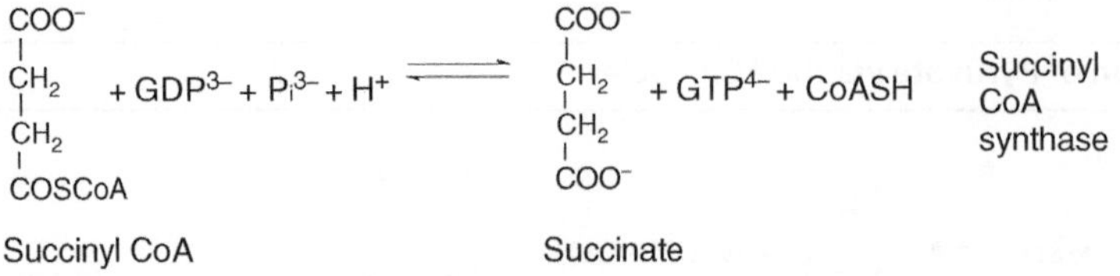

$$
\begin{array}{l}
\text{COO}^- \\
|\\
\text{CH}_2 \\
|\\
\text{CH}_2 \\
|\\
\text{COSCoA}
\end{array}
+ \text{GDP}^{3-} + \text{P}_i^{3-} + \text{H}^+ \rightleftharpoons
\begin{array}{l}
\text{COO}^- \\
|\\
\text{CH}_2 \\
|\\
\text{CH}_2 \\
|\\
\text{COO}^-
\end{array}
+ \text{GTP}^{4-} + \text{CoASH}
\qquad
\begin{array}{l}
\text{Succinyl} \\
\text{CoA} \\
\text{synthase}
\end{array}
\qquad (3.23)
$$

Succinyl CoA   Succinate

Panel 3.9 Recall the difference between FAD and NAD$^+$ (nucleic acid section)

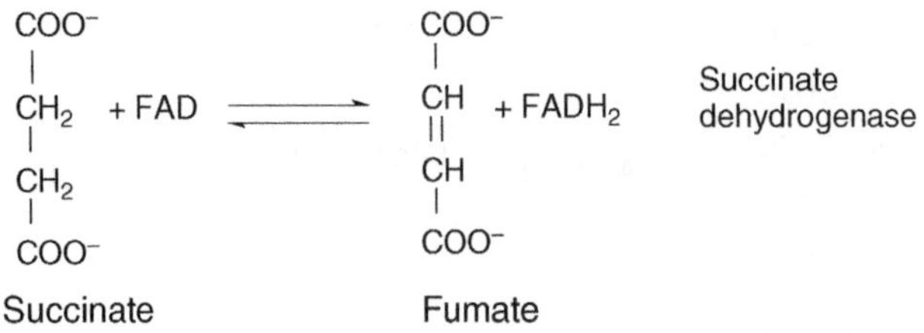

$$
\begin{array}{l}
\text{COO}^- \\
|\\
\text{CH}_2 \\
|\\
\text{CH}_2 \\
|\\
\text{COO}^-
\end{array}
+ \text{FAD} \rightleftharpoons
\begin{array}{l}
\text{COO}^- \\
|\\
\text{CH} \\
||\\
\text{CH} \\
|\\
\text{COO}^-
\end{array}
+ \text{FADH}_2
\qquad
\begin{array}{l}
\text{Succinate} \\
\text{dehydrogenase}
\end{array}
\qquad (3.24)
$$

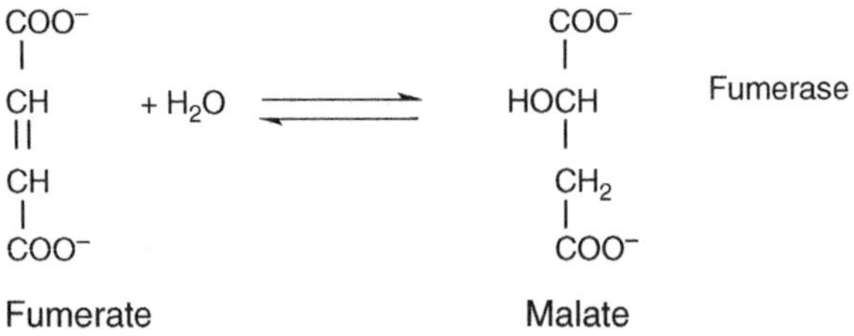

Succinate   Fumate

$$
\begin{array}{l}
\text{COO}^- \\
|\\
\text{CH} \\
||\\
\text{CH} \\
|\\
\text{COO}^-
\end{array}
+ \text{H}_2\text{O} \rightleftharpoons
\begin{array}{l}
\text{COO}^- \\
|\\
\text{HOCH} \\
|\\
\text{CH}_2 \\
|\\
\text{COO}^-
\end{array}
\qquad \text{Fumerase} \qquad (3.25)
$$

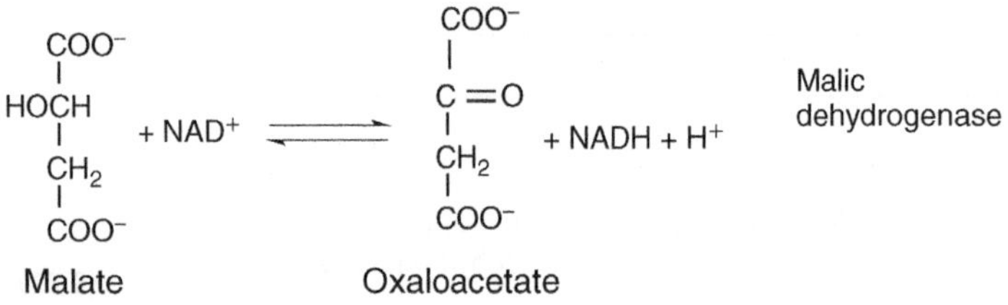

Fumerate   Malate

$$
\begin{array}{l}
\text{COO}^- \\
|\\
\text{HOCH} \\
|\\
\text{CH}_2 \\
|\\
\text{COO}^-
\end{array}
+ \text{NAD}^+ \rightleftharpoons
\begin{array}{l}
\text{COO}^- \\
|\\
\text{C} = \text{O} \\
|\\
\text{CH}_2 \\
|\\
\text{COO}^-
\end{array}
+ \text{NADH} + \text{H}^+
\qquad
\begin{array}{l}
\text{Malic} \\
\text{dehydrogenase}
\end{array}
\qquad (3.26)
$$

Malate   Oxaloacetate

Panel 3.10

Overall reaction of pyruvate oxidation via the TCA cycle:

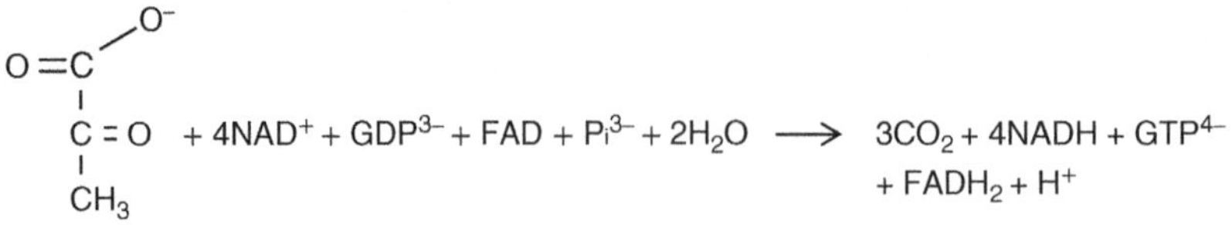

$$
\begin{array}{l}
\text{O} = \text{C}^{-\text{O}^-} \\
|\\
\text{C} = \text{O} \\
|\\
\text{CH}_3
\end{array}
+ 4\text{NAD}^+ + \text{GDP}^{3-} + \text{FAD} + \text{P}_i^{3-} + 2\text{H}_2\text{O} \longrightarrow
\begin{array}{l}
3\text{CO}_2 + 4\text{NADH} + \text{GTP}^{4-} \\
+ \text{FADH}_2 + \text{H}^+
\end{array}
\qquad (3.27)
$$

Overall reaction of glucose oxidation (glycolysis and the TCA cycle):

$$\begin{array}{l} \text{CHO} \\ | \\ \text{HCOH} \\ | \\ \text{HOCOH} \\ | \\ \text{HCOH} \\ | \\ \text{HCOH} \\ | \\ \text{CH}_2\text{OH} \end{array} \quad \begin{array}{l} + \ 2ADP^{3-} + 10NAD^+ + 4P_i^{3-} + 2GDP^{3-} + FAD \\ \quad + 2H_2O \end{array} \longrightarrow \quad \begin{array}{l} 6CO_2 + 2ATP^{4-} \\ + 10NADH + 2GTP^{4-} \\ + 2FADH_2 + 2H^+ \end{array} \quad (3.28)$$

Panel 3.11

Electron transfer/oxidative phosphorylation:

$$NADH + H^+ + \tfrac{1}{2}O_2 \longrightarrow NAD^+ + H_2O \qquad\qquad (3.29)$$

$$FADH_2 + \tfrac{1}{2}O_2 \longrightarrow FAD + H_2O \qquad\qquad (3.30)$$

The reactions on mitochrondrial membrane couple to ATP formation
 Overall Electron Transfer Reaction:

$$10\,NADH + 10\,H^+ + 2\,FADH_2 + 6\,O_2 \longrightarrow 10\,NAD^+ + 2\,FAD + 12\,H_2O$$
$$+ (nATP + nH_2O)$$
$$(+nADP + nP_i + 2\,nH^+) \qquad\qquad (3.31)$$

n: efficiency of ATP synthase varies with species, $\sim$30–33.

While the discussion here centers on aerobic glucose metabolism and anaerobic glycolysis, the two pathways are often both active in many cell types, and the relative contribution of each varies with culture conditions.

3.1.2 Maximum Conversion of a Metabolic Product

Many biotechnology products are raw materials or substrate intensive. This means that the cost of producing a fixed amount of product is highly affected by the cost of the raw materials. For such products, it is important to know the conversion efficiency or the yield of the production process. It is also important to know the theoretical maximal conversion (i.e., the maximal amount of product that can be generated from a unit amount of substrate) (Panel 3.12).

One can generate the overall stoichiometric equation for the formation of a metabolite by compiling all of the component reactions that detail the path from substrate to product. If the product is a nitrogen-containing compound, for instance, the substrates will include both carbon and nitrogen sources, such as glucose and ammonium. In some cases, the synthesis of the product also incurs the generation of a side product, in addition to CO_2, as a result of metabolism (shown as the compound with the stoichiometric coefficient f in Eq. 3.32). This is seen especially frequently in the production of solvent in anaerobic microorganisms. For example, acetone and butanol are co-produced in the fermentation of glucose by *Clostridium acetobutylicum*.

Panel 3.12 Conversion Yield of Metabolic Product – Energetically Favorable

---

The producer confers a biochemical pathway to convert the raw materials to the product with a net gain of ATP or NADH.

In the production of ethanol by yeast, 1 mole glucose is converted to 2 moles of ethanol, and generates 2 moles of ATP.

The reaction is energetically favorable.

The theoretical maximal yield is 2 mol ethanol/mol glucose or 0.51 Kg ethanol/Kg glucose.

The theoretical conversion, or yield, is expressed as the stoichiometric ratio of product to substrate (Eq. 3.32 in Panel 3.13). Often, it is expressed in either a molar ratio or mass ratio. Mass ratio is obtained by multiplying the molar ratio by the molecular weights of the product and substrate. As an example, the conversion of glucose to ethanol in yeast has an overall stoichiometric equation that is a combination of all of the reactions in glycolysis and the decarboxylation and reduction reactions that convert pyruvate into ethanol (Panel 3.12). The overall equation shows that an input of 1 mole of glucose and 2 moles of ADP and phosphate are converted into 2 moles each of ethanol, CO_2, and ATP. The theoretical yield of ethanol from glucose, often written as $Y_{p/s}$ (where p represents the product and s represents the substrate, in this case glucose), is 2 moles of ethanol per mole of glucose, or 0.51 kg ethanol/kg glucose.

Panel 3.13 Overall Reaction for Product Formation

Metabolic product – energetically favorable:

$$C_6H_{12}O_6 + aO_2 + bNH_3 \rightarrow cC_{\alpha'}H_{\beta'}N_{\gamma'}O_{\delta'} + dCO_2 + eH_2O$$
$$fC_{\alpha''}H_{\beta''}N_{\gamma''}O_{\delta''} + nATP \tag{3.32}$$

Synthetic product – requiring energy input:

1) Product synthesis:

$$C_6H_{12}O_6 + aNH_3 + xATP \rightarrow bC_{\alpha'}H_{\beta'}N_{\gamma'}O_{\delta'} + nADP \tag{3.33}$$

2) Energy generation

$$C_6H_{12}O_6 + 6O_2 \rightarrow 6CO_2 + 6H_2O + nATP \tag{3.34}$$

Overall reaction: $[\psi \cdot (\text{Eq. 3.33})] + [(1\text{-}\psi \cdot (\text{Eq. 3.34})]$, so that $\psi \cdot x = (1\text{-}\psi) \cdot n$, no net ATP consumption

$$C_6H_{12}O_6 + a'O_2 + b'NH_3 \rightarrow c'C_{\alpha'}H_{\beta'}N_{\gamma'}O_{\delta'} + d'CO_2 + e'H_2O \tag{3.35}$$

Note: ADP, P_i, and H_2O are not shown. NADH, NADPH, and $FADH_2$ are implicitly expressed in ATP equivalent.

Such stoichiometry of a biochemical pathway-derived conversion is a theoretical maximum. It assumes that no substrates are diverted to other pathways. In reality, some substrates are required for growth or other energetic needs, so the observed conversion is always lower than the theoretical conversion.

In the ethanol example given here, the stoichiometric equation indicates that 2 ATP are generated. The Gibbs free energy change of the reaction also indicates that the reaction is thermodynamically favorable. Many biosynthetic products require extensive energy input, in the form of ATP (Panel 3.13). Some products, such as glycerol and butandiol, are more reduced than glucose. In those cases, there must be a sufficient input of high-energy compounds (e.g., ATP, NADH, NADPH, etc.) on the reactant side of the stoichiometric equation in order to have a negative Gibbs free energy change and for the reaction to proceed (Eq. 3.33 in Panel 3.13). In such cases, a sufficient amount of substrate must be used to generate energy (ATP or NADPH) through a catabolic pathway.

Panel 3.14 Conversion Yield of Synthetic Product – Requiring Energy Input

- Many products are more reduced than the substrate or require net inputs of ATP, NADPH, or other high-energy compounds.
- A fraction of substrate is catabolized to make the overall reaction energetically favorable (i.e., to negate the net ATP/NADPH requirements).
- For first approximation, assume 1 mole of NADH is equivalent to 3 moles of ATP.

An overall reaction equation for generating the required high-energy carriers can be written to describe the energy-generating portion of substrate utilization (Eq. 3.34 in Panel 3.13). The theoretical yield is then obtained by combining the two stoichiometric equations. One reaction describes the product formation with the consumption of ATP and NADPH; the other describes the substrate metabolism needed to generate a sufficient amount of ATP and NADPH (Eq. 3.35 in Panel 3.13). In the overall reaction, there is no ATP or NADPH as reactants.

Example 3.1   The Conversion Yield of Glutamic Acid from Glucose and Ammonia in *Corynebacterium glutamicum*

Solution
The pathway of glutamic acid biosynthesis is shown here:

$$\text{glucose} + 2\text{ADP} + 2\text{P}_i \rightarrow 2 \text{ pyruvate} + 2 \text{ ATP}$$
$$\text{pyruvate} + \text{NAD}^+ \rightarrow \text{acetyl CoA} + \text{CO}_2 + \text{NADH}$$
$$\text{pyruvate} + \text{CO}_2 + \text{ATP} + \text{H}_2\text{O} \rightarrow \text{oxaloacetate} + \text{ADP} + \text{Pi}$$
$$\text{acetyl CoA} + \text{oxaloacetate} + \text{NAD}^+ \rightarrow \alpha\text{-ketoglutarate} + \text{CO}_2 + \text{NADH}$$
$$\alpha\text{-ketoglutarate} + \text{NADPH} + \text{NH}_3 \rightarrow \text{glutamate} + \text{NADP} + \text{H}_2\text{O}$$

Note that it is not sufficient to use a portion of the TCA cycle to produce the precursor α-ketoglutarate. As α-ketoglutarate is drawn out from the cycle to produce glutamic acid, there is nothing that can be used to produce oxaloacetate. Therefore, a reaction to supply oxaloacetate is necessary. The pyruvate is split into two supply streams: one to produce acetyl CoA, and the other to produce oxaloacetate.

The overall reaction is obtained by combining all the reactions shown:

$$glucose + NH_3 + ADP + Pi + NADPH + 2NAD^+ \rightarrow glutamate + CO_2 + ATP$$
$$+ 2NADH + NADP^+$$

Overall, it is energetically feasible because there is a net production of 1 ATP and 1 NADH.

The theoretical maximum conversion base is 1 mole glutamate/mole glucose, or on a mass basis $143/180 = 0.79$.

3.2 Stoichiometry for Cell Growth

Most processes used to produce biochemicals involve the cultivation of large quantities of cells in order to synthesize products. Performing a material balance on the process thus requires balances on both the product and the cell mass. In principle, a stoichiometric equation for cell biomass formation can be obtained by combining all of the component reactions, according to the degree in which they contribute to biomass generation. In reality, not all reactions are known, and the contribution of each reaction is often hard to measure. Nevertheless, by collecting experimental measurements of the inputs required for cell growth (i.e., the nutrients consumed while cells grow) and the amount of new biomass produced, one can derive the biomass stoichiometric equation.

3.2.1 Cell Composition and Material Flow to Make Cell Mass

3.2.1.1 Composition and Chemical Formula of Cells

To establish a stoichiometric equation, the first step is to obtain the molecular formula of the reactants and products. This is relatively straightforward for pathways, since all of the compounds involved have a defined chemical formula. It is more complicated for cells. Cells are made up of proteins, carbohydrates, polysaccharides, lipids, nucleic acids (DNA and RNA), small molecules, and electrolytes (Panel 3.15). Even if the precise composition of cells were known, cells are anything but a mixture of those components. It is the organization and interaction of these components, and many other physiological features, that make a cell a living entity. The organization and integration of these components cannot be represented by a simple chemical formula. Nevertheless, in order to perform material balance on the process of cell growth, an operational chemical formula for cells is needed.

The formula for cell mass can be of different degrees of detail. It may entail chemical compositions, such as percentage of proteins, lipids, nucleic acids, and so on. Alternatively, it may only consider elemental composition, where the number of carbon atoms from lipids, carbohydrates, and proteins are all considered to be equal. In our discussion of stoichiometric equations for biomass, we will only consider elemental composition.

Nearly 80% of the volume or mass of a typical bacterium is water. Much of the cell's water is taken up from the environment as the cell expands its volume during growth. Water is also a metabolic product in biochemical reactions; for each mole of glucose oxidized, 6 moles of water are produced. However, the quantity of water produced from metabolism is relatively small compared to the amount of water taken up from the

---

Panel 3.15 Representative Composition of a Cell

	Mass (pg/cell)		Dry mass (%)	
	Animal cell	Bacterium	Animal cell	Bacterium
Wet weight	3500	1.5		
Dry weight	600	0.3	100	100
Protein	250	0.17	41	55
Carbohydrate	150	0.015	20	5
Lipid	120	0.015	1.5	5
DNA	10	0.015	1.5	5
RNA	25	0.075	2.5	25
Water		1.2	500	400
Volume	4×10^{-9} cm^3	1.5×10^{-12} cm^3		

---

environment to generate biomass. Because cells grow in an aqueous environment, the amount of water produced from metabolism cannot be easily quantified. In discussing and quantifying biomass, one typically uses dry biomass and excludes water.

Considering dry biomass only, proteins typically constitute about half or a bit more than half of the cell mass (Panel 3.15). The lipid content is typically a few percent of the cell's dry mass and is higher in eukaryotic cells because of an abundance of membrane-wrapped organelles. Nucleic acids contribute about 5% of the dry biomass in a cell, with the vast majority being RNA. The carbohydrate content varies widely with cells. Polysaccharides are a major constituent of cell walls in both bacteria and plant cells. The carbohydrate content is lower in animal cells, except for those cell types that store carbohydrate polymer (glycogen) as an energy reservoir.

The major elements that make up proteins, carbohydrates, nucleic acids, and lipids are carbon, hydrogen, nitrogen, and oxygen (Panel 3.16). These are also major elemental constituents in cell mass. Carbon is the most abundant element in terms of mass in the cell, while hydrogen is the most abundant in terms of molar amount. These four elements contribute 92–95% of the cell dry mass in a typical bacterium.

Panel 3.16 Chemical Composition of Cells

- Most abundant component is water; often not included in biomass formulation
- Carbon compounds: proteins, carbohydrates, lipids, and nucleic acids
- Noncarbon compounds; electrolytes
- Major elemental components: C, N, H, O, P, and S

Typical cellular formula:

Bacterium	$CH_{1.6}N_{0.20}O_{0.27}$ to $CH_2N_{0.26}O_{0.45}$
Yeast	$CH_{1.64}N_{0.16}O_{0.52}P_{0.01}S_{0.005}$
Animal cells	$CH_{1.98}N_{0.26}O_{0.49}$

Besides the four major elements, phosphorus constitutes a significant fraction of the dry biomass because phosphate is a major component of nucleic acids. Another element that has a significant presence in cell mass is sulfur. Sulfur is found in proteins as parts of cysteine and methionine. It is also present as sulfate in some polysaccharides. In addition to phosphorus and sulfur, some other inorganic elements are also present in small quantities in the cell as electrolytes, including sodium, potassium, and chlorine ions. Magnesium ion chelates with nucleotides and nucleic acids. Many trace metal elements (e.g., iron, calcium, zinc, copper, and selenium) are also cofactors of metalloproteins, which are so named because of the presence of those metal ions. The contribution of these major and trace elements to cell mass is typically determined by placing a known amount of cells in a high-temperature furnace (~450 °C) until a constant weight is reached, to ensure the complete removal of the combustible portion of the cells. The remaining cellular materials are often referred to as "ashes" and typically occupy about 5–8% of total cell mass.

Using the elemental composition of cells, one can develop an empirical chemical formula of cells (Panel 3.17). A chemical formula for cell mass does not carry much physiological meaning except to indicate the elemental composition of the cell. Nevertheless, a formula is useful for setting up a stoichiometric equation of cell biomass synthesis from known substrates during cell cultivation. A chemical formula for cell mass can be obtained by dividing the mass contribution of each element to its atomic weight.

Panel 3.17 Biomass Formula

Example:

Element	C	H	N	O	Ash
% of Dry weight	53.0	7.3	12.0	19.0	8
Atomic weight of the element	12	1	14	16	
No. of moles per 100 g	4.42	7.3	0.857	1.19	
Normalize to C	1	1.60	0.194	0.269	

$CH_{1.66}$, $N_{0.194}$, and $O_{0.269}$, constitute 0.92 of mass.

Formula weight $= (12 + 1 \times 1.66 + 14 \times 0.194 + 16 \times 0.26)/(1 - \text{ash fraction})$

$= 20.8$

There are multiple ways of presenting the cell mass formula. Conventionally, one can set the coefficient of an element (reference element; usually carbon) to be 1.0 and normalize the coefficients of all of the other elements to the reference element. Another frequently seen approach is to set the formula weight to be equal to a convenient number, such as equivalent to 100 g of biomass. It is important to recall that the elements carbon, hydrogen, nitrogen, and oxygen only constitute biomass, and not the ashes. Consequently, the formula weight of the cell accounts for only a fraction of the total weight, calculated as: $(1 - \text{mass fraction of ashes})$.

3.2.1.2 Material Flow for Biomass Formation

Cells in culture take up nutrients from medium, and then expand their cell mass and number. To perform mass balance on this process of cell growth, we can consider a culture volume to be separated into two phases, a biotic phase and an abiotic phase. The biotic phase constitutes all of the cells in the system. In spite of the discrete nature of individual cells, we treat all cells together as an entity. The abiotic phase, conversely, constitutes all of the mass in the fluid, including medium components and metabolites excreted by cells. Material balance is performed on the boundary between the biotic and abiotic phases. Nutrients taken up by cells are considered to be inputs. They undergo a biochemical transformation to generate biomass, energy, and metabolites. The biomass generated and metabolites excreted are considered to be outputs (Figure 3.3).

A culture medium may be enriched ("complete") or minimal with respect to its nutritional content. A minimal medium contains the least number of required chemical species as nutrient sources. The further omission of any component from its composition will abolish its capability of supporting the growth of the organism. Furthermore, each component is in its simplest structural form. For example, if ammonium and glutamic acid are both able to serve as the nitrogen source for cell growth, ammonium, as the simpler structure, will be the one that is used in a minimal medium. When growing in a minimal medium, cells need to devote much of their energy to synthesizing building blocks. A complete medium, on the other hand, provides many of the cellular building blocks (various amino acids, lipids, etc.), so that cells do not need to spend energy synthesizing those cellular components and are able to grow at a faster rate.

Cell mass is mostly made of biopolymers (e.g., proteins, nucleic acids, and polysaccharides). The energetic cost of making these biopolymers from building blocks of amino acids, nucleotides, and monosaccharides is rather high. The incorporation of each nucleotide into DNA or RNA requires the input of 2 ATP. Furthermore, for each peptide bond within a protein, 4 ATP are required for its formation. Considering that each glucose molecule that is oxidized through glycolysis and the TCA cycle roughly generates only 30 ATP, the energetic cost of making biomass is truly very high.

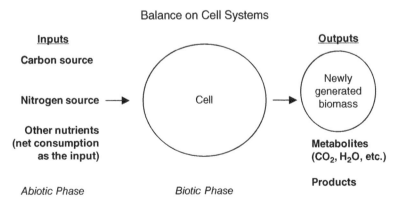

Figure 3.3 Material inputs and outputs in a cell cultivation system.

In addition to assembling the materials to make cell mass, cells also maintain a membrane potential (through a pH and small-electric gradient) and an osmotic pressure across their membranes. They also take up or excrete some compounds against a concentration gradient. All of these processes require energy. Thus, a large portion of nutrients taken up by cells is used to generate energy, simply to sustain cellular functions.

Conceptually, one can divide the nutrients taken up by cells for growth into two portions: one for generating energy through energy metabolism, and the other for synthesizing building blocks and biopolymers to generate biomass. Physiologically, however, the biochemical reactions and pathways are intermingled and cannot completely be segregated into catabolism and anabolism.

3.2.2 Stoichiometric Equation for Cell Growth

The mass balance of cell growth can be written as a stoichiometric equation describing the conversion of nutrients to growth or to excreted metabolites. We will consider a general case and account for only the four major elements (carbon, hydrogen, nitrogen, and oxygen). The inputs will include a carbon source, a nitrogen source, and oxygen. The biomass product is represented by the biomass formula. A general equation for "synthesizing" cell mass from multiple substrate inputs (with subscript i for different nutrients) is shown in Eq. 3.36 (Panel 3.18). In addition to CO_2 and H_2O, other metabolites (shown with a general formula) may also be produced. We have stated above that for assessing the mass balance of cell growth, intracellular water cannot be easily measured and is neglected. Thus, the H_2O in the stoichiometric equation only represents the water molecules generated as a result of the biochemical reactions. It does not consider the H_2O taken up by cells from medium during growth.

Panel 3.18 Stoichiometric Equation for Biomass Formation

General equation for biomass formation:

$$C_6H_{12}O_6 + \sum_i \eta_i C_{\alpha i}H_{\beta i}N_{\gamma i}O_{\delta i} + aO_2 \rightarrow bC_{\alpha'}H_{\beta'}N_{\gamma'}O_{\delta'} + cCO_2 + dH_2O + eNH_3$$

(3.36)

Substrates consumed   Biomass produced

Biomass formation in glucose/ammonium medium:

$$C_6H_{12}O_6 + aNH_3 + bO_2 \rightarrow cC_{\alpha'}H_{\beta'}N_{\gamma'}O_{\delta'} + dCO_2 + eH_2O$$

(3.37)

The stoichiometric equation for cells grown in a minimal medium using glucose and ammonia as carbon and nitrogen sources and producing no other metabolite aside from CO_2 is shown in Eq. 3.37 (Panel 3.18). The composition (i.e., chemical formula) of all terms in the cell growth equation is known or experimentally measurable. The stoichiometric equation contains four elements (carbon, hydrogen, nitrogen, and oxygen) but has five stoichiometric coefficients. The equation by itself is thus undetermined. The stoichiometric coefficients of the participating components in the reaction must be experimentally determined, but cannot be calculated on a theoretical basis. Typically, to calculate the stoichiometric coefficients, one measures the amount of carbon and nitrogen sources, oxygen consumed during growth, and the amount of cell mass and CO_2 produced. After experimentally determining these quantities, the stoichiometric coefficient of water in the equation can be calculated.

The number of stoichiometric coefficients in a general biomass formation equation is larger than the number of equations for elemental balances. Therefore, the value of stoichiometric coefficients for the biomass equation cannot be fixed without further information. One has to rely on experimentally measured values to obtain the stoichiometric coefficients. The undetermined nature of the stoichiometric coefficients in the biomass formation equation reflects the divergent ways of utilizing the same substrate in nature. Even if the cell composition and substrates are the same, cells from different organisms have different efficiencies of converting substrates to cell mass. Furthermore, an organism may use the same substrate differently under different conditions. The differing efficiencies in substrate utilization give rise to different sets of stoichiometric coefficients.

3.2.2.1 Yield Coefficient

In the equation of biomass, the stoichiometric ratio of biomass formation to substrate consumption is the molar form of the yield coefficient of biomass formation. Since the molar cell mass is only operationally defined, molar yield is not used. Rather, the yield coefficient for biomass is often expressed using mass units. From the stoichiometric equation, the yield coefficient is the ratio of the multiplication products of the formula weight and the stoichiometric coefficients of cell mass to substrate.

The term "yield coefficient" is used in two ways. First, when a stoichiometric equation is established, the yield coefficient is the stoichiometric ratio normalized to mass units. Second, when a complete stoichiometric relationship is not established, but the increase in cell mass and the decrease in available substrates are measured, the yield coefficient is the ratio of the two measured quantities (Figure 3.4 and Panel 3.19).

The yield coefficient or stoichiometric ratio can be established for any pair of components, from substrate to cell mass. For example, the yield coefficient of biomass may be based on the source of nitrogen or oxygen, in addition to the carbon substrate. The yield coefficient can also be expressed by the elemental input, rather than be based on the substrate compound. For example, as the ratio of the mass of new biomass per moles

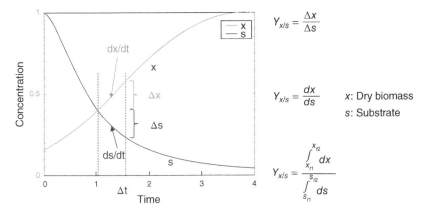

Figure 3.4 Biomass yield coefficient.

Panel 3.19 Yield Coefficient on Biomass

Yield on biomass = (mass of biomass produced)/(mass of substrate consumed)

- Many different ways of expressing yield
- Examples of units: g dry biomass/g glucose; g dry biomass/mole glucose

---

of oxygen is consumed, the yield coefficient based on glucose can be expressed as the moles of carbon atoms in new biomass per mole of carbon atoms in glucose.

Expressing the yield coefficient in the carbon molar ratio allows one to clearly delineate how much of the carbon intake is incorporated into biomass and how much is excreted as CO_2. The yield coefficient based on moles of carbon (i.e., the number of moles of carbon that are taken up versus the number of moles of carbon that get incorporated into biomass) is usually less than 0.5. A larger portion of the carbons that are taken up by cells are used for energy generation, rather than for incorporation into biomass.

In contrast to carbon substrates and oxygen, the elemental yield for nitrogen atoms using ammonia as a nitrogen source would be 1.0, unless some nitrogen-containing compound is excreted as a metabolic product. Nitrogen atoms are not diverted for energy formation and are retained in cells. The vast majority of inorganic components in the medium (including phosphate, sulfur, potassium, and sodium) is only incorporated into biomass and thus has an elemental yield coefficient (the mass of the element in biomass versus the mass of the element consumed) of 1.0.

Another frequently used parameter in biomass formation is the respiratory quotient (RQ), which is the molar ratio of CO_2 produced and O_2 consumed, or the stoichiometric ratio of CO_2 to O_2 as shown in Eqs. 3.36 and 3.37 (Panel 3.18). When glucose is used as the carbon source, the RQ is close to 1.0 but, for hydrocarbon, the RQ is less than 1.

Most organisms can grow on a variety of substrates; however, the yield coefficient on different substrates can vary widely. In general, when multiple substrates can serve as a carbon source and support cell growth equally well, the compound that is more reduced (i.e., having less oxygen per carbon) has a higher yield coefficient. Both the biomass composition and the yield on a substrate vary with the physical and chemical environment. For most microorganisms, the yield coefficient for biomass is higher under optimal culture conditions than in suboptimal conditions.

The yield coefficient of biomass is generally rather low for anaerobic growth. Typically, in these cases, over 90% of carbon is excreted as fermentation products, like lactate and ethanol. For an organism that can grow either aerobically or anaerobically, the yield coefficient based on glucose under aerobic and anaerobic conditions may differ by an order of magnitude. We discussed that the carbon source consumed by a cell can be conceptually divided into two parts: one for incorporation into biomass, and the other for energy generation to support biomass synthesis.

The energetic cost for synthesizing a given amount of biomass is approximately the same, regardless of whether cells are grown aerobically or anaerobically. To generate the same amount of energy, however, the amount of glucose that must be consumed is much larger in the anaerobic process, reflecting efficiency differences of energy generation. In anaerobic processes, 2 ATP are generated per mole of glucose. Conversely, in aerobic processes, each mole of glucose generates over 30 moles of ATP. Also, in anaerobic processes, a large amount of metabolites (e.g., lactate or ethanol and CO_2) are generated as a part of metabolism.

3.3 Hypothetical Partition of a Substrate for Biomass and Product Formation

The stoichiometric equation for products with a known synthetic pathway can always be established. The stoichiometric ratio of the product and the substrate is its yield coefficient. Like biomass, the product yield coefficients can be based on different substrates, such as carbon or nitrogen sources. The stoichiometric equations for biomass formation and product synthesis can be combined to create an equation for overall product formation in a bioprocess, but only if the ratio of the amount of the substrates used for each component reaction is known. The relative weight of a substrate diverted to a product (relative to that used for biomass formation) is denoted as ω (Panel 3.20). ω is thus a measure of the overall efficiency with which a product is formed. The overall (observed) product yield based on substrate is the theoretical yield multiplied by $[\omega/(1+\omega)]$. In developing a process for product synthesis, the objective is to direct the maximum amount of substrate to the product (i.e., to maximize ω). While it is important to generate a sufficient amount of cells for producing the product, an excessive diversion of the substrate for cell growth decreases the overall product yield.

Panel 3.20 Combined Growth and Product Formation

Biomass growth:

$$C_6H_{12}O_6 + aNH_3 + bO_2 \rightarrow cC_{\alpha'}H_{\beta'}N_{\gamma'}O_{\delta'} + dCO_2 + eH_2O \qquad (3.37)$$

Product synthesis:

$$C_6H_{12}O_6 + a'NH_3 + b'O_2 \rightarrow c'C_{\alpha'}H_{\beta'}N_{\gamma'}O_{\delta'} + d'CO_2 + e'H_2O$$
$$\text{\textit{Theoretical maximal molar product yield is } } c' \qquad (3.35)$$

$$\omega = \frac{\textit{glucose consumed for product synthesis}}{\textit{glucose consumed for biomass}}$$

The observed molar yield

$$Y = \frac{\omega c'}{1 + \omega}$$

In experiments and in production, one measures the process variables, including cell mass and product produced, as well as the amounts of substrate consumed. However, from substrate consumption and product accumulation measurements alone, one cannot tell how much of the consumed substrate is used for biomass formation and how much for product synthesis. In other words, ω is not explicitly measured. Instead, one can assume that the stoichiometric equation and theoretical yield obtained from the known (or even postulated) pathway comprise an accurate depiction of product biosynthesis and thereby determine ω.

Many industrial processes for biochemical production employ two stages in manufacturing: a fast-growth stage for biomass accumulation, and a slow-growth stage for maximal product synthesis. This allows more of the total carbon source to be channeled to product than to biomass (i.e., ω ≫ 1). In most cases, the concentration of product accumulated in the medium is substantially higher than its intracellular concentration. In these instances, the secretion of the product, against its concentration gradient, would have to be energy dependent. It is necessary to take the energy expenditure of the secretion process into account for determining the theoretical yield. Also, since the fraction of substrate channeled to product is large, failure to consider substrate transport will result in significant underestimation of the energetic cost.

3.4 Metabolic Flux Analysis

Most of the reactions of interest to biotechnologists form networks, with convergent and divergent branches. Substrates distribute through the network to participate in reactions that result in the generation of biomass, metabolites, and product. In most cases, only some of the inputs and outputs and a limited number of the internal distribution of fluxes are measured. And yet, it is highly desirable to quantify the distribution of materials at various branching points. With a measurement of these inputs, outputs, and so on, and with a system of equations derived from their stoichiometric relationship, the flux distribution through the involved reactions can be determined or estimated based on reasonable assumptions. When so based on material balance, flux determination is called metabolic flux analysis (MFA). It is applied to biochemical reaction networks, or even to the entire cellular network of bioreactions as a whole.

In this section, we will discuss the systematic solution of the material balance equations for deriving metabolic fluxes. The solution for biochemical reactions or a cellular system is no different from solving material balance equations of simple chemical reactions. We will use a simple chemical reaction system to review the solution of material balance equations and will then conclude with a discussion on biochemical and cellular reaction systems.

3.4.1 Analysis of a Chemical Reaction System

We will use methane combustion in a burner as an example to illustrate how MFA equations are set up and solved (Panels 3.21 and 3.22). A fraction of the input of methane is completely oxidized by oxygen to form CO_2 and H_2O; the rest is converted to CO and H_2O. By measuring the input of methane and output of CO_2 and CO, one can determine the fraction of methane that is utilized in each reaction. This problem can be easily solved, but we will use a systematic approach, as the method is also applicable to calculating fluxes of larger reaction systems.

3.4.1.1 Setting Up Material Balance Equations
In order to determine the flux of each of the two reactions involved, we will first perform a balance on the materials that are consumed and produced. We also

---

Panel 3.21 Solving Flux Distribution in a Simple Chemical Reaction System

Example: incomplete combustion of methane:

$$CH_4, O_2 \rightarrow \begin{bmatrix} 2CH_4 + 3O_2 \xrightarrow{J_1} 2CO + 4H_2O \\ CO + \tfrac{1}{2}O_2 \xrightarrow{J_2} CO_2 \end{bmatrix} \rightarrow CO, CO_2, H_2O$$

Establish material balance on chemical species involved:

$$\frac{dCH_4}{dt} = q_{CH_4} - 2J_1$$

$$\frac{dO_2}{dt} = q_{O_2} - 3J_1 - 0.5J_2$$

$$\frac{dCO}{dt} = -q_{CO} + 2J_1 - J_2 \qquad \rightarrow \qquad \frac{dY}{dt} = AX - Q$$

$$\frac{dCO_2}{dt} = -q_{CO_2} + J_2$$

$$\frac{dH_2O}{dt} = -q_{H_2O} + 4J_1$$

Panel 3.22 Assuming the System is at a Steady State

Assumed steady state:

$$\frac{dY}{dt} = 0 \rightarrow \begin{aligned} -2J_1 &= -q_{CH_4} \\ -3J_1 - 0.5J_2 &= -q_{O_2} \\ -2J_1 + J_2 &= q_{CO} \\ -J_2 &= q_{CO_2} \\ -4J_1 &= q_{H_2O} \end{aligned}$$

Rewrite into stoichiometric matrix, flux vector, and input/output vector:

$$\begin{array}{ccc} A & X = & Q \\ \begin{bmatrix} -2 & 0 \\ -3 & -0.5 \\ 2 & -1 \\ 0 & 1 \\ 4 & 0 \end{bmatrix} & \begin{bmatrix} J_1 \\ J_2 \end{bmatrix} = & \begin{bmatrix} -q_{CH_4} \\ -q_{O_2} \\ q_{CO} \\ q_{CO_2} \\ q_{H_2O} \end{bmatrix} \end{array}$$

Eliminate dependent equations. Check whether system is determined, undetermined, or overspecified.

---

need to define the system and identify the inputs, outputs, and chemical reactions. In this case, two reactions are involved for which the chemical formula of the compounds and the stoichiometric coefficients of the reactions are known.

The rate of inputs and outputs is given a symbol q. In a cellular system, those q's are called the specific rates; they will be discussed further in Chapter 5. The input (consumption) is given a negative value, and the output (production) has a positive value.

We denote the flux of each reaction as J. By denoting the flux of reaction 1 as J_1, the fluxes of methane, oxygen, carbon monoxide, and water reacted in reaction 1 are $-2J_1$, $-3J_1$, $+2J_1$, and $4J_1$, respectively. Their fluxes are related by the stoichiometric coefficients -2, -3, 2, and 4.

Next, we set up material balance equations. Each compound involved in the system is given a differential equation describing the change of its concentration in the reactor as the balance of its rates of consumption, production, and other inputs and outputs. The balance equation for CH_4 thus includes input (q_{cH4}) and reaction fluxes ($-2J_1$) [i.e., the net of $-2J_1$ ($+0J_2$) and $-q_{CH4}$]. Note that the volume of the system is kept constant, thus allowing material balance to be conducted on the concentration of different compounds, instead of on the total mass in the system.

3.4.1.2 Quasi–Steady State
The equations set up above are differential equations. For a short duration of time, the concentrations of all species in the reactor (or in the cell) are relatively constant. The system is thus assumed to be at a steady state. With this assumption, the left-hand side of all differential equations becomes zero, and the equations become a simple system of linear algebraic equations (Panel 3.22).

3.4.1.3 Stoichiometric Matrix, Flux Vectors, and Solution
To solve a large set of equations, it is more efficient to perform matrix operations. For material balance on a reaction system, the matrix is called a stoichiometric matrix. This refers to the collection of stoichiometric coefficients organized by reactions in one dimension, and by species involved in reactions in the other dimension. For the example of methane combustion, the matrix has two columns; each consists of the stoichiometric coefficients of reactions 1 and 2.

Each row in the matrix represents the stoichiometric coefficients in the balance of a compound in the system. In the example, column 1 of the matrix includes fluxes in reaction 1, and row 1 includes the balances of methane. J_1 and J_2 are presented in a vector. The inputs and outputs are collected in a vector on the right-hand side. For a compound that is generated in the reactor but absent in the input or output, the corresponding element in the vector on the right-hand side is zero. This is often encountered when analyzing metabolic flux, since many metabolic intermediates are present only intracellularly.

Next, the equations that are not independent are removed from the system of equations. The resulting system of equations may be overspecified, be undetermined, or have a unique solution. An overspecified system would have more independent equations than unknowns. Undetermined would have more unknowns than independent equations. And, finally, a unique solution would have the same number of independent equations as unknowns.

For an undetermined system, one seeks to perform more measurements to constrain the system. For an overspecified system, one has to use knowledge to remove less significant reactions, or find a set of solutions that give rise to less error. In general, for biological metabolic reactions, the system is almost always overspecified. In the above example, one needs to measure only two out of the five inputs and outputs. The system of equations has a 2×2 stoichiometric matrix, has two unknown fluxes, and is determined.

Example 3.2 Metabolic Flux of Yeast Nonaerobic Growth

Calculate the metabolic flux of anaerobic glycolysis in *Saccharomyces cerevisiae*. Consider the case that 100 moles of glucose are consumed. Assume that all the ATP generated is used for work (denoted as W) that converts ATP to ADP and phosphate. Show that without ethanol production, the glycolysis cannot proceed.

Solution

We use J to denote the flux of each reaction in glycolysis using the same numbering as shown in Panels 3.1, 3.2, 3.3, and 3.4 (and Figure 3.1) for glycolysis reactions. For example, J_1 is conversion of glucose to G6P, and J_2 is conversion of G6P to fructose-6-phosphate (F6P).

The material balance on G6P at steady state is $J_1 - J_2 = 0$. At steady state, $J_1 = J_2 = J_3 = J_4$. In the flux analysis, we will just use J_1 to represent all four fluxes. Similarly, $J_6 = J_7 = J_8 = J_9 = J_{10}$, and J_6 represents all equal fluxes in that segment of linear pathway. We thus will use J_1 to represent the flux of combine reactions 1–4 (from glucose to DHAP and GAP) and J_6 to represent combined reactions 6–10 (from GAP to Pyr). Similarly, $J_{11} = J_{12}$.

The material balance equations for the compounds and the reactions involved are written in matrices form as shown here. Each column is for a reaction whose number is shown at the top, and each row is the balance. For reversible reactions, the flux toward downstream is designated as positive. The material balance equations are:

$$
\begin{pmatrix}
-1 & 0 & 0 & 0 \\
1 & 1 & -1 & 0 \\
1 & -1 & 0 & 0 \\
0 & 0 & 1 & -1 \\
0 & 0 & 0 & 1 \\
2 & 0 & -1 & 0
\end{pmatrix}
\begin{pmatrix}
J_1 \\ J_5 \\ J_6 \\ J_{11}
\end{pmatrix}
=
\begin{pmatrix}
q_{glc} \\ 0 \\ 0 \\ 0 \\ q_{EtOH} \\ q_{ADP}
\end{pmatrix}
$$

Through Gaussian elimination:

$$
\begin{pmatrix}
-1 & 0 & 0 & 0 \\
0 & 2 & -1 & 0 \\
1 & -1 & 0 & 0 \\
0 & 0 & 0 & 1
\end{pmatrix}
\begin{pmatrix}
J_1 \\ J_5 \\ J_6 \\ J_{11}
\end{pmatrix}
=
\begin{pmatrix}
q_{glc} \\ 0 \\ 0 \\ q_{EtOH}
\end{pmatrix}
$$

Written as a system of equations, we see:

$$J_1 = -q_{glc}$$
$$2J_5 = J_{11} = J_6 = q_{EtOH}$$

It can be seen that the only meaningful solution ($q_{glc} > 0$) is when ethanol is produced.

3.4.2 Analysis of Fluxes in a Bioreaction Network

The approach described above can be applied both to biochemical reaction networks and to cellular systems involving cell growth and product formation. In this section, we will perform material balance on the PPP (Figure 3.5). The PPP is an important branch of glucose metabolism, generating the ribose needed for nucleotide synthesis and the

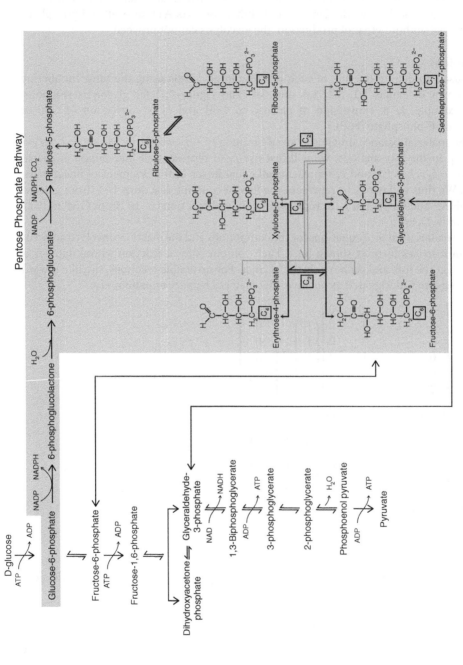

Figure 3.5 Pentose phosphate pathway and its biochemical reactions.

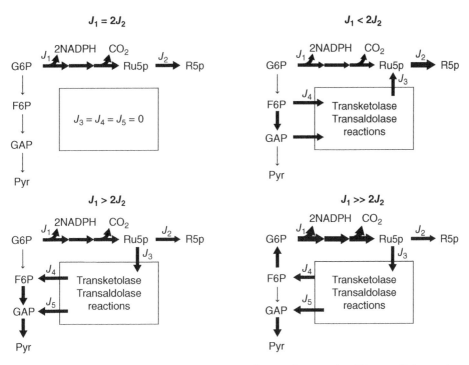

Figure 3.6 Flux distribution through a pentose phosphate pathway under different cellular needs.

NADPH needed for reductive biosynthesis. In actively dividing cells and tissue cells active in lipid synthesis, a significant portion of glucose (~5%) is diverted to the PPP to supply ribose and NADPH.

The PPP can be divided into two segments or stages (Figure 3.6 in Panel 3.23). In the first oxidative decarboxylation stage, it generates NADPH and pentose phosphate. In the second stage, through a series of reversible molecular interconversion reactions catalyzed by transadolase and transketolase, a 2-carbon hydroxyl acetyl group or a 3-carbon dihydroxyl acetonyl moiety is translocated between two monosaccharide-phosphates with 3–7 carbons. Those phosphosugars allow the ribose-phosphate generated in the first stage to reenter glycolysis as F6P or G3P. In the second stage, four enzymes catalyze five reversible reactions of isomerization or aldose-ketose translocation.

In the first stage of the PPP, NADPH and ribose are produced at a 2:1 molar ratio. However, NADPH and ribose are not always needed in a 2:1 ratio for all physiological conditions. Rather, cells need these metabolites at different ratios under different growth conditions, so the second stage of the PPP provides cells with the ability to alter that ratio (Figure 3.6). Through a combination of the five reactions in the second stage of the PPP, ribulose-phosphate can be transformed into F6P and G3P in the second stage. While G3P continues in the glycolysis pathway to be catabolized, F6P, upon reentry into glycolysis, can either be further catabolized through glycolysis, or be returned to the oxidative segment of PPP. To return to the oxidative segment, F6P is converted through the reversible reaction to G6P, which then reenters the decarboxylation branch of PPP to generate more NADPH.

Panel 3.23 Pentose Phosphate Pathway

Reaction	Enzyme
Oxidative segment	
Glucose 6-phosphate + NADP⁺ → 6-phosphoglucono-δ-lactone + NADPH + H⁺	Glucose 6-phosphate dehydrogenase
6-Phosphoglucono-δ-lactone + H_2O → 6-phosphogluconate + H⁺	Lactonase
6-Phosphogluconate + NADP⁺ → ribuilose-5-phosphat + CO_2 + NADPH	6-Phosphogluconate dehydrogenase
Molecular transformation segment	
Isomeration	
Ribulose 5-phosphate ⟺ ribose 5-phosphate	Phosphopentose isomerase
Ribulose 5-phosphate ⟺ xylulose 5-phosphate	Phosphopentose epimerase
Aldose-ketose translocation	
Xylulose 5-phosphate + ribose 5-phosphate ⟺ sedoheptulose 7-phosphate + glyceraldehyde 3-phosphate	Transketolase
Sedoheptulose 7-phosphate + glyceraldehyde 3-phosphate ⟺ fructose 6-phosphate + erythrose 4-phosphate	Transketolase
Xylolose 5-phosphate + erythrose 4-phosphate ⟺ fructose 6-phosphate + glyceraldehyde 3-phosphate	Transketolase

---

Under different physiological needs, cells operate the second stage of the PPP differently to alter the production ratio of NADPH:5-carbon sugar. By setting up material balance equations for this pathway, one can determine the overall flow of molecules. A systematic way of approaching such a reaction system is to first set up a reaction matrix. The matrix consists of all the reactions and reactants, products, and reaction intermediates (Figures 3.7 and 3.8). Each row represents a reaction, and each column represents the material balance on a compound involved in at least one reaction.

In the example shown in Figure 3.8a, row 1 of the matrix lists the stoichiometric coefficients from the G6P dehydrogenase reaction as -1, -1, 1, and 1 for G6P, NADP, 6-phosphoglucono-lactone, and NADPH, respectively. The first column summarizes the balance of G6P, depicting its consumption by two reactions with stoichiometric coefficients of 1 for both reactions.

As you may recall from our example above, the rate of each reaction is typically represented by a J with a subscript depicting the identity of reaction. These J's are presented as a vector. The matrix is then transposed into a stoichiometric coefficient matrix (Figure 3.8). The inputs and outputs of the system consist of G6P, NADP, NADPH, and CO_2 and pentose-phosphate, G6P, and fructose-6-phosphate. The inputs and outputs form another vector on the right side of the equation. All intracellular components that do not have an input or output from outside are set to 0. For those components that have inputs or outputs, the balance should equal the rate of input and output (i.e., the q's).

The reaction equations can then be solved using matrix operations.

Stoichiometric Equations for PPP

Glucose 6-phosphate	NADP+	6-phosphoglucono-δ-lactone	NADPH+H	6-phosphogluconate	ribulose-5-phosphate	carbon dioxide	ribose-5-phosphate	xylulose-5-phosphate	sedoheptulose-7-phosphate	glyceraldehyde-3-phosphate	fructose-6-phosphate	erythrose-4-phosphate	fructose-1,6-bisphosphate		
−1	−1	−1	−1											=0	Glucose-6-phosphate dehydrogenase
		−1		1										=0	Lactonase
	−1		1	−1	1	1								=0	6-phosphogluconate dehydrogenase
					−1		1							=0	Phosphopentose isomerase
					−1			1						=0	Phosphopentose epimerase
							−1	−1	1	1				=0	Transketolase
									−1	−1	1	1	1	=0	Transaldolase
								−1		1	1	−1	−1	=0	Transketolase II
−1											1			=0	Glucose-6-phosphate isomerase
											−1		1	=0	Phosphofructokinase
										2			−1	=0	Aldolase

1. Each row is a reaction to be considered in MFA.
2. In this example, reactants have a negative sign, products have a positive sign.

Figure 3.7 Stoichiometric coefficients of reactions in a pentose phosphate pathway.

3.4.3 Metabolic Flux Analysis on a Cellular System

When MFA is performed on a reaction network, in general, the reaction pathways are well defined. MFA is also applied to cellular systems by encompassing all biochemical reactions in the cell, including the formation of cell biomass and products. It is a rather complex task requiring important assumptions and approximations. This effort is worthwhile, however. When used to ask the right questions, it can yield very useful information.

The first step is to compile all of the reactions involved and set up material balance equations for all intracellular components. Then, solve the system of equations. In most cases, the dilution of materials due to cell volume enlargement (i.e., growth) is neglected. This is regarded as acceptable because the duration that is considered in the MFA is usually much shorter than the doubling time of cells.

MFA provides an additional dimension with which to analyze experimental data. It gives a bird's-eye view of the distribution of nutrients throughout different pathways. Comparing the MFA results of different physiological conditions can reveal significant insights into the dynamics of cell metabolism, not otherwise easily seen.

3.4.3.1 Selecting Reactions for Analysis

There are over a thousand total biochemical reactions in a bacterial cell. That number is even higher for plant cells, which are capable of synthesizing secondary metabolites. However, only a fraction of these reactions has a large flux. Overall, the reactions in the glycolytic pathway have the highest flux for cultured cells, followed by those in the TCA cycle, amino acid and lipid metabolism, protein synthesis, and so on. There are perhaps only a hundred or so reactions with a reaction rate >5% of that of glycolysis. The vast remainders have only a small fraction of the reaction rate of glycolysis.

Matrix Operation for PPP

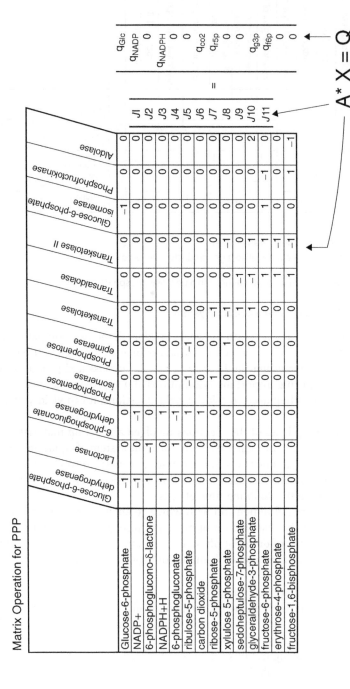

Figure 3.8 Stoichiometric equations of a pentose phosphate pathway for solving flux distribution.

When used to analyze the experimental results of a cellular system, MFA provides more accurate information on major pathways of carbon flow. Because of measurement errors, MFA does not provide accurate estimates for minor reactions in the cellular system. To apply MFA to reactions with a very small flux, it is necessary to accurately quantify the specific formation rate of key intermediates in the particular pathway. This is often performed by tracking the distribution of isotope tracer-labeled compounds to various intermediates.

To apply MFA to a cell system, the first step is to focus on the most relevant and materially significant pathways. In general, those pathways include glucose and amino acid metabolism, and the biosynthesis of building blocks for biomass and product formation. The pathways that are known to have only small fluxes are lumped together and represented by a flux. For example, the synthesis of fatty acids for biomass formation may be lumped into one reaction. During the course of simplifying the reaction network, assumptions are made and reactions are selected for inclusion into the analysis. The selection of pathways for inclusion will inevitably affect the results of the flux analysis.

3.4.3.2 Compartmentalization

Eukaryotic cells have different organelles that serve different functions. Some biochemical reactions occur within a specific organelle and are segregated by space. For instance, glycolysis takes place in the cytosol, while the TCA cycle, oxidative phosphorylation, and fatty acid oxidation occur in mitochondria. Cholesterol biosynthesis occurs across many compartments, including the cytosol, mitochondria, peroxisomes, and the endoplasmic reticulum. Reaction intermediates must cross the membrane barriers of different organelles to complete a pathway or to join another pathway. For example, pyruvate crosses from the cytosol into the mitochondria at a high flux. The NADH generated in glycolysis does not cross the mitochondrial membrane to enter electron transfer. Rather, it transfers its reducing equivalent into the mitochondrion for the generation of ATP through the malate-aspartate shuttle. The flux of the malate-aspartate shuttle to transport the reducing equivalent of NADH is very high, at the same level as the flux of pyruvate into mitochondria.

These shuttles and transferring events across organelle membranes pose further constraints to the material flow. For example, if the malate-aspartate shuttle is hindered by the drainage of malate, the regeneration rate of NAD may be affected, which may ultimately affect the flux of glycolysis. Imposing those constraints on MFA is important for obtaining a realistic estimate of energy metabolism fluxes. Therefore, building a flux analysis model with necessary compartmentalization is a worthwhile effort. When dealing with a compound that is present in multiple compartments, its concentration in each compartment is treated as a separate variable. The concentrations in different compartments are then linked through transport fluxes across compartment boundaries.

3.4.3.3 Biomass

With the exception of cell mass, all outputs of a cell system have a defined chemical composition. The cellular composition is seldom fully characterized. Typically, one resorts to the literature value of similar species to develop a biomass formula for use in the development of a biomass formation equation. In some cases, a more detailed composition (such as the amino acid composition of cellular proteins), in addition to elemental makeup, is measured or estimated. Thus, one can assign the stoichiometric coefficient of various amino acids and other building blocks to assemble to a unit amount of cell mass in a biomass formation reaction equation. This will give a greater confidence in the contribution of metabolic pathways to making biomass.

The contribution of biomass in the total material output is relatively small in anaerobic fermentation and in mammalian cell culture. Thus, MFA on those systems is less prone to error caused by the uncertainty of biomass formula. The vast majority of carbon consumed by cells is converted to metabolites, products, and carbon dioxide. Similarly, for some highly productive industrial processes, the fraction of substrates that is channeled to cell mass is small, with most substrates being used for product synthesis. In the case that the amount of carbon and nitrogen channeled to biomass is a significant portion of the overall material flow, a careful measurement of the composition of cell mass and cell yield is necessary to ensure reliable flux analysis.

3.4.3.4 Limitations on Accounting of Materials

Like material balance in chemical reactions systems, MFA is hindered by the lack of knowledge pertaining to all of the reactions involved and the accounting of all materials. The complete reaction set in a cell is not entirely known, and it varies depending on the cell's tissue of origin and species, and under different conditions. Another challenge in applying MFA for experimental analysis is accounting of material flows in the system that is being analyzed. In principle, a complete accounting of all of the material inputs and outputs in the system is necessary.

Ideally, materials in a reaction system should be balanced according to every major element (C, N, H, and O). The balance of H atoms is sometimes obscured by the difficulty of determining the turnover rate of NAD/NADH and the metabolic production rate of H_2O. Practically, we can perform balances only on carbon- and nitrogen-containing species by measuring the consumption of all carbon and nitrogen sources, and the production of all metabolites and products. Although in some cases this is possible, in many cases, a major metabolic product (i.e., CO_2) is not measured. One may use an estimated value of CO_2 production, if oxygen consumption is measured and the value of the respiratory quotient is available. With measurement errors and uncertainty in determining biomass composition, the closure of balances on C and N is not always easy.

Some culture media contain complex components (e.g., peptones, serum, and plant hydrolysates). Those components provide peptides and lipids for growth. The extent of their consumption is difficult to assess and poses a high degree of uncertainty when performing MFA in these culture conditions.

3.4.3.5 Solution and Analysis

After setting up the reaction model for a cellular system, the number of equations easily reaches 30–50, if not more. Unfortunately, the number of experimentally measured quantities [including inputs, outputs (or specific rates, as will be discussed in Chapter 4), cell mass, and intracellular reaction intermediates] is usually small. The number of unknown fluxes and the number of equations in the system are rarely the same, and the system is often undetermined or overspecified.

After relaxing or further imposing constraints, a best-fit solution is obtained. The solution gives a set of values of the unknown fluxes and the newly estimated specific rates. Some of those specific rates have been measured. One should compare the value of the specific rates given by the solution and the measured value.

It is also prudent to check the material balance based on solution values. For further verification, one can examine the value of the solution set with the biological bound, or the range that is biologically feasible. For example, a respiratory quotient of 5 should arouse caution. The MFA results involve tens of variables depicting the fluxes in the reaction network. Such data is difficult to comprehend without a proper graphic presentation. For further insight, it is helpful to link the results of MFA to a metabolic map for visualization.

Flux balance analysis has also been applied to *in silico* analysis of pathway construction for metabolic engineering. In a cellular reaction system, the number of fluxes and the number of material balance equations that describe the system are not equal. If they are equal, there will be a unique solution that implies that any individual cell, under all conditions in which the reaction system operates, will have the same set of fluxes. If this is the case, all cells will have the same material utilization efficiency. In nature, the biological reaction systems have many degrees of freedom. Under different conditions, the same reaction system may operate differently, and even have different efficiency of energy generation or resource utilization. Imagine that one of the outputs from the system is an industrial product. One can then perform optimization analysis by imposing different constraints and identify the conditions that will give the highest product conversion efficiency. This approach can be used to evaluate, and even design, a new synthetic pathway, as will be briefly described in the Synthetic Biotechnology chapter (Chapter 12).

3.5   Concluding Remarks

Cell growth and metabolite production are processes of material transformation. Performing balances on the material flow in the process is important, whether it is for the cultivation of tissues for regenerative medicine, or for the manufacturing of biochemicals derived from metabolites. The design of medium for cell cultivation, the optimization of manufacturing yield, the reallocation of raw material resources, and many other aspects of process enhancement all require a balance on the materials involved.

In this chapter, we presented material balance on metabolic pathways and on the cell growth system. We also described an MFA that determined a map of the distribution of material flow both in a reaction network and in a cellular system. A simple mass-balanced approach lacks the sophistication of an MFA, but provides a clear view of process efficiency. MFA, on the other hand, gives a detailed flux map, but its conclusions are greatly affected by the given assumptions. The two approaches are complementary; a process engineer should be proficient in using both for process analysis.

Further Reading

Banner, T, Fosmer, A, Jessen, H, Marasco, E, Rush, B, Veldhouse, J & de Souza, M 2011, 'Microbial bioprocesses for industrial-scale chemical production'. In: *Biocatalysis for green chemistry and chemical process development*. Eds. J Tao, RJ Kazlaukas. John Wiley & Sons, Hoboken, NJ.

Geankoplis, CJ 2003, *Transport process and separation process principles*. Chapter 1, pp. 3–27. Prentice Hall, Upper Saddle River, NJ.

Kim, BH & Gadd, GM 2008, *Bacterial physiology and metabolism*. Cambridge University Press, New York.

Park, JM, Kim, TY & Lee, SY 2009, 'Constraints-based genome-scale metabolic simulation of systems metabolic engineering', *Biotechnology Advances*, vol. 27, pp. 979–988.

White, D, Drummond, J & Fuqua, C 2012, 'Central metabolic pathways'. In: *Physiology and biochemistry of prokaryotes*, 4th edn. Oxford University Press, New York.

Nomenclature

A	stoichiometric matrix	
$a, a', b, b', c,$ $c', d, d', e, e',$ m, n	stoichiometry coefficient	
subscript i	chemical species i	
J_i	molar flux of reaction i	mole/cell·t or mole/L^3·t
Q	vector of the inputs	
q_i	specific uptake rate of compound i	M/L^3·t mole/L·t
s	concentration of substrate	
t	time	
X	reaction flux vector	
x	dry biomass concentration	M/L^3
Y	yield coefficient	M/M, M/mole
ψ	mole fraction of carbon source directly incorporated into the product	mole/mole
$\alpha, \alpha', \beta, \beta',$ $\gamma, \gamma', \delta, \delta'$	ratio in molecular formula	
η_i	stoichiometric coefficient of compound i	

Note: A "cell" may be mass of cell or each cell as a unit (e.g., mole per g-cell per hour or mole per cell per hour).

Problems

A. Stoichiometry of Biochemical Reactions

A1 Calculate the theoretical yield of L-lysine production based on glucose. Assume that the microorganism uses glucose and NH_3 as its raw materials and the glycolysis and TCA cycle pathways for synthesis.

A2 The biosynthetic pathway of phenylalanine in *Escherichia coli* is available in a standard biochemistry textbook. Estimate the theoretical yield on phenylalanine based on glucose when a wild-type *E. coli* is grown in a medium containing glucose and ammonia.

A3 Calculate the theoretical maximum conversion yield of producing L-proline from glucose and NH_3 in *E. coli.*

A4 A cell line in culture consumes 100 mmoles of glucose, all of which go to energy catabolism (glycolysis and TCA cycle) to generate lactate and carbon dioxide. The amount of lactate produced is 130 mmoles. Calculate how many moles of carbon dioxide are produced, and how many moles of pyruvate enter mitochondria and how many moles of NADH is generated.

B. Cell Growth and Product Formation

B1 A bacterium uses glucose and NH_3 as carbon and nitrogen sources, respectively. It converts all of the nitrogen atoms they take up from NH_3 to cellular nitrogen (in proteins, nucleic acids, etc.). 50% and 7% of its dry biomass are carbon and nitrogen, respectively. The yield based on nitrogen is the inverse of the nitrogen content in the cells. Why is the yield coefficient based on the glucose not the inverse of carbon content?

B2 A microorganism is capable of utilizing glucose, methanol, and hexadecane as the carbon source. Its average cell composition (by weight) was analyzed to be 47% carbon, 6.5% hydrogen, 31% oxygen, 10% nitrogen, and the remainder ash. During cultivation, this microorganism converts the substrates O_2 and ammonia into cells, carbon dioxide, and water. A batch culture was carried out to estimate the cell yield based on substrate and oxygen (kg cells/kg substrate or oxygen). Air was continuously supplied, and the exhaust gas was vented from the top of the bioreactor. A mass spectrometer was available for gas composition analysis. The water vapor was removed from both inlet and outlet gases before analysis. The inlet gas was composed of 21% oxygen and 79% nitrogen. The composition of the gas at the exit (on a volumetric basis) is listed in Table P.3.1.

Table P.3.1 Gas composition exiting from a bioreactor.

Substrate	% Nitrogen	% Carbon dioxide	% Oxygen
Glucose	78.8	10.2	11.0
Hexadecane	85.6	9.0	5.4
Methanol	85.3	8.5	6.1

a) Write down the stoichiometric reaction equation for each substrate.
b) What are the estimated cell yields, based on the substrates and oxygen, for each of the three substrates? What are the RQ's? (RQ is respiratory quotient = mole carbon dioxide produced/mole of oxygen consumed.)

c) For a bacterium with 8% of its dry mass as nitrogen, what is the yield coefficient for NH_3?

B3 Yeast can be grown in medium containing glucose and hexane. A student performed a series of experiments to determine the yield coefficients. Unfortunately, he forgot to label the results properly and has only the three values for biomass yield based on glucose, hexane, and ammonia (as the sole nitrogen source), as listed below. The ammonia yield is the same for both carbon sources. Identify which value is most likely to be for glucose, hexane, and ammonia, and explain your answer. Of note, 45% of the biomass is elemental carbon, and 7.0% is nitrogen.
a) 0.5 g cell/g
b) 0.8 g cell/g
c) 11.8 g cell/g

B4 The experimental yields of yeast *Candida* grown on glucose, methanol, and hexadecane are 0.51, 0.41, and 0.84 (g dry weight/g substrate), respectively. The nitrogen source is NH_3. The biomass is determined to consist of 47.3% C (by mass), 6.7% H, 10.0% N, and 28.8% O, and the remaining balance is ashes. Estimate the yield coefficient of biomass based on oxygen ($Y_{x/o}$). How is the yield coefficient for biomass based on substrate and oxygen related to the degree of the oxidative state of the substrate? Is there a general relationship?

B5 A batch culture of *E. coli* was initiated with a medium containing 10 g/L of glucose, 2 g/L of ammonia, and 0.1 g/L of cells. At the end of culture, 0.1 g/L of glucose remained in the medium. A total of 5.1 g/L of *E. coli* were recovered. The elemental composition of the cells was determined to be (by mass) 53% C, 7.3% H, 12.0% N, and 19.0% O, with the remaining balance being ashes.
a) Calculate the yield (by mass) of biomass based on glucose and nitrogen.
b) Write down the stoichiometric equation for biomass formation from glucose and ammonia.
c) Assume all the carbon that was not incorporated into biomass was used for energy generation and was completely oxidized to carbon dioxide. Also, assume that each mole of glucose produces 38 moles of ATP when completely oxidized to carbon dioxide. Calculate the biomass yield coefficient based on ATP.

B6 A microorganism can use both CH_3OH (methanol) and CH_3CH_2OH (ethanol) as a carbon source and grow at an equal rate. Which one do you expect to have a higher yield coefficient (g cell/g substrate), and why?

B7 The theoretical maximum of the conversion yield of glucose ($C_6H_{12}O_6$) to ethanol (C_2H_5OH) is 0.51 (g/g). In a fermentation process using yeast, the yield was 0.46 g ethanol/g glucose. It is known that all the carbons in glucose end up in ethanol, CO_2, and biomass and that 50% (by mass) of the biomass is carbon. How much cell biomass is produced per gram of glucose?

B8 The biomass formula for a yeast is $CH_{1.66}O_{0.269}N_{0.194}$. The formula accounts for 92% of dry mass; the other 8% is ashes. When grown on C_6H_{14}, the yield on biomass is

1.4 g cell/g substrate. What is the respiratory quotient (mole CO_2 produced/mole O_2 consumed)? Do you expect the RQ to be higher or lower when the yeast grows on glucose?

B9 Note that the genome sizes are described as Kbp or Mbp (kilo- or mega-base pairs of nucleotides) in a haploid cell of the organism. The size of a human genome is 3.3 Gbp. What is the DNA content (g/cell) in a human somatic cell? You can estimate the average molecular size of each human nucleotide as 280 g/mole.

C. Flux Analysis

C1 The transaldolase and transketolase reactions of the pentose phosphate pathway (PPP) can be considered to be "redistribution reactions," since they can distribute the intermediates back to glycolysis when the cell's needs for ribose-5-phosphate and for NADPH are not at the stoichiometric ratio of 2.0, as dictated by the oxidative section of PPP. How do the fluxes of carbons look, under those conditions that the need of NADPH and pentose phosphate is greater than a molar ratio of 2.0? Draw the "distribution reactions" as a box with ribulose-5-phosphate from the oxidative section as the input and with outputs of fructose-6-phosphate returning to glycolysis. Show qualitatively the relative magnitude of fluxes in the oxidative section and the input and output of the redistribution reactions.

C2 100 mmoles of glucose are taken up by cells and shuttled into the glycolysis and the PPPs. Consider two cases of different growth conditions:
a) Cells need 200 mmoles of NADPH but no ribulose- (or ribose-)5-phosphate.
b) Cells need 100 mmoles of NADPH and 50 mmoles of ribulose- (or ribose-)5-phosphate.
   Determine the distribution of fluxes at the node of glucose-6-phosphate, fructose-6-phosphate, and glyceraldehyde-3-phosphate for both cases.

C3 Refer to glycolysis and the PPP. A cell consumes 5 mmol/h of glucose, and it requires 1 mmol/h of ribose-5 phosphate and 3 mmol/h of NADPH.
a) How much glyceraldehyde-3-phosphate will become pyruvate?
b) Now, the cell's needs have changed; it now requires only 0.1 mmol/h of ribose-5-phosphate, while the glucose input and NADPH requirements are unchanged. What is the new flux of pyruvate generation?

C4 Use flux analysis to determine the relative flux of the molecular transformation portion of the PPP. With 100 moles of ribulose-5-phosphate as the sole input, consider the following cases, and clearly identify the flux for each reaction:
a) 50 moles of fructose-6-phosphate is generated as the outflow of materials (output); the balance goes to triose-phosphate
b) All converted to fructose-6-phosphate.
c) All converted to triose-phosphate.
d) Compare the output of NADPH, ribose, and the flux downstream to pyruvate to generate energy for each of the three cases.

C5 1 mole of glucose is fed to the glycolysis/pentose phosphate pathways. 0.8 moles of ribose-5-phosphate and 1 mole of NADPH are produced. The rest of the reaction intermediates continue in the glycolysis to become pyruvate. What is the amount of pyruvate and NADH produced?

C6 A cell line consumes 100 mmoles of glucose, all of which goes to energy catabolism and generates lactate and carbon dioxide. The amount of lactate produced is 130 mmoles. Calculate how many moles of carbon dioxide are produced, and how many moles of pyruvate and NADH enter the mitochondria. This problem is identical to Problem A4. Solve this problem by setting up the stoichiometric matrices and flux vector.

C7 A yeast culture consumes 10 mmol/h of glucose and produces 18 mmol/h of ethanol. Using isotopic tracer, it was determined that the glucose is either fermented to ethanol or converted to pentose through the PPP. Calculate the amount of carbon dioxide, NADPH, and pentose phosphate produced.

D. Advanced Problems

D1 The yeast *Saccharomyces cerevisiae* and bacterium *Zymomonas mobilis* can both ferment glucose to produce ethanol. The ethanol fermentation process in both microorganisms is growth associated (i.e., the production of ethanol is an energy-generating process that accompanies cell growth). After cell growth ceases, ethanol production slows down and the production process is terminated when cell growth stops. For ethanol fermentation, *Saccharomyces* uses EMP, which produces 2 mol ATP/mol glucose. Conversely, *Zymomonas* uses the Entero–Doudoroff pathway, which produces only 1 mol ATP/mol glucose. The biomass yield based on ATP can be assumed to be 100 mmol ATP/g biomass for both. The cellular composition (percentage of dry biomass) for both microorganisms is listed in Table P.3.2. In a medium containing 80 g/L of glucose, 3 g/L of NH_3, 3 g/L of KH_2PO_4, and plenty of other inorganic nutrients, how much ethanol (g/L) can be produced by each microorganism? Does the medium actually contain sufficient elemental sources for producing the biomass and ethanol (*note*: atomic weight $K = 39$, $P = 31$)? From the perspective of carbon conversion to the product, which organism converts more sugar to ethanol? Assume C, N, O, and H constitute 92% of all dry biomass, while the other 8% are minerals.

Table P.3.2 Cell composition of two microorganisms.

	Saccharomyces	*Zymomonas*
C	50	45
N	8	9
O	30	31
H	7	11
P	1	1.5

D2   A genetically engineered bacterium is used to produce fine cellulose. It grows on glucose and converts glucose to cellulose. The reaction to synthesize cellulose is:

$$nC_6H_{12}O_6 + 2(n\text{-}1)ATP \rightarrow C_6H_{11}O_6(C_6H_{10}O_5)_{n\text{-}2}C_6H_{11}O_5 + 2(n\text{-}1)ADP$$
$$+ 2(n\text{-}1)P_i + (n\text{-}1)H_2O$$

The average chain length, n, of the cellulose molecules is 100. Assume one mole of glucose can be completely oxidized to generate 36 moles of ATP. Calculate the theoretical maximum yield (accounting for the energy requirement for its synthesis) of cellulose (g cellulose/g glucose).

This bacterium is grown in a medium containing 40 g/L of glucose. The initial cell concentration (inoculum) is 0.05 g/L. Cells grow at a maximal rate until cell concentration reaches 5.0 g/L. At that point, the gene for cellulose synthesis is "turned on" (or induced), and the growth immediately stops. The final cellulose concentration is 21 g/L. Assuming that there is no maintenance energy requirement, calculate the yield coefficient of biomass (g cells/g glucose).

D3   Cells excised from plants, especially from embryos in the seeds, can be cultured in bioreactors. These cells can further be propagated to increase biomass, and be induced to form embryos using growth regulators. The embryos can then be encapsulated to become artificial (or synthetic) seeds that give rise to virtually identical clones.

Cells of Douglas fir are grown in culture. The medium contains 200 kg glucose and 25 kg NH_3 per m^3.

Initially, 0.05 m^3 of cells were inoculated into 0.95 m^3 of medium to initiate the culture. After two weeks, glucose was depleted and 0.35 m^3 of biomass was recovered. At that point, the NH_3 concentration was 0.001 kg/m^3. The specific density of the wet biomass is 1020 kg/m^3, and the water content of the biomass is 90%. Calculate the yield of the biomass based on glucose and NH_3 (express the biomass in dry weight).

D4   An *E. coli* strain has been metabolically engineered so that it converts glucose to 1,3-propandiol. This was accomplished by expressing four exogenous genes to create a new pathway. In this new pathway, dihydroxyacetone phosphate (DHAP) from the aldolase reaction in glycolysis is converted to glycerol at the expense of one NADH. The glycerol is then further reduced to 1,3-propandiol using one more mole of NADH. Note that glyceraldehyde-3-phosphate (GAP) still goes to pyruvate.

Using glucose as the carbon source, how much 1,3-propandiol can be produced anaerobically from each mole of glucose? Will there be a side product? How much?

D5   Poly-β-hydroxybutyric acid (PHB) can be used as a biodegradable plastic material. This optically active substance is synthesized by many different bacteria.

Under special growth conditions, the soil bacterium *Alcaligenes eutrophus* is able to store the polymer $(C_4H_6O_2)_n$ at up to 80% of its ash-free dry weight. Elemental analysis of a biomass that contains 5% w/w of this storage product gave the following composition formula for the biomass: $CH_{1.83}O_{0.55}N_{0.25}$.

a) What is the composition formula for the rest of the biomass (i.e., the biomass not included in the storage material)?

b) What is the composition formula for the biomass that contains 80% w/w PHB?

D6 Some yeast cells growing on glucose can mix aerobic and anaerobic metabolism (i.e., they consume oxygen and produce ethanol simultaneously when the O_2 supply is restricted). Consider a cell population that consumes 10 mmol/L-h of glucose and also consumes 30 mmol/L-h of O_2.

a) Calculate the ethanol production rate and CO_2 production rate. Also specify the flux, in mmol/L-h, of pyruvate to acetyl CoA, NAD to NADH, and pyruvate to ethanol.

These cells must maintain a constant total ATP production rate, regardless of whether they grow aerobically or anaerobically. The ATP generation rate has been estimated to be 32 mmoles for each mole of glucose completely oxidized (including 2 mmoles of the ATP produced from the conversion of glucose to pyruvate).

b) Now the supply of oxygen is shut off. How much ethanol will be produced after the oxygen supply is shut off, if the same ATP generation rate is to be sustained? What is the CO_2 production rate?

D7 100 mmoles of pyruvate are used by yeast to generate ATP through the TCA cycle and to convert portions of pyruvate to glucose through gluconeogenesis. Look up gluconeogenesis reactions in a metabolic map. The energy generated in pyruvate oxidation is used for gluconeogenesis. You can assume that each mole of NADH generates 2.5 moles of ATP and that each mole of $FADH_2$ generates 1.5 moles of ADP through oxidative phosphorylation. How many moles of glucose can be generated? Show the fluxes of the reactions converting pyruvate to PEP. To simplify the analysis, consider all reactions are in the same cellular compartment (i.e., cytosol).

D8 A yeast strain is metabolically engineered to convert glucose to polylactate. Two heterologous genes (*Pct* and *PhaC*) have been mutated and cloned in to the strain to complete the engineered pathway. Pct converts propionate to propionyl-CoA, and PhaC converts 3-hydroxylalokanoate to polyhydroxylalkanoate.

a) Briefly describe how those two genes are able to complete the conversion of glucose to polylactate.

b) If the product consists of 20 lactate monomers, calculate the theoretical maximum yield based on glucose.

D9 *E. coli* is used to synthesize succinic acid. The compound can be synthesized in the direction of a typical TCA cycle or through the reverse direction of certain cellular reactions. Mark the pathway that you would use (include all steps needed on the chart shown) to synthesize succinic acid, and briefly explain your reasoning. Calculate the theoretical yield of succinic acid from glucose.

D10 What is the typical volume fraction of mitochondria and nucleus in a eukaryotic cell? Of the 10–13 mmol/h of glucose consumed by a muscle cell, 5% goes to the PPP to generate ribose for nucleic acid. 3% of the total is diverted to generate phosphor-glycerol through glyceraldehyde phosphate. The rest enters the mitochondria after glycolysis. Not all of the pyruvate that enters the mitochondria is oxidized into CO_2. 30% of the pyruvate carbon entering the TCA cycle is diverted (as oxaloacetic acid, alpha-ketoglutarate, and citrate) and used for fatty acid, cholesterol, and amino acid synthesis. Nevertheless, it is easy to see that the molar reaction rate (moles of substrate reacted/volume time) in mitochondria is faster than that in cytoplasm.

a) Estimate the molar reaction rate of pyruvate dehydrogenase and the oxygen consumption rate in mitochondria. Compare the reaction rate to pyruvate kinase in cytoplasm. Which is higher on a per-volume basis?

b) The mitochondrial metabolism can be considered as being at a steady state. Draw a diagram to indicate the fluxes across the mitochondrial inner membrane. Use arrows to indicate transport of all molecular species between the mitochondrial matrix and cytoplasm. Calculate the magnitude of those fluxes.

c) During exercise, the ATP demand of muscle cells increases fourfold. This is accomplished by taking up glucose faster. However, oxygen consumption can only be increased threefold because of the limitations of oxygen supply. How do cells respond, then, to generate enough ATP? Be as quantitative as you can, and state your assumption clearly. Assume that the blood glucose level remains constant.

D11 A bacterial population is growing on lactic acid, consuming it at a rate of 100 mmol/h. It is first converted to pyruvate and then a portion enters the TCA cycle, while 20 mmol/h of pyruvate is used is convert to glucose-6-phophate via the gluconeogenesis pathway to enter PPP to generate NADPH and ribulose-5-phosphate at a 2:1 molar ratio. 90% of the acetyl CoA is used to synthesize lipids, while the remainder is used to generate energy in the TCA cycle.

a) How much CO_2 is generated in mitochondria? How much $NADH/FADH_2$ is generated in the TCA cycle? How much O_2 is consumed?

b) Assuming that each mole of NADH and $FADH_2$ can generate 3 moles and 2 moles of ATP, respectively, what is the net ATP generation rate?

4

Kinetics of Biochemical Reactions

4.1 Enzymes and Biochemical Reactions

Enzymes are fundamental to all aspects of life; they are at the core of bioprocesses. Enzymes are the workhorses of the cell; they break down food and other materials that the organisms take up from the environment to derive energy and convert them to cellular components. Thousands of different enzymes in a cell catalyze virtually all the reactions that are carried out in each cell. They not only make building blocks like amino acids and nucleotides, but also assemble them into DNA molecules base by base, and into proteins amino acid by amino acid, in the correct order. Importantly, they do this fast; a few molecules of DNA synthesis enzymes (and their associated enzymes) in an *Esherichia coli* cell can complete making a new copy of its genome of about four million base pairs in just over half an hour, fast enough for *E. coli* to grow very rapidly. More impressively, they do this with an extremely high fidelity, with only a few mistakes in each new copy of genome synthesized.

We find enzymes everywhere life is in nature. They not only reside inside the cell but also are secreted into the extracellular environment, where they help to degrade materials into a usable form. For example, some microorganisms secrete cellulases that degrade cellulose, converting it to sugars that can be used by cells. Many secrete proteases that degrade proteins to amino acids or small peptides so that they can be taken up as nutrients.

In the human body, extracellular enzymes are present in the bloodstream. They help to clot blood when bleeding occurs and to break up tissue material. They even build a path for cells to move into an injured area, to enable healing of the wound. Enzymes are also in our lungs. Carbonate anhydrase helps us to breathe. Enzymes increase the rate of conversion from bicarbonate ion to CO_2; this allows the exchange of CO_2 between the blood and the air. Enzymes are even in our tears, helping to dissolve or attack invading bacteria.

With their versatility, the catalytic capacity of enzymes has been harnessed by humans, first unwittingly in food making and, since the last century, purposefully in industrial applications. Today, enzymes are used systematically throughout industrial food production. For example, amylases are used to convert starch into glucose. Glucose isomerase is used to convert glucose into a mixture of glucose and fructose (called high-fructose corn syrup). This is used as a sweetener in soft drinks and other processed foods.

Engineering Principles in Biotechnology, First Edition. Wei-Shou Hu.
© 2018 John Wiley & Sons Ltd. Published 2018 by John Wiley & Sons Ltd.
Companion Website: www.wiley.com/go/hu/engineering_fundamentals_of_biotechnology

Enzymes are also used in the pharmaceutical industry for the manufacturing of pharmaceuticals, as well as for direct use as medicine. For example, they are used to convert penicillin produced in bioprocesses to semisynthetic penicillin, in which a side chain of the native penicillin is substituted with another chemical group. This class of derivatives has better clinical characteristics and a different antibacterial spectrum. Enzymes are used directly in the treatment of patients. Some disease states (e.g., congenic diseases) are caused by mutation or deficiency of a physiologic enzyme. Treatment can involve administering this otherwise deficient enzyme to the patient. DNase, for example, is used to treat patients with cystic fibrosis. It digests the DNA that accumulates in their lungs and improves breathing. Similarly, tPA (tissue-type plasminogen activator) is injected into patients who have had a stroke, to help them dissolve residual blood clots. Many patients who lack any one of a number of lysosomal enzymes also benefit from enzyme replacement therapy.

For all their multifaceted roles in nature, enzymes work with other enzymes to link their reactions into a network. Importantly, eventually these enzyme networks and pathways require proteins to move the compounds from the feed into the pathways or are produced by these pathways across the cell membrane. Except for a few small compounds like oxygen and water, only a few molecules can pass through the cell membrane fast enough to sustain cell growth. Virtually all nutrients and metabolites are shuffled through cell membranes by transporter proteins. Enzymes of pathways and their relevant transporters are thus an integrated network. The reaction rate of enzymes and the transport rate are thus coordinated or regulated. An uncontrolled import may result in an accidental surge of intracellular level of a compound that the cellular enzymes cannot cope with. Imagine if the compound is toxic at high levels.

The coordination or regulation of the reaction rate is exerted by controlling the activity of enzymes in a pathway, as will be discussed in this chapter. Importantly, it is also controlled by the amount of each enzyme and/or transporter that a cell makes for a pathway that can meet its need. The control of enzyme synthesis through the expression of the genes coding for the enzyme typically entails binding reactions of protein–protein, protein–nucleic acid (DNA or RNA), or protein–regulatory molecule pairs. Binding reactions are also important in intercellular communication and in the coordination of differentiation and development in multicellular organisms. The binding reactions, enzymatic reactions, and transport functions are three pivotal reaction types fundamental to biochemical reactions. To exploit the synthetic capability of biological systems, we ultimately aim to manipulate these three types of reactions involved in the target biological process. In this chapter, we will learn how to describe the reactions of enzyme, transporter, and molecular binding using kinetic expressions. These quantitative expressions form the basis of systematic analysis of a biological reaction system. Ultimately, a quantitative mechanistic description holds the key to the optimization and design of a synthetic biological reaction system.

4.2 Mechanics of Enzyme Reactions

Enzymes catalyze reactions by lowering the activation energy of the chemical reaction. They have binding sites, which provide the molecular specificity that only allows for binding with the "correct" substrate. Enzymes also have an active site, where the transformation of substrate to product takes place. At the onset of a reaction, the substrate binds to the enzyme, forming an enzyme–substrate complex. The configuration of the enzyme–substrate complex increases the probability of the substrate being elevated to an activated state. As an enzyme-activated substrate complex, the substrate is transformed to the product. The chemical potential of the enzyme–substrate complex is much lower than the activated substrate that must be formed if a conversion reaction from substrate to product were to occur. The participation of the enzyme thus reduces the activation energy required for the conversion.

The basic steps of an enzyme reaction are illustrated using the active site of chymotrypsin during the hydrolysis of a peptide bond (Figure 4.1). Chymotrypsin, like trypsin and many other proteases, is a type of serine protease, so named to highlight the key role of a serine residue in its catalytic activity. In chymotrypsin, a serine and a histidine in the active site are critical for the cleavage of peptide bonds. These two amino acid residues, although distant from each other in the primary peptide sequence (serine at position 195 and a histidine at position 57 from the amino terminus of the protein), are close in the protein's tertiary structure. The 3-OH group in Ser^{195} and the −NH group in the imidazole side chain of His^{57} form hydrogen bonds, thus making the oxygen in Ser^{195} a very strong nucleophile.

In the first step of reaction, a substrate binds to the chymotrypsin-binding site. The carbonyl carbon of the peptide substrate is brought close to the −OH group of Ser^{195}, and the amide nitrogen is brought close to the −NH of His^{57}. A transient-state structure is formed, wherein the carboxylic side of the peptide bond forms an ester bond with Ser^{195}. The amide group of the peptide bond becomes the amino terminus of the other part of the broken protein. The second step of catalysis involves the participation of the

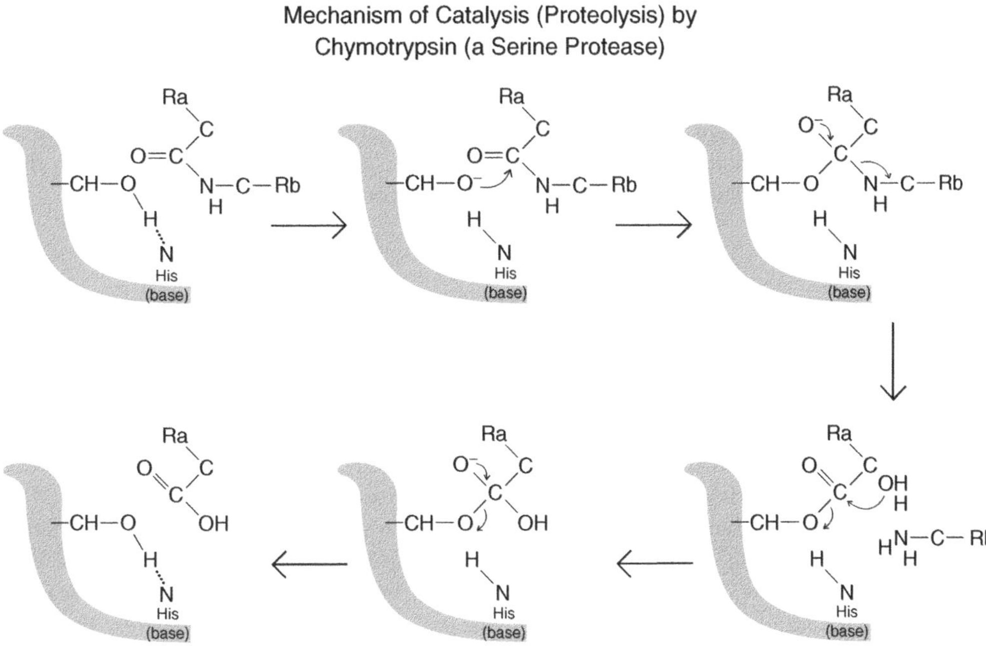

Figure 4.1 Mechanism of peptide bond cleavage by chymotrypsin.

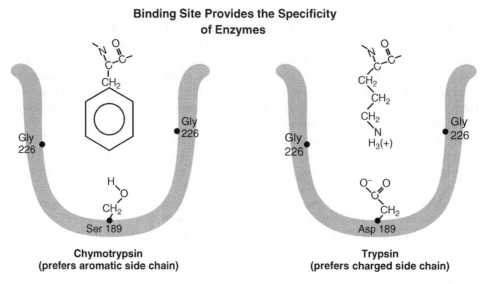

Figure 4.2 Amino acid identity in the substrate binding site determines the substrate specificity of an enzyme. Comparison of two serine proteases, chymotrypsin and trypsin.

H_2O molecule that hydrolyzes the ester bond to free the protein's carboxyl terminus. It is important to note that the generation of the product from the transition-state enzyme–substrate complex also brings the enzyme back to its original state, ready to accept another substrate.

The recognition of substrate through the binding site interaction is illustrated by a comparison of chymotrypsin and trypsin. Chymotrypsin cleaves peptide bonds formed by an aromatic amino acid. Trypsin cleaves a peptide formed by lysine–arginine that has a positively charged amino acid (Figure 4.2). The aspartic acid at position 189, which is a serine in chymotrypsin, renders trypsin preferential to positively charged substrates. Overall, the specificity and activity of an enzyme to a substrate are determined by the binding characteristics of that substrate to the enzyme.

4.3 Michaelis–Menten Kinetics

In 1913, while they were studying the kinetics of invertase that converts sucrose to glucose and fructose, Leonor Michaelis and Maud Menten proposed a mathematical model to describe the relationship between sucrose concentration and the reaction rate. The model has since been known as the Michaelis–Menten equation. In this section, we will examine the underlying principles behind this equation.

Consider an enzyme that catalyzes a reaction to irreversibly convert a single substrate S to a single product P. First, the substrate binds to the enzyme to form the enzyme–substrate complex ES. The substrate in the ES complex can then be released to return ES to its original state (free enzyme and free substrate). Alternatively, the substrate in the ES complex may be converted to form the product before being released. In this depiction, we therefore assume that the formation of the ES complex from the enzyme and substrate is a reversible reaction, whereas the conversion of the *ES* complex to a free enzyme and product is an irreversible reaction. This series of reactions can thus be described as Eq. 4.1 (Panel 4.1), or as individual reactions in Eqs. 4.2 and 4.3 (Panel 4.1).

Panel 4.1

$$E + S \underset{k_{-1}}{\overset{k_1}{\rightleftharpoons}} ES \overset{k_2}{\longrightarrow} E + P \tag{4.1}$$

$$E + S \underset{k_{-1}}{\overset{k_1}{\rightleftharpoons}} ES \tag{4.2}$$

$$ES \overset{k_2}{\longrightarrow} E + P \tag{4.3}$$

In this reaction system, there are four species of compounds (E, S, ES, and P). A balance equation is set up for each species (Eqs. 4.4~4.7 in Panel 4.2), which includes the production and consumption terms. For example, the material balance for the enzyme substrate complex consists of its formation from the binding of substrate to enzyme, as well as the depletion due to the dissociation reaction of the enzyme–substrate complex and the reaction to form the product (Eq. 4.7 in Panel 4.2).

Panel 4.2

$$\frac{dS}{dt} = -k_1 \cdot E \cdot S + k_{-1} \cdot ES \tag{4.4}$$

$$\frac{dE}{dt} = -k_1 \cdot E \cdot S + k_{-1} \cdot ES + k_2 \cdot ES \tag{4.5}$$

$$\frac{dP}{dt} = k_2 \cdot ES \tag{4.6}$$

$$\frac{dES}{dt} = k_1 \cdot E \cdot S - k_{-1} \cdot ES - k_2 \cdot ES \tag{4.7}$$

One can see $\dfrac{dE}{dt} = -\dfrac{dES}{dt}$

Assume that we have an ideal system and that we can operate at a steady state with a fixed amount of enzyme. As the reaction proceeds, substrate is consumed. By adding substrate continuously (at a rate F_s), we can replenish the amount consumed to maintain the substrate concentration at the initial level, thus making Eq. 4.8 equal 0. After a short while, S, E, and ES will be at a steady state (Eqs. 4.4, 4.5, and 4.7 become 0), while P accumulates in the system. The reaction rate will soon equal the rate of product formation and is assumed to follow first-order kinetics with respect to the reactants (Eq. 4.9 in Panel 4.3). Let the enzyme concentration initially be equal to E_o. After the reaction starts, the enzyme exists as either a free enzyme or an enzyme–substrate complex. The sum of the free enzyme and the enzyme–substrate complex is thus equal to the initial enzyme concentration (Eq. 4.10 in Panel 4.3).

Panel 4.3

$$\frac{dS}{dt} = W_s - k_1 \cdot E \cdot S + k_{-1} \cdot ES$$

$$= \text{substrate feeding rate} - \text{substrate net consumption rate} \tag{4.8}$$

$$\frac{dP}{dt} = k_2 \cdot ES = r \tag{4.9}$$

$$E_0 = E + ES \tag{4.10}$$

At steady state, the concentration of every species in the system (except for P) is constant. Rearranging and collecting all ES terms from the steady-state equation of ES balance (4.11) and substituting in Eq. 4.10 (Panel 4.3) give Eqs. 4.12 and 4.13 (Panel 4.4). From Eq. 4.9 (Panel 4.3), one obtains the expression of reaction rate r at a substrate concentration S (4.14). Lumping the three rate constants together as K_m [$= (k_2 + k_{-1})/k_1$] and defining k_2 as k_{cat} lead to the Michaelis–Menten equation (Eq. 4.14 in Panel 4.4).

Panel 4.4

$$k_{-1} \cdot ES = k_1 \cdot E \cdot S - k_2 \cdot ES$$

$$= k_1(E_0 - ES) \cdot S - k_2 \cdot ES$$

$$= k_1 \cdot E_0 \cdot S - ES \cdot (k_1 \cdot S + k_2) \tag{4.11}$$

$$ES \cdot (k_{-1} + k_1 \cdot S + k_2) = k_1 \cdot E_0 \cdot S \tag{4.12}$$

$$ES = \frac{E_0 \cdot S}{\dfrac{k_{-1} + k_2}{k_1} + S} \tag{4.13}$$

$$r = k_2 \cdot ES = \frac{k_2 \cdot E_0 \cdot S}{\dfrac{k_{-1} + k_2}{k_1} + S} = \frac{k_{cat} \cdot E_0 \cdot S}{K_m + S} = \frac{r_{max} \cdot S}{K_m + S} \tag{4.14}$$

---

k_{cat} is often called the "turnover number." It is the maximum rate at which a unit of enzyme can convert the reactant to the product. E_0 is the total enzyme concentration. The maximum reaction rate is achieved when all enzyme molecules become ES complexes. By carrying the experiment over a wide range of substrate concentrations, one can obtain a graphic representation of the relationship between the reaction rate and substrate concentration as described by Eq. 4.14 (Figure 4.3). K_m is called the Michaelis–Menten constant and has units of substrate concentration. At $r = 0.5\, r_{max}$, K_m is equal to S; so K_m is the substrate concentration when $r = 0.5\, r_{max}$. For this reason, it is also sometimes called the half-saturation constant.

We derive the Michaelis–Menten equation using a steady-state situation. S is fed continuously to keep the substrate concentration constant, allowing ES to reach a steady state. The equation can also be used to describe reaction kinetics when the substrate is decreasing, because the change of ES is relatively slow and virtually approaches a steady state.

4.4 Determining the Value of Kinetic Parameters

The Michaelis–Menten equation describes saturation-type kinetics. As the substrate concentration increases to high levels, the reaction rate asymptotically approaches a maximum value. In that range of substrate concentrations (where $S \gg K_m$), the reaction rate is approximated by the constant value of r_{max}. The reactor is zero order with respect to the substrate concentration (Eq. 4.15 in Figure 4.3). At low ranges of substrate concentration, where $K_m \gg S$, the reaction rate can be approximated by $r_{max}S/K_m$ (Eq. 4.16 in Figure 4.3). This reaction follows first-order kinetics as the reaction rate increases linearly with substrate concentration. Between the first- and zero-order regions, there is a transition region. Here, one needs the full Michaelis–Menten equation to properly describe the reaction kinetics.

The kinetic equation can be used to calculate the time profiles of substrate consumption and product formation. In different organisms, or even in different tissues of the same organism, the enzymes catalyzing the same reaction may have different kinetic behaviors. The values of kinetic parameters (k_{cat}, K_m) allow their behavior to be compared quantitatively.

To determine the value of kinetic parameters, one obtains a relationship between the reaction rate and substrate concentration by measuring the increase of product concentration over time at different initial substrate concentrations. The initial reaction rate under different substrate concentrations is then plotted against the initial substrate concentrations to obtain the plot seen in Figure 4.3. It can be shown that by taking the inverse of both the substrate concentration and reaction rate, a linear relationship emerges (Eq. 4.17 in Figure 4.4). In such a $1/r$ versus $1/S$ double reciprocal plot, the slope of the plot is K_m/r_{max}. The intercepts for the axis are $1/r_{max}$ and $-1/K_m$, respectively. This double reciprocal plot is often called a Lineweaver–Burk plot.

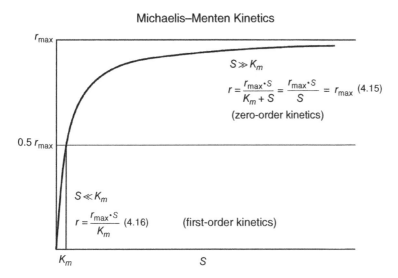

Figure 4.3 The dependence of the enzyme reaction rate on the substrate concentration for an enzyme with Michaelis–Menten kinetics.

$$\frac{1}{r} = \frac{1}{r_{max}} + \frac{K_m}{r_{max}} \cdot \frac{1}{S} \quad (4.17)$$

Lineweaver–Burk Plot

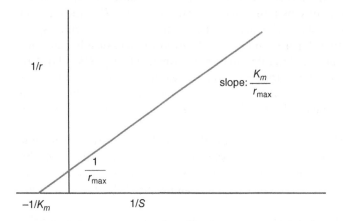

Figure 4.4 Using a Lineweaver–Burk plot to determine the kinetic parameters for the Michaelis–Menten type of enzyme kinetics.

The reaction rate of enzymes varies with operating conditions. Within a range of optimal operating conditions, the reaction rate is highest. Other than substrate concentration, two major factors that affect enzyme activity are temperature and pH. If we measure the reaction rate at a constant enzyme concentration, with a saturation-level substrate concentration while varying temperature or pH, we are likely to obtain a curve as shown in Figure 4.5. The optimal range of an enzyme is affected by its source of origin. Enzymes isolated from a thermophilic organism are likely to have a higher optimal temperature than their counterparts isolated from mesophiles.

In determining kinetic parameters of enzymes, it is common practice to carry out the measurement at the enzyme's optimal conditions. The product formation rate in the initial phase, before the substrate concentration changes substantially, is then used to construct a Lineweaver–Burk plot. Another way of estimating the Michaelis–Menten

Optimal Range of Enzyme Activity

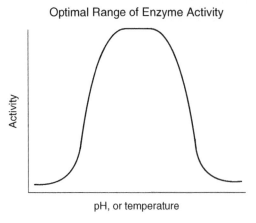

Figure 4.5 Activity profile of an enzyme as a function of an environmental parameter such as temperature or pH.

constant is to carry out the enzymatic reaction and measure the substrate and product concentrations over time. From the slope of the product concentration over time, the reaction rate $r(t)$ is calculated at different times and plotted against the substrate concentration at the corresponding time $[S(t)]$.

The reaction rate versus substrate concentration plot resembles that which is obtained from the initial rate experiment. The data can then be used to calculate the kinetic parameters. Such an approach is valid only when the reaction does not cause the reaction conditions to deviate from the optimal conditions. In many cases, the consumption of the substrate or the accumulation of product can cause the pH to change. Sometimes, the product causes feedback inhibition. In those cases, one must resort to measuring the initial rate to determine the kinetic value.

Example 4.1   Enzyme Reaction Completion Time

A unit of enzyme is the amount of enzyme that converts 1 μmole/min of its substrate to the product. We are developing an assay to measure the lactose level in milk using β-galactosidase. β-galactosidase converts lactose to glucose and galactose. The released glucose is then measured by a standard glucose assay using an enzymatic or chemical method. In this assay, it is important that at least 95% of the lactose in milk is hydrolyzed.

To prepare for the assay, the milk sample will first be diluted to a lactose concentration in the range of 100 μM to 1.0 mM. An aliquot of 0.5 units of enzyme in 2 μL is then added to 1 mL of the diluted milk sample to initiate the enzymatic reaction. The K_m of the enzyme is 20 μM. The reaction is then carried out at 30 °C. How long should the reaction be carried out so that at least 95% of lactose is hydrolyzed?

Solution

The equation describing the consumption of lactose and the production of glucose is:

$$\frac{dS}{dt} = -\frac{r_{max}S}{K_m + S}$$

Separating variables S and t to different sides of the equation:

$$\frac{K_m + S}{r_{max}S}dS = -dt$$

Integrating from $(t=0, S=S_0)$ to (t_f, S_f):

$$\int_{S_0}^{S_f} \frac{K_m + S}{r_{max}S}dS = \int_{t_0}^{t_f} -dt$$

$$\frac{K_m}{r_{max}}(\ln S_f - \ln S_0) + \frac{1}{r_{max}}(S_f - S_0) = t_0 - t_f$$

0.5 units of enzyme are added to 1 mL. The activity of an enzyme is measured in a substrate concentration range where the reaction follows zero-order kinetics. The maximal reaction rate is:

$$r_{max} = 1\frac{\mu mole}{min \cdot unit} \times \frac{0.5 unit}{2\mu L} \times \frac{2\mu L}{1mL + 2\mu L} = 0.0049\frac{\mu mole}{\mu L \cdot min} \approx 500\mu M/min$$

The relationship between the substrate concentration and the reaction time is thus:

$$\frac{20\mu M}{500\mu M/min}(\ln S_f - \ln S_0) + \frac{1}{500\mu M/min}(S_f - S_0) = t_0 - t_f$$

We determine the time in which the reaction goes to 95% of completion for the substrate concentration of 1.0 mM. At that concentration, it will take the longest time. The time is:

$$\frac{20\mu M}{500\mu M/min}(\ln 50\mu M - \ln 1000\mu M) + \frac{1}{500\mu M/min}(50\mu M - 1000\mu M)$$

$$= 0 - t_f$$

$$t_f = 2.02\,min = 121\,s$$

In the case that the sample has 100 µM of lactose:

$$\frac{20\mu M}{500\mu M/min}(\ln 5\mu M - \ln 100\mu M) + \frac{1}{500\mu M/min}(5\mu M - 100\mu M) = 0 - t_f$$

$$t_f = 0.31\,min = 19\,s$$

The reaction should be carried out for 2.02 min.

Note: When 95% of the initial 1.0 mM has reacted, the residual substrate concentration is 50 µM. This level is still 2.5 times of K_m. The hydrolysis reaction is thus mostly in a concentration region of zero-order kinetics. We can apply zero-order kinetics to estimate the reaction time. The reaction time is the initial concentration divided by the maximal reaction rate:

$$\frac{1000\mu M}{500\dfrac{\mu M}{min}} = 2\,min$$

The estimated time is very close to that obtained by the detailed solution.

4.5 Other Kinetic Expressions

The single-substrate/single-product irreversible reaction used to derive the Michaelis–Menten equation is only one of many reaction schemes catalyzed by enzymes. Many reactions are reversible. Most involve more than one substrate and/or more than one product.

Even for single-substrate/single-product reactions, different mechanisms exist. A product may be released from the enzyme complex first, while the enzyme remains in a different state (EAc). The enzyme, now in an intermediate state, is subsequently converted to its native state to start another catalytic cycle. This reaction mechanism is described in Panel 4.5 (Eq. 4.18).

Panel 4.5

$$E + S \underset{k_{-1}}{\overset{k_1}{\rightleftharpoons}} ES \underset{k_{-2}}{\overset{k_2}{\rightleftharpoons}} P + EA_c \overset{k_3}{\longrightarrow} E \qquad (4.18)$$

The reaction can be described by the following equation:

$$r = \frac{\dfrac{k_2 k_3}{(k_2 + k_3)} \cdot E_0 \cdot S}{\dfrac{k_{-1}k_3}{k_1(k_2 + k_3)} + S} \qquad (4.19)$$

We can define K_m, K_{cat}, and K_s as:

$$K_m = K_s \frac{k_3}{k_2 + k_3} \tag{4.20}$$

$$K_{cat} = \frac{k_2 k_3}{k_2 + k_3} \tag{4.21}$$

$$K_s = \frac{k_{-1}}{k_1} \tag{4.22}$$

Using the same approach as for deriving Michaelis–Menten kinetics, we can derive Eq. 4.19 (Panel 4.5) to describe the reaction rate. We can see that by defining K_m, k_{cat}, and K_s as shown in Eqs. 4.20–4.22 (Panel 4.5), the expression of the reaction rate (4.19) is the same as Michaelis–Menten kinetics. It should be emphasized that the definitions of k_{cat} and K_m are different from those in Michaelis–Menten kinetics. In general, for different reaction mechanisms of single-substrate reactions, the basic form of Michaelis–Menten kinetics still applies, but the definitions of K_m and k_{cat} vary.

The majority of reactions in cell metabolism involve more than one substrate and often produce more than one product. Different enzymes utilize different mechanisms to catalyze reactions involving two substrates (Figure 4.6). For instance, in a Ping-Pong two-substrate/two-product (bi-bi) mechanism, a substrate is first bound to its binding site. Then, the product is released (which might be derived from part of the substrate). The binding of the second substrate follows, and the last product is released. Conversely, in order-kinetic mechanisms, both substrates are bound to the enzyme and then both substrates are released.

There are mechanistic kinetic equations for these and other reactions (Eqs. 4.23 and 4.24 in Figure 4.6). These mechanistic equations all have multiple kinetic parameters. Although these kinetic equations accurately reflect the reaction mechanism, the value of their kinetic parameters is often not easily determined. One can determine the

Figure 4.6 Examples of enzymatic, bi-substrate, bi-product reactions.

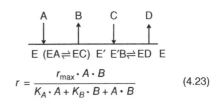

Ping-Pong Bi-Bi

A   B   C   D

E (EA⇌EC) E' E'B⇌ED E

$$r = \frac{r_{max} \cdot A \cdot B}{K_A \cdot A + K_B \cdot B + A \cdot B} \tag{4.23}$$

Ordered Kinetic

A B   C D

E EA EAB⇌ECD ED E

$$r = \frac{r_{max} \cdot AB}{K_A \cdot K_B + K_A \cdot A + K_B \cdot B + A \cdot B} \tag{4.24}$$

value of kinetic parameters experimentally by measuring the reaction rate at different concentrations of A and B. However, the limited number of datasets still makes accurate data fitting difficult. It is common practice to use an empirical form of saturation-type kinetics to describe the kinetics of an enzymatic reaction involving two substrates and two products. A general reaction rate equation for bi-substrate/bi-product reaction (Eq. 4.25) is shown in Eq. 4.26 (Panel 4.6). For process modeling and design, such empirical equations are usually adequate.

Panel 4.6

A reaction with two substrates and two products:

$$A + B \rightarrow C + D \tag{4.25}$$

An empirical rate expression for an irreversible bi-substrate enzyme reaction:

$$r = \frac{r_{max} \cdot A \cdot B}{(K_{mA} + A)(K_{mB} + B)} \tag{4.26}$$

4.6 Inhibition of Enzymatic Reactions

Enzymes recognize their substrates with a very high molecular specificity. However, the enzyme–substrate pairing is not absolute. Some enzymes have evolved to catalyze reactions using a broad range of compounds with similar structural characteristics. Even enzymes with very high substrate specificities may encounter compounds of very similar binding characteristics, capable of binding to the substrate-binding site. Some of those compounds are nonspecific substrates (i.e., not the native substrate of the enzyme) and can be converted by the enzyme to a different product. Others are inhibitors, meaning they bind to the substrate-binding site but are not converted to a product. The binding by those inhibitors reduces the conversion rate of the native substrate to product.

Depending on the characteristics of the inhibition kinetics, enzyme inhibition is generally fall into competitive or noncompetitive type. Competitive inhibitors reversibly bind to the binding site. The effect of a competitive inhibitor can be overcome by employing a higher concentration of substrate to outcompete the inhibitor. Its apparent effect on the kinetic behavior of enzyme–substrate is thus imparted on the value of K_m. In contrast, a noncompetitive inhibitor binds the enzyme at a location other than the active site. The binding of inhibitor cannot be relieved by a high concentration of substrate.

These two different types of inhibition have different kinetic behaviors under different combinations of substrate and inhibitor concentrations. In a Lineweaver–Burk plot, the difference is apparent. At different concentrations of a competitive inhibitor, the intercept of the $1/S$ axis shifts. This is due to the effect of the inhibitor on K_m. Conversely, with a noncompetitive inhibitor, the intercept on the $1/r$ axis shifts due to a changing r_{max} value (Figure 4.7).

Inhibition of Enzyme Reactions

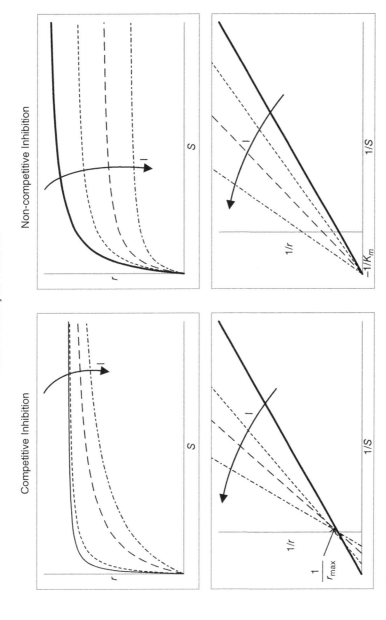

Figure 4.7 Effect of inhibitor concentrations on enzyme reaction rate. (Left) Competitive inhibitor; (right) noncompetitive inhibition.

4.7 Biochemical Pathways

In nature, enzymes work in concert. A group of enzymes with coordinated functions becomes a pathway. With cooperative effort, they channel the flow of materials to interconvert cellular components and dispense the cells' energy carriers. Multiple pathways then form a biochemical network. In a network, a substrate is often used by many reactions. An enzyme often catalyzes a reaction step involved in more than one pathway.

Very few enzymes work in complete isolation. Rather, different enzymes interact through shared substrates or common pathways. To explore or exploit the biochemical rhythm of cells, it is important to understand the interactions of enzymes and pathway kinetics.

4.7.1 Kinetic Representation of a Reaction Pathway

Enzymatic reactions are traditionally grouped together by the serial nature of their reactions and by their common functionality. For instance, reactions that synthesize lysine, threonine, and leucine form a pathway of aspartate family amino acids. Consecutive reactions may form a linear pathway, wherein the product of one reaction is first generated and then becomes the substrate of a subsequent reaction.

A pathway may also be branched, with the product of a reaction being split between subsequent reactions. It is also possible that two reaction paths converge to combine two fluxes into one. In this section, we will use a linear pathway and a branched pathway to illustrate the basics of analyzing pathways. We will also highlight an important feature of biochemical pathways: that their response to substrates and other environmental perturbations is not linear.

Consider a series of reactions catalyzed by enzymes E_1, E_2, and E_3 that convert A to B and then to C and to D. The stoichiometry of these reactions is shown in Panel 4.7. Let the reactions be carried out in a closed system, like a batch reactor. We assume all reactions follow Michaelis–Menten kinetics. The material balance equations for A, B, C, and D are as shown in Eqs. 4.27 to 4.30 of Panel 4.7. The reaction rates for the three reactions are r_1, r_2, and r_3. The material balance for each compound consists of a generation rate and a consumption rate, related by a stoichiometric coefficient. If we know the initial concentrations of A_0, B_0, C_0, and D_0, the system of equations can be solved; the time profile of the concentrations of each compound can be obtained. In such a closed system, the reactions may reach completion and stop entirely.

Panel 4.7 Rate Expression of Reactions in Series

A series of reactions catalyzed by enzymes E_1, E_2, and E_3, converting A to D:

$$A \xrightarrow[r_1]{E_1} \beta B \quad ; \quad B \xrightarrow[r_2]{E_2} \gamma C \quad ; \quad C \xrightarrow[r_3]{E_3} \varsigma D$$

All reactions follow Michaelis–Menten kinetics.
  The concentrations of A, B, C, and D are described by:

$$\frac{dA}{dt} = -r_1 = \frac{-k_{cat,1} \cdot E_{10} \cdot A}{K_{m,1} + A} \tag{4.27}$$

$$\frac{dB}{dt} = \beta r_1 - r_2 = \beta \frac{k_{cat,1} \cdot E_{10} \cdot A}{K_{m,1} + A} - \frac{k_{cat,2} \cdot E_{20} \cdot B}{K_{m,2} + [B]} \qquad (4.28)$$

$$\frac{dC}{dt} = \gamma r_2 - r_3 = \gamma \frac{k_{cat,2} \cdot E_{20} \cdot B}{K_{m,2} + B} - \frac{k_{cat,3} \cdot E_{30} \cdot C}{K_{m,3} + C} \qquad (4.29)$$

$$\frac{dD}{dt} = \varsigma r_3 = \varsigma \frac{k_{cat,3} \cdot E_{30} \cdot C}{K_{m,3} + C} \qquad (4.30)$$

Now, let us consider a case that the first reactant (A) is supplied continuously, such that its concentration is maintained at a constant level. This is much like a nutrient being taken up continuously over a period of time. The downstream reactions in the pathway continuously convert the reactant to the product. The product generated in the pathway may accumulate in the system and have no effect on the progression of the reactions. Otherwise, it can be continuously withdrawn from the system, much like a metabolic product that is secreted out of the cell. We denote the system input and output with q.

The balance equations now include the input and output terms, q_A and q_D. The stoichiometric coefficients for the reactants are given a negative sign, while products are given a positive sign. q for product excretion (i.e., output) is given a negative sign. The reactant taken up from the outside is positive. The balance equations are listed in Eqs. 4.31 to 4.34 (Panel 4.8). Recall that each reaction rate r is represented by a Michaelis–Menten kinetic expression, as shown in Panel 4.7. At steady state, these differential equations become algebraic equations. All of the reaction rates and the input/output rates (q) are related by stoichiometric coefficients (Eq. 4.35 in Panel 4.9). If the value of the kinetic parameters is known, and any of the rates are measured or otherwise known, the steady-state concentrations of each compound can be determined (Eqs. 4.35 to 4.39 in Panel 4.9)

Panel 4.8

If the system has an input of A and excretes D at q_A and q_D, then a steady state can be achieved. At steady state:

$$A \to \boxed{\begin{matrix} A \to \beta B \\ B \to \gamma C \\ C \to \varsigma D \end{matrix}} \to D \qquad \frac{dA}{dt} = -r_1 + q_A \qquad (4.31)$$

$$\frac{dB}{dt} = \beta r_1 - r_2 \qquad (4.32)$$

$$\frac{dC}{dt} = \gamma r_2 - r_3 \qquad (4.33)$$

$$\frac{dD}{dt} = \varsigma r_3 - q_D \qquad (4.34)$$

Panel 4.9

At steady state, the reaction rate can be described based on A, B, C, or D. Different descriptions of rates are related by the stoichiometric coefficient.

$$q_A = r_1 = \frac{r_2}{\beta} = \frac{r_3}{\beta\gamma} = \frac{q_D}{\beta\gamma\varsigma} \tag{4.35}$$

$$q_A = \frac{k_{cat,1} \cdot E_{10} \cdot A}{K_{m,1} + A} \tag{4.36}$$

$$0 = \beta \frac{k_{cat,1} \cdot E_{10} \cdot A}{K_{m,1} + A} - \frac{k_{cat,2} \cdot E_{20} \cdot B}{K_{m,2} + B} \tag{4.37}$$

$$0 = \gamma \frac{k_{cat,2} \cdot E_{20} \cdot B}{K_{m,2} + B} - \frac{k_{cat,3} \cdot E_{30} \cdot C}{K_{m,3} + C} \tag{4.38}$$

$$q_D = \varsigma \frac{k_{cat,3} \cdot E_{30} \cdot C}{K_{m,3} + C} \tag{4.39}$$

From Eq. 4.35, we see that the flux of the pathway from A to D can be described based on any compound in the pathway. These different measures of flux are related by stoichiometric coefficients. The pathway is often written by setting the stoichiometric coefficient of the reactant (or one of them) of the first reaction in the pathway to 1. By convention, the rate is expressed as mole per unit volume per unit time of that first compound. These units are then used to describe the flux (J) of the pathway. It should be stressed that there are many other ways to describe flux. One should always clearly specify the definition of fluxes when reporting.

4.7.2 Linearity of Fluxes in Biochemical Pathways

Using the molar flux of r_1 as the flux of the pathway J (Eq. 4.40 in Panel 4.10), we next examine the condition that all concentrations of reactants (A, B, and C) are at low levels. In this case, all reactions occur in the first-order region (Eq. 4.41 in Panel 4.10). If the reaction conditions change, for example the supply rate q_A or the concentrations of A change, the fluxes at steady state also change. At the new steady state, the concentrations of A, B, and C will change in the same ratio as that of the two steady-state fluxes (Eq. 4.42 in Panel 4.10) (Figure 4.8). The response of the system is thus "linear" when the reaction kinetics are first order.

Now, consider that the reactions are not first order, but described by Michaelis–Menten kinetics as shown in Eq. 4.40. In this scenario, the simple proportionality between concentration change and rate variation does not hold true anymore. Thus, the response of the biochemical systems is rather nonlinear for nonlinear (non-first-order) kinetics.

Next, we examine a branched pathway wherein the reactant A is converted to B in reaction 1. Then, B is used as a substrate by reactions 2 and 3 simultaneously. It is converted to C and D, respectively (Panel 4.11). The material balances are shown in Eqs. 4.43

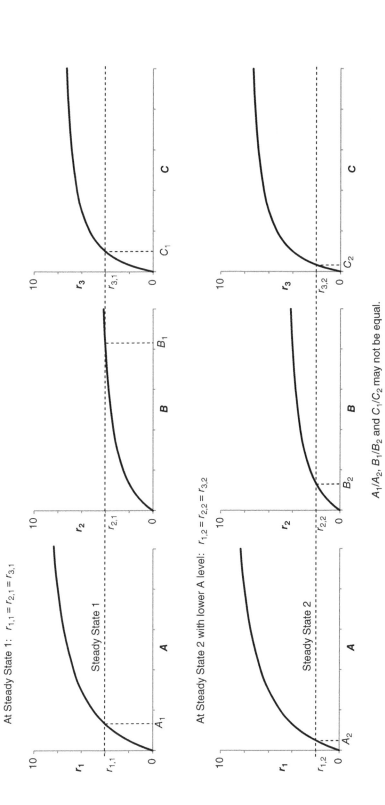

Figure 4.8 Enzymatic reactions forming a linear pathway.

to 4.46 in Panel 4.11. Reactions 2 and 3 share a common substrate, B, and the total consumption of B by reactions 2 and 3 must be equal to the production rate of B from reaction 1 at steady state (Eq. 4.44 becomes 0) (Figure 4.9).

Panel 4.10

$$J = r_1 = \frac{r_2}{\beta} = \frac{r_3}{\beta\gamma}$$

$$= \frac{r_{max,1}A}{K_{m1} + A} = \frac{1}{\beta}\left(\frac{r_{max,2}B}{K_{m2} + B}\right) = \frac{1}{\beta\gamma}\frac{r_{max,3}C}{K_{m,3} + C} \tag{4.40}$$

If the reaction kinetics are in the first-order region:

$$J = \frac{r_{max,1}}{K_{m,1}}A = \frac{1}{\beta}\frac{r_{max,2}}{K_{m,2}}B = \frac{1}{\beta\gamma}\frac{r_{max,3}}{K_{m,3}}C \tag{4.41}$$

The fluxes at two steady states (i, j):

$$\frac{J_i}{J_j} = \frac{\dfrac{r_{max,1}}{K_{m,1}}A_i}{\dfrac{r_{max,2}}{K_m}A_j} = \frac{\dfrac{r_{max,2}}{K_{m,2}}B_i}{\dfrac{r_{max,2}}{K_m}B_j} = \frac{\dfrac{r_{max,3}}{K_{m,3}}C_i}{\dfrac{r_{max,3}}{K_{m,3}}C_j}$$

$$= \frac{A_i}{A_j} = \frac{B_i}{B_j} = \frac{C_i}{C_j} \tag{4.42}$$

The flux and the concentration change in the same proportion.

Panel 4.11 Reactions with Branches

A series of reactions catalyzed by enzymes E_1, E_2, E_3, converting A to C and D:

$$A \rightarrow \boxed{A \xrightarrow[r_1]{E_1} \beta B; \ B \xrightarrow[r_2]{E_2} \gamma C; \ B \xrightarrow[r_3]{E_3} \varsigma D} \rightarrow C, D$$

The material balance equations are:

$$\frac{dA}{dt} = q_A - r_1 = \frac{-k_{cat,1}E_{10} \cdot A}{K_{m,1} + A} + q_A \tag{4.43}$$

$$\frac{dB}{dt} = \beta r_1 - r_2 - r_3 = \beta\frac{-k_{cat,1}E_{10} \cdot A}{K_{m,1} + A} - \frac{k_{cat,2}E_{20} \cdot B}{K_{m,2} + B} - \frac{k_{cat,3}E_{30} \cdot B}{K_{m,3} + B} \tag{4.44}$$

$$\frac{dC}{dt} = \gamma r_2 - q_c = \gamma\frac{k_{cat,2}E_{20} \cdot B}{K_{m,2} + B} - q_c \tag{4.45}$$

$$\frac{dD}{dt} = \varsigma r_3 - q_D = \varsigma\frac{k_{cat,3}E_{30} \cdot B}{K_{m,3} + B} - q_D \tag{4.46}$$

When all reactions follow first-order kinetics, the change in the concentration of reactants A and B is proportional to the change in the flux. The fluxes of the two

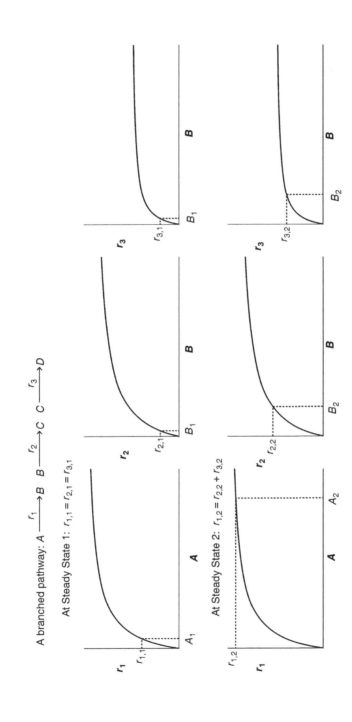

Figure 4.9 Enzymatic reactions forming a branched pathway.

branch reactions (reactions 2 and 3) also change at the same proportion. Such linear responses are not seen when the reaction kinetics are nonlinear, as in the case of Michaelis–Menten kinetics.

4.8   Reaction Network

Hundreds of enzymes participate in cellular metabolism. The enzymes, which work in concert to complete a chemical transformation process, are grouped into pathways. The concentration of metabolic enzymes in an *E. coli* cell is perhaps in the order of 0.1–1 μM, or about 100 to 1000 molecules per cell. Many metabolites are present in the cells in the range of 10–500 μM. Although the enzymes and reaction intermediates of a pathway may be functionally associated, they are usually not spatially isolated from all other enzymes of different pathways. Many enzymes are involved in more than one pathway. In a cellular compartment, enzymes and reaction intermediates of different pathways are mixed together.

There are more reactions that utilize two or more substrates and produce multiple products, compared to the single-substrate/single-product reactions described by Michaelis–Menten kinetics. For easier visualization, biochemical pathways are depicted in straight, orderly paths, showing only the main carbon flow of substrates and products along the path. Reactants common to many cellular reactions, including energy carriers (ATP/ADP, NADH/NAD, and $FADH_2$/FAD) and co-substrates (such as CO_2, H_2O, acetyl CoA, etc.), are shown on the side. These co-substrates participate in a large number of biochemical reactions.

In addition to those cofactors, many substrates are involved in different reactions of multiple pathways. For example, glucose-6-phosphate is used in both glycolysis and the pentose phosphate pathway (PPP). Pyruvate is not only converted to acetyl CoA for the tricarboxylic acid (TCA) cycle and to lactate for the regeneration of NADH; it also leads to alanine, and is converted to oxaloacetate. Fluctuations in the level of a compound that is a common substrate of many reactions may affect the rate of all the pathways that use the compound as a substrate or intermediate. Many pathways are therefore intimately connected to one another through shared enzymes, substrates, and products.

4.9   Regulation of Reaction Rates

Microbial cells must use their resources efficiently in order to compete and thrive in their native environments. Different types of cells in a multicellular organism have complementary biochemical functions, and adjust their activities to enable the organism to utilize the resources optimally. To accomplish this, cells need to adjust the activity of various pathways according to growth needs under different environmental conditions. The regulation of cells' biochemical activities is fundamental to their well-being and is handled on multiple levels.

4.9.1   Flux Modulation by K_m

At the most basic level, the material flux at a branching node is distributed to the different reaction paths, according to a cell's needs. As shown in Figure 4.9 and Panel 4.11,

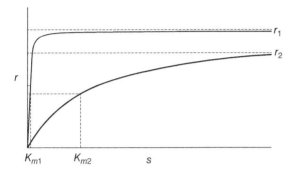

Regulation/Partition of Resources by K_m

Enzyme Competition for Substrate: Regulation by K_m

The reactions critical for cell growth have a lower K_m than others that compete for the same substrate.

Figure 4.10 Competition of reactions sharing the same substrate. Regulation by binding affinity of the substrate.

the distribution of material flow at a divergent branching node is determined by the r_{max} and K_m of each reaction. The amount of enzyme present (E_o, since $r_{max} = k_{cat}E_o$) and the affinity to the substrate ($1/K_m$) together determine the distribution of the common substrate to the two competing reactions. At a given concentration of the common substrate, the enzyme that has a higher $k_{cat}E_o/K_m$ will be more competitive in seizing the substrate. The enzyme of the branch that has a lower priority in the distribution of cellular resources often has a higher K_m for the common substrate; the substrate can be preferentially directed to the path of higher need (Figure 4.10).

For example, α-ketoglutarate is an intermediate in the TCA cycle and is also used as a cofactor in β-lactam antibiotic cephamycin C biosynthesis in some bacteria. The K_m for α-ketoglutarate dehydrogenase in the TCA cycle is around 0.1 mM. This value is about ten times higher in the reaction, where it serves as a co-substrate in cephamycin synthesis. During a period of rapid growth, α-ketoglutarate dehydrogenase has a tenfold higher affinity for α-ketoglutarate than for the enzyme that makes cephamycin. This difference in the K_m values of the two enzymes ensures that α-ketoglutarate is preferentially used for energy generation in the TCA cycle, rather than for antibiotic synthesis during rapid growth. After the growth rate and the consumption of α-ketoglutarate in the TCA cycle slow down, the accumulation of α-ketoglutarate allows it to become more available for secondary metabolism.

4.9.2 Allosteric Regulation of Enzyme Activities

Cells control their distribution of resources, both globally and locally. At a global level, as the growth rate changes, the metabolic activity of all catalytic and anabolic pathways adjusts to demand. At a local level, by exerting control on one or more key enzymes, the flux of a pathway (or even just a branching node) is modulated to respond to changing

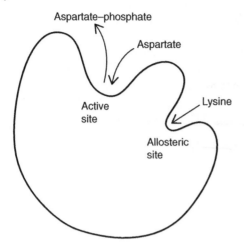

Aspartate–phosphate

Aspartate

Active site

Lysine

Allosteric site

Figure 4.11 Illustration of allosteric regulation of an enzyme.

needs. If regulation is at the level of a branching node within a pathway, it is important to have the capability to adjust the flux of each branch, independent of one another.

Feedback inhibition is the most direct way to control metabolic flux of a pathway and avoid wasteful excessive synthesis. With feedback inhibition, an overaccumulation of a pathway's product (or reaction intermediate) causes the product/intermediate to bind to the pathway's key enzyme and reduce its activity. This reduces the flux of the pathway and, in turn, normalizes the concentration of the product/intermediate. In most cases, the effector (or the inhibitory metabolite) binds to a different site (called the allosteric site) than the usual substrate-binding site of the enzyme (Figure 4.11). Such allosteric inhibition is used extensively in glycolysis and during the biosynthesis of a cell's building blocks, to regulate their production under different growth conditions.

In a metabolic pathway, the node of metabolic regulation is frequently at the head of a pathway (i.e., the first reaction after the branching point). This is illustrated with the pathways that describe the synthesis of aspartic acid and its derivatives in *Corynebacterium glutamicum*. This industrial organism produces a number of amino acids, including glutamic acid and lysine (Figure 4.12).

Four amino acids (lysine, methionine, threonine, and isoleucine) are derived from aspartic acid. The pathway branches into two routes: one leading to lysine, and one leading to the other three amino acids. The second branch has two subbranches: one leading to methionine, and the other to threonine and isoleucine. Extensive feed inhibition and repression regulate the synthesis of these four amino acids. Lysine and threonine both exert multivalent inhibition on aspartate kinase, the first enzyme in the pathway. Similarly, the first enzyme in the second branch is inhibited by threonine and isoleucine and repressed by methionine. The first enzyme after the subbranching node (homoserine) is repressed by threonine.

By affecting the activity and synthesis of these pivotal enzymes, the flux of metabolites is restricted during the times when different amino acids are in abundance and the biosynthetic pathway does not need to be in operation. This prevents wasteful production.

For industrial purposes, this type of regulation can be a hindrance to high productivity. As we will discuss in the chapter on synthetic biology (Chapter 12), a key to

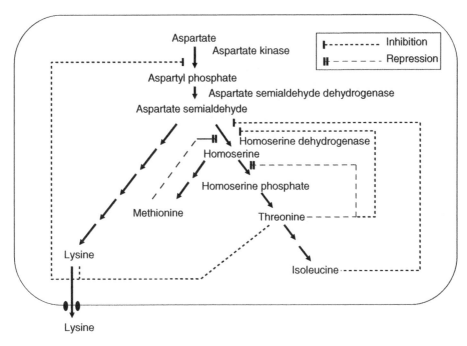

Figure 4.12 Regulation of the flux of biochemical pathways by feedback inhibition and repression. Biosynthesis of aspartate family amino acids as an example.

overproducing microbial metabolites for biotechnological applications is to deregulate the pathway(s) involved. This essentially transforms the producing organism into a "wasteful" mode of production, at least as far as the microbe is concerned.

The allosteric regulation at the enzyme activity level is not limited to feedback inhibition alone. Activation is also seen in biochemical reactions, although not as frequently as feedback inhibition. Activation may act on a downstream enzyme, in which the accumulation of a reaction intermediate upstream in a pathway causes the intermediate to bind to a downstream enzyme and increase its activity. A notable example is the activation by fructose-1,6-bisphosphate on the pyruvate kinase in glycolysis. An accumulation of fructose-1,6-bisphosphate indicates that the subsequent reactions in glycolysis are restricting flux. Activation of pyruvate kinase increases the flux and alleviates the accumulation of upstream fructose-1,6-bisphosphate.

4.9.3 Regulation at Transcriptional and Posttranslational Levels

In addition to the feedback allosteric regulation of enzyme activity, cells also respond to their changing metabolic needs by increasing or decreasing the level of enzymes through transcription and translation. For example, enzymes of a pathway utilizing a particular sugar are not needed unless the sugar is present. Hence, those enzymes are not synthesized until the presence of the particular sugar that triggers the transcription and translation of the enzymes. This phenomenon is called "induction." Feedback repression, on the other hand, suppresses the flux of a pathway by reducing the synthesis of the enzyme in response to the excessive accumulation of some metabolite(s) of the pathway.

There are two important distinctions between feedback inhibition and feedback repression: (1) their contrasting purposes, and (2) their response times. First, note their differing purposes. Feedback inhibition regulates enzymatic activity. Feedback repression, on the other hand, affects the level of the enzyme. Second, note their comparative response times. In the case of feedback inhibition, response time is relatively fast. When the inhibitor accumulates, it quickly reduces the activity of the target enzyme, and rapidly restores it once the inhibitory compound drops. Feedback repression, in contrast, takes a longer time before we can observe its effects. A decrease in the intracellular level of an enzyme, by enzyme turnover or by dilution due to cell growth, is a slower process. In some cases, cells may invoke additional regulatory mechanisms to degrade a particular enzyme and to remove it from the cellular environment.

4.9.4 Modulation of Resource Distribution through Reversible Reactions

Many reactions in biochemical pathways are reversible. In fact, cell metabolism involves more reversible reactions than irreversible ones. Cells have evolved this way to efficiently grow and persist under very diverse conditions. For instance, consider a cell that usually relies upon a pathway to convert a given substrate to a desired product. Under some circumstances, the substrate may be limited but the product is in excess. In these cases, it may become necessary for the cell to reverse the direction of the pathway, and instead utilize the product to synthesize the substrate.

One such example is glucose, the most abundant and available substrate of most cells. It is metabolized through glycolysis to provide energy and to supply the precursors for PPP and lipid synthesis. If glucose is depleted but pyruvate or lactate is available, cells must then use the pyruvate to generate energy (through the TCA cycle) and also to generate glucose-6-phophaste for PPP and the synthesis of other cellular components. The process of synthesizing glucose from the downstream product of energy metabolism, like pyruvate, lactate, and acetyl CoA, is called gluconeogenesis. Since the majority of glycolytic reactions are reversible, gluconeogenesis can occur when the need arises. Cells only need to make a few enzymes to circumvent a few irreversible steps.

A typical reversible, single-substrate/single-product reaction and its kinetic expression are shown in Panel 4.12 (Eq. 4.47). The reaction mechanism is the same as that for Michaelis–Menten kinetics, with one exception: the last step of dissociation of the enzyme–product complex to the free enzyme and product is a reversible step. The kinetic expression is a combination of two Michaelis–Menten equations: one is the forward reaction from S to P, and the other is the reverse reaction from P to S. In the limiting case that P is zero, the equation becomes a Michaelis–Menten equation. In another limiting case that S is zero, the expression is a Michaelis–Menten equation for the conversion of P to S. The reaction rate may have a positive value for forward reaction or a negative value for the conversion of P to S (Figure 4.13).

An important characteristic of a reversible reaction is that the reaction rate depends not only on the substrate concentration, but also on both S and P. Multiple pairs of S and P can give the same reaction rate. In an irreversible reaction, a higher reaction rate is associated with a higher substrate concentration, as described by simple Michaelis–Menten kinetics. However, in a reversible reaction when the reaction rate is lower, the substrate concentration may even be higher, albeit with a higher concentration of the product (and thus a higher rate of reverse reaction). In the case that a

Panel 4.12 A general single-substrate reversible reaction:

$$S \rightleftharpoons P \quad r = \frac{k_s E_0 S - k_p E_0 P}{1 + \dfrac{S}{K_{ms}} + \dfrac{P}{K_{mp}}} \tag{4.47}$$

A general bi-substrate reversible reaction:

$$A + B \rightleftharpoons C + D$$

$$r = \frac{(r_{max,f} \cdot A \cdot B - r_{max,r} \cdot C \cdot D)}{(K_{m,A} + A)(K_{m,B} + B) + (K_{m,C} + C)(K_{m,D} + D)} \tag{4.48}$$

An example of a mechanistically derived equation (for pyruvate kinase):

$$r = \frac{\left(\dfrac{r_{max,f} \cdot A \cdot B}{K_{m,A} \cdot K_{m,B}} - \dfrac{r_{max,r} \cdot C \cdot D}{K_{m,C} \cdot K_{m,D}} \right)}{\left(1 + \dfrac{A}{K_{m,A}} \right)\left(1 + \dfrac{B}{K_{m,B}} \right) + \left(1 + \dfrac{C}{K_{m,C}} \right)\left(1 + \dfrac{D}{K_{m,D}} \right) - 1} \tag{4.49}$$

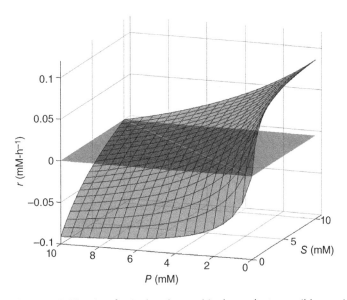

Figure 4.13 Kinetics of a single-substrate/single-product reversible reaction.

branching node has a reversible branch, a reduction in its flux may be accompanied by many possibilities of new S and P pairs, depending on the kinetic behavior of the pathway. Thus, changes in the flux of the reversible node may or may not affect the flux of the other branch. The dependence of reaction rate on a substrate–product pair gives one more degree of freedom in the control of reaction rate. This is important for regulating the distribution of fluxes in a network.

Many bi-substrate reactions are also reversible. A general form of the bi-substrate reversible reaction and an example of a mechanistic equation are shown in Eqs. 4.48

and 4.49 (Panel 4.12). The kinetics of inhibition or activation by an effector molecule is described through a multiplicative regulatory term (N) that depicts the action of the inhibition or activation (Eq. 4.50 in Panel 4.13). We can verify that inhibition is described by N with a positive value of exponent n; activation is by a negative exponent. The strength of the inhibition or activation is affected by the magnitude of n. In the case that the regulatory effect is exerted by changing K_m instead of by modulating k_{cat}, the effect is described by multiplying the corresponding K_m by N.

Panel 4.13 Description of Allosteric Regulation on Enzyme Activity

Inhibition or activation by an allosteric compound L:
  Multiplying the rate equation r by an allosteric regulation term N.

$$N = \frac{1}{\left(1 + \dfrac{L}{K_{I,L}}\right)^n} \tag{4.50}$$

The exponent n is positive for allosteric inhibition, and positive for allosteric activation.
  When the allosteric regulation affects K_m rather than r_{max}, K_m is multiplied by the N term. N can be determined by fitting experimental data.

4.10 Transport across Membrane and Transporters

So far, we have focused on reactions that involve the conversion of biochemicals. Another type of reaction that is very important to cellular function is the transport of molecules across the cell membrane. In Chapter 2, we discussed the lipid bilayer membrane acting as a barrier separating a cell's interior from its surroundings. Aside from oxygen and water, few nutrients can pass through the cell membrane fast enough to meet the need for cell growth. Essentially, all nutrients are transported into the cell through their transporters. Metabolites and products are also excreted through transporters.

A cell may encounter very different chemical environments at different times, but it must maintain its intracellular environment within a relatively narrow range. For example, the intracellular pH, concentration of major ions, and osmolarity are all relatively constant. This is achieved by balancing intracellular reactions through an exchange with the extracellular environment. Transporters are therefore involved not only in the transport of nutrients and products, but also in the influx and outflux of many other solutes to sustain the cell's homeostasis. Indeed, transporters constitute a very large fraction of all proteins in the cell membrane.

4.10.1 Transport across the Cell Membrane

To evaluate solute diffusion across a membrane, we will treat the membrane as a stagnant lipid bilayer film that is in contact with an aqueous solution on each side. The diffusion of a solute in a medium is described by Fick's law, as shown in Eq. 4.50 in Figure 4.14.

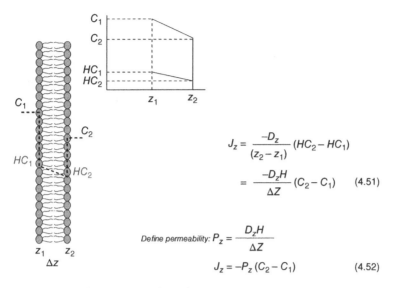

$$J_z = \frac{-D_z}{(z_2 - z_1)}(HC_2 - HC_1)$$

$$= \frac{-D_z H}{\Delta Z}(C_2 - C_1) \quad (4.51)$$

Define permeability: $P_z = \dfrac{D_z H}{\Delta Z}$

$$J_z = -P_z(C_2 - C_1) \quad (4.52)$$

Figure 4.14 Diffusion across cell membrane.

Consider the diffusion of a solute from one side of a membrane to the other side. The flux is proportional to the difference in the concentration of the solute on the two sides of the membrane and the diffusion coefficient of the solute in the lipid bilayer. We can assume that at the interface, the solute is at equilibrium between the two phases of the lipid membrane and the aqueous solution. The equilibrium relationship between two phases is described by a partition coefficient, H. If the concentration at the interface is C in the aqueous phase, then that in the membrane is HC. The flux can then be expressed in terms of the concentration in the aqueous phase by lumping the partition coefficient, the diffusion coefficient, and the thickness of the membrane into one parameter. This parameter is called the permeability (Eq. 4.52).

Permeability is thus a quantitative indicator of how fast a solute diffuses across a membrane. It is affected by the diffusion coefficient, which is inversely proportional to the cubic root of molecular weight of the solute. For molecules with a similar diffusion coefficient, the permeability is proportional to the partition coefficient. In other words, the compound that is more soluble in the membrane has a higher flux. The diffusion rate is determined by the permeability of the solute and the concentration difference of the solute across the membrane. The permeability has units of velocity (cm/s), while the diffusion coefficient has units of cm^2/s.

4.10.2 Transport of Electrolytes

Biological solutions contain many electrolytes, including Na^+, K^+, PO_4^{3-}, and Cl^-. In addition, they contain charged organic compounds, such as those that have a carboxylic or phosphate group that dissociates at neutral pH. The diffusion of electrolytes between two points in a solution is affected by the concentration difference and by the electric potential difference. In other words, their movement is affected by other electrolytes that contribute to the balance of charge neutrality.

The flux for an electrolyte can be described by Eq. 4.53 in Panel 4.14. The first term on the right-hand side accounts for the contribution of chemical potential, or the concentration gradient. The second term is the contribution of electropotential difference. ψ is the electric potential, z is the valence of charge (e.g., +1 for Na^+), and $\mathfrak{F}$ is Faraday's constant. Faraday's constant is essentially a unit conversion factor relating electrostatic potential to chemical potential, with a value of 96,500 coulomb/volt. The C/RT converts the electrostatic potential gradient to the chemical concentration gradient. This equation is often referred to as the Nernst–Planck equation.

Panel 4.14 Diffusion of Electrolytes

The flux equation for an electrolyte can be written as (Nernst–Planck equation):

$$J = D \left(\frac{\Delta C}{\Delta z} + Cz\mathfrak{F} \frac{\Delta \psi}{RT \Delta z} \right) \tag{4.53}$$

- The flux is affected by both the concentration gradient and electric potential gradient $\Delta\Psi$.

Consider the diffusion of a strong, single monovalent electrolyte, like HCl, in a stagnant solution in which its concentration is not uniform. To maintain electroneutrality, the concentrations of the two dissociated cations and anions must be the same at any location in the solution. Thus, the two ions will have the same concentration gradient. This is in spite of the difference between their mass and diffusion coefficients. The diffusion of each ion is constrained by the diffusion of its counter ion. For example, HCl completely disassociates in water to form H^+ and Cl^-. They have different diffusion coefficients of 9.3×10^{-5} and 2.0×10^{-5} cm^2/s, respectively. Because the two ions must diffuse together to maintain electroneutrality throughout, the effective diffusion coefficient of the two ions in a one-electrolyte solution is the harmonic average of their diffusion coefficient.

4.10.3 Transport of Charged Molecules across Membrane

The principle of maintaining electroneutrality in the diffusion of electrolytes also holds true for diffusion across the cell membrane. Consider two electrolyte solutions that are separated by a membrane. If different numbers of an ion and its counter ion are transferred from one side of the membrane to the other side, a net charge difference across the membrane will be created. An electropotential will be built up across the membrane.

A membrane potential exists in all cytoplasmic membranes and mitochondrial inner membranes. The quantity of charges that need to be transferred across a membrane to create a fixed level of electropotential is inversely proportional to the capacitance of the membrane. Because of the large capacitance of the cellular membrane, only a small number of ions need to be transferred across the cell membrane in order to generate an -80 mV potential. The intracellular concentration of both Na^+ and K^+ is in multiples of millimolars; the amount of ions transferred to create the membrane potential (in the order of 10^{-18} moles) is too small to change their concentration. But it should be noted that electropotential across the cell membrane has a major effect on the transport of electrolytes like H^+, Na^+, K^+, and Cl^- across the membrane.

Like their neutral counterpart, ions and other charged compounds are translocated across the cellular membrane by transporters. A number of transporters, including Na-K-ATPase, K-channel, and Na-channel proteins, do not transport ions in a charge-neutral fashion. They are involved in maintaining membrane potential. Other transporters of charged molecules, particularly those for metabolites that carry a net charge under neutral pH (such as lactate), transfer their solute in a charge-neutral fashion. The transport of an organic solute is often coupled to the transport of another species. The co-transported species (possibly another organic metabolite or an ion like H^+ or Cl^-) may carry an opposite charge and is transported in the same direction. Or, it may carry the same charge and move in the opposite direction. In both cases, charge neutrality is maintained through the transport process.

4.10.4 Types of Transporters

With the diversity of nutrients, there is a large number of transporters in each organism. Many solutes have multiple transporters to meet different physiological needs. For example, a low-affinity transporter may be expressed under normal growth conditions, and another, high-affinity transporter is expressed when the nutrient level is low.

These transporters are classified in different ways. A transporter can be either a uniporter or a co-transporter; a uniporter transports a single solute, while a co-transporter must simultaneously transport a pair of solutes. The two solutes in a pair can be transported in the same direction (symporter) or in the opposite direction (antiporter) (Figure 4.15).

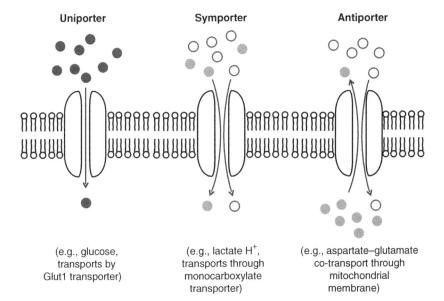

Uniporter	Symporter	Antiporter
(e.g., glucose, transports by Glut1 transporter)	(e.g., lactate H^+, transports through monocarboxylate transporter)	(e.g., aspartate–glutamate co-transport through mitochondrial membrane)

Figure 4.15 Different types of membrane transporter: a channel protein, a facilitated diffusion transporter, and two transporters for active transport. One uses an energy source such as ATP; the other uses the concentration gradient of Na+ and the electropotential gradient that helps drive a positively charged Na+ into cytosol.

Transporters may also be categorized according to their energy dependence. Many rely on the chemical potential difference or concentration difference of the solute that they transport. Such transporters include channel proteins, like a potassium channel. Once it is open, K^+ quickly passes from the high-concentration side to the low-concentration side. Another example is the facilitated transporter of glucose, glut1, which allows glucose to pass across the membrane along its concentration gradient.

The other type of transporter requires another energy source (usually ATP) to drive the transport of a solute against its concentration gradient. For example, driven by the power of ATP, the multidrug resistance transporter pumps intracellular cytotoxic drugs from the cytoplasm to the extracellular space. Some co-transporters use the chemical potential energy of a solute as their own energy source, relying upon the potential of a favorable gradient of the co-transported solute (such as Na^+ and H^+). However, these transporters must transport the target solute simultaneously in order to proceed. This allows the target solute to be transported against its concentration gradient.

4.10.5 Kinetics of a Facilitated Transporter

The transport of a solute by a facilitated transporter can be assumed to start with the reversible binding of the solute at the outside surface (at a concentration S_o) of the transporter protein (T) to form a transporter–solute complex (TS) (Eq. 4.54 in Panel 4.15). Upon a conformational change of the protein, the solute is exposed to the interior on the other side. Subsequently, the solute is released to the interior that has a lower solute concentration (when a concentration is S_i), and the protein reverts to its original conformation.

Panel 4.15 Panel

$$T + S_o \underset{k_{-1}}{\overset{k_1}{\rightleftharpoons}} TS \underset{k_{-2}}{\overset{k_2}{\rightleftharpoons}} T + S_i \tag{4.54}$$

$$T_0 = T + TS \tag{4.55}$$

Balance on TS:

$$\frac{dTS}{dt} = k_1 \cdot T \cdot S_o + k_{-2} \cdot T \cdot S_i - (k_{-1} + k_{-2}) \cdot TS \tag{4.56}$$

Assume steady state:

$$TS = \frac{k_1 \cdot T_0 \cdot S_o + k_{-2} \cdot T_0 \cdot S_o}{k_{-1} + k_2 + k_1 \cdot S_o + k_{-2}S_i} \tag{4.57}$$

Substituting TS into rate expression:

$$r = k_2 \cdot TS - k_{-2} \cdot T \cdot S_i$$
$$= \frac{k_1 \cdot k_2 \cdot T_0 \cdot S_o - k_{-1} \cdot k_{-2} \cdot T_0 \cdot S_o}{k_{-1} + k_2 + k_1 \cdot S_o + k_{-2} \cdot S_i} \tag{4.58}$$

Define:

$$\frac{k_{-1} + k_2}{k_1} = K_{mo} \tag{4.59}$$

$$\frac{k_{-1} + k_2}{k_{-2}} = K_{mi} \tag{4.60}$$

$$\frac{k_1 k_2}{k_{-1} + k_2} = k_o \tag{4.61}$$

$$\frac{k_{-1} k_{-2}}{k_{-1} + k_2} = k_i \tag{4.62}$$

$$r = \frac{k_o \cdot T_0 \cdot S_o - k_i \cdot T_o \cdot S_o}{1 + \dfrac{1}{K_{mo}} \cdot S_o + \dfrac{1}{K_{mi}} \cdot S_i} \tag{4.63}$$

when:

$$S_i \to 0$$

$$r = \frac{K_{mo} \cdot k_o \cdot T_0 \cdot S_o}{K_{mo} + S_o} = \frac{r_{max} \cdot S_o}{K_{mo} + S_o} \tag{4.64}$$

This mechanistic model for transport is similar to that for an enzyme-catalyzed reversible reaction. We will assume that (1) the total concentration of transporters (T_o) in the membrane is constant, and (2) a transporter can be either free (T) or solute bound (TS). We will perform a material balance on TS, assuming that the system quickly reaches a pseudo–steady state with a constant concentration of TS, just as we derived the Michaelis–Menten kinetics by assuming ES to be constant. By lumping the rate constants together and defining parameters akin to the half-saturation constant in the Michaelis–Menten equation, the resulting rate expression is very similar to a reversible single-substrate enzymatic reaction. In the case that intracellular concentration of the solute is small, the transport rate expression has the same form as Michaelis–Menten kinetics. The half-saturation constant has units of concentration. In the absence of a transport rate in the opposite direction, the half-saturation constant is the concentration at which the transport rate is at half of its maximum. Since most transporters are for transporting nutrients from the external environment, the half-saturation constant of the extracellular domain is often referred to as K_m.

The specificity of a transporter toward its solute is not absolute. Some transporters move a number of structurally similar molecules across the membrane, albeit often at different affinities. For example, the glut1 transporter can transport both glucose and galactose, but the K_m for galactose is about ten times higher than for glucose. It is also not uncommon that multiple transporters can transport the same solute. They serve different physiological needs. For example, cells of the liver and intestine express both glut1 and glut2 transporters for glucose. The former has a very low K_m of ~1 mM, while the latter has a K_m of ~15 mM. Glut2 is not active when the glucose concentration is

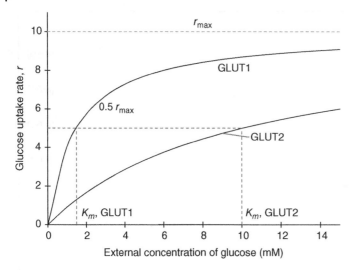

Figure 4.16 The kinetic behavior of two glucose transporters, Glut1 and Glut2, with a low and a high Km, respectively.

normal; it becomes active when the glucose level is raised. This allows for the rapid absorption of glucose (Figure 4.16).

4.11 Kinetics of Binding Reactions

4.11.1 Binding Reactions in Biological Systems

Many important biological reactions involve the binding of two molecules to form a complex, but do not entail the formation or breakage of a covalent chemical bond. For example, an antibody molecule binds to its specific antigen to form the antibody–antigen complex, which can be cleared from circulation by tissue cells (Figure 4.17). Cells also produce various carriers that bind to nutrients to facilitate their uptake. Some microorganisms secrete siderophores that have a very strong binding for iron, which helps the cells to scavenge iron when its concentrations are low. Serum albumin produced in our liver binds to various lipids after a meal and carries them in circulation to where they are used.

The low permeability of the lipid bilayer membrane insulates cells from various environmental stimuli. To transmit signals from other cells and the environment, elaborate signaling receptors are embedded in the cell membrane. The signaling molecule binds to the external region of the receptor that transduces the signal into the intracellular environment and triggers a cell's responses. The single cells of an amoeba *Dictyostellium discoideum* (commonly called slime mold) respond to the binding of a cAMP signal to its receptor by inducing cell–cell adhesion and the differentiation into multicellular-organism-like fruit bodies. Insulin binds to an insulin receptor on the cell surface that results in the phosphorylation of insulin receptor substrate. This triggers a growth-stimulating response through the mTOR signaling pathway.

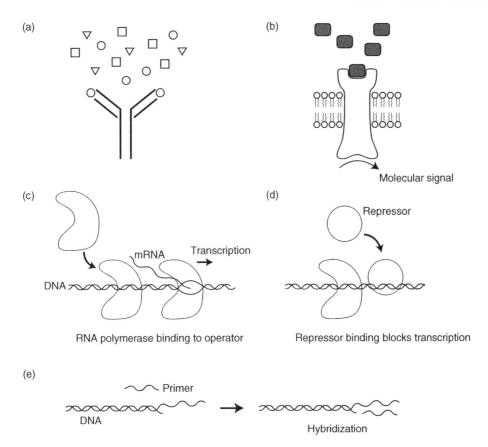

Figure 4.17 Examples of binding reactions important in biological systems.

Binding reactions involving DNA molecules are of great importance both biologically and technologically. Nucleic acid hybridization is the binding between a single strand of DNA and another single-stranded DNA (or RNA) through the formation of hydrogen bonds between their complementary bases. It is the foundation of all transcription, translation, and DNA replication. It is also the basis of polymerase chain reaction (PCR) and DNA microarray. Protein binding to a specific site(s) of DNA, like repressor protein binding to the operator region, or RNA polymerase binding to the promoter region, controls the expression of a gene.

In a binding reaction, specificity is critical and hinges on the difference of binding affinities between the proper target and mimetics.

4.11.2 Dissociation Constant

In this section, we will focus on a monovalent binding (i.e., one effector molecule binds to a binding receptor that has only one binding site). Binding of an effector molecule to its receptor to form an effector–receptor complex is depicted as a reversible process (Eq. 4.65 in Panel 4.16). At equilibrium, the formation rate and the dissociation rate are equal. The equilibrium is expressed as either a dissociation constant or association constant, with a unit of concentration or the inverse of concentration (Eq. 4.68 in Panel 4.16). The association/dissociation constant is thus a quantitative measure of the binding affinity of an effector–receptor pair. The dissociation constant of different effector–receptor pairs varies over a very wide range, from a biotin–avidin pair of 10^{-15} M to some with a weak binding of 10^{-6} M (Panel 4.17).

Panel 4.16 Binding of an Effector to a Receptor

$$A + B \underset{k_{off}}{\overset{k_{on}}{\rightleftharpoons}} AB \tag{4.65}$$

$$B_0 = B + AB \tag{4.66}$$

At equilibrium: $k_{on} \cdot A \cdot B = k_{off} \cdot AB$ $\tag{4.67}$

Substitution: $k_{on} \cdot A \cdot (B_0 - AB) = k_{off} \cdot AB$

$$AB = \frac{k_{on} \cdot A \cdot B_o}{k_{on} \cdot A + k_{off}} = \frac{A \cdot B_o}{\dfrac{k_{off}}{k_{on}} + A}$$

$$AB = \frac{A \cdot B_o}{K_d + A} \tag{4.68}$$

Panel 4.17 Dissociation Constant of Some Effector–Receptor Pairs (K_d, M)

Effector–receptor pair	K_d(M)
Biotin–avidin	10^{-14}
Antibody–antigen	10^{-10}–10^{-12}
Signal molecule–receptor	10^{-9}–10^{-11}
Insulin–insulin receptor (pocket 1/pocket 2)	$10^{-11}/10^{-9}$
IgG–protein A	10^{-8}

4.11.3 Saturation Kinetics

Many binding reactions involve a fixed number of receptors. For example, a finite number of receptors are present on the cell surface for the binding of signaling molecules. The expression of some bacterial genes is regulated by repressor protein–operator binding. For these genes, the repressor protein binds to the operator site upstream of the region coding for the target gene in the genome. The binding state of the operator governs whether the target gene is transcribed to RNA molecules (transcription "on") or not (transcription "off"). There may be only one specific operator site in the bacterium's genome. If the gene is encoded on a plasmid, there will be multiple copies of plasmids and, thus, multiple operator sites, but the number is still finite. In either case, the total number of a specific operator is fixed.

The receptor exists as either effector occupied or unoccupied. One can expect the equilibrium concentration of the occupied receptor sites to follow a saturation type of kinetics with respect to the effector concentration (Panel 4.16). The equation is analogous to the Michaelis–Menten equation (Eq. 4.69). However, Michaelis–Menten kinetics describe the reaction rate of an enzymatic reaction; the binding kinetics depict the concentration of effector-bound receptors at equilibrium. In the equation, the kinetic constant K_d (the dissociation constant) is the concentration when the effector-bound receptor concentration is half of the maximum.

4.11.4 Operator Binding and Transcriptional Regulation

As we have learned, enzymes do not work in isolation. They form networks to carry out biochemical functions. The biochemical networks are further linked to the extracellular environment through transporters. Binding reactions, also, do not work in isolation. They are linked to other biological reactions to derive functionality. The binding of a signaling molecule to its receptor triggers receptor-mediated reactions to elicit downstream effects. In many cases, the intracellular chain of events leads to the binding of a transcription factor or regulatory protein to the regulatory region of DNA, to initiate the transcription of a target gene.

An *E. coli* genome encodes about 4000 genes in total. However, fewer than 3000 are being transcribed at any given time. To conserve the cell's resources, genes that are not needed for a particular environment are not expressed. These genes are "repressed" when not needed. They are "induced" once the inducer is present and the protein encoded by the gene is therefore needed. The regulation of transcription is often executed by controlling the binding state of the operator upstream of the promoter.

A promoter is a segment of DNA upstream of a gene that the RNA polymerase binds. The RNA polymerase can then move downstream to the gene-coding region to begin to transcribe the gene into RNA. A classic example of gene expression regulation is the lactose operon (lac operon) in *E. coli* (Figure 4.18). It is called an operon because three genes are encoded on the DNA consecutively and are transcribed as a single piece of transcript (called a multicistronic transcript) and then translated into three proteins. The proteins are all involved in the hydrolysis of lactose to glucose and galactose. The

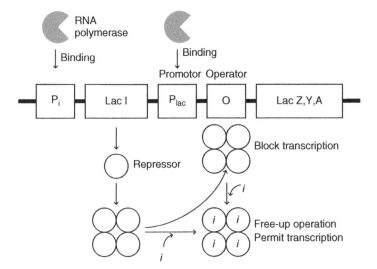

Figure 4.18 The organization of a lactose operon.

genes are not transcribed in the absence of lactose. Once the lactose is available, the operon is induced and the genes are made into proteins.

Upstream of the gene, and typically downstream of the promoter, sits a region called the operator. To the operator, a tetrameric protein (called the repressor) can bind and create a "roadblock" along the path of the RNA polymerase to prevent it from transcribing the gene. Further upstream is another gene (and its promoter) that codes for the repressor protein. The repressor protein is transcribed and translated at a relatively constant level all the time (i.e., constitutively expressed). In the absence of lactose, the repressor protein binds to the operator and prevents the lac operon from being transcribed. When lactose (the inducer) is present, it binds to the repressor protein, causing its binding affinity to the operator to reduce drastically. The repressor protein is released from the operator, thus allowing the transcription to commence.

The interaction between the regulator protein and the DNA binding site is a very important mechanism of gene regulation in all organisms. The architecture of a typical gene in a bacterial genome is shown in Figure 4.19. To transcribe a gene, the RNA polymerase moves along the coding region of the DNA to read the template and to add complementary bases, one by one. This process synthesizes the growing mRNA strand.

The lac operon and the system shown in Figure 4.19 are under a negative control. The binding of a repressor protein to the operator (or the regulatory region of the DNA) suppresses transcription. In some cases, the binding of the regulatory protein or the transcription factor triggers or enhances the transcription, whereas in the absence of the binding of the transcription factor the gene expression is diminished. In one scenario,

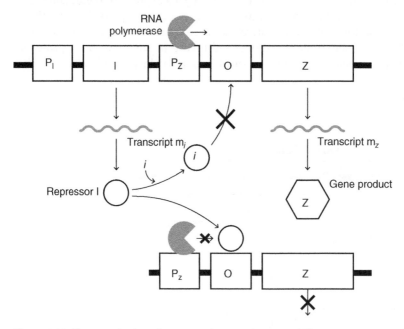

Figure 4.19 The organization of a gene under negative control. The repressor protein binds to the operator to suppress transcription. Binding of an inducer to the repressor releases it from the operator and induces the transcription.

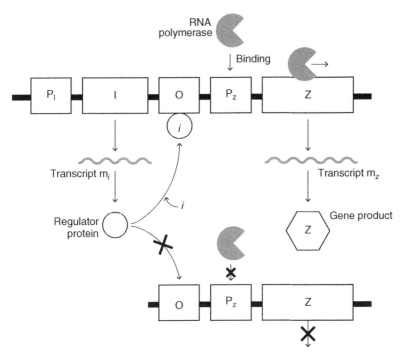

Figure 4.20 The organization of a gene under positive control. Binding of the regulatory protein to the operator is necessary for starting the transcription. In this example, the binding of an inducer to the regulatory protein enhances the binding affinity of the inducer–regulatory protein complex to the regulatory region of the DNA, and thus induces the transcription.

the binding affinity of the regulatory protein to the operator is low; thus, the default state is the suppression of transcription. Once an inducer is available, the binding protein binds to the regulatory region to initiate the transcription (Figure 4.20).

4.11.5 Kinetics of Transcription and Translation

Different genes are transcribed and translated at different rates. Thus, cellular proteins have different levels of abundance to meet different cellular needs. An abundant protein usually has a higher level of corresponding mRNA. The abundance level of the transcript depends on how frequently transcription is initiated by an RNA polymerase, but does not depend on its transcript length. To initiate the transcription of a gene, RNA polymerase binds to the promoter region and occupies a physical length (~50 bps) of DNA. The RNA polymerase then unwinds the two strands of DNA, forms a transcription initiation unit, and moves downstream to start transcription.

Incoming RNA polymerase can bind to the promoter only after the preceding RNA polymerase moves out of it. This limits how fast transcription can be initiated. The strongest promoters have a transcription initiation rate of approximately 1 transcript per second. Most promoters are initiated at a slower rate. Along a segment of the coding region, there may be multiple RNA polymerases moving downstream in tandem. For genes that are under regulatory control like induction and repression, the repressor protein binding to the operator influences the transcription rate.

In an *E. coli* cell, the transcription of a gene is coordinated with translation. Some regions in the transcript, notably segments in the 5′ and 3′ ends, are not translated. To initiate translation, the ribosome binds to the ribosome-binding site near the 5′ end of the transcript. In *E. coli*, the transcription starts even before the entire transcript molecule is synthesized. A ribosome binds to the nascent mRNA at the ribosome-binding site and begins to translate the mRNA into a protein molecule. The ribosome then moves along the mRNA strand at a velocity of about 15 amino acids per second. The transcription rate by RNA polymerase is about 45–50 bases per second. A codon coding for an amino acid consists of three bases. Thus, the transcription rate balances the translation rate. It can take a period of time to then complete translation, depending on the length of the gene.

In addition to the initiation rate, the abundance levels of a transcript and its protein product are also affected by their degradation rate. Proteins and mRNAs have a limited lifespan; they are "turned over" to prevent the accumulation of any one molecule that is damaged during its regular functioning. Some proteins are specifically targeted for rapid degradation after their activities are no longer needed. In some cases, the continuing presence of their activity may interfere with other cellular functions. Most cellular proteins are thus continuously synthesized through transcription and translation, and are constantly turned over. Additionally, when a cell grows to become larger in volume, the concentration of any particular protein molecules is also diluted.

We can use a system of equations to describe the transcription of a gene under repressor control and its translation into protein molecules (Panel 4.18). Consider the cell volume change due to cell growth (Eq. 4.69). A fast-growing bacterial cell can double its volume every 20 minutes. Some mRNAs are transcribed only once every minute or longer. The intracellular concentration is constantly diluted by the increase in cell volume. This is in contrast with the material balance for biochemical compounds in metabolic flux analysis (MFA). For enzyme-catalyzed reactions, the k_{cat} is in the order of thousands per second or higher. Thus, the intracellular metabolite concentration is not significantly changed by the much slower cell volume expansion, and it reaches a steady-state value in a relatively short time.

Panel 4.18

$$\frac{dV}{dt} = \mu \cdot V \tag{4.69}$$

Transcript of *z*:

$$\frac{d(V \cdot m_z)}{dt} = \eta_z \cdot V - \delta_z \cdot V \cdot m_z \tag{4.70}$$

$$m_z \frac{dV}{dt} + V \frac{d(m_z)}{dt} = \eta_z \cdot V - \delta_z \cdot V \cdot m_z \tag{4.71}$$

$$\frac{d(m_z)}{dt} = \eta_z - \delta_z \cdot m_z - \mu \cdot m_z \tag{4.72}$$

Transcript of *l*:

$$\frac{d(m_l)}{dt} = \eta_l - \delta_l \cdot m_l - \mu \cdot m_l \tag{4.73}$$

Proteins of z, I:

$$\frac{dZ}{dt} = \xi_z m_z - \lambda_z \cdot Z - \mu \cdot Z \tag{4.74}$$

$$\frac{dI}{dt} = \xi_I m_I - \lambda_I \cdot I - \mu \cdot I - k_1 \cdot i \cdot I + k_{-1} \cdot iI \tag{4.75}$$

$$\frac{diI}{dt} = k_1 \cdot i \cdot I - k_{-1} \cdot iI \tag{4.76}$$

The balance equation for the transcript of the inducible gene, z, is written for the total transcript (transcript concentration multiplied by cell volume), rather than for its intracellular concentration (Eq. 4.70). We define the volume expansion rate on per-cell volume as the specific growth rate, as will be discussed in Chapter 5. In the material-balance equation for the concentration of mRNA of gene z, the μm_z term describes the effect of dilution caused by cell volume expansion (Eq. 4.72). Similarly, the material balance of the concentration for the proteins Z, I, and the mRNA of I also includes a dilution term.

The initial conditions for all species can be set to zero, as they will be under uninduced conditions. The exception is the repressor protein, which is constitutively expressed and can be set at a steady-state level. The inducer concentration (i) can be set to have a step change, as in the case of the addition of inducer.

The transcription initiation rate depends on the binding of a regulator–inducer complex to the operator (Figure 4.20). Two binding reactions are involved in induction: (1) inducer i binding to regulator protein I to form the inducer–regulator complex (iI) (Eq. 4.76 in Panel 4.18); and (2) iI binding to the operator (O) to allow for the transcription to initiate (Eq. 4.77 in Panel 4.19). The inducer–regulator complex binding to the operator is assumed to reach equilibrium very rapidly following a saturation type of kinetics, because the total concentration of the operator-binding site is fixed (Eq. 4.79). The nonspecific binding of regulatory protein (without inducer) to the operator is assumed to be negligible and does not induce transcription. The transcription initiation rate from the promoter P_z is thus proportional to the fraction that the operator is occupied by the inducer–regulator complex (Eq. 4.80). Eq. 4.80 confers a saturation type of binding characteristic with respect to iI.

Panel 4.19

$$O + iI \underset{k_{-1}}{\overset{k_1}{\rightleftharpoons}} OiI \tag{4.77}$$

Total concentration of operator is fixed:

$$O_o = O + OiI \tag{4.78}$$

At equilibrium:

$$OiI = \frac{iI \cdot O_o}{K_{dO} + iI} \tag{4.79}$$

(Continued)

Panel 4.19 (Continued)

Transcription initiation rate proportional to the fraction of "induced" operator:

$$\eta_z = \eta_{zo} \cdot \frac{Oil}{O_o} = \frac{\eta_{zo} \cdot il}{K_{dO} + il} \tag{4.80}$$

We further examine the case that the total concentration of the regulator protein is fixed. This scenario is highly relevant because often the regulator protein is only present at a very low concentration. For example, the lacI protein was estimated to be present only at about 5–10 copies per cell. In this case, the binding of the inducer i to regulator I follows saturation-type kinetics (Eq. 4.81 in Panel 4.20). We also assume that the binding of the inducer to the regulator rapidly reaches equilibrium.

Panel 4.20

i and I binding reaches equilibrium rapidly:

$$il = \frac{k_1 \cdot i \cdot l_o}{K_{d,I} + i} \tag{4.81}$$

Combine Eqs. 4.79 and 4.81:

$$Oil = \frac{il \cdot O_o}{K_{dO} + il} = \frac{\left(\dfrac{k_1 \cdot i \cdot l_o}{K_{d,I} + i}\right) O_o}{K_{dO} + \left(\dfrac{k_1 \cdot i \cdot l_o}{K_{d,I} + i}\right)} = \frac{k_1 \cdot i \cdot l \cdot O_o}{K_{dO}(K_{d,I} + i) + k_1 \cdot i \cdot l_o}$$

$$= \frac{k_1 \cdot i \cdot l_o \cdot O_o}{K_{dO}K_{d,I} + K_{dO} \cdot i + k_1 \cdot i \cdot l_o} = \frac{(k_1 \cdot l_o \cdot O_o)i}{K_{dO}K_{d,I} + (K_{dO} + k_1 \cdot l_o) \cdot i} \tag{4.82}$$

$$\frac{Oil}{O_o} = \psi = \frac{\dfrac{1}{(K_{dO} + k_1 \cdot l_o)} \cdot (k_1 \cdot l_o) \cdot i}{\dfrac{K_{dO}K_{d,I}}{(K_{dO} + k_1 \cdot l_o)} + i} \tag{4.83}$$

$$\psi = \frac{\psi_o \cdot i}{K_{OI} + i} \tag{4.84}$$

Define: $\psi_o = \dfrac{(k_1 \cdot l_o)}{(K_{dO} + k_1 \cdot l_o)}$ \hfill (4.85)

$$K_{OI} = \frac{K_{dO}K_{d,I}}{(K_{dO} + k_1 \cdot l_o)} \tag{4.86}$$

Transcription rate:

$$\eta_x = \eta_{xo} \cdot \frac{\psi_o \cdot i}{K_{OI} + i} \tag{4.87}$$

The equations describing the two binding events (inducer to regulator, and inducer–regulator complex to the operator) can be combined. This gives an equation

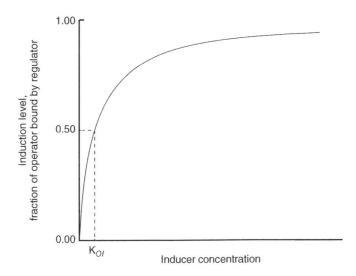

Figure 4.21 The effect of inducer concentration on the transcription rate of an inducible gene.

for the transcription rate as a function of the inducer concentration (Panel 4.20 in Eq. 4.87). Figure 4.21 shows the effect of inducer concentration on the transcription rate of the gene that it induces. The resulting behavior depends on the value of the kinetic parameters of the two binding events (the inducer to the regulator, and the inducer–regulator complex to the operator).

A very small K_{OI} can elicit a very sharp response in induction; even a very low concentration of inducer can trigger a very rapid increase in transcription. Regulation by inducer is thus an almost on-off type of control. With a large K_{OI}, the response to the inducer concentration is more gradual.

4.12 Concluding Remarks

Enzymes are the engines of biochemical processes in living systems and biotechnology. They work in concert, forming pathways and reaction networks. To understand and design biological processes, it is necessary to have a good grasp on their kinetic behavior.

To understand the dynamic behavior of a pathway or a network, it is necessary to establish a model of the reactions and perform a kinetic analysis. Kinetic analysis complements MFA by incorporating kinetic constraints and considering the regulations at activity and enzyme-synthesis levels. The extension of the reaction networks to the cellular physiological space is bridged by two other types of biochemical reactions: (1) transport across the cellular membrane, and (2) the binding of regulators to receptors.

Although the kinetics of the cellular reaction network, transport, and gene expression regulation based on binding reactions are discussed separately, kinetic analysis also aims to integrate those individual aspects into a full description of cell physiology. Ultimately, this quantitative description of cellular physiology will help us to understand, describe, and evaluate reactor performance, as described in subsequent chapters.

Further Reading

Ehlert, FJ 2015, 'Analysis of drug-receptor interactions using radioligand binding assays on G protein-coupled receptors', in *Affinity and efficacy: the components of drug-receptor interactions*. World Scientific, Hackensack, NJ.

Mulukutla, B, Yongky, A, Daoutidis, P & Hu, WS 2014, 'Bistability in glycolysis pathway as a physiological switch in energy metabolism', *PLOS ONE*, **9**, 6. Available from: http://journals.plos.org/plosone/article?id=10.1371%2Fjournal.pone.0098756 [19 July 2016].

Segel, IH 1975, *Enzyme kinetics: behavior and analysis of rapid equilibrium and steady state enzyme systems*. John Wiley & Sons, New York.

Stein, WD & Litman, T 2015, *Channels, carriers, and pumps: an introduction to membrane transport*, 2nd edn. Academic Press, London.

Tao, JA & Kazlauskas, RJ (eds) 2011, 'Gene expression, receptor model', in *Biocatalysis for green chemistry and chemical process development*. John Wiley & Sons, Hoboken, NJ.

Nomenclature

A, B, AB, C, D	chemical species participating in reaction, and its concentration; subscript denotes different conditions	mole/L^3
B_0	total receptor concentration	mole/L^3
ΔC	concentration difference of c across the membrane	mole/L^3
D_z	diffusion coefficient	L^2/t
E	enzyme, enzyme concentration (when used in an equation)	mole/L^3 M/L^3
E'	enzyme at an activated or intermediate state	mole/L^3 M/L^3
E_0	initial enzyme concentration	mole/L^3 M/L^3
E_1, E_2, E_3	enzyme for reaction 1, 2, 3	mole/L^3 M/L^3
E_{10}, E_{20}, E_{30}	initial concentration of enzyme 1, 2, 3	mole/L^3 M/L^3
ES	enzyme–substrate complex	mole/L^3 M/L^3
H	Henry's law constant	(mole/L^3)/(mole/L^3)
I	concentration of repressor protein I	mole/L^3
I_o	initial concentration of repressor protein I	mole/L^3
i	(concentration of) inducer	mole/L^3
iI	concentration of complex of i and I	mole/L^3
J	flux of a metabolic reaction	mole/L^3·t mole/cell·t
J_i, J_j	flux of a metabolic reaction at state i or reaction j	mole/L^3·t mole/cell·t

J_z	solute flux across a membrane	mole/L^3·t mole/cell·t
K_d	dissociation constant of a binding reaction	mole
K_{dO}	dissociation constant of il binding to operator	mole
$K_{I,E}$	inhibition constant of an allosteric regulator E	mole, (for first order kinetics)
K_m	Michaelis–Menten constant, half-saturation constant for an enzyme or a transporter	mole, mole/L^3
K_{m1}, K_{m2}	half-saturation constant for enzyme 1, 2	mole/L^3
$K_{mA}, K_{mB}, K_{mC}, K_{mD}$	half-saturation constant of substrate A, B, C, D	mole/L^3
K_{mi}, K_{mo}	half-saturation constant of export and import for a transporter of facilitated diffusion	mole/L^3
K_{mp}, K_{ms}	half-saturation constant of substrate and product of a reversible reaction	mole/L^3
K_{OI}	half-saturation constant for induction	mole/L^3
K_s	equilibrium constant	depends on nature of reaction
$k_1, k_2, k_3, k_{-1}, k_{-2}$	reaction rate constant	t^{-1}, 1/mole·t, dependent on the kinetics of reaction
k_{cat}	turnover number for an enzyme	mole/mole·t
k_i, k_o	turnover number of import and export of a facilitated diffusion transporter	mole/mole·t
k_{on}	rate constant for binding reaction	1/mole·t
k_{off}	rate constant for unbinding reaction	1/t
k_s, k_p	turnover number of forward and reverse reactions of a reversible reaction	1/mole·t
L	allosteric regulatory compound	
m_i, m_z	intracellular concentration of mRNA of I and z	mole/L^3
N	allosteric regulation term	
O	operator concentration	mole/L^3
O_o	total operator concentration	mole/L^3
OiI	operator occupied by il	mole/L^3
P	permeability	1/t
q_A, q_C, q_D	specific uptake rate of A, B, C	mole/cell·t, M/cell·t
R	universal gas constant	
r, r_1, r_2, r_3	reaction rate or transport rate	mole/L^3·t
r_{max}	maximum reaction rate	mole/L^3·t
S	substrate	mole/L^3, M/L^3
$S_i, S_o, \Delta x$	internal and external (outside) substrate concentration	mole/L^3, M/L^3
T	transporter concentration	mole/L^3

T	(in Eq. 4.53) absolute temperature	K
T_0	total transporter concentration	mole/L^3
V	cell volume	L^3
W_s	substrate supply rate to balance the consumption by the enzyme	mole/L^3·t
Z	concentration of protein z	mole/L^3
z	valence of charge	
$z_1, z_2, \Delta z$	position 1, 2, and distance	L
β, γ, ς	stoichiometric coefficient	
$\Delta \psi$	electric potential gradient	V
ψ, ψ_o	fraction and maximum fraction of O bound by il	
$\Im$	Faraday's constant	coulomb/volt
μ	specific growth rate	t^{-1}
λ_I, λ_z	degradation rate constant of protein I and Z	t^{-1}
δ_I, δ_z	degradation rate constant of transcript m_I and m_Z	t^{-1}
η_I, η_x, η_z	transcription initiation rate of gene I, x, z	1/mole·t
η_{xo}, η_{zo}	maximal transcription initiation rate of gene x, t	1/mole·t
ξ_I, ξ_z	protein translation rate of I and z	1/mole·t

Problems

A. Enzyme Kinetics

A1 **True or False**: When converting a substrate to a product, enzymes sometimes form covalent bonds with a reaction intermediate.

A2 **True or False**: A noncompetitive inhibitor will only change the slope of the Lineweaver–Burk plot.

A3 **True or False**: In competitive inhibition, the presence of inhibitors increases the K_m but does not affect the r_{max}.

A4 **True or False**: While cells in most tissues have only hexokinase for energy metabolism, liver cells also use glucokinase to take up glucose for glycogen synthesis. Hexokinase has a higher K_m for glucose than glucokinase.

A5 **True or False**: Some vitamins, such as thiamine, play a major role in an enzyme's catalytic function. In fact, some vitamins are parts of the protein and are referred to as an enzyme's "prosthetic groups."

A6 **True or False**: The substrate-binding site in an enzyme is like a pocket; it discriminates a specific substrate from other compounds primarily by size. Therefore, all the compounds that are small enough to get in are reacted.

A7 **True or False**: The binding of a substrate causes the secondary structure of an enzyme to change, prompting the substrate and the enzyme to form a complex and triggering the catalytic reaction.

A8 For an enzyme that exhibits Michaelis–Menten kinetics, when the substrate concentration is much less than K_m, the reaction is approximately:
   a) zero order.
   b) first order.
   c) second order.
   d) none of the above.

A9 A competitive inhibitor affects:
   a) the value of K_m.
   b) the value of r_{max}.
   c) the value of k_{cat}.
   d) all of the above.
   e) none of the above.

A10 In Michaelis–Menten kinetics, at what point is the reaction rate equal to half the maximum reaction rate?
   a) $K_m \ll S$
   b) $K_m = S$
   c) $K_m \gg S$
   d) None of the above

A11 In the presence of a competitive inhibitor, the slope of the Lineweaver–Burk plot:
   a) increases.
   b) decreases.
   c) stays the same.

A12 The presence of a competitive inhibitor in an enzyme assay affects the intercept of the trendline and the axis $1/r$ in a Lineweaver–Burk plot. In which direction does the intercept change with increasing concentration of a competitive inhibitor?
   a) Increase
   b) Decrease
   c) Stay the same

A13 If we want to be able to control the rate of a Michaelis–Menten reaction by changing the substrate concentration, it is best to work in the regime where:
   a) $S \gg K_m$.
   b) $S \ll K_m$.

A14 Which of the following is not an advantage of enzyme catalysis?
   a) Substrate specificity
   b) Substrate flexibility
   c) Regioselectivity
   d) Relatively mild reaction conditions
   e) None of the above

A15 Chymotrypsin and trypsin are both serine proteases because a serine residue plays a critical role in their similar mechanism of catalytic function. Chymotrypsin hydrolyzes a peptide bond that has a large hydrophobic amino acid at its carboxyl side, while trypsin hydrolyzes a peptide bond that has a lysine or arginine. The binding sites of the two proteases are well studied; one of them has glycine, serine, and glycine as the key amino acid residues contributing to substrate specificity, while the other has glycine, aspartic acid, and glycine. Which one belongs to trypsin?

A16 Compound A is the substrate for two enzymes, E_1 and E_2, involved in the synthesis of an amino acid and an antibiotic, respectively. Their reaction rates, r_1 and r_2, at an arbitrary concentration of E_1 and E_2 were measured at different concentrations of A. However, the data were not labeled properly (shown in Table P.4.1), and the student who did the measurement needs help to sort out which dataset is for E_1. Determine the K_m and r_{max} for both enzymes, with respect to the concentration of A. Which set of data is more likely to be for E_1 and which for E_2, and why?

Table P.4.1 Reaction kinetics of a common substrate A for two enzymes.

Concentration of A (mM)	0.2	0.6	1.2	2	3	4	5	6	8	9	12	15
Reaction rate (r_x) (mmol/L*min)	3.33	4.29	4.62	4.76	4.84	4.88	4.9	4.92	4.94	4.95	4.96	4.97
Reaction rate (r_y) (mmol/L*min)	0.09	0.23	0.38	0.5	0.6	0.67	0.71	0.75	0.8	0.82	0.86	0.80

A17 An enzyme that converts A to B was added at 1 μM to a solution of A for activity measurement. At concentrations of 10 mM, 20 mM, and 50 mM, the reaction rate was constant at 1×10^{-10} mol/L-s. When the concentration of A decreased to 0.2 mM, the reaction rate dropped to 5×10^{-11} mol/L-s. Calculate k_{cat} and K_m for the enzyme. Show units in the answer.

A18 The initial reaction rate for the enzyme that converts A to B was measured at different enzyme concentrations and different initial substrate concentrations. The data were then used to perform a Lineweaver–Burk plot to determine the K_m of the substrate for the enzyme. Sketch a plot for different enzyme concentrations; label the direction that the enzyme concentration increases.

A19 Glucose-6-phosphate is converted to fructose-6-phosphate by phosphohexomutase to generate ATP in glycolysis. It is also converted by phosphoglucomutase to glucose-1-phosphate for glycogen synthesis. Glycogen is a major energy storage compound in cells. Which of the two reactions is likely to have a higher K_m for glucose-6-phosphate? Explain.

A20 The kinetic behavior of a reversible reaction, $S \rightleftharpoons P$, can be described as:

$$r = \frac{(0.5\mathrm{sec}^{-1})S - (0.2\mathrm{sec}^{-1})P}{1 + \left(10\dfrac{L}{mole}\right)S + \left(20\dfrac{L}{mole}\right)P}$$

A series of experiments were carried out at different P, but with $S = 0$. The initial reaction rate (r) was measured in order to obtain the data to plot $1/r$ versus $1/P$. Sketch the plot, be quantitative, and specify the values at intercepts of the $1/r$ and $1/P$ axes.

A21 Which one of the following properties of an enzyme gives rise to its saturation behavior observed in Michaelis–Menten kinetics (i.e., the maximum rate is relatively insensitive to increasing concentrations of substrate)?
a) The enzyme reaction reaches an equilibrium.
b) The enzyme lowers the activation energy of a chemical reaction.
c) At a given enzyme concentration, only a fixed number of substrate binding sites are available.
d) The product of the enzyme usually inhibits the enzyme.

A22 A reaction converting A to B follows Michaelis–Menten kinetics. After measuring the reaction rate at different substrate concentrations of A, the plot in Figure P.4.1 is obtained.
a) Write down the kinetic expression of the reaction rate, with numerical values and units.
b) Now, the enzyme concentration is increased to three times of that in part (a). Plot the new results, and rewrite the equation as in (a).

Figure P.4.1 Lineweaver–Burk plot of an enzymatic reaction.

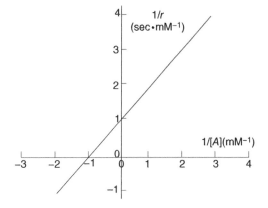

A23 A Lineweaver–Burk plot of a single-substrate irreversible reaction is shown in Figure P.4.2. Propose a regulatory action that can give rise to such a behavior. Propose a possible kinetic expression (e.g., a modified Michaelis–Menten kinetic equation) that describes the kinetic behavior.

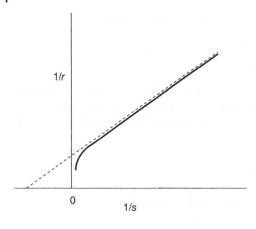

Figure P.4.2 Lineweaver–Burk plot of an enzyme with an allosteric regulation.

B. Biochemical Reaction Network

B1 Consider the biochemical pathway shown in Figure P.4.3. The enzyme and reaction rate are labeled with E and r, respectively. The supply rate of A is q_A, and the production rates of D and F are q_D and q_F, respectively. The K_m and r_{max} for different enzymes are shown in Table P.4.2.

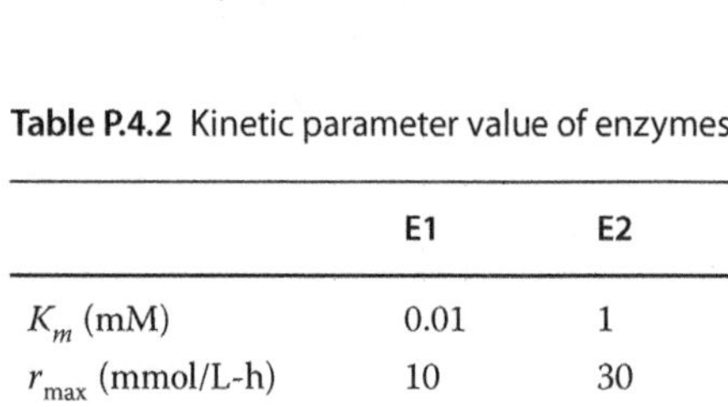

$$A \xrightarrow[r_1]{E_1} B \xrightarrow[r_2]{E_2} C \xrightarrow[r_3]{E_3} D$$

$$E_4 \downarrow r_4$$

$$E$$

$$E_5 \downarrow r_5$$

$$F$$

Figure P.4.3 A pathway with branched reactions.

Table P.4.2 Kinetic parameter value of enzymes in the pathway.

	E1	E2	E3	E4	E5
K_m (mM)	0.01	1	2	0.2	1
r_{max} (mmol/L-h)	10	30	15	10	15

The pathway is operating at a steady state, and the production rate of D is 1 mmol/L-h. What is the supply rate of A and the production rate of F? What is the concentration of B?

a) The consumption rate of A is now increased three times. What are the new production rates of D and F? Do they change in the same proportion? Why or why not?

b) Make a table to list the concentration of all of the species under the conditions of (a). Do the concentrations of all species change in the same proportions after the rate increase?

c) If the production rates of both D and F are to be increased fourfold from the initial conditions described in (a), how can this be accomplished? Assume that the supply of A can be increased to meet the demand.

B2 Consider a linear pathway A is converted to B by enzyme E_1, then B is converted to C by enzyme E_2. The reaction rates of the two reactions are denoted as r_1 and r_2, respectively.

The K_m and r_{max} for E_1 are 1 mM and 3 mmol/L-s, respectively, and the K_m and r_{max} for E_2 are 0.1 mM and 10 mmol/L-s, respectively.

a) Determine the concentrations of A and B when the production rate of C is 2 mmol/L-s. Assume the system is at a steady state.

b) Now, the concentration of A is increased three times higher. What is the new production rate of C? What is the new concentration of B?

B3 A yeast grows on glucose by simultaneous fermentative and aerobic metabolism when oxygen is available. The pyruvate generated in glycolysis is split into two fluxes: the pyruvate decarboxylase (PDC) reaction channels it to ethanol fermentation, while the mitochondrial pyruvate carrier (MPC) transfers pyruvate into mitochondria for oxidation in the TCA cycle. The kinetics of the two reactions in the cell have been determined. The maximal reaction rates are 50 and 20 mmol/L-h for PDC and MPC, respectively, while the K_m's are 2 mM and 6 mM, respectively. The intracellular concentration of pyruvate is 3 mM, while the metabolic fluxes are at a steady state. To increase ethanol production, the supply of oxygen is to be shut off to drive pyruvate to ethanol fermentation. If the glucose consumption can be sustained at the same level as before oxygen is shut off, what will the new pyruvate concentration be? What will the ethanol production rate be?

B4 Two enzymes, E_1 and E_2, respectively catalyze the conversion of A to B, and B to C in two reactions in series. r_1 and r_2 denote reaction rates. Both reactions follow Michaelis–Menten kinetics. The (r_{max}, K_m) for the E_1 and E_2 are (5 mmol/L-s, 2.5 mM) and (10 mmol/L-s, 0.1 mM), respectively. Note that the reaction rate is based on the cellular volume. The product, C, is then secreted out of the cell.

a) At a steady state, the cell converts 2 mmol/L-s of A to C. What are the intra-cellular concentrations of A and B?

b) Now, the supply rate of A is increased to 4 mmol/L-s, but E_1 and E_2 remain at the same levels. What are the new concentrations of A and B?

C. Advanced Problems

C1 Consider problem B4, but change the reaction r_1 to a reversible reaction. The $r_{max,1}$ for forward and reverse reactions (i.e., $k_s E_o$ and $k_p E_o$) are the same. $K_{ms,1}$ and $K_{mp,1}$ are 1 mM and 2.5 mM, respectively. Compare the results with those of problem B4.

C2 An *E. coli* culture consumes glucose at 1 mmol/h. Upon uptake by cells, glucose is phosphorylated to become glucose-6-phosphate (G6P). 90% of this glycolysis is converted by G6P isomerase to become fructose-6-phosphate (F6P) and continues on the glycolysis pathway. The remaining 10% enters the pentose phosphate pathway (PPP) via G6P dehydrogenase to generate NADPH and pentose phosphate. The reaction rates of the two reactions are shown here:

$$\text{G6P isomerase} \quad r = \frac{r_{max}(G6P - F6P)}{1 + \dfrac{G6P}{0.1} + \dfrac{F6P}{1}} = \frac{k_{cat1}E_{10}(G6P - F6P)}{1 + \dfrac{G6P}{0.1} + \dfrac{F6P}{1}}$$

$$\text{G6P dehydrogenase} \quad r = \frac{r_{max2}G6P}{0.5mM + G6P} = \frac{k_{cat2}E_{2,0}G6P}{0.5mM + G6P}$$

Since the F6P concentration is relatively low, the rate of the reverse reaction can be neglected. What is the ratio between the two enzymes? (One can assume that the metabolism is at a steady state and that the k_{cat} of G6P isomerase is ten times the k_{cat} of G6P dehydrogenase.)

5

Kinetics of Cell Growth Processes

5.1   Cell Growth and Growth Kinetics

The term "growth" often refers to an increase in the mass or size of an individual organism. It can also refer to an increase in the number of cells in a population. In the context of bioengineering, growth refers to an increase in the coordinated biosynthesis needed to produce new cells. In the process of reproduction or proliferation, a cell increases its cellular content by duplicating its genome twice and then dividing into two daughter cells. Therefore, growth entails cyclical increases of cell mass and genome copies followed by cell division.

For eukaryotic cells, the process is divided into four stages (Figure 5.1). The end of mitosis (M phase) also marks the beginning of a period called the first gap (G1) phase. During G1, a cell increases its mass and size, but not its DNA content. Upon reaching a critical size, DNA replication occurs in the S phase, and the cell's genome is doubled. Following that, there is another gap phase (G2). The M phase occurs only if cells complete the G2 checkpoints, and the cell's readiness for division (i.e., essential cellular components have been doubled) is verified.

If we examine a single cell moving through the cell cycle, the cellular content of different materials will progress along one cycle, as shown in Figure 5.2. As cells grow, some aspects of the cell change continually. This is the case for cell volume, cell mass, and some cell components like proteins. However, changes in the DNA content appear to be discontinuous, and only occur during the S phase and cell division.

If we extend this to a population of cells whose growth is synchronized, the profile will be the same as shown for individual cells. However, if these same properties are measured in an unsynchronized population over time, the discrete increase in some materials (e.g., DNA) will be averaged out and thus will not be apparent. Although the total content of each material will increase proportionally as cells grow, the material content per cell will remain relatively constant with a consistent cell growth rate.

When measuring the content of a given cellular material, we typically take a sample of the population (consisting of a large number of cells) to perform the measurement. The measured result is divided by the number of cells in the sample. This way, we can estimate the amount of material per cell. Measured this way, cellular content is considered as an average of the entire population.

Alternatively, we can use other methods to quantify the material content of each individual cell. In that case, we will observe that some cells have a higher value and others

Engineering Principles in Biotechnology, First Edition. Wei-Shou Hu.
© 2018 John Wiley & Sons Ltd. Published 2018 by John Wiley & Sons Ltd.
Companion Website: www.wiley.com/go/hu/engineering_fundamentals_of_biotechnology

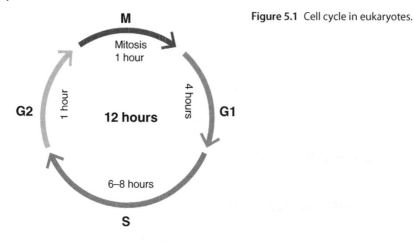

Figure 5.1 Cell cycle in eukaryotes.

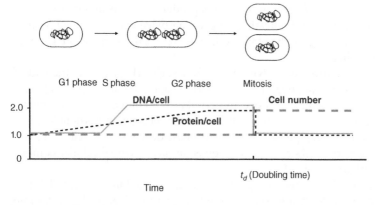

Figure 5.2 Cell number, protein, and DNA content change during growth in a synchronized population or in a single cell.

have a lower value. Each property (e.g., mass, volume, or DNA content per cell) will be distributed over a range and can be displayed as a frequency distribution curve.

If one measures DNA content for each cell of a population, the distribution curve would look as shown in Figure 5.3. The population is not synchronized; thus, cells are at different stages of the cell cycle. Cells that are immediately before division and in G2/M will have a higher DNA content, as identified by the peak to the right side. Immediately after cell division, each cell is left with only half of the content of the parent cell. In the G1 phase and before DNA synthesis commences, the DNA content remains at this same low level, as represented by the peak to the left. During the S phase, the DNA content increases gradually until it is doubled. Cells in S phase are thus represented in the relatively flat region between two peaks. After S phase, the DNA content remains doubled throughout G2 until the cell enters the M phase. The area under the distribution curve is the fraction of the cell population at different stages in their cell cycle. If the size of each cell is measured, we can see that the cell size varies within a range. Immediately before

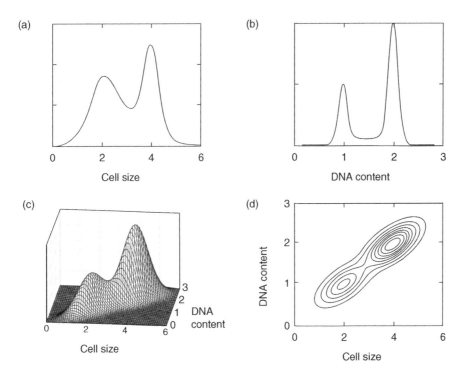

Figure 5.3 Distribution of cellular DNA content and cell size of a eukaryotic cell population. (a, b) Separate measurements of DNA and size; (c) distribution of simultaneous measurements of DNA content and cell size; (d) contour plot of the same data in (c).

cell division, cells will have a larger size than immediately after cell division (Figure 5.3). Since the increase in cell size largely occurs throughout the cell cycle, its distribution curve may or may not be bimodal, unlike DNA content.

Property distribution in a cell population is only partially caused by the cells' progression through the cell cycle. Even in a synchronized cell population, most cell properties are subject to random variation. For example, the cell cycle time may not be uniform given a set of nutrient concentrations; the cycling time will likely be somewhat shorter or longer in some cells. This is also true for the increase in cellular protein content and in cell size. Immediately before mitosis, all cells may not have the same size. Some may grow to somewhat larger sizes than others.

DNA content is an exception; it behaves differently. For normal cells, DNA synthesis is tightly regulated. In each division of the cell cycle, the entire genome is replicated once and only once. The DNA content per cell in the G2 phase is precisely twice that after the M phase. However, in the DNA content distribution plot, instead of uniform discrete values, we see small variations in the DNA content of cells in the G2 phase and in the postmitotic phase. This is largely caused by the limited precision of DNA content measurement. The distribution of properties as experimentally observed arises from intrinsic population distribution and random measurement error. With sound measurement, the error rate should be small compared to the intrinsic variation.

5.2 Population Distribution

To conduct the DNA content and cell size measurements, we typically employ two samples: one for DNA content and the other for cell size. Increasingly, it is preferable to use multiparametric measurement. In this case, one would make DNA content and cell size measurements simultaneously on each cell; the data is plotted as cell size versus DNA content. On such a chart, a high density of dots in a region indicates a cluster of cells with the same combination of size and DNA content. We can also use a 3D plot, with the third dimension being the frequency of occurrences (Figure 5.3c). In this way, we can observe that cells of the same DNA content have a range of sizes; cells with the same size also have very different cell content.

By projecting the data onto a plot of population density versus cell size, one can generate the cell volume distribution curve as seen in Figure 5.3. Likewise, one can obtain the DNA content distribution curve by projecting the data onto the population density versus DNA content plane. A contour plot can be generated from the DNA content–cell size density plot, wherein each contour line represents equal frequency. Each inner contour has a higher frequency than those outside of it (Figure 5.3d). The innermost contour, thus, corresponds to the peaks in the graph. The peak of the G2 phase on the DNA content plot is also represented by larger cells.

Property distribution within a population is inherent to living systems (Panel 5.1). This is caused by the "randomness" of various events in a population. Cell division is not naturally synchronized. Thus, at a given time, cells in a population are in different stages of the cell cycle. Other randomness (or stochasticity) occurs during the making of various cellular contents. This leads to cells having slightly different amounts of a particular component.

Panel 5.1 Cell Properties Heterogeneity

- A typical population is not synchronized. Some properties will vary from one cell to another. They are distributed properties.
- An averaged value is more commonly used to describe such a property.
- A distributed function can describe the population property more precisely.

Knowing the distribution of properties in a cell population is sometimes important for biotechnological processes. For example, the settling velocity of large cells and small cells in a population may be different depending on the size distribution. This information could be important for developing a cell sedimentation protocol.

However, a distributed property is also less convenient for use in design calculations or other data analysis (Panel 5.2). For this reason, it is more desirable to use a population-averaged value. If we measure a population density distribution over a property, θ, and obtain a histogram, we can then generate a population density distribution function, $f(\theta)$. The average of the population could then be described by Eqs. 5.1 and 5.2 (Panel 5.2) using a histogram or the population density function, respectively.

Panel 5.2 Average of a Distributed Property

Average value from distribution data (e.g., histogram):

$$\bar{\theta} = \frac{\sum_i \theta_i N_i}{\sum_i N_i} \tag{5.1}$$

Average value from a population distribution function:

$$\bar{\theta} = \frac{\int_{\theta_1}^{\theta_2} \theta f(\theta) d\theta}{\int_{\theta_1}^{\theta_2} f(\theta) d\theta} \tag{5.2}$$

5.3 Description of Growth Rate

In this chapter's following discussion of cell growth kinetics, we will describe a cell population using its averaged properties. However, it is important to keep in mind that not all cells are equal. Growth is an autocatalytic process, meaning the rate of change of cell concentration (dx/dt) over time (t) is dependent on the cell concentration (x) present at the given time point, as shown in Eqs. 5.3 and 5.4 (Panel 5.3). Cell concentration may be expressed as cell mass per volume or other units, although dry cell mass per volume is most frequently used. Hence, x is the concentration of cells in terms of dry biomass.

Panel 5.3 Specific Growth Rate and Doubling Time

1) In terms of biomass concentration, x:

$$\frac{dx}{dt} = \mu x \tag{5.3}$$

$$\mu = \frac{1}{x}\frac{dx}{dt} \tag{5.4}$$

2) In terms of cell number per volume, n:

$$\frac{dn}{dt} = \mu_n n \tag{5.5}$$

For microorganisms that divide by fission, the increase in cell number can be described as in Eq. 5.5 (Panel 5.3), where n is the number of cells per unit volume. In this equation, μ and μ_n denote the specific growth rate based on dry mass and cell number, respectively. The specific growth rate is a measure of how quickly the cells grow. The time interval necessary for the cell number or cell biomass concentration to increase by twofold ($n_2 = 2n_1$) is called the doubling time (t_d).

The composition of cells is not necessarily constant during the course of growth, because not all cellular materials increase at an equal rate at all times. In the case that all cellular materials increase proportionately with the number of cells, cells are considered to be at a state of balanced growth. During that time, the specific growth rate can be estimated from the mass change of any cellular component (Eq. 5.6 in Panel 5.4).

Panel 5.4 Balanced Growth

- At some stages of growth, all cellular components remain in the same proportion:

$$\frac{1}{n}\frac{dn}{dt} = \frac{1}{x}\frac{dx}{dt} = \frac{1}{\xi_i}\frac{d\xi_i}{dt} = \mu \tag{5.6}$$

5.4   Growth Stage in a Culture

As cells grow, they increase in number, mass, and size. In a natural environment or in culture medium, their growth rate can change in response to fluctuations in the concentration of nutrients, metabolites, oxygen, temperature, pH, and other factors. Cell growth in culture is divided into stages (Figure 5.4).

For example, consider the cultivation of a pure population of a microorganism in a liquid nutrient mixture. If the inoculum is from a dormant culture, cells may experience a short period of lag phase immediately after inoculation. During this lag phase, they first become metabolically active, increase in mass and size, and synthesize the machinery required for active growth. They may begin to divide when their cellular state is ready for active growth. Subsequently, they enter a period of rapid growth, called the exponential growth phase or log phase. During log phase, cells have sufficient nutrients to grow at the maximum rate achievable in the chemical and physical conditions provided.

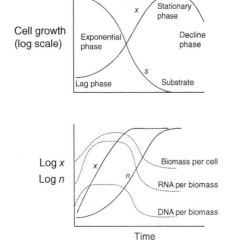

Growth Stages and Cell Composition

Figure 5.4 Growth phases and variation in cellular properties in a batch culture.

After a period of rapid growth, some nutrients may decrease to a level too low to sustain the maximum growth rate. In that case, cell growth slows. Eventually, cell concentration reaches a plateau and the culture enters a stationary phase. Some cells may enter a decline phase, in which they lose viability and the cell concentration begins to decrease.

Throughout the cultivation period, cell size and composition may change. As described above for the lag phase, cells may increase in mass first before they divide. In general, actively growing cells have a higher content of RNA per cell due to the large amount of rRNA that is needed for protein synthesis. Under very rapid growth conditions, *E. coli* cells double their number every 20 minutes, almost twice as fast as its genome can be replicated. Because of this, the genome replication for a given daughter cell starts two or more generations earlier than its inception. In a rich medium, the rapidly growing *E. coli* cells in the exponential phase thus have higher DNA content per cell than in the stationary phase.

Conversely, other nondividing cells in late stationary stages may have more than one copy of a genome per cell. They may cease growing before cell division and thus have a higher DNA content per cell than cells in the exponential stages. In the stationary phase, many mammalian cells are small. In most cases, cultured mammalian cells begin to lose their viability in the stationary phase. The dead cells are often smaller than nongrowing cells in the stationary phase.

For most microbial cells and many animal cells capable of growing in suspension, this process of growth can be repeated. By harvesting a small portion of the old culture around the end of the exponential phase, we can use that portion to inoculate a new culture. Thus, the cycle of cell growth and culturing can begin again.

5.5 Quantitative Description of Growth Kinetics

Microorganisms respond to the availability of nutrients or substrates. In the presence of abundant nutrients, they grow quickly and take up more nutrients, assembling the materials necessary to grow cell mass and to generate more energy. As nutrients diminish, cells correspondingly reduce their growth and consumption rates.

Some nutrients provide vital functions and are essential to the cell, especially those that are the sole source of an elemental component of the cell or those that are component compounds that the cell cannot synthesize. For example, some nutrients can supply carbon for making biomass. Others might supply nitrogen, phosphate, or sulfur. In most cases, the organic carbon source also serves as the energy source. Among the carbon-containing substrates taken up by cells, a portion is converted to biomass. The remainder is used to generate energy for biomass synthesis.

When growing cells, all of the essential nutrients must be provided. A medium that contains the minimal set of nutrients is referred to as minimal medium. The inclusion of additional nutrients can often alleviate a cell's need to synthesize cellular components and can facilitate cell growth. For example, many microbial cells can grow on inorganic nitrogen, such as ammonium or nitrate, to make amino acids for making proteins. However, if amino acids are provided, they will grow faster. Often, the supply of a larger number of amino acids will facilitate growth.

In other words, a medium that is richer in nutrient variety can support faster growth. As cells grow, the concentration of nutrients decreases. At some point, a nutrient may become low and cause the cell growth to slow down; that nutrient is then the "limiting substrate."

Let us consider a case where the growth rate of a microorganism depends on the concentration of a limiting substrate. At very high levels of this rate-limiting substrate, cells grow at a maximum specific growth rate (μ_{max}). At low levels of the substrate, the specific growth rate decreases until the point where no nutrient is available ($s = 0$) and no growth can occur ($\mu = 0$). Such a biological response can be described with an equation (Eq. 5.7 in Panel 5.5), and was first used by J. Monod. It is therefore commonly referred to as the "Monod model" of microbial growth (Figure 5.5).

Panel 5.5 Monod Model for Cell Growth

$$\mu = \frac{\mu_{max} \cdot s}{K_s + s} \tag{5.7}$$

$$\frac{1}{\mu} = \frac{1}{\mu_{max}} + \frac{K_s}{\mu_{max}} \cdot \frac{1}{s} \tag{5.8}$$

With the Monod model, the growth rate dependence on the substrate concentration can be viewed as a zero-order region, and a first-order region bridged by a transition region. At a very high substrate concentration ($s \gg K_s$), the growth rate can be approximated by μ_{max} and is not affected by substrate concentration. At low substrate concentrations ($K_s \gg s$), the growth rate is linearly dependent on the substrate concentration and is said to follow first-order kinetics. At these times, the growth rate has a saturation-type dependence on substrate concentration (i.e., μ asymptotically approaches a maximum value as s continues to increase).

This is an almost universal feature of microbial growth. When nutrient resources are abundant, cells maximize their growth rate to expand their population. Conversely, when resources are scarce, cells conserve their resources and minimize their expansion of biomass.

Monod Model of Growth Kinetics

- When $s \gg K_s$, $\mu \rightarrow \mu_{max}$
- When $s \ll K_s$, $\mu \rightarrow \mu_{max} s / K_s$

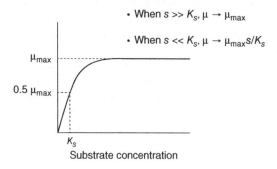

Figure 5.5 Dependence of growth rate on substrate concentration. The relationship can be described by the Monod model.

Figure 5.6 A double reciprocal plot (1/μ vs. 1/s) of the Monod model for estimating μ_{max} and K_m.

Kinetic Parameters in Monod Model

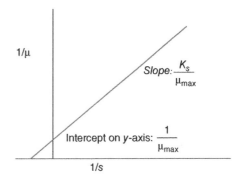

The Monod model employs an identical equation, known as the Michaelis–Menten equation, for single-substrate enzyme kinetics. However, there is a major distinction between the two models. While the Michaelis–Menten equation was derived mechanistically, the Monod model is an empirical model and does not have a mechanistic foundation. Like the Michaelis–Menten equation, the model has two parameters: a maximum rate (μ_{max}) and a half-saturation constant (K_s). K_s has units of substrate concentration. At a substrate concentration of K_s, the specific growth rate is $0.5\ \mu_{max}$.

As in Michaelis–Menten kinetics, the Monod equation can be rearranged to give a relationship among the reciprocals of *s* and μ (Eq. 5.8 in Panel 5.5), as shown in Figure 5.6. When growth rates at different substrate levels are experimentally measured, the data can be plotted in a double reciprocal plot for the estimation of μ_{max} and K_s values.

For higher organisms, or cells derived from them and cultured *in vitro*, the phenomenon of growth is more complex. In a multicellular organism, the growth of individual cells is regulated by developmental cues. These cues direct cells to grow, differentiate, and develop into different tissues and organs of the organism. Therefore, although those cells do grow faster when nutrients are abundant, their growth is more profoundly affected by the concentration of growth factors or other regulators. Their growth rate still responds to nutrient concentration, especially when substrate levels are low. However, even at high substrate levels, cells may still require some growth regulators to proliferate. A Monod model may not be adequate for describing the growth of cells derived from higher organisms.

5.5.1 Kinetic Description of Substrate Utilization

Consider a batch culture of a microorganism. As cells proliferate, the specific growth rate (Eq. 5.4 in Panel 5.3) can be calculated by taking the slope of the cell concentration plot divided by the cell concentration at the time point of interest. Similarly, we can define a specific substrate utilization rate that is the substrate consumption rate divided by cell concentration (Eq. 5.9 in Panel 5.6). The specific substrate consumption rate is a measure of how fast cells consume a nutrient. We can also define a specific production formation rate (Eq. 5.10 in Panel 5.6).

Panel 5.6 Specific Rates for Growth, Consumption, and Production

$$\mu = \frac{1}{x}\frac{dx}{dt} \tag{5.4}$$

$$q_s = \frac{-1}{x}\frac{ds}{dt} \tag{5.9}$$

$$q_p = \frac{1}{x}\frac{dp}{dt} \tag{5.10}$$

Recall that we previously defined a yield coefficient ($Y_{x/s}$) of the conversion of substrate to biomass (3.2.2.1 in Chapter 3). Consider a scenario where no substrate is used to make the product. In that case, the rates of "new" biomass synthesis (μx) and of current biomass synthesis are related by the yield coefficient (Eq. 5.11 in Panel 5.7). Assuming that $Y_{x/s}$ is constant, then a plot of specific nutrient consumption rates against the specific growth rate should yield a straight line that passes through the origin (Figure 5.7).

Panel 5.7

If all the substrate consumed goes to make cell mass, the material balance on substrate gives:

$$\frac{ds}{dt} = -\frac{\mu x}{Y_{x/s}} = -q_s x \qquad\qquad \rightarrow q_s = \frac{\mu}{Y_{x/s}} \tag{5.11}$$

Specific Maintenance Substrate Consumption Rate

- Cells consume energy source even when not growing.
- Hypothesis: a requirement of "maintenance energy"

$$\frac{ds}{dt} = -\frac{\mu x}{Y_{x/s}} - mx = -\left(\frac{\mu}{Y_{x/s}} + m\right)x = -q_s x$$

$$q_s = \frac{\mu}{Y_{x/s}} + m \tag{5.12}$$

Some experimental data from microbial cultures, although giving a relatively straight line, show a positive intercept on the y-axis (i.e., axis of the substrate consumption rate) (Figure 5.7). This is rationalized by the fact that the nutrients consumed by cells are not

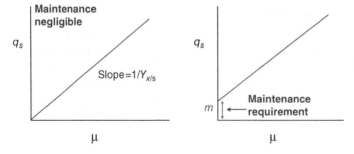

Figure 5.7 Maintenance requirements and specific substrate consumption rate.

solely used to synthesize biomass, as was previously assumed. Additional nutrients need to be consumed to "maintain" the cells, even when they are not growing. This is more obvious when the substrate serves as an energy source, as in most cases when the line gives a positive intercept on the *y*-axis (Figure 5.7).

Even when cells are not actively growing (i.e., $\mu = 0$), cellular materials are still being turned over and the membrane potential and concentration gradients of cellular components across the cell membrane are still being maintained. All of this "work" requires energy. Therefore, there is always a net consumption of the substrate for energy. Such a nonzero intercept on the graph for specific nutrient consumption is referred to as "maintenance substrate consumption" and is described by adding a maintenance term to the substrate utilization equation (Eq. 5.12 in Panel 5.7).

5.5.2 Using the Monod Model to Describe Growth in Culture

The growth of a microorganism in a batch culture with constant volume can be described using balance equations of the concentrations of cell mass and substrate (Eqs. 5.13 and 5.14, in Panel 5.8). In this case, we assume that the specific growth rate follows Monod kinetics and that no substrates are being used for product synthesis or maintenance (Panel 5.8). The two balance equations (biomass and substrate concentrations) are thus related by the dependence of the specific growth rate on the substrate concentration, and by the amount of substrate consumed per unit amount of biomass produced. Given the initial conditions of the concentrations of cells and substrate, this set of simultaneous equations can be solved to give the time profile of cell and substrate concentrations. These time profiles resemble what are typically seen in batch cultures.

Panel 5.8 Monod Model Describes a Batch Culture Kinetics

$$\frac{dx}{dt} = \mu x = \frac{\mu_{max} s}{K_s + s} x \tag{5.13}$$

$$\frac{ds}{dt} = -\frac{\mu x}{Y_{x/s}} = -\frac{\mu_{max} s}{K_s + s} \frac{x}{Y_{x/s}} \tag{5.14}$$

Assume *m* is negligible:

- Initial conditions: $x = x_0$, $s = s_0$
- Assume constant $Y_{x/s}$.

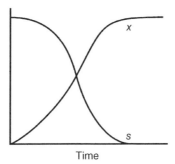

Time

In the case that a maintenance term is incorporated into the substrate balance equation (Eq. 5.15 in Panel 5.9), a negative substrate concentration will be predicted by the model. The model would depict a nonnegative specific nutrient consumption rate due to maintenance, even as both the specific growth rate and substrate concentration approach zero. This situation is certainly unrealizable and illustrates one of the shortcomings of using a simple empirical model. Nevertheless, the Monod model can approximate the growth of many organisms, especially those under less extreme conditions (Panel 5.9).

Panel 5.9 Simulation of Bath Growth Kinetics

$$\frac{dx}{dt} = \mu x = \frac{\mu_{max}s}{K_s + s}x \tag{5.3}$$

$$\frac{ds}{dt} = -\frac{\mu x}{Y_{x/s}} - mx = -\frac{\mu_{max}s}{K_s + s}\frac{x}{Y_{x/s}} - mx \tag{5.15}$$

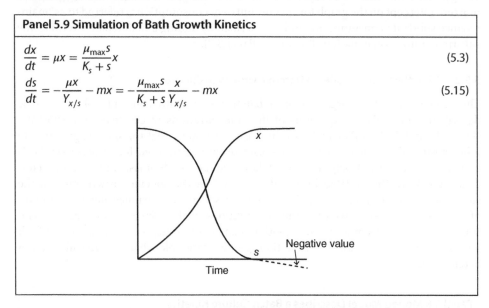

To use a Monod model, we first need to determine the values of μ_{max} and K_s. It is common practice to plot the specific growth rate obtained from the batch culture growth curve against the concentration of the rate-limiting substrate. By curve fitting or using a double reciprocal plot, one can then obtain the μ_{max} and K_s values. When obtained in this way, however, these values may not be accurate due to the impact of other factors [e.g., pH and accumulated metabolite(s)] on growth rate.

5.6 Optimal Growth

In addition to substrate concentration, many other factors affect the growth rate. Such factors include pH, osmolality, and temperature. Aside from temperature, the growth rate dependence on most other parameters falls into three regions: suboptimal, optimal, or inhibitory. In the suboptimal region, increasing the level of a given parameter increases the growth rate until it reaches an optimal region. In the optimal region, the level of parameters is equivalent to an optimal steady state of growth. In the inhibitory region, the further increase of a given parameter will reduce the growth rate.

Using the optimal growth temperature as a criterion, microbes can be classified as psychophiles (with an optimal temperature range of subzero to 15 °C), mesophiles (15–45 °C), thermophiles (50–60 °C), and extreme thermophiles (65 °C and above).

Figure 5.8 Effect of substrate and product inhibition on cell growth rate.

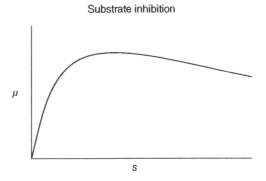

Substrate inhibition

Product inhibition

Microbes also have different levels of optimal osmolality for growth. Halophiles and extreme halophiles are those growing in high salt concentrations (i.e., 3–5 M or higher). Similarly, there are acidophilic microorganisms that grow in very acidic conditions (i.e., pH 0–0.7).

Cell growth may be inhibited by high concentrations of a substrate or product (Figure 5.8). For some nutrients, the substrate inhibitory concentration is low. For example, some yeast can utilize methanol as a carbon source. However, the methanol concentration range for optimal growth is narrow, and it is inhibitory at rather low levels. Some products are inhibitory to the producing cells when accumulated at high levels. Examples include ethanol, acetic acid, butanol, and lactic acid. One way to describe the negative dependence of the specific growth rate on the concentration of substrate and product is to add an inhibitory term to the Monod equation (Eqs. 5.16 and 5.17, in Panel 5.10), where K_I is the inhibition constant and n is an exponent with a positive value. K_I and n are obtained, along with K_s and μ_{max}, by fitting the experimental data using the equation.

Panel 5.10 Modifications of the Monod Model

- Substrate inhibition:

$$\mu = \frac{\mu_{max}s}{K_s + s\left(1 + \dfrac{s}{K_I}\right)} \tag{5.16}$$

(Continued)

Panel 5.10 (Continued)
• Product inhibition:

$$\mu = \frac{\mu_{max}S}{(K_s + s)\left(1 + \dfrac{(P)}{K_I}\right)^n} \tag{5.17}$$

5.7 Product Formation

The accumulation of the product in culture is dependent on the specific production rate of the producing cell, q_p, and the cell concentration (Eq. 5.18 in Panel 5.11). The specific production rate is a quantitative measure of the activity of cells to produce a product. Under different culture conditions, the specific production rate changes. A kinetic description of the specific production rate as a function of the key culture variable that influences it is useful to predict the production characteristics of a culture.

Panel 5.11 Description of Product Formation

$$\frac{dp}{dt} = q_p x \tag{5.18}$$

$$q_p = \alpha\mu + \beta \tag{5.19}$$

The nature of products is very diverse, as are the key variables influencing the specific production rate. Some products are metabolites of energy metabolism, such as lactate, ethanol, and acetic acid, while others are compounds in biosynthetic pathways, such as amino acids and nucleotides. Still others are secondary metabolites for which productivity is related to cell differentiation. With recombinant DNA technology, more products are produced only after induction. The productivity of these products may be affected by the inducer concentration. With a wide variety of microbial products, it is not surprising that productivity is also affected by many diverse factors, including the physical and chemical environments, as well as the biological nature of the product.

The biosynthesis of many metabolic products is affected by the growth of microorganisms. Their kinetic behavior is frequently classified according to its relation to growth rate (Figure 5.9). Some products are produced more when cells are growing faster. These products' productivity is often modeled as being proportional to the specific growth rate, and is categorized as positively growth associated. For example, ethanol production by yeast provides the energy for cell growth and is positively growth associated. The synthesis of some products is not affected by growth rate (non–growth associated), while others are negatively affected by growth (negatively growth associated). Many secondary metabolites fall into the category of being negatively growth associated.

A simple equation with a growth-dependent term and a growth-independent term can be used to describe the specific production rate (Eq. 5.19 in Panel 5.11). The symbol α can be a positive or negative value or zero. This expression can serve as a first approximation of a specific product formation rate. The material balance equation for product formation (Eq. 5.20 in Panel 5.12) can be combined with the growth model for the simulation of a product time profile in a batch culture (Panel 5.12).

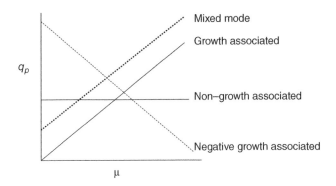

Figure 5.9 Relationship between specific product formation rate and specific growth rate.

Panel 5.12 Product Formation in a Batch Culture

$$\frac{dx}{dt} = \mu x = \frac{\mu_{max}s}{K_s + s}x \tag{5.14}$$

$$\frac{ds}{dt} = -\frac{\mu x}{Y_{x/s}} - \frac{q_p x}{Y_{p/s}} = -\frac{\mu_{max}s}{(K_s + s)} \cdot \frac{x}{Y_{x/s}} - \frac{q_p x}{Y_{p/s}} \tag{5.20}$$

$$\frac{dp}{dt} = q_p x \tag{5.18}$$

With recombinant DNA technology, one often prefers to control the time profile of production by manipulating the expression of the gene(s) or product formation using an inducible promoter. Upon reaching the desired culture stage, one can then "turn on" the product formation gene(s) to switch production. In this case, productivity may depend on the concentration of the inducer. On other occasions, a precursor is added for its conversion to the product. In that case, productivity depends on the concentration of the precursor.

5.8 Anchorage-Dependent Vertebrate Cell Growth

Cells from multicellular organisms, such as plants, insects, and vertebrate cells, have been isolated from tissues after enzymatic dissociation. With appropriate culture conditions, cells can be continually cultivated and expanded over a long period of time, and can be used for production purposes. They are frequently used to produce metabolites, proteins, or viruses. Mammalian cells cultured *in vitro* have been used for the production of viral vaccines and recombinant protein therapeutic agents.

Depending on the tissue of origin and the nature of the cells, a few types of animal cells can grow suspended in medium, just like most microorganisms. The vast majority of cells isolated from vertebrate animals, however, are anchorage dependent, meaning they require a surface to where they can attach and grow. Some cells, especially those that are derived from normal tissue, grow only for a limited number of cell divisions before they enter senescence. Other cells that have abnormal growth control mechanisms can grow in culture indefinitely.

Cultivation of Normal Diploid Cells

Figure 5.10 Derivation of cells from tissues and surface attachment–dependent growth.

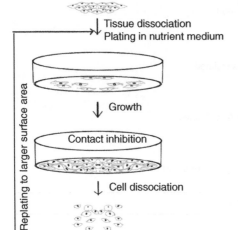

Mammalian cells are isolated from tissues by enzyme digestion. This removes the extracellular materials that bind them together. They are then plated on a compatible surface, supplied with medium, and allowed to attach to the surface and grow (Figure 5.10). Once they grow to cover the entire surface area and form a monolayer, cell growth ceases due to contact inhibition (i.e., the cells do not like to become too crowded and do not grow to overlap each other). Although cells require nutrients and growth regulators (such as insulin) to proliferate, contact inhibition is independent of these requirements. Cellular growth can only be resumed after cells are detached from the surface (using a proteolytic enzyme) and replated on a new surface with a larger area, where they again grow to reach confluence. This process can be repeated until the cell line reaches senescence. For transformed cell lines, the process can be repeated indefinitely.

The phenomenon of contact inhibition is seen in normal cells originally isolated from normal tissues. It is less strict for continuous cell lines (i.e., lines that can indefinitely grow in culture). However, even for continuous cell lines, the growth rate decreases substantially once cells reach confluence.

The equation describing the growth rate of anchorage-dependent cells will have a parameter related to the cell density per surface area. We may use a logistic equation (Eq. 5.21 in Panel 5.13) with x_m as the confluent cell density (number of cells/surface

---

Panel 5.13 Two Empirical Equations for Density-Dependent Cell Growth

$$\mu = \mu_{max}\left(1 - \frac{x}{x_m}\right) \tag{5.21}$$

$$\mu = \mu_{max}\left\{1 - Exp\left[(-B)\left(\frac{x_m - x}{x_m}\right)\right]\right\} \tag{5.22}$$

area). An alternative equation uses one more parameter and allows for a better fit of the experimental data (Eq. 5.22 in Panel 5.13). As in the Monod model, these equations are also empirical expressions, so the value of the parameter B is determined by data fitting.

5.9   Other Types of Growth Kinetics

Most culture conditions used for production have one substrate that is limiting, typically the most expensive component of the medium. All other substrates are provided in excess. A Monod model is well suited to describe these conditions. However, in some cases, the Monod model cannot adequately describe how growth kinetics depend on substrate concentration.

One example is with some mycelial microorganisms, which grow as pellets or dense cell clumps. Nutrient transport into this cell mass is difficult and can be a limiting factor. As a result, only a portion of cells near the exterior surface grows at the maximum rate, even with abundant nutrients. Another example is with some hydrocarbon-utilizing microbes, which grow on the interface of oil droplets. The available surface area of the oil droplets may limit their growth. In the cultivation of plant tissue, such as the hair root or meristem, growth may be limited to the tip region of the tissue. In those cases, the apparent specific growth rate (i.e., the rate of biomass increase divided by the total biomass) will decrease with time, and the increase in biomass over time will not be an exponential function.

In some cultures, an organism may use two major substrates that are both growth limiting, but in a sequential manner rather than a concurrent one. One example is the growth of *E. coli* in the presence of glucose and lactose. *E. coli* cells prefer to utilize glucose first. As glucose is depleted, cell growth is temporarily paused. At this point, the absence of glucose "de-represses" β-galactosidase, which then allows cells to use lactose to resume growth until it is also depleted. Such biphasic growth is called diauxic growth (Figure 5.11 and Panel 5.14).

Figure 5.11 Diauxic growth of *E. coli* on a mixture of glucose and lactose.

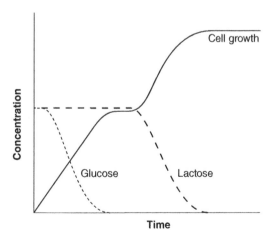

Panel 5.14 Growth of *E. coli* in Glucose and Lactose

- Diauxic growth:
  - *E. coli* uses only glucose, until glucose is exhausted.
  - After glucose depletion, a short lag period is seen in which cells express enzymes needed for lactose utilization.
  - Lactose-utilizing enzymes (β-galactosidase) are "repressed" in the presence of glucose and "induced" by the presence of lactose.

In some cases, the cell growth rate is affected by more than one substrate. One example is the simultaneous utilization of glucose and glutamine by animal cells. Both nutrients must be present simultaneously; cells utilize both while growing. The kinetic expression of growth will involve both substrates (s_1 and s_2) and their specific consumption rates. A frequently used formula for the specific growth rate uses a multiplicative product of the Monod equations for s_1 and s_2, written as: $\mu = \mu_{max} s_1 s_2 / [(K_1 + s_1)(K_2 + s_s)]$.

5.10 Kinetic Characterization of Biochemical Processes

Kinetic descriptors, the specific rates, and yield coefficients are important for characterizing and comparing processes involving cell growth. Once an experiment is completed, the value of model parameters should be determined. A comparison of the results of different experimental conditions is performed on both direct measurements and kinetic parameters. Direct measurements would include cell and nutrient concentration profiles. Kinetic parameters would include the specific rates of growth, nutrient consumption, and product formation. The consistency of manufacturing runs is evaluated also at the levels of both measured variables and kinetic model parameters (Panel 5.15).

Panel 5.15 Analysis of Experimental Data

In-process data analysis:

- Plot measurement data for data quality evaluation.
- Calculate derived variables (specific rates, stoichiometric ratios, yields, etc.).

Post-process data analysis:

- Data regression on entire time profiles of data
- Calculate derived variables based on regressed data.

In most industrial bioprocesses, the values of different kinetic parameters do not remain constant. Instead, they change over the course of cell cultivation. The process is typically more complex than being limited by a single substrate. In addition to substrate limitation, the adverse effects of metabolite accumulation often lead to a reduction in the growth rate.

In general, analyses should be performed during and after processing. Performing analyses during the process allows for real-time diagnosis. Postprocess analyses provide more data to identify outliers and perform regressions and statistical analysis. This can yield a more accurate assessment.

5.11 Applications of a Growth Model

The simplest use of a growth kinetic model is to summarize a large set of kinetic data. After collecting a large dataset for a given process (e.g., after a number of manufacturing runs), the accumulated sets of time profile data can be difficult to describe, visualize, and compare. We can fit time profile data using a suitable model to better distill the data into a set of parameters. The smaller the number of parameters, the easier it is to compare the time profiles of kinetic data (Panel 5.16).

Panel 5.16 Kinetic Models

Why use a process model?

- To summarize a large amount of data
- To simulate system behavior to facilitate process analysis
- To predict process performance

At least two classes of models:

- Empirical models
  - Monod model and its variants
- Mechanistic models
  - Based on mechanism; more predictive

While this is a useful application, the ultimate value of a growth model is for performance prediction or process control. Many recombinant protein productions in microorganisms use inducible systems. In that case, the protein is induced only after the cell concentration reaches a threshold. Upon induction, cells are then switched from a rapid growth mode to a more controlled growth rate. This minimizes the possibility that exceedingly high rates of protein synthesis will overwhelm the cell's capacity for protein folding.

Many other processes employ similar strategies for controlling cell growth rate and optimizing process performance. Examples include baker's yeast fermentation and secondary metabolite production. With a growth model, one can control the supply rate of a key nutrient to manipulate the overall growth rate.

A model for cell growth and product formation must include relevant "state" variables (Panel 5.17). State variables typically include the concentrations of cells, key substrate(s), and metabolites. All of these may affect process performance. We also need to define the relationship among state variables. For example, in a simple model, the dependence of cell growth on substrate concentration is described by a Monod relationship. This relationship joins the balance equations for cell mass and substrate.

Panel 5.17 Setting Up a Kinetic Model for Cell Growth

- Define "state" variables (e.g., concentrations of cells, limiting nutrients, products).
- Determine the kinetic relationship among state variables.
- Set up balance equations for state variables.
- Obtain the value of kinetic parameters.
- Simulate, and refine the model; apply the model to test various hypotheses and to optimize the process.

For any application of a kinetic model, the kinetic parameters must be accurate. These parameters, including the specific rates and yield coefficients, are obtained from experimental data. They can be refined continually as more experimental results accumulate. Once these are all in place, we can use this data to generate a simulation or to calculate the substrate feeding rate profile that can yield the target growth rate. Once a model is verified with experimental data, it can be used to explore different process scenarios, test hypotheses, or even optimize the process.

5.12 The Physiological State of Cells

Cells under different environments or at different stages of culture may have different growth rates. They may also metabolize nutrients differently, or even have various differentiation states. Under different growth conditions, a producing organism may have a different product synthetic rate and/or produce different products. For example, a microorganism may be capable of producing a family of related secondary metabolites. Under different chemical environments or growth conditions, different antibiotics would be produced. In other words, cells have different physiological states under different conditions. Understanding the impact of their physiological state on product synthesis, and being able to control the physiological state of cells, are important for many processes.

In some cases, the physiological state that is important for product formation is the specific growth rate. For example, in the production of antibiotics and other secondary metabolites, biosynthesis starts only after rapid cell growth ceases. This phenomenon is sometimes referred to as switching from trophophase to idiophase. In some other cases, the physiological state of concern may be related to different types of metabolism or to different "metabolic states." For example, yeast cells in culture may convert most of the glucose they consume to ethanol through fermentation, or they may oxidize most of the glucose to CO_2 through aerobic respiration.

These two types of metabolism will have very different cell yields. Given an equal amount of glucose as the substrate, the resulting cell concentration is lower when most of the glucose carbons are converted to ethanol than when respiration is the primary mechanism for glucose utilization. In ethanol fermentation, the desired metabolic state is fermentative. In the production of baker's yeast, an aerobic metabolic state is desirable.

In some processes, our objective may be to grow cells and direct them to differentiate to a particular cell type. For example, some plant cell culture processes are designed to direct somatic cells to differentiate into somatic embryos.

Stem cell cultivation processes usually entail two stages. The first stage is the expansion of cells and maintaining stem cells at their "potent" state. This means that they are capable of self-renewal and have the capacity to differentiate. In the second stage, the stem cell is guided to differentiate to a particular type of cell (e.g., liver, nerve, or muscle cells). The process is thus rather complicated in terms of maintaining and controlling a cell's potent state.

When the physiological state of the cell is important to the process outcome, the state variables must also include the "state" of the cell. The "state" of interest is very diverse and depends on the process objective. It may be as simple as the growth rate or metabolic state, or as complex as the differentiation state or the degree of morphological transformation. For those processes, it is important to develop a quantitative description of the state of the cell. One could argue that a comprehensive description of a cell's physiological state is the combination of descriptions of its transcriptome, proteome, and metabolome. However, from an engineering perspective, the description only needs to include the cell characteristics that are critical to the process' objective.

Once a physiological state function (g) is defined, cells in the population can then be categorized according to their physiological state, g_i (where the subscript i denotes each state). The concentration of cells at each state is accounted for by $x(g_i)$. Such a discrete description of state (categorizing cells using a number of distinct classes) is well suited in some cases, such as vegetative cells versus spores. Sometimes, there may not be discrete "states" but rather a continuous distribution of cells at different values of the "state" properties.

In the process to direct a population of cells to change their state, we must usually control the concentration of an external factor(s) in the medium. This enables us to influence a cell's transition from its initial state to the "target" state. For example, in the case of yeast fermentation, glucose and oxygen concentration is controlled to influence cell metabolism. A high concentration of glucose and a low level of oxygen favor fermentation; the opposite promotes aerobic respiration. For embryonic stem cell differentiation to liver cells, a combination of signaling molecules is used to guide undifferentiated cells to become endodermal cells, and then hepatic cells.

A complete model for describing a cell population undergoing state transition will thus have: (1) balance equations for the concentrations of each subpopulation in each state; (2) a balance equation for the external factor that drives the change of the state; and (3) a function that describes how the external factor directs the switch from one state to another.

An example of metabolic state shifting in a population is the switch between oxidative respiration (converting glucose to CO_2) and glycolytic metabolism (fermenting glucose to ethanol) in yeast. Neglecting the flux to the pentose phosphate pathway, the glucose consumed by cells is split between two paths at the pyruvate node: toward ethanol production or toward the tricarboxylic acid (TCA) cycle. The energy needs [adenosine triphosphate (ATP)] required for cell growth are relatively constant, regardless of whether the yeast cell is at an oxidative metabolic state or a fermentative metabolic state. Therefore, cells with a fermentative metabolism consume far more glucose than cells with an oxidative metabolism. The two fluxes, the specific glucose consumption rate (q_{glc}) and the specific ethanol production rate (q_{EtOH}), can be used to describe the metabolic state. A fermentative state is represented by high q_{glc} and high q_{EtOH}, whereas the oxidative state will be low in both specific rates.

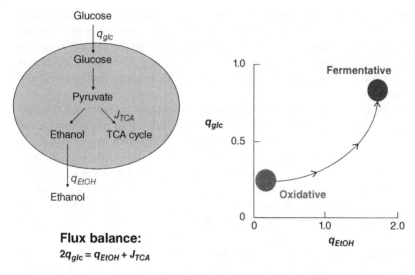

Flux balance:

$$2q_{glc} = q_{EtOH} + J_{TCA}$$

Figure 5.12 Fermentative and aerobic metabolic states of yeast cells.

On a plot of q_{glc} versus q_{EtOH}, cells at the two states are well separated (Figure 5.12). On such a plot, one can chart the trajectory of a population that is originally in one state and moving toward the other. If we use arrows to mark the state change over a unit of time (e.g., each arrow indicates the change of state in an hour), we can also visualize the rate of change. By plotting the trajectories under different external factors (such as a step change of glucose concentration to different levels), we can compare and attempt to derive a function describing the effect of the external factor on state transition.

5.12.1 Multiscale Model Linking Biotic and Abiotic Phases

As discussed in Section 5.12, the physiological state of the cell determines its growth and production. A complete model for a biochemical process should thus connect the chemical environment of the culture (including the concentration of substrates, metabolites, and product) to the intracellular biochemical reactions that define a cell's physiological state (Figure 5.13). In such a model, the intracellular reactions and the

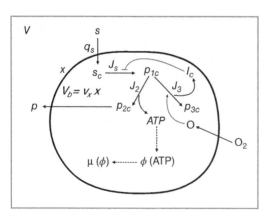

Figure 5.13 Schematic of a bioprocess system. Model variables include abiotic reactor phase and biotic intracellular reactions.

extracellular environment interact and mutually influence each other. The magnitude of the rate constants of the biochemical reactions is in the order of a millisecond or lower. The specific rates (growth and substrate consumption) are in the order of hours. These types of models thus include parameters whose values span over a wide range of time and length scales.

Such a model can be considered to be a combination of two submodels: (1) an intracellular submodel of the network of major biochemical reactions in the cell, and (2) a reactor submodel of the concentrations of cells and nutrients. The cell growth submodel consists of the balance equations of state variables, and cell and substrate concentrations (Eqs. 5.23 and 5.24, in Panel 5.18). Linking these two submodels together is the material exchange between the cellular compartment and the extracellular compartment (Eq. 5.25 in Panel 5.18). The consumption of a substrate (i.e., the decrease in the abiotic phase; see the right-hand side of Eq. 5.25) is equal to its flux into the intracellular reaction network (i.e. the intracellular substrate flux). In the example shown in Panel 5.18, another link between the two submodels is the dependence of the specific growth rate on the intracellular level of ATP (Eqs. 5.25 and 5.26).

Panel 5.18 Reactor Model (Abiotic Phase)

Assumptions:
    Constant V; biomass occupies a very small fraction of total volume.
    Constant cell density

$$\frac{dx}{dt} = \mu x = \mu(\phi) \cdot x \tag{5.23}$$

$$\frac{ds}{dt} = -q_s x \tag{5.24}$$

Intracellular Reaction Model

$$J_s = q_s \cdot v \tag{5.25}$$

$$\phi = \phi(J_{ATP}) \tag{5.26}$$

Example: $\phi = \phi(J_{ATP})$, intracellular substrate converted to an intermediate that is split into two ATP generation pathways. One is inhibited by oxygen concentration.

$$J_s = \frac{r_{max,1} \cdot s_c}{(K_{m,1} + s_c)\left(1 + \frac{I_c}{K_I}\right)} \tag{5.27}$$

$$\gamma_{s,p1} J_s = J_2 + J_3 \tag{5.28}$$

$$J_2 = \frac{r_{max,2} \cdot p_{1c}}{K_{m,2} + p_{1c}} \tag{5.29}$$

$$J_3 = \frac{r_{max,3} \cdot p_{1c}}{K_{m,3} + p_{1c}} \cdot \frac{O}{K_{m,3O} + O} \tag{5.30}$$

$$J_{ATP} = \gamma_{2,ATP} \cdot J_2 + \gamma_{3,ATP} \cdot J_3 \tag{5.31}$$

In general, the model considers the biotic phase to be single compartment, collapsing a large number of individual cells into a single phase. For simplicity, the volume occupied by cells is assumed to be negligible, so the total volume of the abiotic phase is constant.

That allows one to neglect the abiotic volume change due to the uptake of water for the expansion of cellular volume. In balancing the substrate and metabolites transferred between the biotic and abiotic phases, a specific cell volume (v_x) (volume of cell/mass of dry biomass) is used to relate the intracellular metabolic flux (based on intracellular volume) to the extracellular rate (based on culture volume) (Eq. 5.25).

The intracellular bioreaction model incorporates the reactions that describe substrate utilization, metabolite production, and chemical potential energy formation. All of these variables may affect the physiological state ($\varnothing$) of the cell. The effect of the physiological state on the growth rate (Eq. 5.23) is then established to provide another link between the intracellular submodel and the extracellular submodel.

Panel 5.18 illustrates a possible scenario in which the intracellular reactions in yeast cells determine the physiological state. The intracellular substrate is depicted as consumed in glycolysis with a flux J_s (Eq. 5.27) that is split into fermentative (J_2) and oxidative (J_3) branches (Eqs. 5.28, 5.29, and 5.30). The reaction in the oxidative branch is highly dependent on oxygen concentration. It also produces a metabolite (I) that inhibits glycolysis. Both branches generate ATP. The total ATP level is the physiological variable that determines the specific growth rate (Eq. 5.31).

Through intracellular biochemical reactions, the intracellular level of ATP dictates the growth rate. With the provision of substrate in the abiotic phase, the chemical environment in the reactor also affects the biochemical reactions and ATP generation rate, and thus the cell's physiological states.

The example in Panel 5.18 uses simplified reaction equations for glycolysis (Eq. 5.27). It is possible to employ mechanistic biochemical reactions to describe the biochemical reactions. However, the model will become very complex, with many parameters to be estimated. Often, mechanistic reaction equations are available for only major biochemical reactions. Furthermore, it is very difficult to make reliable estimates of the enzyme levels (i.e., the r_{max}'s) for all of the reactions involved.

When using an integrated system model, it is important to examine the simulated intracellular concentrations of various reaction intermediates and metabolites to check that they are all in the physiological range. With proper scrutiny and verification, a multiscale model is an attractive tool for better understanding and even controlling biosystems.

5.13 Kinetics of Cell Death

As a culture enters the stationary phase, cells may lose viability due to nutrient depletion or metabolite accumulation. Cell death inevitably causes a decrease in productivity. In some cases, the intracellular enzymes released by dead cells can cause product degradation or product quality deterioration (e.g., by altering the glycan structure on glycoprotein products). It is thus important to assess cell death in a biological process.

For this purpose, we introduce variables that separately describe the concentrations of viable (x) and dead cells (x_d). The death of cells is generally considered to occur according to first-order kinetics with respect to viable cell concentration (Eqs. 5.32 and 5.33, in Panel 5.19). The specific death rate (μ_d) may be a constant value or a function of the effector that causes cell death. For example, the death rate constant can be a function of ethanol concentration if the fermentation product is the key factor causing cell death.

Panel 5.19

$$\frac{dx}{dt} = \mu x - \mu_d x = \frac{\mu_{max} S}{K_s + s} x - \mu_d x \tag{5.32}$$

$$\frac{dx_d}{dt} = \mu_d x \tag{5.33}$$

Example 5.1   Inactivation of Virus
Recombinant cells derived from mouse myeloma are used to produce a monoclonal antibody. These cells harbor endogenous retroviruses. Even though these viral genome sequences in the producing cell have no known effects on humans, they must be eliminated from the purified product to keep the potential public health risk to a minimum.

The antibodies produced in cell culture are purified by affinity chromatography. Subsequently, the antibody solution is kept at pH 3.0 with acetate buffer for 2 h. The incubation reduces the virus titer by five logarithms. Calculate the first-order virus inactivation rate constant.

Solution
The inactivation of virus follows first-order kinetics:

$$\frac{dv}{dt} = -k_i v$$

Integrating from $t=0$ $v=v_o$ to (t, v) gives:

$$\ln\left(\frac{v}{v_o}\right) = -k_i t$$

$$\ln(10^{-5}) = -k_i(2h)$$

The inactivation rate constant is:

$$k_i = 5.76 \; h^{-1}$$

5.14   Cell Death and the Sterilization of Medium

Sterilization is highly important in bioprocesses. This includes the sterilization of both the medium and the other materials that are in contact with cultured cells. Sterilization, or the elimination of viable bacterial cells, is accomplished in two ways. One approach is removal, such as filtration using a membrane with small pores. Another method is destruction, such as using dry heat, wet heat, a chemical agent, or irradiation.

In this section, we consider only the destructive means of sterilization. As in the case of cell death in culture, the process of killing cells by sterilization is typically considered to follow first-order kinetics with respect to live cells (Eq.5.34 in Panel 5.20). In the case of sterilization, one is more concerned about the total number of viable cells that remain after sterilization, rather than the mass quantity of cells. The objective of sterilization is to eliminate live bacteria in the entire volume. Reducing the live bacteria to less than 1/mL is not sufficient, because even a single survivor in the entire bioreactor may be enough to cause the process to fail. Thus, the calculation of death kinetics in sterilization is typically written as total bacterial count rather than cell concentration. This is reflected in the use of N (total viable bacterial count), instead of n (viable bacterial concentration).

Panel 5.20 Thermal Sterilization of Medium

$$\frac{dN}{dt} = \frac{d(nV)}{dt} = -k_K \cdot N \tag{5.34}$$

Thermal death rate constant as a function of temperature:

$$k_K = A \cdot Exp(-E/RT) \tag{5.35}$$

The log killing is the integral of the death rate constant over time.

$$\int_{N_0}^{N} \frac{dN}{N} = -\int_{t_0}^{t} k_K dt \tag{5.36}$$

$$\ln \frac{N}{N_0} = -\int_{t_0}^{t} A \cdot Exp(-E/RT)dt \tag{5.37}$$

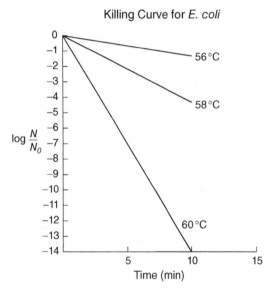

Killing Curve for *E. coli*

Figure 5.14 Effect of temperature on the killing of *E. coli* cells.

The rate at which cells die under different treatments varies widely among different microorganisms and their physiological states. Vegetative cells of mesophilic bacteria lose their viability quickly, even at only moderately high temperatures (e.g., 60 °C) (Figure 5.14). In comparison, spores of thermophilic rod-shaped bacteria (i.e., *Geobacillus stearothermophilus*) must be exposed to a very high temperature (121 °C) for their destruction (Figure 5.15). For a given microorganism, the killing rate constant (k_K) increases with temperature. It is often observed that the increase in the killing rate constant is an exponential function of temperature. In that case, it can be described by an Arrhenius plot or by an Arrhenius constant and activation energy (Eq. 5.35). Again, this is an empirical correlation and should be used within the observed range

Figure 5.15 Effect of temperature on the killing of *Bacillus stearothermophilus* spores.

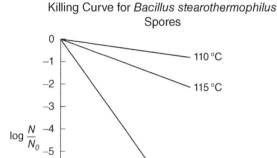

Killing Curve for *Bacillus stearothermophilus* Spores

of temperature. From the temperature profile of a medium sterilization cycle, one can calculate the degree of killing (Eqs. 5.36 and 5.37).

Prior to sterilization, the bacteria present in the medium consist of a variety of species carried over from different medium components and water. In designing the sterilization procedure, we typically adopt the worst-case scenario, assuming that the raw medium is contaminated by a certain level of thermoresistant spores. We then calculate the time required to reach a very low level of total viable cells. Typically, an acceptable total viable bacteria level after sterilization is $N=0.01$. This level of sterilization is equivalent to one in a hundred runs having one surviving bacterium. In other words, it equivalent to a sterilization failure rate of 0.01. This approach demands that, given the same initial level of bacterial concentration in the medium, the degree of reduction of viable bacteria (and thus the duration of thermal sterilization) is higher for a larger scale operation. In general, given the uncertainty of the nature of contaminating microorganisms, overdesign of the sterilization process is a common practice.

5.15 Concluding Remarks

Almost all biochemical productions involve growing cells in at least one segment of their overall process. Even if the immediate process may not involve any cells, such as in enzyme processes, the process may still depend on cell growth to produce its raw materials or key ingredients. The behavior and the physiological state of the producing cell in culture influence the ultimate measures of productivity and product quality. In this chapter, we discussed the quantitative description of cell growth (and briefly cell death), as well as the mathematical description and the growth model that can be used to predict the kinetic behavior of cells in culture. The principles of growth kinetics and stoichiometry form the foundation for the discussion on bioreactors and their operations, as well as scale-up reactions in the following chapters.

Further Reading

Macklin, DN, Ruggero, NA & Coverts, MW 2014, 'The future of whole-cell modeling', *Current Opinion in Biotechnology*, **28**, pp. 111–115.

Shuler, ML & Kargi, F 2002, 'Sterilization of process fluids' in *Bioprocess Engineering*, 2nd edn, pp. 314–323. Prentice Hall, Upper Saddle River, NJ.

Tyson, JJ & Novak, B 2013, 'Irreversible transitions, bistability and checkpoint controls in the eukaryotic cell cycle: a systems-level understanding' in *Handbook of Systems Biology Concepts and Insights*. Academic Press/Elsevier, Amsterdam.

Nomenclature

A	Arrhenius constant	
B	a constant obtained from data fitting	
E	activation energy	
f	population density function	
I_c	intracellular inhibitor concentration	mole/L^3
J	molar flux of a metabolic reaction or a linear segment of pathway. Subscript denotes the reaction or pathway.	mole/L^3·t
K_I	substrate or product inhibition constant	varies with inhibition kinetics
K_m	half-saturation constant of a reaction	mole/L^3
$K_{m,1}, K_{m,2}, K_{m,3}$	half saturation constant for substrate p_{1c}, p_{2c}, p_{3c} and for oxygen (O) in reaction 1, 2, 3	mole/L^3
K_s	half-saturation constant of substrate for growth	mole/L^3
k_K	killing rate constant	t^{-1}
m	maintenance coefficient, specific substrate consumption rate for maintenance	M/cell·t
N	total number of viable cells	cell
N_0	initial number of viable cells	cell
N_i	number of cells with value i of a distributed property	cell
n	viable cell number per volume	cell/L^3
O	oxygen concentration	mole/L^3
p	product concentration	mole/L^3 M/L^3
p_{1c}, p_{2c}, p_{3c}	intracellular metabolic intermediate concentration	mole/L^3
q_{EtOH}	specific ethanol production rate	mole/cell·t M/cell·t
q_{glc}	specific glucose consumption rate	mole/cell·t M/cell·t
q_p	specific product formation rate	mole/cell·t M/cell·t

q_s	specific substrate consumption rate	M/cell·t
R	universal gas constant	
r_1, r_2, r_3	reaction rate of reaction 1, 2, 3	mole/L^3·t
r_{max}	maximum reaction rate	mole/L^3·t
s	substrate concentration	mole/L^3
s_0	initial substrate concentration	mole/L^3
s_c	intracellular substrate concentration	mole/L^3
T	absolute temperature	K
t_d	doubling time	t
V	total volume of medium	L^3
V_b	total volume of biomass	L^3
v_x	specific cell volume	L^3/M
x	biomass concentration	M/L^3
x_0	initial biomass concentration	M/L^3
x_d	dead cell concentration	M/L^3
x_m	maximum cell density	M/L^3
$Y_{p/s}$	yield coefficient of product based on substrate	M/M
$Y_{x/s}$	yield coefficient of growth based on substrate	M/M
α	growth-associated specific productivity	mole/cell M/cell
β	non growth-associated specific productivity	mole/cell·t M/cell·t
θ	a variable that describes a cell property	
μ	specific growth rate	t^{-1}
μ_d	specific death rate	t^{-1}
μ_{max}	maximum specific growth rate	t^{-1}
μ_n	specific growth rate based on cell number	t^{-1}
ξ_i	concentration of cellular component i	M/cell
ϕ	cellular energetic state function	
γ	stoichiometry coefficient	

Problems

A. Basic Growth Kinetics

A1 The specific glucose consumption rates for a microorganism at different specific growth rates are shown in Table P.5.1. Assume that the yield coefficient on biomass is constant. What are the yield coefficient and specific maintenance coefficient?

A2 The biomass formula of a microorganism using glucose and NH_3 as carbon and nitrogen sources is $C_{4.4}H_{7.3}N_{0.86}O_{1.2}$, which accounts for 93% of dry biomass; the balance is ash. The respiratory quotient is 1.25.

Table P.5.1 Specific glucose consumption rate of a microorganism at different specific growth rates.

μ (h^{-1})	0.1	0.2	0.3	0.4
q_s (g/gDW-h)	0.12	0.14	0.16	0.18

Calculate the yield coefficient for the biomass, based on glucose and NH_3 (g cells/g substrate).

In a 1 L batch culture that has a cell concentration of 20 g/L and a doubling time of 1 h, how much glucose (in grams) has to be fed to the culture over the next hour in order to maintain the concentration of glucose at a constant level? Assume that glucose can be added in a very concentrated solution, so that the culture volume does not noticeably change.

A3 Figure P.5.1 is a graph of a microorganism's growth kinetics. Give a quantitative estimate of the specific growth rate and specific substrate consumption rate during this organism's exponential growth phase.

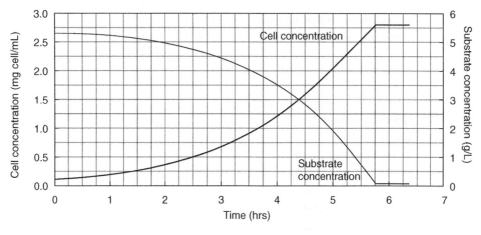

Figure P.5.1 Kinetics of cell growth and substrate consumption of a microorganism in culture.

A4 A bioreactor is used to grow a methanotroph by continuously bubbling methane into the culture. At a given point, the methane concentration is being maintained at 0.1 mmol/L in the culture broth and the cell concentration is 10 g/L. After a mechanical failure, the methane supply was halted, resulting in a linear reduction of the methane concentration to 0.01 mmol/L in 5 min. What is the specific methane consumption rate of the cell?

A5 Viral inactivation is a crucial step in the purification of recombinant proteins produced by mammalian cells. While microwave treatment can inactivate live viruses, it may also damage the protein product. Table P.5.2 shows the first-order inactivation rate constant for virus and protein, as a function of the input level of

Table P.5.2 Inactivation rate constant at different energy input levels.

Energy input	k_{virus} (1/s)	$K_{protein}$ (1/s)
1	1×10^{-3}	0.01×10^{-3}
10	5×10^{-3}	0.08×10^{-3}
100	25×10^{-3}	0.62×10^{-3}
1000	125×10^{-3}	2.08×10^{-3}
10,000	625×10^{-3}	1.08×10^{-3}

microwave energy. Mark the level you would use, and explain why. At that level, how long of an exposure time is needed in order to reduce the virus content by 10,000-fold?

B. Kinetic Model for Growth

B1 An experiment was carried out to measure a cell's growth rate at different glucose concentrations. The results were plotted in a standard double reciprocal plot (Figure P.5.2). The behavior is obviously not typical Monod-type kinetics.
Explain the results, and propose an alternative equation for this cell's growth dependence on glucose.

B2 A sample of algae uses light as its energy source and carbon dioxide as its carbon source. With abundant carbon dioxide in the culture, its growth rate increases with light intensity until it reaches a maximum growth rate. However, as the intensity becomes too high, the light bleaches the chloroplasts and the growth rate declines. Write a possible model that describes the dependence of specific growth rate (μ) on light intensity (I).

B3 A scientist has isolated a new unicellular microorganism! Most microorganisms grow faster with a shorter doubling time (t_d) when the environment is favorable.

Figure P.5.2 Double-reciprocal plot of the specific growth rate and substrate (glucose) concentation.

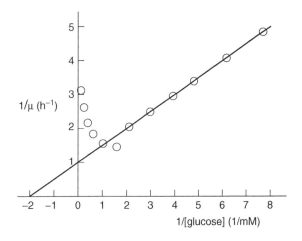

However, this new creature behaves very differently. Instead of dividing more frequently, it keeps the same cell division time, but divides into three daughter cells as opposed to two daughter cells for each round of cell division. In other words, after a generation time t_g, under suboptimal conditions it divides into two cells; but under optimal conditions, it divides into three cells.

a) Use N as the cell number, N_o as the initial cell number, and t_g as the cell division time. Write down the differential equations to describe the change of N due to growth under both conditions. Plot N/N_o versus t/t_g (where t is time) from $N=N_o$ until $N=100\,N_o$. Assume that the size of the daughter cell after cell division is the same in both cases.

b) What is the ratio of their doubling time (the time interval in which cell number or cell biomass increases twofold) under the two conditions?

B4 Table P.5.3 shows a set of experimental data on yeast growth kinetics. Plot the data to show the time course of kinetics, including cells, substrates, and specific rates.

B5 Methanol is used to grow a bacterium. The dependence of the specific growth rate on methanol concentration follows Monod kinetics with substrate inhibition, as shown here:

$$\mu = \frac{\mu_{max}s}{K_S + s + K_I s^2}$$

Table P.5.3 Concentration profiles of a yeast batch culture using glucose as the carbon source.

Time (h)	Cell conc. (g/L)	Glucose conc. (g/L)
0	0.6	8
1	0.7	7.9
2	0.76	7.71
4	0.86	7.6
6	1	7.4
8	1.56	6.45
10	2.1	5.14
12	2.5	4.04
14	3.24	3.54
16	3.76	2.10
18	4.00	1.23
20	4.20	0.84
24	5.00	0.32
28	5.40	0.17
30	50	0.05
40	5.60	0
60	5.80	0
80	5.50	0

Consider a scenario where the μ_{max} is 0.693 h^{-1}, $K_S = 0.2$ g/L, and $K_I = 0.5$ L/g. To avoid inhibiting cell growth, methanol is fed continuously to keep its concentration below the inhibitory range. In this case, it will be kept at five times of K_S. The elemental composition of the yeast is 48.6% carbon, 9.1% nitrogen, 33.7% oxygen, 1.3% phosphorus, and 0.6% sulfur. The yield coefficient based on methanol is 0.6 g dry mass/g methanol.

    a) Design a minimal medium to generate a final yeast cell concentration of 10 g/L. You can use NH_3, K_2HPO_4, and $MgSO_4$ as the sources of N, P, and S, respectively. Discuss other nutrients that are not included in the elemental composition that may need to be included in the medium.

    b) You are starting a culture at an initial cell concentration of 0.5 g/L and a methanol level of 1 g/L. You also want to grow cells at their maximal growth rate as long as possible. Calculate the initial feeding rate of methanol. Derive an equation that can be used to feed methanol (in g methanol/L culture-h).

B6 A yogurt drink product is pasteurized using a batch reactor. The initial bacterial count is 10^7 cells/mL, which can be reduced to 10^3 cells/mL in 30 sec at 80 °C. The heating and cooling are so rapid that the contribution of the heating and cooling period to killing can be neglected. It is now decided to change the process to a continuous operation operated at a higher temperature that has a killing rate constant that is three times higher. You are given a 50 cm long tube with a 2 cm diameter for this purpose. Again, assume that heating and cooling are rapid. What flow rate should you use to achieve the same degree of reduction in bacterial count?

C. Advanced Problems

C1 Cells taken from the skin of a patient can be cultivated *in vitro* to expand their number and then be transplanted back into the patient to cover an area of the body where the skin has been damaged. These cells grow on the surface of Petri dishes to form a patchy monolayer (one single layer of cells on the surface) where only cells that are located at the periphery grow – the cells located on the interior are "contact inhibited." Consider the case that there are N_0 cells initially that form a single patch. The shape of the patch can be considered to be circular. There are plenty of nutrients, and the growth rate is not limited by substrate concentration. The doubling time is t_d for those cells that are on the periphery and dividing. Write an expression for total cell number (N) in terms of N_0, t_d, and t (time). Sketch a plot of the growth curve (N/N_0 vs. t/t_d) for the following values of N/N_0: 1, 2, 4, 16, and 64. How does the apparent specific growth rate ($dN/dt = \mu_{app}N$) vary with time? Write down the expression for μ_{app} as a function of time.

C2 For the production of rDNA proteins using *E. coli*, it is important to attain a very high cell concentration in the bioreactor to ensure a high level of productivity. However, at a very high cell concentration, the oxygen transfer capacity of the reactor cannot support the maximal growth rate. The specific growth rate of *E. coli* is dependent on both glucose and oxygen concentration, as described by Monod kinetics. The K_s for glucose and oxygen are 10 μM and 5 μM, respectively. While the K_s and yield ($Y_{x/s}$) are independent of temperature, the μ_{max} decreases with temperature. At 28 °C, the μ_{max} decreases to 60% of what it was at 37 °C. The μ_{max} at 37 °C is $2\,h^{-1}$, and $Y_{x/glucose}$ and $Y_{x/O2}$ are 0.4 g/g and 0.7 g/g, respectively.

C3 A production strategy is to reduce the temperature to 28 °C once the cell concentration reaches the critical value at which oxygen supply becomes limiting. From then on, the oxygen concentration will be maintained at 80 μM, and glucose will be fed continuously according to a prescribed algorithm so that the oxygen consumption rate remains constant.

C4 Derive an expression to show how the specific growth will change as a function of time after switching the temperature to 28 °C. Then, derive an expression of glucose concentration, and glucose feeding rate versus time for sustaining such a growth-rate profile when glucose is used as the growth rate-limiting substrate.

C5 A recombinant bacterium is transformed with a Bio100 plasmid. The transformant was selected by antibiotic resistance using LB medium supplemented with ampicillin, and has been shown to produce compound A in LB medium. When it grows in LB medium without ampicillin, this bacterium loses the plasmid gradually and, along with it, the ability to produce A. Cells that do not have to replicate the plasmid and to produce A grow faster than those that have retained the plasmid. Cells that have lost the plasmid also have a net energy saving, so they have a higher yield coefficient. The relevant parameters are listed here:
For the transformed cells (carrying the plasmid):

$$\mu_{max} = 0.69\ h^{-1}$$
$$Y_{x/s} = 0.4\ g\ dry\ cell\ mass/g\ glucose$$
$$K_s = 0.1\ g\ glucose/L$$

The frequency of losing the plasmid (the specific "plasmid segregation rate") follows first-order kinetics with respect to plasmid-containing cells. The specific plasmid segregation rate is $k_1 = 0.069\,h^{-1}$.
For the plasmid-free cells:

$$\mu'_{max} = 0.92\ h^{-1}$$
$$Y'_{x/s} = 0.5\ g\ dry\ cell\ mass/g\ glucose$$
$$K'_s = 0.1\ g\ glucose/L$$

A batch culture is started with pure transformed cells (i.e., all of them have the plasmid) at 1 g cell/L. You are asked to produce the maximum quantity of cells using an LB medium containing 20 g/L of glucose but without ampicillin. You can assume that at glucose concentrations greater than 15 times K_s, the growth rate is practically the same as the maximum growth rate. When should you harvest the cells? What percent of the cells harvested will have plasmid?
What can you do to change the process to grow a cell population that can be assured of a pure population of cells with plasmid?

C6 Sometimes it is desirable to control the growth rate in a cell culture at a low level to prevent excessively fast metabolism. In a study that you are designing, you plan to use the strategy of limiting nutrient supply to restrict the growth rate. You want to allow cell number increase linearly over time by controlling the substrate level. The growth kinetics of the cell is described by Monod kinetics. The culture is started at a cell concentration of x_0, an initial volume of V_0, and a limiting substrate concentration of s_0. You are to feed a concentrated limiting nutrient at concentration s_f. The concentration s_f is very high, so the amount of feed does not change the culture volume appreciably. Develop an expression that gives the feeding rate, F, as a function of time that can be used to accomplish this. The conversion yield of substrate to cell mass is Y. Plot the concentrations of cells, substrate, the specific growth rate, and feeding rate over time. You do not need to obtain an analytical solution. You can express your results in terms of $x(t)$ and $s(t)$, but you must be clear on your reasoning and reflect your solution in the plot.

C7   One of the advantages of continuous sterilization over batch sterilization is that the destruction of heat-sensitive nutrients can be minimized. The strategy is to use an HTST (high-temperature short-time) process that rapidly increases the temperature to a high holding temperature for a shorter time in the sterilizer and quickly cools it down afterward. It may be decreased by increasing the holding temperature.
Given the thermal killing rate constant (k_K) for bacterial spores and the thermal destruction rate constant (k_v) for an essential vitamin, discuss why temperature affects the relative destruction of the spores and vitamin. Note that both k_K and k_v increase exponentially with increasing temperature as described by the Arrhenius equation. Discuss three conditions: when the activation energy for the destruction of the nutrient is (1) greater than, (2) equal to, or (3) smaller than that of the destruction of spores.

C8   A batch bioreactor is to be sterilized with heat. The volume is 1500 L, and the initial contamination level is 5×10^6 microorganisms/mL. Table P.5.4 shows the temperature profile during heating and cooling has been obtained along with the thermal killing rate constant determined at some temperature. The sterilization is to reduce the viable bacterial count in the reactor to at most 0.01. Is the holding time of the sterilization process long enough to achieve the sterilization criterion? (The degree of killing can be determined by integrating the killing rate constant over time. Note that at low temperatures, the killing is negligible.)
The medium contains a vitamin that has to be present at 15 mg/L (MW = 346) to support cell growth. This vitamin is heat labile. The specific decomposition rate is 0.012 min^{-1} at 120 °C, and the activation energy of the decomposition is 26 kcal/mol. How much vitamin do you have to add to the medium before the sterilization to make sure there is a sufficient quantity after sterilization?

C9   A bacterium can be infected by a virus. The genome size (i.e., the DNA content) of the virus is only 0.05% of the bacterium, and the mass of the virus is an even smaller fraction of bacterial biomass. After the bacterium is infected, it ceases growing while the DNA and protein synthesis machinery are taken over by the virus. It takes 10 min to make 100 copies of virus DNA and another 5 min to make the viral proteins after a bacterium is infected. After the completion of

Table P.5.4 Temperate and death rate constant profile during heat sterilization.

Time (min)	Temperature (°C)	Specific death rate (1/min)
0	30	Negligible
5	45	Negligible
10	60	Negligible
15	70	Negligible
20	80	Negligible
25	88	Negligible
30	95	0.017
35	100	0.041
40	105	0.098
45	110	0.22
50	114	0.50
55	118	1.15
60	120	1.77
65	120	1.77
90	120	1.77
95	120	1.77
100	117	0.98
110	103	0.071
120	90	Negligible
130	76	Negligible
140	62	Negligible
150	50	Negligible
160	42	Negligible
170	33	Negligible

viral replication, the infected bacterium is lysed, and 100 newly made viruses are released to infect other not-yet-infected bacteria. Each virus, after being released from the lysed cell, attaches to another uninfected bacterium and, after 2 min, enters the newly infected bacterium and disappears from the culture medium. The growth kinetics of uninfected bacteria are not affected by other infected cells. You are asked to construct a model to describe the dynamics of such a bacterium–virus system. You can assume that there are plenty of nutrients for bacterial growth and that the bacterium is essentially growing at the maximum rate supported by the medium throughout the infection period. The infection process is over when all the cells haven been infected and lysed, and the viruses are released.

First, list the state variables that are needed to describe the system. Then, write down their balance equations and the initial conditions. You can assume that only one virus can infect one bacterium, and the efficiency of infection is 100%. In other words, once viruses are released from an infected cell, it will find another bacterium to infect unless there are no more uninfected bacteria to be infected.

C10 You are asked sterilize (by heat) 1 L medium containing bacterium at 10^5 cells/mL. The bacterium carries a plasmid that encodes a heat shock protein. The expression of heat shock protein protects the cell from high-temperature stress and delays thermal death. However, because the cells can carry different numbers of plasmids, there is a range of susceptibility to heat treatment, as shown in Table P.5.5.

Table P.5.5 Percentage of bacteria carrying different number of plasmids.

No. of plasmids	Bacteria in the population (%)	k (thermal death rate constant) (s^{-1})
1	2	6
2	80	2
3	10	1
4	7	0.5
5	1	0.1

You are asked to achieve a sterilization criterion of 10^{-2} (with a survival rate of 0.01 cell in the 1 L medium). What is the sterilization time (the length of time held at the sterilization temperature) that you will use?

C11 A microorganism capable of utilizing hexadecane was isolated from an oil well. It produces an emulsifier to stabilize the hydrocarbon substrate as droplets. When it adheres to the surface of an emulsified droplet, it grows at a maximum specific growth rate with a μ_{max} and a yield coefficient Y. This organism normally grows adhered, but when the cells reach a maximum coverage on the surface of the droplet [denoted as χ (gram cells/cm^2)], they begin to shed into the aqueous phase. Because of the low solubility of the substrate, the growth rate when they are suspended in the aqueous phase can be assumed to be negligible.

Consider a case with an initial substrate concentration of S_o and substrate density of ρ_s. Enough emulsifier is to be added so that the substrate can be dispersed as uniform droplets of diameter (D_o) from the beginning of the experiment. Also, an inoculum cell concentration (X_o) is to be used so that all the droplets have cells on them. Assume that the droplets can be stabilized so that the total number of droplets does not change. The droplet diameter decreases over time due to substrate consumption. Plot the concentration of cells and substrate, and the diameter of the droplets as a function of cultivation time.

6

Kinetics of Continuous Culture

6.1   Introduction

Many industrial manufacturing processes are operated in a continuous mode. This means that except for a short startup period, the process runs continuously with steady inputs and outputs over an extended time period (Figure 6.1).

A batch mode process is subject to frequent cycles of startup, production, product recovery, and reactor cleanup. In contrast, a continuous process has constant through-put. For this reason, a continuous operation uses a reactor more efficiently than a batch process. Continuous cultures, like other continuous processes, have potential for higher productivity than batch cultures. They are also versatile research tools for studying cell growth kinetics and population dynamics (Panel 6.1).

Panel 6.1 Advantages of Continuous Processes
• Reduced process equipment turnaround time • Higher throughput • Steady operation

A continuous culture typically begins as a batch process. When the cell concentration reaches a level near the desired steady state, the medium flow is started. The culture fluid is withdrawn to maintain a constant volume in the bioreactor (Panel 6.2). Through this process, the concentrations of cell, nutrients, metabolites, and product will gradually become constant. At this point, the system is considered to be at a "steady state." The culture can be maintained at the steady state over a long time, and reliably produce a desired product.

Panel 6.2 General Operation of Continuous Culture

• Initiate the culture in a batch mode.
• Once approaching anticipated steady-state cell concentration, start continuous flow.
    • Keep feed and effluent flow rates the same.
    • Maintain constant volume.
• Reach steady state (at least 3 holding times).

Engineering Principles in Biotechnology, First Edition. Wei-Shou Hu.
© 2018 John Wiley & Sons Ltd. Published 2018 by John Wiley & Sons Ltd.
Companion Website: www.wiley.com/go/hu/engineering_fundamentals_of_biotechnology

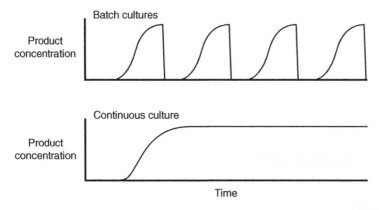

Figure 6.1 Comparison of batch and continuous processes.

In practice, it is not easy to keep every process parameter at a constant value operationally. The medium flow rate may fluctuate, causing the reactor volume to change; or the pH monitoring may suffer from noise interference, causing pH control actions to deviate from the norm. These fluctuations may cause the cell growth to vary.

Most continuous cultures keep the influx and output flow rates equal and constant to maintain the reactor volume constant and to sustain a steady-state operation. A steady state of a continuous culture can also be controlled by other means. For example, an on-line turbidity sensor can be used to measure the cell concentration of a continuous culture. The flow rates are then adjusted to maintain the cell concentration at a constant level. Such a system is called a turbidostat. When the chemical composition (especially the rate-limiting nutrient) is measured and kept at a steady-state level, the system is called a chemostat.

Continuous culture has become a very useful research tool for understanding cellular kinetic behavior because, at a steady state, the relationship between cell growth rate and the limiting nutrient concentration can be clearly defined. This makes it a useful tool for determining the kinetic equation for cell growth. Continuous cultures are also used to observe the relationships, such as competition and complementation, between different microorganisms in a mixed population over time.

In the mid-twentieth century, many chemical processes were switched from batch to continuous processes. Continuous operations allow industrial plants to have a high, sustained output over a long period of time. With their increased productivity and decreased equipment downtime, continuous cultures are an attractive choice for manufacturing bioproducts.

However, it should be noted that for industrial manufacturing using fermentation and cell culture processes, fed-batch and batch operation are still the predominant methods. In many processes, the bulk of product is produced only after a period of rapid growth has occurred. The production process thus has two segments: (1) a cell expansion period of rapid growth to reach a high cell concentration in the bioreactor, and (2) a production period in which the culture conditions are switched to favor product formation (such as by changing temperature, pH, or nutrient composition). Such separation of the cell growth period and product formation stage makes a continuous process less attractive.

Furthermore, the risks of mechanical failure and microbial contamination increase exponentially with a prolonged operating period (Panel 6.3). Also, in some cases, the productivity of the producing organism decreases over a long culture period. In other cases, the quality of the product (especially the flavor) may change in a continuous process. In a long-term operation, there may also be other changes over time, such as a drift in the glycosylation pattern of a protein.

Panel 6.3 Drawbacks of Continuous Cultures

- Microbial contamination
- long-term mechanical reliability of equipment
- Instability of producing organisms
- Possible change of product quality (e.g., beer, yogurt, or glycosylation profile of therapeutic proteins)

Continuous culture is often used for processes that have to manage a very large-volume throughput, such as in the treatment of municipal waste (Panel 6.4). It is also commonly used when the product is labile and subject to degradation in a batch culture. However, in many of those operations, cell recycling or a multistage operation is used instead of a simple continuous culture system. In this chapter, we will first describe the continuous culture system. Then, we will discuss why modified continuous cultures are employed.

Panel 6.4 Where Are Continuous Cultures Used?

- Biocatalysis using immobilized enzyme/microbial cells
- Very high-throughput process (e.g., waste treatment)
- Low substrate concentration
- Labile product
- Very low-productivity process

6.2 Kinetic Description of a Continuous Culture

6.2.1 Balance Equations for Continuous Culture

In this chapter, our discussion will center on an idealized, continuous, well-mixed bioreactor. The system entails a stirred-tank bioreactor with a continuous influx of nutrients (referred to as the "feed stream"), as well as a continuous effluent stream to remove spent medium and cells from the bioreactor (Figure 6.2). The "well-mixed" assumption states that the substrate from the feed stream is instantaneously mixed with the content of the bioreactor, and that the effluent stream has the same composition as what is in the bioreactor. We will also assume that the flow rates of the feed and effluent streams are the same (i.e., $F_i = F_o = F$) to keep the volume of the reactor constant. Furthermore, the

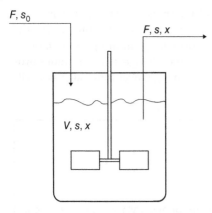

F, s_0

F, s, x

V, s, x

Figure 6.2 A continuous culture system.

specific density of the fluid is kept constant, so we can perform material balance based on the culture fluid volume, instead of mass.

The material balance equation for cells as described in Eq. 6.1, Panel 6.5 is a balance of (1) the amount of cell mass being carried into and out of the reactor, and (2) the change caused by cell growth and cell death. To account for the dead cells, a second equation is written. It describes the balance of dead cells carried into and out of the reactor, and the increase arising from the death of viable cells (Eq. 6.2). In subsequent discussions, the volume of the reactor will be kept constant. This allows us to perform material balance on concentrations, rather than the total quantities in the reactor. In mammalian cell processes, the nonviable cells often constitute a significant portion of the cell population. In most microbial processes, the dead cell concentration is often negligible. The biomass balance equation for the case that cell death is negligible is shown in Eq. 6.3.

Panel 6.5

$$\frac{dVx}{dt} = Fx_0 - Fx + \mu xV - \mu x \tag{6.1}$$

Reactor volume is kept constant:

$$\frac{dx_d}{dt} = \frac{Fx_{do}}{V} - \frac{Fx_d}{V} + \mu_d x \tag{6.2}$$

$$\frac{dx}{dt} = \frac{Fx_0}{V} - \frac{Fx}{V} + \mu x \tag{6.3}$$

$$\frac{ds}{dt} = \frac{Fs_0}{V} - \frac{Fs}{V} - \frac{\mu x}{Y_{x/s}} \tag{6.4}$$

The substrate concentration in a continuous culture is the balance of the amount of substrate fed into the reactor minus that withdrawn by the outflow streams and that consumed by cells to produce cell mass (Eq. 6.4). The amount of substrate used for "maintenance energy" (as discussed in Chapter 5) is assumed to be negligible. If a product is also formed, the amount of substrate used to synthesize the product should be considered. The equation describing the concentration of product can be similarly written, including terms for the input and output flow streams and the production rate.

We will first focus on cell growth in a continuous culture with high viability (i.e., assuming no dead cells) and no product formation. The system is thus described by Eqs. 6.3 and 6.4 (Panel 6.5). We also consider the case that the feed stream contains no cells (Panel 6.6). We define the ratio of the flow rate to the culture volume (F/V) as the dilution rate, D, which is the inverse of the holding time in the bioreactor. At steady state, the differential terms on the left-hand side of the equations become 0. After substituting F/V with D, it can be seen that, for the conditions wherein the cell concentration is not 0 (i.e., $x > 0$), the dilution rate and the specific growth rate must be equal.

Panel 6.6

Set $x_0 = 0$.
  Define dilution rate, $D = F/V$.
  At steady state:

$$0 = -\frac{F}{V}x + \mu x = (\mu - D)x \qquad \therefore \ \mu = D$$

$$0 = D(s_0 - s) - \frac{\mu x}{Y_{x/s}} \qquad x = \frac{D(s_0 - s)}{\frac{\mu}{Y_{x/s}}} = Y_{x/s}(s_0 - s)$$

$$x = Y_{x/s}(s_0 - s) \tag{6.5}$$

The equality $\mu = D$ is an important characteristic of a continuous culture at steady state. It allows one to maintain the culture at a constant growth rate, as long as the same steady state is maintained. This enables us to investigate the effects of growth rate on different cell behaviors.

From the substrate balance, we see that the steady-state cell concentration, x, is determined by the yield coefficient and the substrate concentration difference between the influx and outflow (Eq. 6.5, Panel 6.6).

6.2.2 Steady-State Behavior of a Continuous Culture

6.2.2.1 Monod Kinetics
In Chapter 5, we discussed using a model to describe the dependence of the specific growth rate of a microorganism on substrate concentration. Consider the case that this relationship is described by the Monod model. At steady state, μ and D are equal. Thus, the Monod relationship is extended to D in a simple continuous culture. Also in this case, the Monod model depicts the growth rate being limited by a substrate, s. This predicts that the residual substrate concentration (i.e., the substrate concentration in the bioreactor) increases with the growth rate or the dilution rate until a certain level of saturation. At this saturation level, the specific growth rate is at its maximum under the given culture conditions. In the Monod model, the substrate s is called the rate-limiting substrate. In these scenarios, it is implicitly assumed that all other required substrates are in excess.

6.2.2.2 Steady-State Concentration Profiles
To examine how cell and substrate concentrations change over different dilution rates, we rearrange the steady-state equations for cells and substrates. First, we begin with the case in which no product is formed.

After substituting μ with D in the Monod equation and rearranging, one obtains the relationship between the dilution rate and substrate concentration (Eq. 6.6, Panel 6.7). This equation can be used to plot the substrate concentration over different dilution rates (Figure 6.3).

Panel 6.7 Using Monod Model to Describe Growth Kinetics

At steady state, μ and D are equal:

$$\mu = \frac{\mu_{max}s}{K_s + s} = D$$

Rearranging:

$$s = \frac{DK_s}{\mu_{max} - D} \tag{6.6}$$

Substituting s in Eq. 6.5 with Eq. 6.6:

$$x = Y_{x/s}(s_0 - s)$$

$$x = Y_{x/s}\left(s_0 - \frac{DK_s}{\mu_{max} - D}\right) \tag{6.7}$$

When $s_0 \gg s$, $x \sim Y_{x/s}\, s_0$.

To obtain an equation relating cell concentration to the dilution rate, we can use the material balance equation of the substrate. By rearranging and substituting s with Eq. 6.6 (Panel 6.7), we can obtain the expression of cell concentration (x) as a function of the dilution rate (D) (Eq. 6.7, Panel 6.7). The corresponding plot is shown in Figure 6.3.

As expected, the concentration of the limiting substrate (s) increases with increasing dilution rate. Over a wide range of dilution rates, the steady-state cell concentration stays relatively constant as the dilution rate increases. In this range of dilution rates,

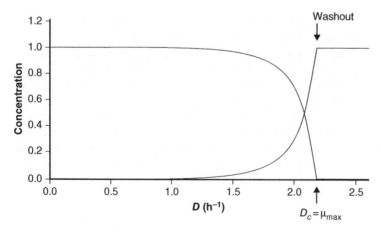

Figure 6.3 Steady-state biomass and limiting substrate concentrations in a continuous culture at different dilution rates.

the substrate concentration in the feed is much greater than the concentration in the reactor ($s_o \gg s$). So, the cell concentration is relatively constant. It can be approximated by the multiplicative product of the yield coefficient and the feed substrate concentration ($Y_x/_s s_o$). As the dilution rate increases to approach the maximal specific growth rate and s becomes large, the steady-state cell concentration begins to decrease (with a corresponding increase of the dilution rate).

Since $\mu = D$, the substrate concentration curve is basically a transposition of the μ versus s curve for the Monod model, in the range $s = s_0$ to $s = 0$ (Figure 6.3). Since $D = \mu_{\max}$ is the point at which the cell concentration becomes zero, this dilution rate is referred to as the critical dilution rate.

6.2.2.3 Washout
At a dilution rate that is higher than the cell's maximum specific growth rate, the rate of biomass generation is not able to replenish the cells being washed out of the reactor. Eventually, all cells will be washed out and the cell concentration will reach zero. In this case, since substrates are no longer being used, their concentrations within the reactor become the same as the feed concentration. The critical dilution rate is the maximum dilution rate at which the system can be operated with cells in the reactor. Beyond this point, cells are washed out and the biomass concentration is zero.

At the critical dilution rate the cell concentration is zero. By rearrangement, one can obtain a relationship between the maximum specific growth rate and the critical dilution rate (Panel 6.8). In most cases, the feed substrate concentration is much higher than K_s ($s_0 \gg K_s$). The only condition that the balance equation for $x = 0$ has a solution is $D = \mu_{\max}$. Therefore, the critical dilution rate equals the maximum specific growth rate. We will limit our discussion to only the regions are below washing out.

Panel 6.8 Critical Dilution Rate

At washout $x = 0$:

$$0 = Y_{x.s} \left(s_0 - \frac{DK_s}{\mu_{\max} - D_c} \right)$$

Rearrange:

$$s_0 = \frac{D_c K_s}{\mu_{\max} - D_c}$$

$$\mu_{\max} s_0 - D_c(s_0 - K_s) = 0$$

$$s_0 \gg K_s$$

$$D_c = \mu_{\max}$$

The Monod model defines a unique relationship between μ and s, as seen in Figure 6.3. At a given μ, there is only one corresponding s. Since $\mu = D$ at steady state, each D has also only one corresponding s value. If the feed substrate concentration changes, the residual substrate concentration at a given dilution rate will still be the same, but the cell concentration (which is determined by the feed and the yield coefficients) will vary (Figure 6.4). In regions that are higher than the washout dilution rate, the residual substrate concentration will be the same as the feed.

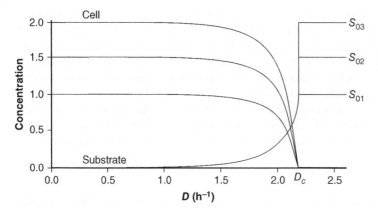

Figure 6.4 Steady-state cell and limiting substrate concentrations at different feed substrate concentrations. Note that all different feed concentrations give the same substrate concentration curve in the reactor until the washout dilution rate. Beyond the washout dilution, the residual substrate concentration is the same as the feed (not shown).

6.2.3 Productivity in Continuous Culture

There have been extensive studies on continuous cultures of bacteria, yeast, and animal cells. In the middle ranges of the dilution rate, experimental data on the relationships between cell, substrate concentrations, and dilution rate show a general agreement with Monod model simulations. In comparison, fewer studies have examined these relationships in the low ranges of the dilution rate. In those regions, the substrate concentration is low, and accurate measurement becomes challenging. Also, operating a continuous culture in a higher dilution rate region is preferable because it has a higher throughput.

In the case that cell mass is the product of interest in a continuous process, the throughput is the multiplicative product of cell concentration and the flow rate (xF). Here, we will explore the case of a continuous culture bioreactor to determine which dilution rate will yield the highest throughput. Since $D = F/V$, we would like to find out at which D the output (xD: cell mass product/unit volume per hour) will be maximal (Panel 6.9). The cell concentration in culture is described with the equation derived from the substrate balance.

Panel 6.9

The throughput is the multiplicative product of cell concentration and dilution rate:

$$xD = DY_{x/s}\left(s_0 - \frac{DK_s}{\mu_{max} - D}\right) \qquad (6.8)$$

To find max(xD), we find the value of D at which dxD/dD is 0:

$$D_m = \mu_{max}\left(1 - \sqrt{\frac{K_s}{K_s + s_0}}\right) \qquad (6.9)$$

When we incorporate the cell concentration expression into xD (Eq. 6.8, Panel 6.9), the maximum throughput is where the derivative, with respect to D, is 0 (Eq. 6.9, Panel 6.9).

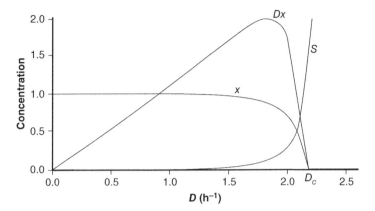

Figure 6.5 The throughput of biomass in a continuous culture.

It can be shown that maximum throughput occurs very closely to the critical dilution rate (somewhat higher than $0.9\ \mu_{max}$ when $K_s \sim 0.01\ s_0$). Beyond this, the throughput decreases sharply due to increased substrate concentrations in the reactor. A higher residual substrate concentration also demonstrates that a lower amount of substrate is being used to generate biomass.

In most cases, $s_0 \gg K_s$, so the term $K_s\ /\ (K_s + s_0)$ is very small. The dilution rate, which gives the maximum productivity (D_m), is close to μ_{max}, as illustrated in Figure 6.5.

For most processes, the products of interest are secreted metabolites, rather than biomass. For product, we can add a balance equation consisting of a production term and a flow-out term (Panel 6.10). In this case, the substrate consumption includes contributions from substrates that are utilized for both biomass generation and product formation.

Panel 6.10

$$\frac{dx}{dt} = (\mu x - Dx) - \mu_d x \tag{6.10}$$

$$\frac{dx_d}{dt} = \mu_d x - Dx_d \tag{6.11}$$

$$\frac{ds}{dt} = D(s_0 - s) - \frac{\mu x}{Y_{x/s}} - \frac{q_p x}{Y_{p/s}} - mx \tag{6.12}$$

$$\frac{dp}{dt} = -Dp + q_p x \tag{6.13}$$

In general, one considers that dead cells do not metabolize substrate. The product formation term, $q_p x$, may be affected by the dilution rate (or the specific growth rate) depending on whether the specific production formation rate (q_p) is non–growth associated, positively growth associated, or negatively growth associated. The relationship between product concentration, productivity, and dilution rate depends on the functional relationship between the specific growth rate and the specific productivity (Eq. 6.14, Panel 6.11).

Panel 6.11

Assume that the feed stream does not contain cells or product (i.e., $x_0 = p_0 = 0$); neglect maintenance energy:

$$0 = (\mu - D - \mu_d)x$$

$$0 = D(s_0 - s) - \frac{\mu x}{Y_{x/s}} - \frac{q_p x}{Y_{p/s}}$$

Then:

$$\mu - \mu_d = D \qquad\qquad (6.14)$$

$$x = \frac{D(s_0 - s)}{\dfrac{\mu}{Y_{x/s}} + \dfrac{q_p}{Y_{p/s}}} \qquad\qquad (6.15)$$

$$p = \frac{q_p x}{D} \qquad\qquad (6.16)$$

Example 6.1   A Continuous Culture

A food-processing plant generates a waste stream at $1\ \mathrm{m^3/h}$. The waste that contains glucose at 5 g/L will be used to grow yeast cells for use in another operation in the plant. Using this waste stream as the medium, the yeast cells grow at a maximal specific growth rate of $0.69\ \mathrm{h^{-1}}$. Its yield coefficient is 0.5 g/g glucose. The doubling time increases to 2 h if the glucose concentration is reduced to 0.1 g/L. Using a continuous culture that is operated at a steady state with a dilution rate of $0.62\ \mathrm{h^{-1}}$, what is the daily output of yeast cells? What is the size of the reactor?

Solution

The relevant parameters are:

$$\mu_{max} = 0.69\ \mathrm{h^{-1}} \text{(the doubling time is thus 1 h)}$$

$$Y_{x/s} = 0.5\ \mathrm{g/g}$$

$K_s = 0.1$ g/L (as the doubling time increases to 2 h, the specific growth rate is $0.345\ \mathrm{h^{-1}}$, which is half of μ_{max}; the substrate concentration at $0.5\ \mu_{max}$ is K_s)

The operating dilution rate is:

$$D = 0.62\ \mathrm{h^{-1}}$$

$$F = 1\ \mathrm{m^3/h}$$

Since $D = \mu$ at steady state, $\mu = 0.62\ \mathrm{h^{-1}} = \frac{0.69\ \mathrm{h^{-1}} \cdot s}{0.19/L + s}$.

Solve for s (or see Eq. 6.6):

$$s = \frac{0.629 \cdot 0.1}{0.69 - 0.62} = 0.88\ \mathrm{(g/L)}$$

The cell concentration in the reactor and the output stream is:

$$x = Y_{x/s}(s_0 - s) = 0.5\ \mathrm{g/g}((5 - 0.88)\ \mathrm{g/L}) = 2.06\ \mathrm{g/L}$$

The output per day is $Fx = 1 \text{ m}^3/\text{h} \cdot 2.06 \text{ g/L} \cdot 10^3 \text{ L/m}^3 \cdot 24\text{h/day} = 49.4 \text{ kg/day}$.
The culture volume in the reactor is:

$$D = F/V \qquad V = F/D = (1 \text{ m}^3/\text{h})/0.62 \text{ h}^{-1} = 1.61 \text{ m}^3$$

6.3 Continuous Culture with Cell Recycling

6.3.1 Increased Productivity with Cell Recycling

One drawback of a simple, continuous, stirred-tank bioreactor is that its maximum throughput is constrained by the washout dilution rate (Panel 6.12). This is especially unfavorable when the achievable cell or product concentration is low, which may be caused by a low feed substrate concentration or low specific productivity.

Panel 6.12 Shortcomings of Simple Continuous Culture

- Feed substrate concentrations are often low in biological processes.
  - This results in low cell concentrations and low throughput.
- In waste treatment, the effluent substrate concentration must be low to meet environmental requirements.
  - Low substrate concentration is accomplished only when specific growth rate is low.

In many bioremediation treatment processes, the discharge from the bioreactor must have a low residual substrate concentration in order to meet environmental regulation standards. In a simple continuous culture, the low substrate concentration is accompanied by a low dilution rate, as described by the Monod model (wherein a low specific growth rate, and thus the dilution rate, keeps substrate concentration low). Therefore, using a simple continuous bioreactor for bioremediation will result in a low throughput.

To improve productivity, a cell-recycling system can be incorporated into the bioreactor. This includes a cell-separating or -concentrating device (Figure 6.6). Instead of discarding the cells into the effluent stream, the cell mass can be separated and a portion of it can be returned to the bioreactor. This results in a higher cell concentration inside of

Figure 6.6 A continuous culture with cell recycling.

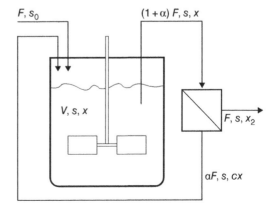

the bioreactor and allows the flow rate to be higher than the maximum specific growth rate. The cell separation device may be a centrifuge, a cell filtration apparatus, a settling tank, or a means of inducing cell agglomeration for settling.

When using such a cell separation device, the stream coming out of the bioreactor is separated into two streams: one with a higher cell concentration, for recycling back into the bioreactor; and one with a lower cell concentration, for purging out of the system and for product recovery. For a steady-state operation, a purging stream for cell removal is necessary. Without cell purging, the cell concentration will continue to increase, unless the specific growth rate is zero or is balanced out by cell lysis.

A schematic of a continuous culture with a cell-recycling system is shown in Figure 6.6. To analyze the system, we perform material balance on the cells and rate-limiting substrate (Panel 6.13). Importantly, the dilution rate is defined as F/V, as it is in the simple continuous culture. F is the flow rate into the entire system, but not the flow entering or exiting the bioreactor. The flow stream into the bioreactor is a combination of F and the recycled flow.

Panel 6.13

Balance on biomass for the bioreactor is:

$$V\frac{dx}{dt} = \alpha Fcx - (1+\alpha)Fx + \mu xV \tag{6.17}$$

Balance on the cell recycle system gives:

$$(1+\alpha)Fx = \alpha Fcx + Fx_2$$

$$(1+\alpha-\alpha c)Fx = Fx_2 \tag{6.18}$$

$$\frac{x_2}{x} = 1+\alpha-\alpha c \qquad c>1 \tag{6.19}$$

The balance on the substrate on the bioreactor:

$$V\frac{ds}{dt} = Fs_0 - \frac{\mu x}{Y}V - F(1+\alpha)s + \alpha Fs \tag{6.20}$$

Defining $F/V = D$. Applying steady-state conditions to Eqs. 6.17 and 6.20.

$$0 = \alpha Dcx - (1+\alpha)Dx + \mu x$$

$$\frac{\mu}{D} = 1+\alpha-\alpha c \tag{6.21}$$

$$c>1, \text{ so } D>\mu$$

$$0 = Fs_0 - \frac{\mu x}{Y_{x/s}}V - Fs$$

$$0 = D(s_0-s) - \frac{\mu x}{Y_{x/s}} \tag{6.22}$$

From material balance on cells one can see that the equality between μ and D in a simple continuous culture does not hold true with cell recycling. As can be seen in Eq. 6.21 (Panel 6.13), because $0 < \alpha < 1$, D is larger than μ, meaning the dilution rate used is greater than the specific growth rate.

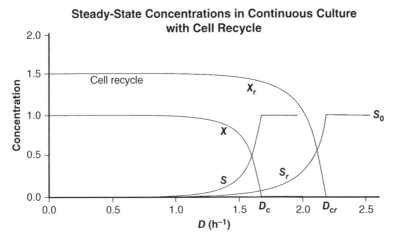

Figure 6.7 The concentrations of biomass and substrate in a continuous culture with cell recycling. A simple continuous culture of the same microorganism is also shown for comparison.

We next use the substrate balance to examine cell concentrations under cell-recycling conditions. Recall that without a cell-recycling system, the cell concentration is described by $x = Y_{x/s}(s_0 - s)$. In a cell-recycling system, the cell concentration in the recycling stream is higher than it is in the bioreactor, thus $c > 1$. The denominator in Eq. 6.19 (Panel 6.13) is thus less than 1. Therefore, the cell concentrations achievable with cell recycling are greater than they would be without (Figure 6.7).

The throughput of the system, in terms of cells produced, is the product of the effluent stream from the cell separator (F) and its cell concentration (x_2). It can be seen that the specific throughput (on a per-reactor volume basis), Dx_2, is equal to μx (Panel 6.14). The expression for throughput is the same as that without cell recycling: the cell concentration multiplied by the specific growth rate. Recall that at the same specific growth rate, the cell concentration is higher with cell recycling than without it. Therefore, the throughput is correspondingly higher. Also, with cell recycling, the critical dilution rate at which washout occurs is also higher, as shown in Eq. 6.24 (Panel 6.15). Ø is the enhancement factor of recycling. Its magnitude depends on the concentration factor and the proportion of medium flow that is recycled.

Panel 6.14
Throughput of the system is the product of the effluent stream from the settler (F) and its cell concentration (x_2): $Fx_2 = (1 + \alpha)Fx - \alpha Fcx$ $Dx_2 = (1 + \alpha)Dx - \alpha Dcx$ $Dx_2 = Dx(1 + \alpha - \alpha c) = \mu x$ (6.23)   It can be seen that the specific throughput (on a per reactor volume basic), Dx_2, equals μx (but x is greater than the x that can be achieved in simple continuous culture).

Panel 6.15

Let $\phi = 1/(1 + \alpha - \alpha c)$, $\phi \mu = D$

Apply the Monod model to describe the growth kinetics. At the critical dilution rate, $s = s_0$.

$$\phi \frac{\mu_{max} s_0}{K_s + s_0} = D_c$$

ϕ: recycle enhancement factor

$$\phi \mu_{max} s_0 = D_c K_s + D_c s_0$$

For the case that $s_0 \gg K_s$:

$$D_c \cong \mu_{max} \cdot \phi \qquad (6.24)$$

So, $D_c > \mu_{max}$ for a cell recycle system. It thus allows a dilution rate higher than μ_{max} to be used.

6.3.2 Applications of Continuous Culture with Cell Recycling

Continuous culture with cell recycling is used in waste treatment and in the cultivation of mammalian cells. In the case of waste treatment, cells in the reactor are often flocculated or immobilized on particles to make them easier for sedimentation. When cell recycling is employed in mammalian cell culture, it is often referred to as "perfusion culture." Both internal and external cell separation devices are used in perfusion cultures, including internal or external centrifugal filters, microporous hollow fiber membranes, and cell settlers.

The majority of continuous culture processes are ongoing for many months. In such long operations, there are several significant challenges. For example, to sustain culture integrity, we must avoid contamination and ensure cell line stability. At the same time, we must minimize mechanical failures. A cell-recycling process is more complex than a simple continuous culture. It entails a very significant effort to develop and to operate. It is not a task to be undertaken lightly, only to increase productivity.

6.3.2.1 Low Substrate Levels in the Feed

To increase the productivity of a continuous culture, one can increase the feed substrate concentration without resorting to the use of a high dilution rate that requires the use of cell recycling. However, in some processes, increasing the feed substrate concentration is not possible. For example, the solubility of the substrate may limit its feed level, or a high feed concentration is simply unavailable (like in waste treatment). In cell culture processes, higher substrate glucose levels also give rise to higher levels of lactate. This inhibits cell growth. For those processes, if we operate the culture as a simple continuous culture, productivity will be too low for the task to be economically viable. Thus, one could resort to cell recycling in order to boost productivity. With a cell-recycling system, the substrate supply rate can be increased by adopting a high dilution rate, instead of using a high substrate concentration.

6.3.2.2   Low Residual Substrate Concentration

In bioremediation processes, the substrate in the feed is metabolized and converted to environmentally neutral compounds before discharge. Their concentrations in the effluent stream from the reactor must be kept at low levels. In a steady-state operation of a continuous process, when the substrate concentration is low, the growth rate will also be low. Using a simple continuous culture, the dilution rate will be low, along with the throughput. With a cell-recycling system, the cell concentration and the throughput can be enhanced while still maintaining a low specific growth rate and a low substrate concentration.

6.3.2.3   Labile Product

Some products are susceptible to degradation after being excreted by cells in the bioreactor. Using a continuous process, the holding time of the product in the reactor can be reduced. Assuming that the degradation of the product follows first-order kinetics, the balanced equation for a product in a continuous culture is $Dp/dt = -Dp + q_p x - \delta p$, where δ is the degradation rate constant. It is clear that for an efficient process, D should be much greater than δ. With a simple continuous culture, the magnitude of D is limited by the cell's maximum specific growth rate. D can be increased using a cell-recycling system.

6.3.2.4   Selective Enrichment of Cell Subpopulation

Some cell separation methods may allow for discriminative retention of a subpopulation of cells. For example, a settling device may provide selectivity based on cell size and density. For some mammalian cells, the size and density of dead cells are smaller than with viable cells. Also, some cells are grown as multicellular aggregates to provide an adhesion surface for one another. Dead cells often fall off of the aggregate. By optimizing the sedimentation conditions, larger cells or aggregates may be selectively retained to enrich the culture with viable cells.

6.3.2.5   High-Intensity Mammalian Cell Culture

Most mammalian cell culture processes are operated as fed-batch cultures. The growth of mammalian cells is typically accompanied by a high excretion of lactate resulting from glucose consumption. The accumulation of lactate in culture prevents the cell concentration from reaching very high levels. In a fed-batch culture, one uses repetitive nutrient feeding to increase cell concentration, but simultaneously also accumulates lactate at high levels, causing the process to terminate. Nevertheless, this mode of operation has been very effective in producing many therapeutic proteins. In fact, only a few products are produced in continuous culture with cell recycling (called perfusion culture in that segment of the industry). Most of them are labile products that cannot be produced in a fed-batch culture.

Recently, the advances in cell separation devices have allowed more efficient cell separation from fluid stream over a long period of time. As a result, these cell retention devices have been applied to continuous cultures to increase the cell concentration to a level much higher than that achievable in fed-batch cultures. The resulting productivity per reactor volume is much higher than in the conventional fed-batch culture.

This combined with the advantage of continuous operation has made continuous culture a very attractive means of manufacturing. With its high productivity, the size of the reactor required for manufacturing the same quantity of therapeutic proteins is substantially reduced. In many cases, the adoption of such intensified continuous culture has allowed thousand-liter-scale disposable bioreactors to replace the traditional stainless-steel reactors of much larger volumes.

Example 6.2   A Continuous Culture with Cell Recycling

This example continues from Example 6.1.

Now the reactor size must be kept smaller, and one with a culture volume of 0.7 m^3 will be used. It is obvious that the feed stream flow rate will result in the washout of cells if a simple continuous culture is used. A centrifuge will be used for recycling the cell. The glucose concentration is to be kept at the same level as in Example 6.1. The centrifuge generates a heavy stream that has five times higher cell concentration than the reactor. The heavy stream is to be recycled into the bioreactor, while cells in the light stream will be harvested. Determine the operating conditions.

Solution

The new dilution rate is:

$$D = (1 \text{ m}^3/\text{h})/0.7 \text{ m}^3 = 1.43 \text{ h}^{-1}$$

This is greater than μ_{max}.

Since glucose at the effluent stream is 0.88 g/L, as in Example 6.1, the specific growth is also unchanged, $\mu = 0.62 \text{ h}^{-1}$.

The ratio of cell concentration between the recycle stream and the reactor is 5. Thus, $c = 5$ (Figure 6.6):

From Eq. 6.21:

$$\mu/D = 0.62/1.43 = 1 + \alpha - 5\alpha$$
$$\alpha = 0.14$$

Therefore, in a steady-state operation, the recycle stream has a flow rate of $0.14 \text{ m}^3/\text{h}$. Once this value is specified, the operation conditions are determined (i.e., there is no more degree of freedom to specify the operating conditions).

It is instructive to examine the value of other variables under the operating condition. The cell concentration in the exit stream of the recycle system (i.e., the harvest stream) can be obtained by overall material balance (Figure 6.6):

$$F(s_0 - s) \cdot Y_{x/s} = Fx_2$$
$$x_2 = (5 - 0.88) \cdot 0.5 = 2.06 \text{ g/L}$$

The cell concentration in the purge stream is the same as the output stream in Example 6.1. However, the same throughput is accomplished using a smaller reactor operating at a dilution rate that is higher than μ_{max}.

It should be noted that cell concentration in the reactor is higher than in Example 6.1. From Eq. 6.19:

$$x = 2.06/(1 + 0.14 - 0.7) = 4.68 \text{ g/L}$$

6.4   Specialty Continuous Cultures

A continuous culture is attractive due to its increased productivity. Over the years, continuous cultures have evolved to meet different special process needs. These specialty continuous cultures are used in industrial processes and provide basic tools for mechanistic studies.

6.4.1   Multiple-Stage Continuous Culture

A continuous culture is constrained by the steady-state equality of the specific growth rate and the dilution rate ($\mu = D$). These optimal growth conditions may not be optimal for product synthesis. To decouple growth and production, continuous bioreactors may be operated in series. By using two continuous reactors in series, cells in the two stages (i.e., reactors) can have different specific growth rates. The first reactor allows for rapid growth to provide the cell biomass necessary for production. The second reactor operates under conditions optimal for product synthesis. A schematic of two continuous bioreactors in a series is shown in Figure 6.8.

With two continuous bioreactors in series, the degrees of freedom are increased. In the second reactor, the influx stream carries cells from the first reactor with it. Thus, its dilution rate and the specific growth rate are not equal. Even if the flow rate and reactor volume of the two reactors are kept the same, their specific growth rates are not equal. By adjusting the volume of each reactor, even if the flow rate is kept the same, different dilution rates can be imposed for each of the two stages. This allows the second stage to be operated under conditions favoring product formation. Also, a feed stream may be introduced. The culture conditions, such as pH and temperature, may be adjusted in the second stage. Through the use of multistage culture, the production process for products with critical quality factors (such as flavors) that affect customer acceptance, including yogurt and beer, can be converted to continuous operations.

Two Continuous Bioreactors in Series

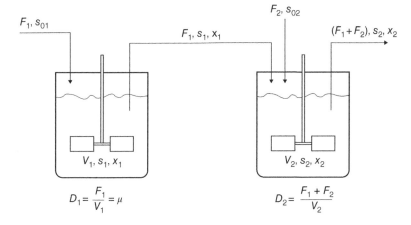

Figure 6.8 A two-stage continuous culture system.

6.4.2 Immobilized Cell Culture System

Another way of retaining cells in a continuous reactor is by physically retaining the cells within large particles that are not washed out with the flow of media. The particles can be formed through cell self-aggregation, or by immobilizing cells inside a solid support. A simple way of cell immobilization is to entrap cells inside agarose beads. Alternatively, various other materials are used to immobilize cells, including Celite®, ceramics, and natural or synthetic polymers.

As cells grow, some cells are shed into the medium. Some animal cells suspended in medium do not multiply, so the cellular growth in these cultures is restricted to cells in the solid phase. Most microbial cells can grow in both solid and liquid phases. While cells in the solids are retained in the reactor, those in the liquid may be purged in the effluent stream, depending on the particle separation method used.

Another frequently encountered form of cell immobilization is the formation of a biofilm. Many microorganisms are capable of excreting extracellular polymeric substances (including polysaccharides and proteins) that allow them to adhere to various surfaces and to each other, forming a layer of cells. Biofilm is seen on teeth, on the surface of the digestive tract, and on the surface of much liquid-contacting equipment that we use daily and are in contact with. A steady state may be established between the biofilm and the liquid surrounding it, sustaining a relatively constant size of the biofilm.

Waste treatment plants using activated sludge are often operated as large-scale cell-immobilized systems. In some cases, particles are added to induce cell immobilization. Other cases allow flocculated cells to form particulate and become a cell-immobilized system. In the example shown (Figure 6.9), a feed stream carrying the substrate containing the carbon, nitrogen, and sulfur sources enters the treatment tank along with the recycled stream. The treatment tank perpetuates faster microbial growth by providing the organic materials from the feed. Some mixing is provided to support cell growth.

In the second tank, organic materials are slowly further degraded. The particulates containing the microbial population (a mixed population of a variety of microorganisms) grow in size. They settle to the bottom of the tank and are recycled to the first tank. The treated effluent is discharged from the treatment plant.

Activated Sludge Plant with Recycle

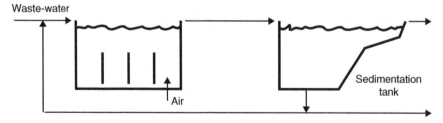

Figure 6.9 An activated sludge waste-water treatment plant with recycling of flocculated cells.

Possible Relationship between Two Microorganisms

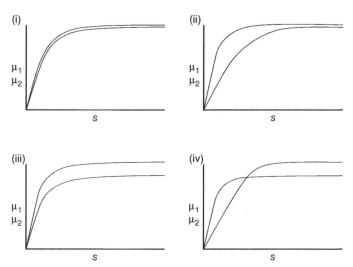

Figure 6.10 Different scenarios of the communal relationship between two microorganisms in a mixed culture.

6.4.3 Continuous Culture with Mixed Populations

Except in bioremediation, virtually all microbial and cell cultures entail a pure culture of a single organism. Inadvertent mixed cultures, containing two or more subpopulations, can arise by contamination or genetic change. Some processes employ microorganisms that harbor plasmids. In these cases, mixed populations may include a population of cells that have lost that plasmid (i.e., plasmid-free cells). Similarly, in a long-term culture of an aneuploid mammalian cell line (whose chromosomes may undergo rearrangement or reorganization), a segment of the chromosome may be lost or duplicated, thus giving rise to a new subpopulation. If the two populations are virtually identical in their growth characteristics (i.e., they have the same μ_{max} and K_s), they can coexist indefinitely. But if their growth characteristics are different, competition for resources is inevitable and the makeup of the overall population may change appreciably.

Consider the case that two organisms compete for the same limiting substrate and their growth kinetics are described by the Monod model. The outcome of the competition depends on the maximum growth rate of each population and how it is affected by the concentration of the substrate. The relationship between the two populations may fall into one of four different classes: (1) $\mu_{max,1} = \mu_{max,2}$, $K_{s,1} = K_{s,2}$; (2) $\mu_{max,1} > \mu_{max,2}$, $K_{s,1} = K_{s,2}$; (3) $\mu_{max,1} > \mu_{max,2}$, $K_{s,1} < K_{s,2}$; or (4) $\mu_{max,1} = \mu_{max,2}$, $K_{s,1} > K_{s,2}$ (Figure 6.10 and Panel 6.16).

Panel 6.16

In the case that two microorganisms compete for the same limiting substrate:

$$\frac{dx_1}{dt} = \mu_1 x_1 - Dx_1$$

$$\frac{dx_2}{dt} = \mu_2 x_2 - Dx_2$$

$$\frac{ds}{dt} = -\frac{\mu_1 x_1}{Y_{x_1/s}} - \frac{\mu_2 x_2}{Y_{x_2/s}} + D(s_0 - s)$$

(a) $\begin{array}{c} \mu_{max_1} = \mu_{max_2} \\ K_{s_1} = K_{s_2} \end{array}$ (b) $\begin{array}{c} \mu_{max_1} = \mu_{max_2} \\ K_{s_1} \neq K_{s_2} \end{array}$ (c) $\begin{array}{c} \mu_{max_1} > \mu_{max_2} \\ K_{s_1} < K_{s_2} \end{array}$ (d) $\begin{array}{c} \mu_{max_1} > \mu_{max_2} \\ K_{s_1} > K_{s_2} \end{array}$

In case 1, the two subpopulations will coexist. In the second and third cases, subpopulation 1 will always have a growth advantage over subpopulation 2. In case 4, depending on the operating region of the dilution rate, one of the subpopulations will have a competitive advantage over the other. At the point where the two Monod curves cross over, the two populations will coexist. However, after crossing over to the other region of the dilution rate, the growth rate competitiveness between the two subpopulations will be reversed.

6.5 Transient Response of a Continuous Culture

In this chapter, we discussed a continuous culture system where the cell growth kinetics followed Monod kinetics. The equations describing the system (i.e., the differential equations for cell and substrate concentrations) can be solved for steady states. For every D, there is a unique solution of steady x and s, as shown in Figure 6.3. A steady state can be either stable or unstable. A stable steady state describes a system that will return to its original steady state after a small perturbation is introduced. An unstable steady state, in contrast, will move away and not return to its original state once the steady state is perturbed.

For a continuous culture with Monod kinetics, the steady state is stable. This can be shown by Eigenvalue analysis. Upon small perturbations of an operating variable, such as the dilution rate or feed substrate concentration, the system will deviate from the steady state. Eventually, it will return to its original steady state. On the other hand, if a permanent change in operating conditions is imposed (such as changes in the substrate concentration in the feed or the dilution rate), then the system will gradually move to a new steady state. The transient behavior of this system is described by the same set of systems of equations used for cell and substrate concentrations. By solving the differential equations, one can see how cell and substrate concentrations change over time after a perturbation or a change in operating conditions. Examining the transient behavior of a simple continuous culture is an excellent way to understand the growth dynamics of microbial systems. How a microorganism responds to nutrient or flow rate perturbation in a continuous culture is affected by their kinetic behavior, including K_s and μ_{max}; nevertheless, we will examine two cases qualitatively.

6.5.1 Pulse Increase at the Substrate Level

Consider a continuous culture system of a microorganism whose growth kinetics are described by Monod kinetics with known K_s and $Y_{x/s}$. When a steady state is reached, the state of the system is defined by the dilution rate, $D\ (=\mu)$; substrate concentration, $s\ [=s(\mu)]$; and cell concentration, $x\ [=Y_{x/s}(s_o-s)]$. The steady state of the system is thus defined by only two variables, D (or μ, or s) and s_o (or x). Any temporary perturbation in operating conditions will cause short-term changes, and the system will return to the original steady state.

Let us consider the following scenarios:

First, consider a system that is initially at a steady state, with parameter values of D_1, μ_1, s_1, s_{o1}, and x_1. This system receives a short-pulse addition of a substrate into the culture, bringing the substrate level to s_2 instantaneously (Figure 6.11). Examining the cell concentration equation of the system, one sees that the increase in s causes

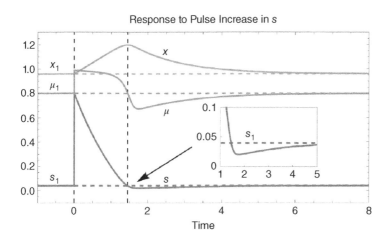

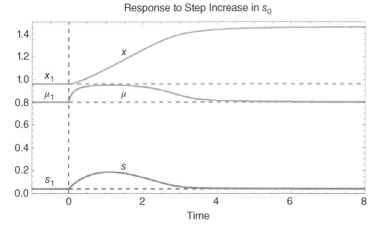

Figure 6.11 The transient responses of cell concentration and rate-limiting substrate concentration to pulse and step perturbations.

μ to increase to μ_2 ($>D$), which then causes the cell mass generation term (μx) to be larger than the cell mass flow-out term (Dx). Consequently, the cell concentration will increase. The increase in cell concentration results in a higher substrate consumption, which causes the slope of the substrate concentration to become negative and the substrate level to decrease. With decreasing substrate concentration, μ also begins to decrease.

At a point, μ becomes equal to D, and so $\mu x - Dx = 0$. So, the slope of $dx/dt = 0$; the rates of biomass increase due to growth and that of biomass washout become equal too, and cell concentration reaches its maximum. Since D is unchanged from that of the original steady state, $D = \mu_1$ still holds. At μ_1, s equals to s_1. From that point on, s decreases to below s_1, and the slope of cell concentration becomes negative.

The drop in both μ and x causes the nutrient consumption to decrease. Gradually the consumption rate of the substrate ($Dx/Y_{x/s}$) is smaller than the subsequent feeding rate (Ds_o), so the substrate level begins to increase (see the inset of Figure 6.11). Eventually, the increased substrate and the corresponding increase of μ take the cell concentration back to the original steady-state level.

As a second scenario, we consider an increase of non-growth rate limiting substrate. Since the growth rate is not limited by this substrate, increasing its concentration will not increase either the growth rate or cell concentration. In this case, regardless of whether the substrate concentration is increased in the feed or added directly into the reactor, no change in the state variables (D, s, x) would be expected.

6.5.2 Step Change in Feed Concentration

We have discussed the effect of feed substrate concentration on steady-state behavior. Now, consider a step increase in the feed concentration while the dilution rate is kept constant (Figure 6.11). We will consider the initial steady state as state 1 with D_1, μ_1, s_1, s_{o1}, and x_1. After introducing the step change, the provision rate of the substrate would be larger than the consumption rate; the residual substrate concentration in the reactor would gradually increase. An increase in the limiting substrate concentration would also cause the specific growth rate to increase. This, in turn, would cause the cell mass generation term to be larger than the washout term. As a consequence, the cell concentration would increase.

In this period of increasing cell and substrate concentrations and specific growth rate, Ds_o would remain constant while Ds and qx terms increase. Eventually, this increasing trend of substrate concentration would come to a stop. We shall call this point state 2, and denote the variables as values of D_1, μ_2, s_2, s_{o2}, and x_2. At this state (which is not a steady state), $D(s_o - s_2)$ and $q_2 x_2$ are equal. The cell concentration is balanced by $\mu_2 x_2$ and $D_1 x_2$. Note that $\mu_2 > D_1$ because $s_2 > s_1$. So, cell concentration would continue to increase. However, with the continued increase in x, qx would become larger than $D(s_{o2} - s)$. The balance of substrate would now tip to greater consumption, and the substrate concentration would begin to decrease, causing the growth rate to also decrease. Eventually, the system would reach a new steady state, state 3. At the new steady state, the dilution rate and substrate concentration would be the same as in state 1, but with a different x_3.

6.6 Concluding Remarks

Continuous cultures have many advantages over batch bioprocesses. In practice, however, only a small number of bioprocesses are operated as continuous cultures. Mechanical failure, cell stability, and contamination are all significant concerns to be addressed with this approach. Advancements in process technology, including more recent devices for cell recycling and the successful implementation of continuous culture for therapeutic biologics manufacturing, continue to highlight the value of continuous culture as an important process technology. Continuous cultures are also valuable for exploring the growth kinetics of various organisms. By observing their behaviors at steady states and during environmental perturbation, one can construct a reliable growth model for describing the process control of cell culture operations.

Further Reading

Basin, MJ (1978) Theory of continuous culture. In: Continuous culture of cells. I. Ed. P. H. Calcott. 1978. CRC Press.

Croughan, MS, Konstantinov, KB and Cooney, C (2015), The future of industrial bioprocessing: Batch or continuous? Biotechnol. Bioeng., **112**: 648–651. doi:10.1002/bit.25529

Hu, WS (2012) Cell retention and perfusion. In: Cell culture bioprocess engineering. Pp. 249–261. Published by author.

Wang, LK, Guss, D and Krofta, M (2010) Kinetics and case histories of activated sludge secondary flotation system. In: Handbook of advanced industrial and hazardous waste treatment. Eds. Wang, L.K., Hung, Y-T. and Shammas, N. Pp. 1155–1189. CRC Press.

Nomenclature

c	concentration factor in cell-recycling device	
D	dilution rate	t^{-1}
D_c	critical dilution rate	t^{-1}
D_m	the dilution rate at which the biomass productivity is maximal	t^{-1}
F	volumetric flow rate	L^3/t
K_s	half-saturation constant for specific growth rate	M/L^3 mole/L^3
m	maintenance coefficient	$M/cell \cdot t$
p	product concentration	M/L^3, mole/L^3
q_p	specific productivity	$M/cell \cdot t$
s	substrate concentration	M/L^3, mole/L^3
s_0	substrate concentration in the feed	M/L^3, mole/L^3

t	time	t
V	volume of the culture	L^3
x	dry biomass concentration	M/L^3
x_0	cell concentration in the feed	M/L^3
x_d	dead cell concentration	M/L^3
Y	yield coefficient	M/M
$Y_{p/s}$	product yield based on substrate	M/M
$Y_{x/s}$	cell yield based on substrate	M/M
α	recycle ratio	
ϕ	recycle enhancement factor	
μ	specific growth rate	t^{-1}
μ_d	specific death rate	t^{-1}
μ_{max}	maximum specific growth rate	t^{-1}

Problems

A. Continuous Culture Kinetics

A1 Chinese hamster ovary (CHO) cells were cultivated in both batch and continuous stirred-tank bioreactors. The growth kinetics in the batch bioreactor are shown in Table P.6.1. After a series of continuous cultures, the steady-state cell and glucose concentrations (at different dilution rates) were tabulated and are shown in Table P.6.2. From those data, the relationship between the specific growth rate and the glucose concentration can be obtained for both the batch culture and the continuous culture. In both cases, the growth rate is limited by glucose concentration. The specific growth rate and glucose concentration can be fitted using the Monod model. Obtain the μ_{max} and K_s for both the batch and continuous cultures. Discuss the similarities or differences between the results obtained with these two different methods.

Table P.6.1 Batch culture data for a CHO cell line.

Time (h)	Biomass (g/L)	Glucose (g/L)
1	0.008	1.80
15	0.005	1.70
24	0.007	1.64
39	0.016	1.48
48	0.025	1.04
60	0.040	0.73
72	0.048	0.24
78	0.050	0.06

Table P.6.2 Steady-state data for continuous cultures of CHO cells, with a feed glucose concentration of 1.0 g/L.

Dilution rate (h^{-1})	Biomass (g/L)	Glucose (g/L)
0.01	0.005	0.0009
0.02	0.005	0.0010
0.025	0.049	0.0013
0.03	0.048	0.002
0.04	0.045	0.005
0.045	0.042	0.011
0.0475	0.035	0.021
0.05	0.025	0.041

A2 A baker's yeast *Saccharomyces cerevisiae* is grown on glucose. The formula for the cell is $C_{4.4}H_{7.3}N_{0.86}O_{1.2}$, and the ash content is 8%. A medium consisting of glucose, ammonia, and mineral salts is used to grow it in a continuous culture at a dilution rate of 0.5 h^{-1}, which is also 80% of its maximum growth rate. At the steady state, glucose is the rate-limiting substrate and the K_s for glucose is 0.01 g/L. NH_3 is to be maintained at 1 g/L in the feed to ensure that it is not limiting. At the steady state, the cell concentration is 10 g/L. The yield coefficient for glucose is 0.5 g cells/g glucose.
a) What is the glucose concentration in the effluent stream?
b) What is the NH_3 concentration in the culture?

A3 *E. coli* cells are grown in a continuous culture at a steady state using glucose as the limiting substrate. The concentration of glucose in the feed is 10 g/L. The maximum specific growth rate is 1.38 h^{-1}. The yield coefficient of biomass is 0.4 g cell/g glucose. The growth rate can be described by Monod kinetics with $K_s = 0.05$ g/L. If the culture is operated at a dilution rate of 1 h^{-1}, what are the steady-state cell and glucose concentrations?

A4 A yeast cell is cultivated in a steady-state, continuous culture to produce baker's yeast. Using glucose as the limiting substrate, the growth kinetics of this yeast strain follow the Monod model. With a maximum growth rate of 0.69 h^{-1} and $K_s = 0.05$ g/L, the culture is to be operated at 90% of its maximum growth rate. The feed glucose is 50 g/L and the yield is 0.3 g cell/g glucose. All of the other nutrients are abundant and in excess. What is the production rate of yeast per day using a 100-L bioreactor? If the yeast contains 7% N in its dry mass and NH_3 is used to supply nitrogen, how much NH_3 is needed per day? The ammonium concentration in the effluent stream is 0.01 g/L.

A5 A baker's yeast production plant employs glucose as the carbon source and operates a steady-state culture continuously over an extended period. The feed glucose concentration is 30 g/L, the half-saturation constant of glucose is 0.05 g/L, and the yield coefficient based on glucose is 0.5 g/g. When growing at its maximum rate, the doubling time of the organism is 1 h. Typically, the culture is operated at 90% of the critical dilution rate to avoid washout. The plant operates a 10,000 L reactor for 330 days a year. How much yeast is produced a year?

A6 A yeast growing on glucose has a maximum specific growth rate of 0.45 h^{-1}. The yield coefficient is 0.4 g cell dry weight/g glucose. Using a feed stream with 80 g/L of glucose, a continuous culture will be established to produce 40,000 kg per year of dry yeast cells. The plant will be operated 340 days a year. What is the total reactor volume needed? Clearly state your assumptions. If necessary, make assumptions on additional conditions to obtain a good estimate.

A7 A continuous culture is operated at steady state using glucose as a limiting substrate. The conditions are s_o (feed glucose) $= 20$ g/L, $s = 0.02$ g/L, $x = 10$ g/L, and $D = 0.25$ h^{-1}. Now, D is changed to 0.3 h^{-1} and s_o to 15 g/L. The new D is 90% of the maximum specific growth rate. What is the cell concentration (x) at the new steady state? Assume that the growth is described by the Monod model.

A8 A microorganism uses glucose aerobically for growth. The relationship between its specific glucose consumption rate and the specific growth rate is shown in Figure P.6.1. The maximum specific growth rate is 0.7 h^{-1}. Given this relationship, sketch the steady-state biomass and glucose concentrations (as a function of the dilution rate) that you would expect to see in a continuous culture that is limited by the glucose concentration and with a feed glucose concentration of 10 g/L. Give an explanation for the behavior at very high and very low dilution rates.

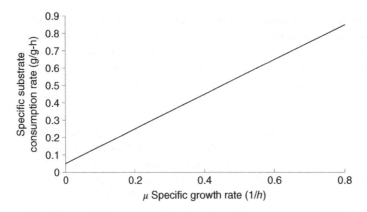

Figure P.6.1 The relationship between a microorganism's specific glucose consumption rate and the specific growth rate.

A9 *Pseudomonas aeruginosa* was grown in a single-stage continuous culture system using NH$_3$ as the growth-limiting substrate. When the inlet feed contained 0.05 M NH$_3$, the maximum throughput of cells was found to occur at a dilution rate of 0.5 h-1, with a cell concentration of 5.8 g/L. The washout dilution rate was 0.6 h-1. From these observations, determine K_s and μ_{max} for this system. No nitrogen-containing compound was excreted by the cell as a product or metabolite. What is the nitrogen content of the cell?

B. Continuous Culture with Recycle

B1 You are asked to design a continuous process to treat a waste stream. This stream contains 1.0 g/L of organic waste that can be used as growth substrate, but the discharge regulation requires that it be reduced to 0.05 g/L. The yield on biomass is 0.5 g for each gram of the substrate. The Monod constant is 0.1 g/L, with a maximum specific growth rate of 0.23 h^{-1}. The feed stream flow rate is 3000 L/h. You are given a 5000-L reactor with which to operate. Specify the operating conditions so that the conditions described above can be achieved at a steady state.

The secondary biological treatment of municipal waste operates in a fashion similar to a continuous culture using cell recycling. A plant is operating at a flow rate of 10^6 L/h. The raw sewage enters this tank at an organic pollutant concentration of 1 g/L. Determine the maximum growth rate.

The treatment vessel operates essentially as a well-mixed reactor. The effluent from the treatment vessel enters a settler where the activated sludge is flocculated for recycling. The supernatant from the settler is void of microorganisms. The flow rate of recycling is 10^5 L/h, the concentration of flocculated cells for recycling is 10 g/L, and the feed flow rate to the treatment tank is 10^6 L/h.

The portion of cells that are flocculated but not recycled is discharged for sludge disposal. The organic pollutant concentration in the effluent stream is 50 mg/L. The cell yield from the pollutant is 0.4 g cell/g organic pollutant. The volume of the vessel is 10^7 L. Assuming that the Monod growth model applies and the K_s for the substrate is 0.2 mg/L, determine the steady-state cell concentration from the treatment vessel and the flow rate of the flocculated activated sludge for final disposal.

By introducing pure oxygen, the same plant can now produce the flocculated sludge in the settler at a concentration of 40 g/L. Assuming the same growth rate can still be attained and the same recycling rate can be used, determine the increased feed flow rate to the tank that the treatment plan can now handle under the new operating conditions. What is the new steady-state cell concentration in the treatment vessel?

B2 A 1 L chemostat, with a cell recycler, is operated at a feed flow rate of 100 mL/h under glucose limitation. The yield coefficient $(Y_{x/s})$ is 0.5 g/g. The glucose concentration in the feed is 10 g/L. The kinetic constants of the organisms are (max specific growth rate) $\mu_m = 0.2\ h^{-1}$, $K_s = 1$ g glucose/L. The value of C (cell concentration ratio of recycle stream to that in the reactor) is 1.5, and the recycle ratio (α, the ratio of flow rate of recycle stream to feed stream) is 0.4. The system is at a steady state.

a) What is the substrate concentration in the effluent stream?
b) What is the specific growth rate (μ) of the organisms?
c) Determine the cell concentration in the recycle stream.

C. Specialty Continuous Culture

C1 Secondary metabolite formation is often dissociated from growth. A reactor cascade is potentially suitable for a continuous process of secondary metabolite production. Consider a two-stage continuous culture system. The volume of the first reactor is 400 L, and the feed that has glucose at a concentration of 5 g/L is fed at 80 L/h. The second reactor has a volume of 240 L. Cell growth can be described by a Monod model with the following parameters: $\mu_{max} = 0.3\ h^{-1}$, $K_s = 0.1$ g/L, $Y_{sx} = 0.4$ g/g. Cell growth is negligible in the second reactor, and product formation occurs only in the second reactor at a specific production rate (q_p) of 0.02 g product/g-cell-h. The yield, $Y_{p/s} = 0.6$ g product/g glucose.
   a) Determine the cell and glucose concentrations in the effluent of the first stage.
   b) Determine the product and glucose concentrations in the effluent of the second stage.

C2 Table P.6.3 lists data of two organisms growing together in a continuous culture and using the same growth rate-limiting substrate.
   a) Which organism will dominate if the feed substrate concentration is 15 g/L and the dilution rate is $1.2\ h^{-1}$?
   b) When the culture reaches a steady state, what are the cell concentrations for each organism?
   c) Is there any condition at which the two organisms can coexist at a steady state? If so, specify those conditions.

Table P.6.3 Kinetic parameters of two microorganisms growing in a continuous mixed culture.

	Organism 1	Organism 2
μ_{max}	1.5 h	2.2 h
K_s	0.005 g/L	0.085 g/L

C3 A CHO cell line that produces antibody has a small subpopulation that has lost its product gene. The heterogeneous populations are cultivated together in a continuous culture with glucose as the growth rate-limiting substrate. The μ_{max} and K_s for glucose for the producer is $0.03\ h^{-1}$ and 0.005 g/L, respectively. For the non-producers, the μ_{max} is $0.04\ h^{-1}$ and K_s is 0.01 g/L. The glucose concentration in the feed is much higher than K_s. Describe the population composition at steady state if the dilution rate is: (1) $0.041\ h^{-1}$; or (2) $0.01\ h^{-1}$.

D. Transient Behavior

D1 In a continuous culture of a bacterium running at a dilution rate of $0.345\ h^{-1}$, glucose is the limiting substrate with a residual concentration of 0.01 g/L. The yield coefficient based on glucose is 0.5 g/g. What will happen if the feed glucose is increased by 0.5 g/L while the dilution rate is maintained at the same level? Plot the response of the culture in terms of x, s, and μ after the feed concentration changes and until a new steady state is reached.

D2 A bacterium is grown in continuous culture at a dilution rate of $D = 0.69\ h^{-1}$, with glucose as the rate-limiting substrate and with a feed glucose concentration of 10 g/L. At steady state, the glucose concentration in the culture is 0.05 g/L and the cell concentration is 4.97 g/L. The dilution rate is then changed to $0.62\ h^{-1}$. After a new steady state is reached, the glucose concentration in the culture is 0.04 g/L.
a) What is the cell concentration at the new steady state?
b) If the dilution rate is further decreased to $0.55\ h^{-1}$, what will the glucose and cell concentrations become?
c) How will the cell concentration, glucose concentration, and growth rate change when the dilution rate is changed from the steady state of $D = 0.62\ h^{-1}$ to $D = 0.55\ h^{-1}$? Sketch the time profile until the new steady state is reached. Set up equations to explain your answer. Be as quantitative as you can. Mark the numerical values of the time points and substrate and cell concentrations.

D3 A continuous culture of a bacterium is running at $D = 0.5\mu_{max} = 1\ h^{-1}$ using glucose as the limiting substrate. The steady-state concentrations of glucose and NH_3 are 10 mg/L and 0.1 mM, respectively. The feed glucose concentration is 5 g/L, and the cell concentration is 2 g/L. To ensure that the nitrogen source, NH_3, is not the actual rate-limiting substrate, the ammonium concentration in the feed is increased twofold. The nitrogen content of the cell is 7% dry mass.
If ammonium is indeed not the rate-limiting substrate, how will the concentrations of cell, glucose, and NH_3 respond to the increasing concentration of NH_3 in the feed? Calculate the new ammonium concentration at the new steady state.
A step change of glucose to 6 g/L in the feed is performed. You are asked to predict the glucose, NH_3, and cell concentrations at the new steady state. Sketch a time profile of glucose and cell concentrations to describe how the culture will reach the new steady state.
The dilution rate is increased from the original steady state, but still at a level lower than the washout dilution rate. Sketch the response of the concentrations of cell, glucose, and NH_3 until the new steady state is reached.

E. Advanced Problems

E1 A bacterium is engineered to carry a plasmid that contains a glucose transport system, which allows it to grow faster at lower glucose concentrations. The non-engineered parent has a K_s of 0.03 g/L for glucose, while the engineered strain has a K_s of 0.01 g/L. However, the parent, not having to make the plasmid, has a higher maximum growth rate. The μ_{max} for parent and transformant are $0.69\ h^{-1}$ and $0.62\ h^{-1}$, respectively. The yield coefficients for both the parent and plasmid-containing cells are the same, at 0.6 g cells/g glucose. A continuous culture is started with a pure population of cells containing the plasmids and operated at a dilution rate of $0.59\ h^{-1}$. The glucose concentration in the feed stream is 5 g/L.

    a) Consider that the engineered cells began to lose their plasmids at a rate of 0.01% per cell doubling. Will the steady-state population of plasmid-containing cells change with time and, if so, how long will it be before the new steady state is reached? You can assume that the culture is essentially at a steady state when it reaches 99% of the new steady-state value.

    b) Consider the case that the culture is running at $D = 0.55$ h^{-1}. Will there be a new steady state? Demonstrate how you derive your conclusion.

E2 An engineered *E. coli* is used to produce ethanol using glucose derived from corn-starch as the raw material. The process will have a capacity of 10^6 kg of ethanol per year, using glucose at a concentration of 20 g/L as the feed. To minimize the environmental impact, the glucose concentration in the waste stream has to be no higher than 0.1 g/L. The maximum specific growth is 0.23 h^{-1}. The K_s for glucose is 0.03 g/L. The specific glucose consumption rate at μ_{max} is 20 g/g h^{-1}. Note that ethanol fermentation is an anaerobic process. From the 20 g of glucose consumed, 9.0 g of ethanol is produced in addition to 1 g of cells. You can assume the specific glucose consumption rate is proportional to the specific growth rate, and the dis-tribution of glucose carbons channeled to biomass and ethanol is unaffected by the growth rate. Determine the size of the reactor, the dilution rate, and the cell and ethanol concentrations. Assume the plant is operated at steady state for 340 days a year.

E3 An aerobic organism is grown in a 10 L continuous culture. The feed containing 50 g/L of glucose is added at 0.1 L/h. The bioreactor is kept at a constant volume of 10 L by using a level controller that triggers a pump to initiate the flow of the effluent stream whenever the liquid level is above the 10 L level. At steady state, the cell concentration is 24.95 g/L while glucose is at 0.1 g/L. Compressed air is passed through the bioreactor at one volume of air per volume of liquid per minute (vvm). The air is first saturated with water by passing it through a humidification column before entering the bioreactor.

Unfortunately, the humidifier malfunctioned. As a result, the entering com-pressed air is completely dry, but it is saturated with water upon exiting the bioreactor. How will this affect the interpretation of analytical data obtained, without accounting for evaporation? Describe how the real dilution rate and specific growth rate will deviate from the values obtained when the incoming air is saturated with water.

E4 A cellular protein is produced from methanol, ammonia, and mineral salts in a 100-L stirred tank operating in a continuous mode and at steady state. The maxi-mum specific growth rate (μ_{max}) is 0.4 h^{-1}, and the half-saturation constants (K_s) for methanol and ammonia are 0.05 g/L and 0.0017 g/L, respectively. The yield coefficients for the two substrates are 0.7 g cell/g methanol and 18 g cell/g ammo-nia. The methanol feed, at 20 g/L, is supplied at 40 L/h. Due to the environmental regulation, methanol concentration in the effluent stream is set at 0.05 g/L. Only methanol will be the rate-limiting substrate, to ensure that all the other substrates will be kept at a level ten times of its half-saturation constant. Apparently, one way to meet these constraints is to employ cell recycling.

a) A cell-settling separator will be used to separate the effluent from the bioreactor into two streams. The more concentrated stream has a cell concentration seven times higher than the cell concentration in the dilute stream. The concentrated stream is recycled, while the dilute stream is purged. Specify the operating conditions (dilution rate, and recycle stream/feed stream ratio) that are sufficient to define the system. Sketch a diagram of your system, and clearly label any symbols that you may use in your explanation.

b) What is the cell concentration in the reactor? What is the throughput (g cells/h)?

c) What is the concentration of ammonia in the feed?

d) The reactor was initiated from a batch culture, with an initial cell concentration of 0.1 g/L. Propose an economic way of starting up the process. (*Hint*: You will most likely use a fed-batch culture startup.) Devise a time profile of operating conditions. There is no single answer to this problem, but be specific. Simulate your results with appropriate equations.

E5 *Perfusion cell culture*: One factor limiting the productivity for mammalian cell culture is the high conversion rate of glucose to lactate. In many industrial continuous cell culture processes, the cell growth rate is limited by the accumulation of lactate but not nutrient supply. These cells convert glucose to lactate at a molar ratio of 1.5 moles of lactate/mole of glucose. In a culture of baby hamster kidney (BHK) cells producing Factor VIII, under nutrient-abundant conditions, the specific growth rate is affected by lactate, as shown here:

$$\mu = \frac{\mu_{max}}{K_I + I^2}$$

where $\mu_{max} = 0.035 \ h^{-1}$, $K_I = 1.2 \ mM$, and I is the concentration of lactate in mM. The specific consumption rate for glucose is 2×10^{-10} mmol/cell-h, while lactate production is 3×10^{-10} mmol/cell-h. A perfusion process has a cell concentration of 2×10^{10} cell/L and a specific growth rate of 0.030 h^{-1}. Specify the feeding rate, given a feed glucose concentration of 8 g/L. The cell retention efficiency in the settling tank (defined as total cells recycled/total cell feed into the settler) is 0.9, with a purge stream to recycle stream volume ratio of 0.3.

E6 *The stability of simple continuous culture*: Consider a continuous culture for which the growth kinetics of the microorganism can be described adequately by Monod kinetics. At a given steady state, is the system stable? If the system at a steady state is perturbed temporarily, will it return to the same steady state? This question is answered by performing a stability analysis. Take the following material balance equations:

$$\frac{dx}{ds} = (\mu - D)x$$

$$\frac{ds}{dt} = D(s_o - s) - \frac{\mu x}{Y_{x/s}}$$

$$\mu = \frac{\mu_m s}{K_s + s}$$

The deviation variables, X and S, are:

$$X = x - \tilde{x}$$
$$S = s - \tilde{s}$$

where $\tilde{x}$ and $\tilde{s}$ are steady-state values.

Rewrite the material balance in terms of x, s. Expand the expression for $\mu(s)$ (i.e., a Monod kinetics equation in terms of deviation variable) in Taylor's series. Linearize it by dropping second-order terms. Substitute it into the balance equations written in deviation variables. It can be argued that in the resulting equations, the terms that include the product of deviation variables are all very small and can be neglected. That gives equations in the following form:

$$\frac{dX}{dt} = a_{11}X - a_{12}S$$

$$\frac{dS}{dt} = a_{21}X - a_{22}S$$

These two linearized equations can be used to obtain analytical solutions for the system. Use the Eigenvalues to discuss the stability of the system.

E7 Factor VIII is produced by recombinant CHO cells for the treatment of hemophiliac patients. It is rather unstable; its half-life in a bioreactor is 120 h. The degradation of the product not only reduces its productivity, but also causes problems in its purification. During the course of cultivation, cells produce lactate that is growth inhibitory. To reduce the degradation of the product and to decrease the concentration of lactate, a continuous process with cell recycling is used. The relevant parameters for the process are listed here:

$$\mu = 0.036 \text{ h}^{-1}$$
$$\mu_{max} = 0.04 \text{ h}^{-1}$$
$$K_s(\text{glucose}) = 1 \text{ mM}$$
$$q_{glc}(\text{glucose}) = 10^{-9} \text{mmol/cell} - \text{h}$$
$$q_{lac}(\text{lactate production}) = 1.5 \times 10^{-9} \text{mmol/cell} - \text{h}$$
$$q_p(\text{Factor VIII}) = 10^{-12} \text{g/cell} - \text{h}$$

The lactate concentration has to be kept below 10 mM, and the degradation product has to be below 5% of the total product. The settler used for cell recycling produces a concentrated and a dilute stream. The concentrated stream has a cell concentration that is five times higher than that in the reactor. Specify the process conditions and product concentration that can be achieved.

E8 A *Flavobacterium* sp. is capable of degrading pentachlorophenol (PCP) and uses it as the sole carbon and energy source. When PCP is the rate-limiting substrate, the specific growth rate is substrate inhibited, as described by:

$$\mu \ (h^{-1}) = \frac{0.21 \cdot s}{3 + s + 0.0125 \ s^2}$$

where (s) has units of mg/L. The yield coefficient for biomass is 1 mg cells per 10 mg of PCP.

a) Water at a contaminated site has a PCP concentration of 50 mg/L. The throughput of the biodegradation treatment plant must be capable of processing 5 m³ of contaminated water per day. What is the reactor size if the substrate concentration in the effluent stream is 5 mg/L? What is the cell concentration in the effluent stream? Assume the reactor is well mixed.

b) After the operation starts, the PCP levels at the effluent stream are sometimes far higher than what is predicted from the calculation. The engineer checks the flow rate and holding volume in the reactor, and found everything normal except the cell concentration and the residual PCP concentration. From the growth kinetics, explain the results and predict the concentrations of cells and residual PCP for this "unusual" case.

c) Using steady-state analysis, an engineer proposes to solve the problem of process inconsistency by diluting the feed stream. Will that work? What will be the consequence of this proposal on the reactor volume? After hearing the proposal, another engineer proposed to combine diluting the feed stream with cell recycling while using the existing reactor. Comment on this proposal. Be as quantitative as you can.

E9 Chinese hamster ovary (CHO) cells are used to produce a phosphorylated glycoprotein in a continuous culture with cell recycling. The μ_{max} is 0.035 h⁻¹. The dilution rate used is 1.5 times of the μ_{max}. Under these operating conditions, cells have a finite death rate, and thus both viable and dead cells are seen in the culture. However, the cell separation device that is used preferentially recycles the viable cells with a 90% recovery and purges 10% of the viable cells. Conversely, for the dead cells, the recovery is only 50% (i.e., purging 50% of cells). The concentrations of viable and dead cells in the outlet stream are 10⁹ cells/L and 2×10^8 cells/L, respectively. The operation recycles 20% of the stream taken from the stirred-tank bioreactor to the bioreactor. Calculate the specific death rate of these cells.

E10 An open pond bioreactor is used to grow photosynthetic algae as feed supplement in fish farming. During the day, it is operated as a continuous culture with an objective of maximizing biomass output. The strategy is to supply sufficient CO_2 and nitrate to allow cells to grow as fast as the light source will support ($\mu_{max} = 0.69$ h⁻¹). However, because of decreased light penetration, the specific growth rate decreases when cell concentration is higher than 1 g/L, as described by:

$$\mu = \mu_{max}(5 - x)/4$$

where x: cell concentration in g/L, and is between 1 and 5.

To enrich the carbon source, sodium carbonate is added as the CO_2 source. The culture is to be maintained at a steady state, and CO_2 concentration in the pond is maintained at 0.03 g/L, a sufficient level to support the maximum growth rate. The conversion efficiency of CO_2 to glucose by photosynthesis is 100% (i.e., all CO_2 taken up by cells is converted to glucose). However, among the glucose assimilated, only half by mass is converted to biomass. Determine the optimal dilution rate to maximize the biomass output.

E11 An anaerobic *Clostridium* sp. is used to degrade cellulose and produce ethanol. It first hydrolyzes cellulose to glucose before converting the glucose to ethanol using only the glucose derived from cellulose as the carbon source. The stoichiometric relationship between the chemical species is measured in an experiment. For 100 g of glucose consumed, 8 g of cells, 46 g of ethanol, and 44 g of CO_2 are produced. When the organism is grown in a reactor, it is necessary to moderately sparge N_2 through the fluid, in order to remove any O_2 that might have leaked into the medium, and also to remove CO_2 to guard against CO_2 inhibition. The doubling time at its maximum growth rate is 1 h. The K_s (half saturation constant) with respect to glucose is 0.05 g/L. You are asked to design a continuous process that will produce an ethanol stream of 40 g/L at a dilution rate of 0.43 h^{-1}. Specify the glucose feeding rate and CO_2 removal rate.

7

Bioreactor Kinetics

7.1 Bioreactors

Soy sauce is a traditional condiment in many East Asian cultures. Some factories still use centuries-old methods, such as fermentation and aging soybeans in jars, to produce it (Figure 7.1). As in modern bioprocesses, where the medium is sterilized before the process begins, soybeans are first cooked to suppress the growth of unwanted microorganisms. Then the inoculum, consisting of fermented *koji*, is added. The mixed-culture fermentation goes on in stages, first dominated by molds of the *Aspergillus* genus, then by yeasts and lactic acid bacteria that become prominent in the later stages.

The fermentation process of soy sauce progresses rather slowly. The oxygen demand of the growing microorganisms is relatively small. The heat released from the metabolism of the microorganisms is also relatively moderate. The jar sits still on the ground, and oxygen reaches the fermenting soybean by diffusion through the solid. As the culture expands, oxygen diffusion into the solid substrate becomes more restricted. The temperature of the content in the jar also increases because the heat released from microbial metabolism becomes harder to dissipate. Using these small jars sitting stationary on the ground, the traditional method has made a high-quality product for centuries. But the jar is kept small. In a large jar, the microbial growth and fermentation cannot be sustained because of difficulties in oxygen delivery and heat dissipation.

Modern operations for soy sauce fermentation present a great contrast to these centuries-old processes. The reactor size and production capacity dwarf the old-fashioned jars. The fermenting soybeans are moved slowly on a conveyor. The temperature and humidity in the incubation chamber are well controlled. Importantly, mechanical mixing is used to enhance oxygen transfer and heat dissipation.

Without mixing, like the fermentation in the jar, increased production capacity is met by using a large number of jars. With mixing, the scale of the reactor can be amplified tremendously. One of the most important consideration in selecting a reactor type is thus mixing. Mixing facilitates the transfer of materials and energy from one location in the reactor to another. How much mixing is necessary depends on the reaction rate in the reactor. If the reaction is slow, the nutrients will have sufficient time to be distributed in the reactor, and a vigorous mixing may not be important. On the other hand, if the reaction rate is fast compared to the nutrient diffusion rate, then mixing becomes more important.

The nutrient diffusion rate of a solute in the gas, liquid, or solid phase differs by orders of magnitude. Few biological processes occur in the gas phase. A large number of food

Engineering Principles in Biotechnology, First Edition. Wei-Shou Hu.
© 2018 John Wiley & Sons Ltd. Published 2018 by John Wiley & Sons Ltd.
Companion Website: www.wiley.com/go/hu/engineering_fundamentals_of_biotechnology

Soy sauce factory, Wu Jen, Jiang Su, China

Figure 7.1 Solid-state fermentation for soy sauce production in a centuries-old production plant in China.

processes, such as soy sauce fermentation, entail solid-state fermentation. However, because of slow diffusion rates in particles, solid-state processes cannot sustain a high reaction rate. Providing adequate mixing to solid material is more difficult than handling liquid. Thus, other than food processing, virtually all other bioprocesses are carried out in the aqueous phase.

Compared to many chemical processes, biological processes are all conducted in relatively mild temperatures. The mild temperature of biological systems limits their reaction rate; the probability of molecular collision and chemical reaction is lower than what occurs at a high temperature. Each adult human consumes about 0.5 mole/kg of body weight-h of oxygen. Small mammals, such as rats, can consume oxygen at a faster rate because they can dissipate body heat more quickly. The order of magnitude of the reaction rate in a living system is around 1 mole/L-h. In comparison, the reaction rate of a combustion engine with a volume similar to that of an adult human is much higher. For example, burning 10 L/h of gasoline would consume approximately 1000 mole/h of oxygen.

However, although biochemical processes have lower reaction rates, they can produce products of far more complex structure. Even with the relatively slow reaction rates in biological systems, as the scale increases, the diffusion and mass transfer time also increases. As a result, mixing is necessary in most processes. We will discuss bioreactors with an emphasis on their mixing characteristics.

7.2 Basic Types of Bioreactors

We will use continuous reactors to illustrate mixing and reaction kinetics in reactors. Here, the reactor volume and density are considered constant. A feed stream and an exit (effluent) stream are operated at the same flow rate. We will use two types of idealized continuous reactors with two extremes of mixing, to demonstrate the effects of mixing on reactor performance.

In one case, the liquid content in the reactor is instantaneously and completely mixed upon the addition of any feed. The composition of fluid in the reactor is uniform; samples taken at any position in the reactor at the same moment will have the same composition. Assuming there is no time delay in withdrawing the effluent stream, the composition of the effluent stream is identical to that of the content in the reactor.

In the other extreme, the liquid content in the reactor is void of mixing. Feed entering at the inlet flows toward the outlet by moving like a plane or a plate. Any fluid element at the same plane of a given distance from the inlet will have the same composition. These two idealized continuous reactors are referred to as (1) well-mixed stirred-tank reactors, and (2) plug flow (tubular) reactors.

7.2.1 Flow Characteristics in Idealized Stirred-Tank (Well-Mixed) and Tubular (Plug Flow) Reactors

The distinction between a well-mixed, continuous stirred-tank reactor (CSTR) and plug flow reactor (PFR) is best shown by a comparison of their behaviors after a step change in the composition of the feed. First, we consider a continuous reactor that has an inlet stream (feed) and an outlet stream at the same volumetric flow rate. Initially, the fluid in both the reactor and the feed stream are colorless. At time $t = 0$, the feed stream is switched to a fluid containing a red dye at a concentration of C_0. We then observe how the dye concentration changes at the outlet.

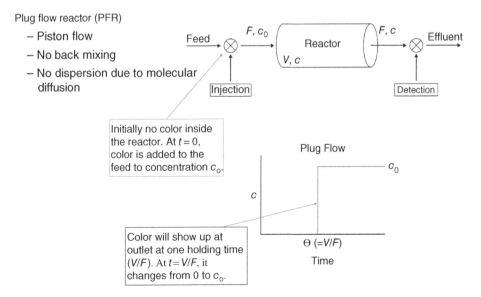

Figure 7.2 An idealized plug flow continuous reactor and the concentration profile resulting from a step change of concentration in the feed.

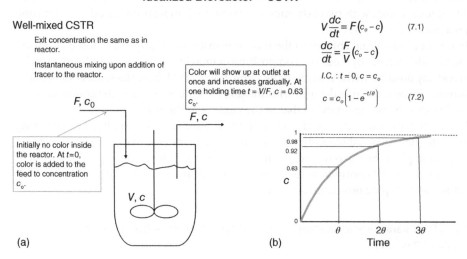

Figure 7.3 (a) An idealized stirred-tank reactor and (b) the concentration profile in the reactor resulting from a step change of concentration in the feed.

In a PFR, the red dye moves downstream like a sharp band, since there is no back-mixing or diffusion to blur the sharp boundary between the color and colorless streams (Figure 7.2). After switching the feed stream to the red color, it takes one holding time for the red color to appear at the exit. After the colored solution begins to exit, its concentration becomes identical to the feed concentration.

In a well-mixed CSTR, the dye will quickly be distributed in the reactor in a uniform manner as soon as the feed stream is switched to the red solution (Figure 7.3(a)). Since the effluent stream carries the red dye at the same concentration as in the reactor, the red color will appear instantaneously in the exit stream. As more colored feed comes into the reactor, the concentration of the color will gradually increase.

The balance equation for the dye concentration is shown in Eq. 7.1, where F is the flow rate, V is the reactor volume, c is the dye concentration, and the subscript o denotes inlet. Because we assume the density of the fluid is constant, we can perform material balance based on the concentration of the dye instead of using the total mass.

By setting the initial condition $c = 0$, one can solve the equation (Eq. 7.2) and plot the profile of the normalized concentration of the dye (c/c_o) over time (Figure 7.3(b)). V/F has units of time and represents the holding time (θ) (i.e., the time it takes for one reactor volume of feed to pass through the reactor). In a continuous culture of microbial cells, the inverse of holding time is called the dilution rate. One can see that, after one holding time, the concentration of the dye in the reactor is 0.63 of that in the feed. It takes three holding times (3θ) for the concentration to approach that of the feed ($\sim$98%).

7.2.2 Reaction in an Idealized CSTR

When a reaction is carried out in reactors with different mixing behaviors, the rate and yield of product formation may be rather different (Figure 7.4). Consider a first-order reaction converting reactant c to product p in a CSTR (Eq. 7.3 in Panel 7.1). The material balance equations for c and p thus include terms for the feed and effluent streams, and a term accounting for the reaction (Eqs. 7.4 and 7.5, in Panel 7.1). The r in Eqs. 7.4 and

When Reactions Occur in the Reactor

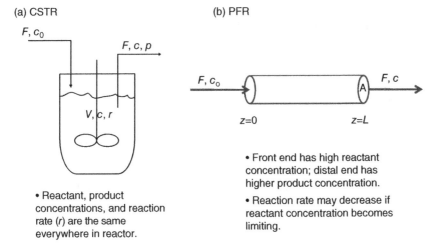

(a) CSTR

(b) PFR

- Reactant, product concentrations, and reaction rate (r) are the same everywhere in reactor.

- Front end has high reactant concentration; distal end has higher product concentration.
- Reaction rate may decrease if reactant concentration becomes limiting.

Figure 7.4 Idealized reactors with reactions, with symbols used in material balance.

7.5 is the reaction rate. Although a first order reaction is used for illustration purposes in Panel 7.1, it can be any mechanism or type of reaction, such as the enzyme reactions (discussed in Chapter 4) or cell growth, substrate consumption, and product formation reactions (discussed in Chapter 5). Again, the volume of the reactor and the density of its fluid content are considered to be constant. So the material balance can be performed on a concentration basis. After a steady state is reached, the reactant concentration c in the reactor decreases with increasing holding time (Eq. 7.7). After defining holding time (Eq. 7.8) the concentration of c can be expressed as a function of holding time (Eq. 7.9).

Panel 7.1 First-Order Reaction in a Stirred-Tank Reactor

Consider a first-order reaction:

$$c \xrightarrow{r} p \quad r = k \cdot c \tag{7.3}$$

$$V\frac{dc}{dt} = F(c_0 - c) - rV \tag{7.4}$$

$$V\frac{dp}{dt} = -Fp + rV \tag{7.5}$$

At steady State:

$$F(c_0 - c) = rV = k \cdot c \cdot V \tag{7.6}$$

$$(c_0 - c) = k \cdot c \cdot \theta \tag{7.7}$$

$$\theta = \frac{V}{F} \text{ (holding time)} \tag{7.8}$$

$$c = c_0 \left(\frac{1}{1 + k\theta}\right) \tag{7.9}$$

$$\text{Yield: } Y = \frac{c_0 - c}{c_0} = \frac{k\theta}{1 + k\theta} \tag{7.10}$$

The yield, or the fraction of reactant converted to the product, can be obtained from the concentration difference between the reactant in the feed and in the effluent streams (Eq. 7.10). Increasing the holding time thus increases the conversion or the yield. However, with a longer holding time, the output flow rate is lower. The overall throughput, or the productivity per reactor volume (i.e., quantities of product formed per reactor volume per unit time), will be affected. The operating condition is then optimized by balancing the cost of raw material and the throughput.

7.2.3 Reaction in an Idealized PFR

In a PFR, the reactant is consumed as the fluid moves downstream toward the exit. Its concentration decreases along the flow direction. We will now examine how the reactant concentration changes in the reactor. Let z be the position along the longitudinal direction of the reactor, and let 0 be the position at the inlet (Figure 7.4(b)).

In this example, the concentration profile has reached steady state at different positions in the reactor. In other words, the concentrations of the reactant and the product, although different in different positions in the reactor, are constant over time at any given position. We perform material balance on the concentration of the reactant over a very thin sectional slice in the reactor from the position z to $z+\Delta z$. The input into this small section is thus Fc evaluated at z, while the output is Fc evaluated at $z+\Delta z$. Since there is no mixing in an ideal PFR, the difference between input and output in the section is caused by the reaction whose rate is represented by r (Eq. 7.11 in Panel 7.2). If the reaction follows first-order kinetics, then r is represented by $-kc$. By taking a differential or reducing Δz and Δc to very small values, the equation is transformed into a differential equation of dc/dz (Eqs. 7.12 and 7.13, in Panel 7.2). The concentration profile of the reactant at the exit of the reactor can be obtained (Eqs. 7.14 and 7.15), and is the yield (Eq. 7.16).

Panel 7.2 First-Order Reaction in a Plug Flow Reactor

Reaction converting c to p: $c \xrightarrow{r} p$ $\quad r = \dfrac{dp}{dt} = -\dfrac{dc}{dt} = kc$

$$Fc\big|_{z+\Delta z} - Fc\big|_z = -A \cdot \Delta z \cdot r \tag{7.11}$$

$$F\frac{dc}{Adz} = -r = -k \cdot c \tag{7.12}$$

$$\frac{dc}{c} = -\frac{A}{F}kdz \tag{7.13}$$

$$\ln c = -\frac{AL}{F}k = -\frac{V}{F}k = -\theta k \tag{7.14}$$

$$c = c_0 Exp(-\theta k) \tag{7.15}$$

$$\text{Yield:} \quad Y = \frac{c_0 - c}{c_0} = 1 - Exp(-\theta k) \tag{7.16}$$

In the case that reaction r follows first-order kinetics (i.e., $r = kc$ at any position z), the concentration of the reactant decreases exponentially as the feed stream moves downstream, while the product concentration increases (Figure 7.5). Concurrently, the product concentration increases as the fluid moves toward the exit of the reactor. Similar to a CSTR, the conversion yield increases with an increasing holding time. If the reaction kinetics are zero order with respect to the reactant concentration,

Reactant Concentration in PFR for First-Order and Zero-Order Reaction

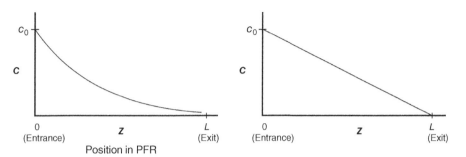

Figure 7.5 The concentration profile of the reactant along a plug flow reactor, for first-order and zero-order kinetics.

Plug Flow Bioreactor (Michaelis–Menten Kinetics)

Consider an enzymatic reaction following Michaelis–Menten kinetics, converting S to P, $(S \rightarrow P)$ taking place in a plug flow reactor in which the fluid flow down stream at a linear velocity (the flow rate divided by the cross sectional area of the reactor) of u. The material balance on the substrate as the fluid moves downstream is shown below.

$$\frac{ds}{dz} = \frac{-r_{max} \cdot S}{K_s + S} \cdot \frac{1}{u} \qquad (7.17)$$

Let the initial conditions be $z=0$, $s=s_0$; the concentration profiles of substrate can be obtained. (*Note: $dP/dz = -ds/dz$*.)

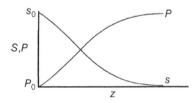

Figure 7.6 The concentration profile of the reactant for an enzyme reactor with Michalis–Menten kinetics in a plug flow bioreactor.

then the reactant concentration decreases linearly in a downstream fashion along the longitudinal direction of the reactor (Figure 7.5).

Now, we will consider a case in which the reactor is used for the enzymatic conversion of a substrate to a product, and the reaction follows Michaelis–Menten kinetics (Eq. 7.17 in Figure 7.6). The substrate concentration in the feed is high, so the reaction rate is in the zero-order region. As the fluid flows downstream, it is then consumed, and the product accumulates.

In the front part of the reactor, the reaction is in the zero-order region, and the substrate concentration decreases linearly with increasing distance from the inlet. The product concentration also increases linearly in a similar matter away from the inlet. As the concentration decreases further downstream, the reaction enters the

Similarity in the Behavior of a PFR and Batch Reactor

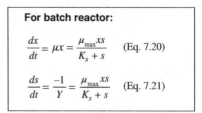

For PFR:

$$\frac{dx}{dz} = \frac{\mu_{max} xs}{K_s + s} \cdot \frac{1}{u} \quad \text{(Eq. 7.18)}$$

$$\frac{ds}{dz} = -\frac{\mu_{max} xs}{Y(K_s + s)} \cdot \frac{1}{u} \quad \text{(Eq. 7.19)}$$

For batch reactor:

$$\frac{dx}{dt} = \mu x = \frac{\mu_{max} xs}{K_s + s} \quad \text{(Eq. 7.20)}$$

$$\frac{ds}{dt} = \frac{-1}{Y}\frac{\mu_{max} xs}{K_s + s} \quad \text{(Eq. 7.21)}$$

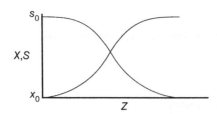
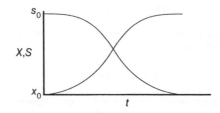

Figure 7.7 The concentration profiles of cell and substrate in a plug flow and a batch bioreactor.

first-order region and the reaction rate begins to decrease, so the slope of the substrate concentration curve also begins to decrease.

Next, consider a case where a PFR is used for growing microbial cells that follow Monod kinetics (Eqs. 7.18 and 7.19, in Figure 7.7). At the entrance of the reactor, the substrate concentration is sufficiently high to support the maximal growth rate. As the fluid moves downstream, the substrate concentration decreases, but the specific growth rate will be constant as long as the substrate is still in the zero-order region. The cell concentration increases exponentially until the substrate concentration decreases to a lower level and the specific growth rate begins to decrease. The cell concentration then reaches its maximum as the substrate gradually becomes depleted (Figure 7.7).

The kinetic behavior of a PFR resembles that of a batch reactor, if one interchanges the time scale of a batch reactor with the time it takes for a fluid element to move along the reactor. For a fluid element entering at the entrance and moving along the longitudinal direction, its position in the reactor is related to the time it spends in the reactor: $t = z/U$, where $U(= F/A)$ is the linear flow rate or the velocity of the stream in the reactor. The model equations for PFRs can be rewritten by substituting z/U with t to become model equations, with t as the independent variable (Eqs. 7.20 and 7.21, in Figure 7.7). The model equations are then identical to the growth kinetic equation for cell growth, discussed in Chapter 5. The growth kinetics described in Chapter 5 are essentially those for a batch culture with a uniform culture environment (i.e., stirred-tank conditions). The concentration plot along the reactor length also depicts those in a batch reactor if z is replaced by $t(= z/U)$ (Figure 7.7).

A stirred-tank bioreactor, when operated in batch mode, thus has the same kinetic behavior as a plug flow bioreactor. The kinetics of biochemical reactions discussed in Chapter 4 and the growth reactions discussed in Chapter 5 also describe kinetics in batch reactors. In Chapter 6, we discussed the kinetic behavior of continuous culture in a stirred-tank bioreactor; we will not specifically discuss it again in this chapter.

7.2.4   Heterogeneous and Multiphasic Bioreactors – Segregation of Holding Time

A bioreactor typically includes three phases: a liquid phase (medium), a solid phase (cells), and a gas phase (aerating bubbles). Often, cell mass represents a very small fraction of the culture volume and is considered negligible. Consequently, even though a bioreactor volume is often considered as having a biotic phase (cells) and an abiotic phase (medium), it is rather common that the volume occupied by the biotic phase is ignored, and the content in the cell cultivation bioreactor is treated solely as a homogeneous liquid phase. In this case, cells are treated like a "soluble" component.

However, in some microbial fermentation and plant cell cultures, cells may make up a large fraction (up to 50%) of the total reactor volume. In these cases, the culture must be treated as a heterogeneous system. The reactor culture volume is thus partitioned into cell volume and medium volume. When describing the cell and nutrient concentrations, one must specify whether the concentrations are based on medium volume or total reactor volume. Most bioprocesses involving cell growth use aerobic microorganisms, thus requiring continuous aeration of a gas phase (air bubbles). However, except in oxygen transfer or in scale-up discussions, the culture is treated as a homogeneous liquid phase, and material balance and kinetic analyses are performed based on the liquid phase component.

Although most bioreactors are assumed to be single phase and homogeneous, heterogeneous bioreactors are employed in many processes, especially in biocatalysts. Cells and enzymes that are used as biocatalysts are often immobilized on a solid support to convert the reactants to products. For example, large-scale, packed-bed bioreactors with an immobilized enzyme, glucose isomerase, are used to convert glucose to a ~50:50 mixture of glucose and fructose. This is used in high-fructose corn syrup.

In a continuous biocatalytic reactor, those immobilized catalysts are retained in the reactor while the substrate and product are continuously supplied and withdrawn. In the cultivation of anchorage-dependent animal cells, microcarriers are frequently used to support cell attachment and growth and are used in a high concentration, sometimes taking up 20–30% of the culture volume.

Many bioreactors used for tissue engineering are also heterogeneous. The reactor chamber usually involves two compartments: one for fluid flow to supply nutrients and remove metabolites, and one to grow tissue cells on a support (e.g., collagen) or on a biodegradable scaffold that can be transplanted. After transplantation, the material is then gradually degraded and leaves the tissue cells behind to integrate into the host tissue. These bioreactors are genuinely heterogeneous and must be considered to have two separate liquid and solid phases in material balance analyses.

7.3   Comparison of CSTR and PFR

7.3.1   CSTR versus PFR in Conversion Yield and Reaction Rate

An important distinction between PFR and CSTR is the degree of uniformity of the constituents in the reactor. In an idealized CSTR, the reactant concentration is uniform throughout the reactor, so is the reaction rate. In a PFR, conversely, the reactant concentration varies from high at the inlet to low at the exit. The reaction rate differs significantly over the flow direction of a PFR, unless the reaction kinetics is zero order.

Consider a CSTR and PFR both giving rise to the same high conversion yield in a first-order reaction, wherein both will have a low exit concentration of c. In the case of a CSTR reactor, the reaction rate will be equally low in all positions in the reactor. In a PFR, although the concentration of c at the exit is the same as in a CSTR, its level can be rather high in the upstream region. The frontal part of a PFR would therefore have a higher reaction rate. As a result, given the same reaction, the size of a PFR required to accomplish the same yield is smaller than a CSTR (or the required holding time is shorter). This is also evident from the expression of yield coefficient (or the conversion) as a function of holding time [Eq. 7.10 (Panel 7.1) and Eq. 7.16 (Panel 7.2)]. For bio-catalysis processes using enzymes or immobilized cells and for reactions with first-order kinetics, a PFR is the reactor of choice.

Panel 7.3 Mass Balance for a Fed-Batch Culture

$$\frac{d(xV)}{dt} = \mu xV = \frac{\mu_{max}s}{K_s + s}xV \tag{7.22}$$

$$\frac{d(sV)}{dt} = F(t)s_0 - \frac{\mu x}{Y_{x/s}}V \tag{7.23}$$

$$\frac{dV}{dt} = F(t) \tag{7.24}$$

$$V\frac{ds}{dt} + s\frac{dV}{dt} = Fs_0 - \frac{\mu x}{Y}V$$

$$V\frac{ds}{dt} = F(s_0 - s) - \frac{\mu x}{Y}V \tag{7.25}$$

7.3.2 CSTR versus PFR in Terms of Nutrient Depletion and Scale-Up

Plug-flow bioreactors are intrinsically more difficult to scale up than mixing vessels, since the concentration of essential nutrients decreases in the longitudinal direction. Eventually, the nutrient becomes limiting in the downstream region. Thus, extending the reactor length beyond the nutrient-limiting region will not be effective. This is in contrast to a CSTR, in which concentrations can be maintained above the critical level and the reactor size can be increased while still maintaining sufficient supply to all regions in the reactor.

To overcome the size limitation in a PFR, one may increase the nutrient supply rate by using a higher nutrient concentration in the feed, or by using a higher flow rate. However, there are practical limits at the level of nutrient concentration and flow rate. A high nutrient concentration may be limited by its solubility or the inhibitory effect on cell growth. A high flow rate will require a higher capacity pump to overcome the higher pressure drop across the reactor. Such a limitation is especially acute for the cultivation of aerobic organisms because of the low solubility of oxygen in water. In a PFR in which oxygen is supplied by the feed stream, oxygen will be quickly depleted after a short distance from the inlet (Example 7.1).

Thus, the size and scalability of a PFR reactor, for the cultivation of aerobic organisms, are rather limited. If one introduced continuous aeration through direct air bubbling (or "sparging") in the reactor, a significant degree of mixing would also need to be

Residence Time Distribution in a PFR and CSTR

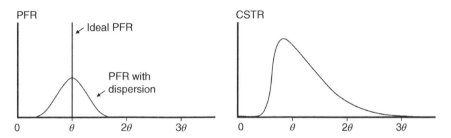

Figure 7.8 The distribution of the residence time of fluid elements in a plug flow bioreactor and a continuous stirred-tank bioreactor.

introduced. This would adversely affect the fluid flow pattern and mixing characteristics of a PFR reactor. This limitation on oxygen supply makes PFRs ill-suited for the cultivation of aerobic organisms, except on small scales.

To employ a PFR for aerobic processes, we can use membrane oxygenation to supply oxygen without resorting to fluid mixing. In a wastewater treatment plant, for example, the bioreactor is operated closer to a PFR than a CSTR; a limited degree of oxygen sparging is used to supply oxygen with minimal mixing.

7.3.3   CSTR versus PFR – A Perspective from Residence Time Distribution

In describing the two idealized bioreactors in this chapter, we assume that the fluid elements in the reactor are either (1) mixed completely and instantaneously, or (2) completely isolated and never mix. In reality, bioreactors do not perform in these absolute, idealized extremes. Rather, they fall on a spectrum between the two, leaning more toward one direction or the other.

By plotting the amount of time each fluid element spends in an idealized PFR, one obtains a delta function. That is, all fluid elements would have spent exactly 1 holding time in the reactor (Figure 7.8). In a real bioreactor that employs tubular flow, the distribution curve will have a spread around 1 holding time. Some fluid elements would have spent <1 holding time before exiting the reactor, while others would have spent >1 holding time. The extent of deviation from an ideal PFR is affected by the flow rate, the geometry of the reactor, and the fluid dispersion, or the extent of fluid mixing due to flow.

Similarly for a CSTR, there will be a distribution over the time each fluid element spends in the reactor. The spread of the distribution for a CSTR is substantially wider than in a PFR; some elements may exit the reactor soon after they enter, while others may spend a long duration before they exit (Figure 7.8). In general, a larger proportion of fluid elements have <1 holding time.

The distribution of the time each fluid element spends in the reactor is often called the "residence time distribution." Residence time distribution is an important way to characterize mixing and is especially important in medium sterilization. Consider an extreme case of chemical reactor operation, where half of the reactor volume has a long residence time with a conversion yield of 95%, and the other half has a short residence time with a conversion yield of 90%. The overall process yield is thus 92.5%. So, while segregation may not be desirable, the outcome may still be acceptable.

Now, consider an operation to sterilize the medium. If half of the reactor volume is sterilized 1.5 times longer than the designated sterilization time, while the other half is sterilized for only half of the time duration required, the outcome may be catastrophic and lead to contamination of the reactor. A very narrow distribution of residence time is thus crucial for sterilization, to avoid oversterilization of nutrients and understerilization of contaminating microbes. Consequently, virtually all continuous sterilization processes are carried out in a tubular flow type of reactor. This ensures a narrow range of residence time distribution. Well-mixed reactors are only used in batch sterilization operations.

Example 7.1 A Plug Flow Bioreactor for Cultivating Aerobic Microorganisms

We will examine the feasibility of using a plug flow bioreactor for growing an aerobic bacterium. The nutrients and oxygen will be flown through the reactor. At the entrance to the reactor, the microorganism is inoculated at a concentration of 1.0 g/L. The oxygen concentration at the inlet is 0.2 mmol/L. The nutrient and oxygen concentrations at the inlet are sufficient to support the bacterium to grow at its maximum specific growth rate of 1.0 hr^{-1}. The consumption of oxygen can be assumed to follow Monod kinetics with a K_s value of 0.01 mmol/L. The yield coefficient of biomass is based on oxygen of 0.033 g cell/mmol O_2. The cross-sectional area of the reactor is 100 cm^2. With a medium flow rate of 1 L/min, what is the maximum length that the reactor can be productively producing the bacterium?

Solution

We will use the maximum oxygen consumption to do the calculation. Basically, we will assume that the oxygen consumption follows zero-order kinetics, rather than using Monod kinetics. Note that the K_s value is 5% of the input oxygen concentration. Most of the cell biomass increase occurs well before the oxygen concentration decreases to K_s. The results obtained using zero order kinetics assumption will be very similar to that obtained with a detailed calculation using Monod kinetics for oxygen consumption.

Oxygen demand [oxygen uptake rate (OUR)] at the given cell concentration:

$$OUR = \frac{\mu x}{Y_{x/O}} = \frac{1\ h^{-1} \cdot 1\ gL^{-1}}{0.033\ g\ mmol^{-1}} = 30\ mmol/L - h$$

Oxygen supply rate (flow rate multiplied by inlet oxygen concentration):

$$Q \cdot C_o = 60\ L/h \cdot 0.2\ mmol/L = 12\ mmol/L - h$$

Next we calculate the critical volume, V_c, that can sustain the OUR:

$$OUR \cdot V_c = Q \cdot C_o$$

$$V_c = \frac{12\ mmol/L - h}{30\ mmol/L - h} = 0.4\ L$$

The length of the reactor is:

$$\ell = \frac{0.4\ L}{100\ cm^2} = 4\ cm$$

Oxygen will be depleted after 4 cm from the inlet.

The allowable reactor size would be too small even for such a low cell concentration.

It is clear that oxygen concentration will be quickly depleted in a PFR used for cell cultivation. Therefore, PFRs are rarely used for large-scale operations of cell cultivation.

7.4   Operating Mode of Bioreactors

The operation of a bioreactor is generally classified as a batch or continuous mode. In a continuous process, the feed is continuously being introduced into the reactor, while the product stream is continuously being withdrawn from the reactor. In a batch process, the medium (including all nutrients with the exception of oxygen) is added at the beginning. No other nutrient addition occurs until the end of cultivation. Continuous culture is not commonly practiced in the cultivation of mammalian cells, unless in conjunction with cell recycling.

7.4.1   Batch Cultures

Batch processes are simple and widely used. Before reaching the production scale, a series of batch cultures (referred to as a "seed train" or "seed cultures") are used to expand the culture volume and cell mass, to provide the quantities of cells needed for starting the production bioreactor. Such batch cultures are mostly operated only in the exponential growth phase, to ensure rapid cell expansion in the seed train. In the final production batch culture, the culture period may be extended into the stationary phase to prolong the period of production (Figure 7.9). The kinetic expression for growth and product formation discussed in the kinetics chapter (Chapter 5) can be readily used as a mathematical model for batch processes.

7.4.2   Fed-Batch Cultures

Batch cultures are often limited by the product concentrations that can be achieved. To achieve a high product concentration, it is necessary to have high cell concentration and a high starting nutrient concentration in the medium. In some cases, a high concentration of nutrients is not attainable due to the solubility limit or growth inhibition. In such cases, a feeding stream can be employed to provide additional substrates or product precursors to sustain continued growth and production.

7.4.2.1   Intermittent Harvest

Fed-batch processes do not differ significantly from batch cultures. The simplest form of fed-batch culture involves intermittent harvest (Figure 7.9(c)). At a late exponential growth stage of the culture, a portion of the cells and product is harvested. The culture is replenished with fresh medium. This minimizes metabolite inhibition of cell growth and replenishes nutrients for continued production. Such harvest and medium replenishment may be repeated several times. This simple strategy is used for the production of viral vaccines by persistent infection, as it allows for an extended production period.

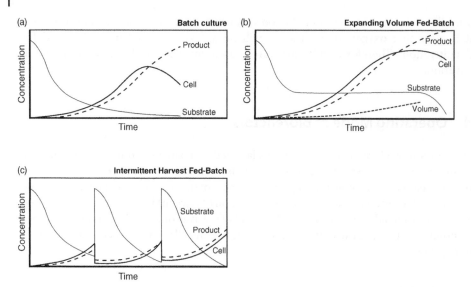

Figure 7.9 Growth kinetics in a stirred-tank bioreactor operated at (a) batch, (b) intermittent fed-batch, and (c) continuous feeding fed-batch mode.

7.4.2.2 Fed-Batch

Fed-batch processes are frequently used in the production of microbial metabolites and recombinant proteins. A fed-batch culture is started with a volume lower than the full capacity of the bioreactor (approximately 60–80% of the maximum volume, or at a level sufficient to allow the impeller to be submerged). Nutrients, usually in a more concentrated form than in basal medium, are added during the cultivation to allow cell and product concentrations to reach much higher levels than can be attained in batch culture (Figure 7.9(b)).

Modeling of fed-batch processes is virtually identical to that of batch processes, except that the material balances are performed on the total mass of each species (thus, a multiplicative product of concentration and volume) instead of the concentrations (Eqs. 7.22 and 7.23 in Panel 7.3). Furthermore, a volume balance should be set up to account for volume increase during fed-batch culture.

Fed-batch cultures are also used to control process dynamics in different stages of cultivation. For example, during the production of recombinant proteins in *Escherichia coli*, a fast-growing stage is first introduced to allow the cell concentration to quickly reach a high level. Inducer is then added to initiate the synthesis of recombinant protein. After the induction of recombinant protein synthesis, the growth rate is controlled at a lower level to reduce the rate of protein synthesis and to avoid protein aggregation. This can be accomplished by slowly feeding a growth rate-limiting substrate (typically glucose) to restrict cell growth. At the same time, nutrient feeding is also closely monitored to avoid starvation. Using kinetic equations for cell growth and substrate consumption, one can prescribe the feeding rate of the limiting nutrient (Eq. 7.24 in Panel 7.4).

Panel 7.4 Strategy for Keeping the Substrate Level Constant in a Fed-Batch Culture

$$\frac{dxV}{dt} = \mu xV = \frac{\mu_{maxs}}{K_s + s} xV \tag{7.22}$$

Set $ds/dt = 0$

$$0 = F(t)(s_0 - s) - \frac{\mu x}{Y_{x/s}} V$$

$$F(t) = \frac{\mu x}{Y_{x/s}} V \cdot \frac{1}{(s_0 - s)} \tag{7.26}$$

7.5 Configuration of Bioreactors

7.5.1 Simple Stirred-Tank Bioreactor

Stirred tanks, or conventional fermenters, have been used widely for growing microorganisms and plant and animal cells since the 1950s. The basic configuration of stirred-tank bioreactors consists of a water-jacketed vessel with an aspect ratio (the ratio of height to diameter) varying from >5, in very large-scale reactors, to only slightly over 1 in small specialty (mostly cell culture) reactors (Figure 7.10a). In general, the aspect ratio is usually smaller in mammalian cell culture bioreactors than it is for microbial fermentation. Most microbial fermenters use Rushton turbines as the main impeller. Their rotational speed ranges from over 500 rpm in small laboratory fermenters to ~100 rpm in very large fermenters.

Depending on the type of cell and product, the fluid may be Newtonian or highly viscous and non-Newtonian. In industrial large fermenters, multiple impellers are used. They are spaced at just over 1 impeller diameter apart. To increase the efficiency of heat and mass transfer, baffle plates protruding from the vessel wall are often installed (Figure 7.10a). These plates break up the liquid moving along the vessel wall rotationally, thus creating more turbulence and providing a better mixing.

At the bottom of the fermenter, beneath the lowest impeller, is the sparger. From this, air enters the reactor. There are a variety of sparger designs. In some small-scale cell culture bioreactors, microspargers with ~100-μm micropores generate fine bubbles. They are used to increase the surface area of the gas phase for enhancing oxygen transfer. In large reactors, a sparger can be as simple as a single, open pipe. Air bubbles released from the sparger move upward and hit the bottom side of the impeller plate of the Rushton turbine. The air bubbles then move outward toward the edge of the plate, where they encounter the quickly rotating vertical blade of the turbine. Here, they are broken into a swarm of smaller bubbles that rise upward and outward in the fluid.

Before the start of the process, steam is injected into the water jacket as well as the interior of the vessel for sterilization purposes. The water jacket surrounding the vessel is also for providing cooling water during cell cultivation. The heat arises from the agitation

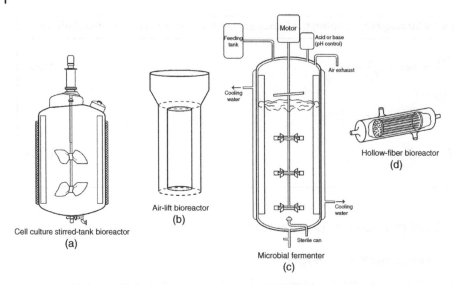

Figure 7.10 Different type of bioreactors: (a) a stirred tank for microbial culture, (b), a stirred tank for cell culture, (c) an air-lift fermenter, and (d) a hollow-fiber bioreactor.

power and the metabolic heat released by growing cells. In industrial fermenters, it is not unusual to see cooling coils installed in the interior of a fermenter to increase the heat-removal capacity.

The general geometry and shape of a stirred-tank bioreactor for mammalian cell culture are very similar to those used for microbial fermentation. However, cell culture bioreactors have several distinct features. For example, the agitation power input per unit volume of a bioreactor is substantially lower in mammalian cell culture bioreactors. Also, while the Rushton-type impeller is the norm in microbial fermenters, mammalian cell culture fermenters typically employ axial flow-type impellers. This difference reflects the differing purposes of agitation in microbial fermentation and in cell culture. In microbial fermentation, agitation is needed at a higher power input to disperse air bubbles and to increase oxygen transfer efficiency. In mammalian cell culture, the primary purpose of agitation is to maintain a relatively uniform distribution of cells and medium components.

Generally, the mixing time in a mammalian cell culture bioreactor is substantially longer than that in a microbial fermenter of similar scale. The oxygen transfer capacity in a cell culture bioreactor is also substantially lower than that in a microbial fermenter. In general, the typical oxygen demand in a mammalian cell culture is 10–50 times lower than that in microbial fermentation.

In addition to the hardware described here, a bioreactor is typically installed with sensors for pH, dissolved oxygen, and other measurements. These sensors are designed to withstand the temperature of steam sterilization. Various ports for withdrawing culture samples and for connecting to transfer pipes are also in place. These sensor, sampling, and transfer ports are designed with an emphasis on creating sufficient barriers for the entry of bacteria from the surrounding environment. The goal is to virtually eliminate the chance of introducing microbial contamination into the interior of the bioreactor.

7.5.2   Airlift Bioreactor

A bubble column reactor can be simply a tubular vessel with an entry port at the bottom for gas flow. Such reactors are used in gas–liquid reactions. Airlift bioreactors are essentially bubble column reactors with an internal circulation loop for the cultivation of mammalian, insect, and plant cells. In these airlift reactors, internal liquid circulation is achieved by sparging through the internal draft tube (Figure 7.10c). The fluid in the draft tube has a lower effective density than the bubble-free section. Along with the upward momentum generated by airflow, this induces liquid circulation. Media flows upward through the sparged section (riser) and downward in the bubble-free section (downcomer). This method of generating circulation has a low energy requirement compared with stirred-tank reactors.

Because there is no mechanical agitation, airlift bioreactors for cell cultivation are considered to be low-shear devices. They have been used successfully with suspension cultures of BHK-21, human lymphoblastoid, Chinese hamster ovary (CHO), hybridomas, and insect cells.

7.5.3   Hollow-Fiber Bioreactor

A hollow-fiber system is mostly/frequently used in kidney dialysis and for bioseparations. For kidney dialysis, the system consists of tens of thousands of hollow fibers encased in a chamber that forms the extracapillary space. Hollow-fiber systems have also been used for both anchorage-dependent and suspension cells. In these cases, they consist of a bundle of capillary fibers sealed inside a cylindrical tube. The basic configuration is similar to the hollow-fiber cartridge used in kidney dialysis (Figure 7.10d).

The hollow fiber entails a porous, polymeric layer (providing mechanical support) covered by a thin membrane that provides selectivity based on the size of molecules. In most cases, an ultrafiltration membrane is used. The molecular weight cutoff (MWCO) of the membrane differs according to the specific application, ranging from a few thousand to a hundred thousand Daltons.

The culture media are pumped through the fiber lumen. In most applications, cells are cultivated in the extracapillary space, or on the shell side. Supply of low-molecular-weight nutrients to the cells and the removal of waste products occur by diffusive transport across the membrane, between the lumen and the shell spaces. The ultrafiltration membrane prevents free diffusion of secreted product molecules from passing through the membrane. This allows them to accumulate in the extracapillary space at a high concentration.

Hollow-fiber microfiltration membranes are not frequently used for cell culture. However, they do appear in various research applications for studying metabolism, and for producing small quantities of materials for research or diagnostic applications.

7.6   Other Bioreactor Applications

The bioreactor is the core of bioprocesses for growing cells, enzyme conversions, and synthesis. Bioprocess-based manufacturing produces various biochemicals and pharmaceuticals. More recently, bioreactors have been used to produce cells for therapeutic applications, or even to generate engineered tissues derived from cultured cells.

Today, bioreactors are being used for innovative purposes beyond manufacturing. Many of those applications are a product of the past 25 years of progress in life sciences. Now, reactions can be used for measurement of various biochemicals for diagnostic, research, or discovery purposes. Large-scale bioreactors are used to maximize economic production for manufacturing. In contrast, small-scale operations have their own advantages: reduced use of reagents, less cost for research and development, and the possibility of high-throughput operations.

Many such applications involve microfluidic systems. They use a reactor chamber that has a volume from tens of microliters to sub-microliters. These devices have microscale fluid delivery systems, which transfer a precise amount of liquid at a precise time, mix the liquids, and adjust the temperature of the device for the reaction to proceed for a given period of time. The content of the reactor is then transferred to a sensor to measure the concentration of a reaction product, or to the next reactor chamber.

In one such application, a population of cells that have undergone some treatment are fed into the system. The fluid distribution system passes the cell suspension through a detector. For each cell that passes through the detector, a bubble is injected. As a result, each cell is preceded and followed by a bubble and can then be directed to a culture chamber that contains a small volume of culture media. A device can be designed to handle thousands of cells in a short time. Cells in each chamber are allowed to grow. While growing, the cells increase in number, and their growth rate can be monitored. At the end of the cultivation period, the culture fluid can be directed to another chamber to quantify the cell product or to measure some other characteristics of the cells.

In another application, individual cells are directed to reaction chambers individually. Reagents for lysing cells and releasing RNA are then added. After being transferred to another chamber, the single-stranded RNAs can then be converted to DNA molecules to amplify their quantity. The content of each chamber can then be transferred to another chamber, where polymerase chain reaction (PCR) is carried out to detect DNA containing specific sequences. This method allows for one to detect the expression of a given messenger RNA (mRNA) through the simultaneous processing of many single cells.

Note that in these nonmanufacturing applications, bioreactors or microreaction chambers are used in tandem or in series. Practically speaking, even in manufacturing processes, reactors are employed in series: from a small scale, and then serially transferred to the next larger scale until they become the manufacturing scale that is discussed in Chapter 1. In enzymatic processes, it is not unusual for the product stream from a reactor to be fed into a subsequent reactor. Because the time scale in between different reactors operating in batch mode is relatively long (e.g., often in hours if not days), each reactor in a series tends to be analyzed separately. For microfluidic systems, especially in applications involving PCR or other nucleic acid manipulations, the time scale has to be short.

Models can be set up to account for reactions in each of the reactors in the series. In between the reaction steps, additional reagents are added. This means that each reactor may have different volumes. The product of a preceding reactor becomes the reactant of the next, while the unreacted product from the first reactor may undergo side reactions or result in adverse effects. A model linking the submodels that describe the behavior of individual reactors can be established and used for process optimization.

7.7   Cellular Processes through the Prism of Bioreactor Analysis

Although the discussion in this chapter centers on the bioreactors for bioprocesses, the concepts introduced are applicable to many biological processes inside the cell or in a multicellular organism. Mixing characteristics in various cellular compartments or tissues can be vastly different. In analyzing their kinetic behavior, we may find that we can model tissue as a CSTR or PFR, or make considerations of residence time distribution.

For example, blood flow through vessels clearly resembles plug flow, whereas food digestion in the stomach resembles a stirred tank. In the secretion of extracellular proteins, the protein molecules synthesized and folded in the endoplasmic reticulum (ER) are taken as cargo into membrane vesicles. These vesicles bud off from the ER membrane and translocate to the Golgi, where they fuse with the Golgi membrane and release the cargo into Golgi apparatus. The maturing protein molecules undergo various posttranslational modifications, including glycosylation, in different functionally distinct compartments of the Golgi apparatus (Figure 7.11). The protein product molecules are eventually loaded as cargo in secretory membrane vesicles. The secretory vesicles move along microtubules to the cytoplasmic membrane, where they fuse with the membrane and release their cargo into the extracellular environment.

Many cell biologists have been curious whether this process of transporting maturing protein molecules resembles a plug flow or a mixing tank. Different mechanisms of

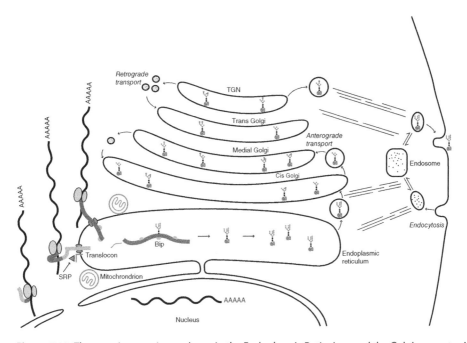

Figure 7.11 The protein secretion pathway in the Endoplasmic Reticulum and the Golgi apparatus in a eukaryotic cell.

transport will have different residence time distributions and will profoundly impact a protein molecule's ability to be posttranslationally modified. As the protein molecules pass through different Golgi compartments, their glycans are lengthened and become more complex. If the process of protein secretion resembles a plug flow reactor, all protein molecules will have a similar holding time, which would yield a more uniform structure.

On the other hand, if the process resembles a stirred tank in each compartment, then the heterogeneity of the glycans will be greater. If mixing occurs (even between different compartments of the Golgi apparatus), then the degree of heterogeneity will be even greater.

In the production of proteins as biological therapeutics, the glycan structure often plays a key role in determining pharmacokinetic behavior (see Chapter 10). A better understanding of the secretory process, from a bioreactor perspective, will help us to devise ways to modulate cellular physiology for a more robust process.

7.8 Concluding Remarks

In this chapter, we discussed the behavior of the two idealized reactors, PFR and CSTR. PFRs find applications in enzymatic processes and medium sterilization because of their high conversion yield and homogeneous residence time distribution. For the vast majority of bioprocesses, stirred-tank bioreactors or other reactors with more extensive mixing are the choice because of the need to deliver oxygen to aerobic organisms in the culture fluid. More recently, miniaturized bioreactors are finding applications other than biomanufacturing in areas like analytics, diagnostics, and high-throughput studies. As the biochemical engineering field increasingly deals with intracellular processes and intercellular or intertissue reactions, concepts in reactor kinetics are also employed in the analysis of those systems.

Further Reading

Hill, CG and Root, TW (2014) Introduction to chemical reaction kinetics and reactor design. 2nd ed. John Wiley & Sons.

Huang, TK and McDonald, KA (2012) Bioreactor systems for in vitro production of foreign proteins using plant cell cultures. Biotechnol. Adv., **30**, 398–409.

Posch, AE, Herwig, C and Spadiut, O (2013) Science-based bioprocess design for filamentous fungi. Trends Biotechnol., **31**, 37–44.

Sikorski, DJ, Caron, NJ, Vaninsberghe, M, Zahn, H, Eaves, CJ, Piret, JM and Hansen, CL (2015) Clonal analysis of individual human embryonic stem cell differentiation patterns in microfluidic cultures. Biotechnol. J., **10**, 1546–1554

Ziongaro, KA, Nicolaou, SA and Papoutsakis, ET (2013) Dissecting the assays to assess microbial tolerance to toxic chemicals in bioprocessing. Trends Biotechnol., **31**, 643–653.

Nomenclature

A	cross-sectional area	L^2
c	concentration of a marker dye or a reactant	$mole/L^3$
c_o	concentration of c in the feed	$mole/L^3$ M/L^3
F	volumetric feeding rate of substance	L^3/t
K_s	half-saturation constant of specific growth rate	M/L^3, $mole/L^3$
k	reaction rate constant	t^{-1} (first-order kinetics)
L	length of reactor	L
p	product concentration	$mole/L^3$ M/L^3
r	reaction rate	$mole/L^3 \cdot t$
s	substrate concentration	
s_o	substrate concentration in the feed	$mole/L^3$ M/L^3
t	time	t
u	linear velocity	L/t
V	volume of reactor	L^3
x	cell concentration	M/L^3
Y	yield coefficient	M/M
$Y_{x/s}$	yield of biomass based on substrate	M/M
z	reactor length coordinate	L
θ	holding time	t
μ	specific growth rate	t^{-1}
μ_{max}	maximum specific growth rate	t^{-1}

Problems

A1 Membrane bioreactors are frequently used in tissue-engineering applications. There are many different configurations. One commonly used flatbed membrane reactor has two flat sheets of permeable membrane that divide the bioreactor into three chambers. The middle chamber is used to place cells in a gel matrix or a scaffold. The other two chambers, one above and the other below the cell chamber, are for medium flow. The medium is supplied from one end of the reactor and flows in the direction parallel to the membrane. When it enters the reactor, it splits into two streams of equal flow rate in the two medium chambers. Oxygen and other nutrients diffuse through the membrane to supply to cells in the cell chamber. The metabolites produced by cells also diffuse from the cell chamber to the medium chamber and are carried away.

The two streams of medium exiting the other end of the reactor are combined in a medium reservoir for recirculation into the bioreactor. However, before recirculating to the cell bioreactor, the medium is pumped through another membrane reactor that is typically of the hollow-fiber membrane type. In the hollow-fiber membrane reactor, the medium passes through the lumen of the hollow fibers while air flows in the extracapillary space of the fibers to enrich the medium with oxygen. The oxygenated medium is then recirculated into the cell bioreactor to cultivate cells.

In the cell chamber, cells at a given concentration (x) consume oxygen at a specific rate (q_{02}).

a) Sketch a drawing of the system, including cell bioreactor, reservoir, and oxygenator. Assign symbols of important variables that are necessary to quantify the kinetics of the system.

b) Set up a model to describe the oxygen concentration in the medium chamber along the longitudinal direction of the bioreactor. The oxygen permeability [1/s(cm²)] is P. Assume that the cells are distributed uniformly in the longitudinal direction and that diffusion is the dominating mechanism of oxygen transfer.

A2 A hollow bioreactor is used to sustain liver cells for use as an artificial liver in a setup similar to the one described in Problem A1, except that the cell bioreactor is also a hollow-fiber system. The cells are kept in the extracapillary space at a concentration of 3×10^{11} cells/L. Their specific oxygen consumption rate is 3×10^{-10} mmol/cell-h, and the total extracapillary volume is 80 mL. The oxygen is supplied by recirculating the medium. The medium, after oxygenation, enters the bioreactor at a dissolved oxygen concentration of 1 mmol/L (which is about five times more saturated than air because pure oxygen is used) and a flow rate of 200 mL/min. The medium flows in from the upstream end of the reactor and exits from the downstream end, and can be assumed to behave like a piston flow (plug flow). What is the oxygen concentration in the medium at the point of exit from the bioreactor?

A3 Glucose is converted to ethanol in a packed-bed reactor using immobilized *S. cerevisiae* cells entrapped in Ca-alginate beads. The specific rate of ethanol formation is 0.15 g ethanol/g cell-h, the average cell concentration in the alginate bead is 25 g/L, and the alginate beads occupy 70% of the bed volume (the other 30% is liquid that the feed flows through). Assume that growth is negligible so that the cell concentration is constant, and that the bead sizes are sufficiently small such that there is no diffusion limitation of substrate and product in the bead. The feed flow rate is $F = 400$ L/h and contains glucose at a concentration of 100 g/L, the bed diameter is 1 m, and the bed length is 5 m. Glucose is converted into ethanol at a yield of 0.5 g ethanol/g glucose. Determine the glucose and ethanol concentrations at the exit of the bed.

A4 The consumption rate of oxygen in a bioreactor is measured using a culture fluid recirculating loop. A small stream of culture broth (containing medium and cells) is withdrawn from the bioreactor through a small tube. As the fluid flows through the tube, the oxygen concentrations at two points are measured. The total fluid volume in the tubing between the two positions of oxygen measurements is 0.1 L, and the flow rate is 6 L/h. In one instance of oxygen consumption rate measurement, the cell concentration is 10 g/L and the oxygen concentrations at the two points are 0.1 mM/L and 0.05 mM/L, respectively.

a) What is the residence time of the fluid in the section between the two oxygen measurement positions? What is the specific oxygen consumption rate?

b) Instead of using the flow system, you take 0.1 L of the cell suspension into a flask to fill it up completely with any gas space in the flask. Then, you measure the oxygen in the same culture broth over a period of exactly a holding time as determined in (a). If the starting oxygen concentration is the same as that in (a), what is the final oxygen concentration, and what is the specific oxygen consumption rate? Remember that (a) employs a plug flow reactor, while (b) has a batch operation mode.

A5 A recombinant *E. coli* is used to produce the variable region of an antibody for use in affinity chromatography. After fermentation, the concentration of the organism has to be reduced to no more than 1 survivor in 100 L fermentation broth. To avoid denaturing the antibody, thermal killing cannot be used. Instead, it was proposed to use acid treatment at moderate temperatures. An experiment showed that at pH 1.0 and 45 °C, cell death corresponded to first-order kinetics and the viable cell concentration decreased by tenfold for every 20 s of exposure to such conditions.

After fermentation, the viable cell concentration is 10^{10} cells/mL, and the reactor volume is 100 L. To reduce the time required for mixing and heating up, a continuous process employing a pipe with 1.5 cm inner diameter will be used. The broth will be mixed with an equal volume of preheated acid; after mixing, you can assume that the temperature and pH will reach the sterilization conditions immediately. After the holding section, the stream will be cooled down to 4 °C and rapidly neutralized. To ensure that the flow resembles a plug flow, the Reynolds number has to be larger than 10^4. Furthermore, all of the broth has to be sterilized within 15 min after completion of fermentation to avoid product degradation. Oversterilization must be avoided to assure a high yield of product recovery. Calculate the length of the holding section.

A6 Fruit yogurt is to be pasteurized by heat treatment with a plate heat exchanger at 100 °C. Because of the limited capacity of the heat exchanger, the holding time cannot be longer than 30 s, but the heating and cooling are almost instantaneous. The killing of the microorganism is not only temperature dependent, but also pH sensitive. The dependence of the death rate constant can be described as:

$$K\ (s^{-1}) = 0.06 \times 10^{(7\text{-pH}) \times 0.8}(s^{-1})$$

This equation is valid for a pH between 3.0 and 7.0. You are asked to determine the highest pH that can be used to reduce the concentration of surviving organisms from 10^8/mL to 10^4/mL under the conditions currently used for pasteurization.

A7 An orange juice processing plant uses a high-temperature short-time (HTST) process to pasteurize its product. A plate heat exchanger is used to provide rapid heating and cooling. In a laboratory, the process parameters were optimized. Heating, holding, and cooling cycles take 15 s, 2 min, and 30 s, respectively. In scale-up testing, the degree of pasteurization was unsatisfactory and a very high spoilage rate was seen. It was found that the holding pressure in the holding section was not set properly and that boiling occurred. Provide a diagnosis of the pasteurization process, and recommend necessary changes as needed.

A8 In the production of recombinant antibody [immunoglobulin G (IgG)], any endogenous retrovirus from the host cells must be inactivated during product recovery. One step of virus inactivation entails incubating the product solution at pH 3.0 after it is eluted from a protein A affinity column for 30 min. This process results in a 1000-fold reduction in virus infectivity (i.e., virus concentration).
a) Assume the inactivation follows first-order kinetics. Calculate the death rate constant.
b) A new method of inactivation by microwave gives a death rate constant that is tenfold higher than the pH 3.0 incubation process. This allows one to use a continuous process in a tubular sterilizer. At a flow rate of 1 L/min in a tube with a cross-sectional area of 5 cm^2, what is the length of the holding tube needed to achieve a 3-log reduction? You may assume that the flow in the holding tube is plug flow and that the microwave energy is uniformly distributed in the fluid phase.

A9 To achieve the desired sterilization criterion of a 10-log reduction, a batch sterilizer is used to sterilize medium at 121 °C for 20 min. Now, the process needs to be changed to a continuous one. Using a perfect plug flow sterilizer and a flow rate of 1 L/min, how large does the sterilizer have to be? What is the death rate constant of the contaminating microorganism?

A10 The holding section of a continuous sterilizer for fermentation media is 0.2 m × 40 m (i.d. × *l*, or inner diameter × length) and operates at a flow rate that gives a 30-s holding time in that section. Now, the throughput of the sterilizer must be increased by fourfold. Two proposals are presented to accommodate the four-times-higher flow rate: (1) increase the diameter by twofold, and keep the same length of the pipe; or (2) keep the same pipe diameter, and increase its length by fourfold. Which option provides a better sterility of the medium under what conditions? Explain your answer clearly and quantitatively.

8

Oxygen Transfer in Bioreactors

8.1 Introduction

Oxygen is one of the most critical nutrients for all aerobic organisms, especially because most organisms have a very low capacity to carry it. Other nutrients can be stored for later use and thus have a much longer time to depletion. In contrast, oxygen must be supplied continuously and is quickly depleted.

Many nutrients can also be temporarily substituted in place of others, at least for a period of time. For example, if glucose becomes depleted, cells can use other sugars or amino acids as an alternative carbon source. When necessary, they can use lipids or other reserve materials to generate energy. In most organisms, however, there is no energy-generating substitute for oxygen.

The low carrying capacity of oxygen is due to its low solubility in an aqueous environment. When 1 mole of glucose is oxidized to generate energy, 6 moles of oxygen are typically consumed. The ratio of the specific oxygen consumption rate to the specific glucose consumption rate can vary with the type of microorganism. It depends on how much glucose is diverted to produce cell mass and other products. In general, the consumption rates of glucose and oxygen are on the same orders of magnitude or are somewhat higher for oxygen. However, while glucose is highly soluble in water, the solubility of oxygen is more than 1000 times lower. Under typical growth conditions, if the supplies for both glucose and oxygen are interrupted, oxygen will be depleted much faster than glucose or any other nutrient. Thus, the question of how to best supply oxygen to the cell is a very important issue in any system that cultures cells (Panel 8.1).

Panel 8.1 Importance of Oxygen as a Nutrient and Oxygen Transfer in Cell Cultivation

Because of its low solubility, oxygen is the first nutrient to be depleted in culture, unless it is continuously supplied.

How to supply?

1) Continuously supply dissolved oxygen by providing soluble oxygen in liquid. For example: circulating blood to bring oxygen from lung to tissues
2) Provide a gas phase (air) to allow oxygen to diffuse from gas into liquid (medium). For example: lung in contact with blood bubbles air into medium

---

Engineering Principles in Biotechnology, First Edition. Wei-Shou Hu.
© 2018 John Wiley & Sons Ltd. Published 2018 by John Wiley & Sons Ltd.
Companion Website: www.wiley.com/go/hu/engineering_fundamentals_of_biotechnology

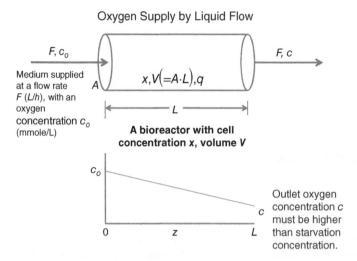

Figure 8.1 Oxygen supply into a tubular flow bioreactor.

8.2 Oxygen Supply to Biological Systems

Cells in culture and in multicellular organisms are always in an aqueous environment and rarely ever in direct contact with a gas phase. They do not obtain oxygen directly from the air. Rather, they extract dissolved oxygen molecules from their aqueous surroundings. Multicellular organisms employ oxygen carriers to increase their oxygen-carrying capacity. In humans, the presence of hemoglobin in red blood cells increases the oxygen-carrying capacity of blood by about 40-fold. This allows blood, flowing at 5 L/min, to supply enough oxygen for all of the roughly 10^{12} cells in the adult body. If the oxygen-carrying capacity of blood were just like water, a much higher flow rate and blood pressure (and a larger heart) would be necessary to meet the oxidative cellular demands.

Performing a material balance on oxygen in a tubular reactor highlights these principles (Figure 8.1). At a steady state, the amount of oxygen transferred into the reactor is the difference of the oxygen concentration at the inlet and the outlet, multiplied by the flow rate (Eq. 8.1 in Panel 8.2). This value, in turn, is equal to the oxygen demand of cells in the reactor. Importantly, the concentration of oxygen at the exit end of the tubular reactor must be higher than the starvation concentration in order to keep cells at the downstream region physiologically active.

Panel 8.2 Henry's Law
$x_A = P_A/H_A$ $\hfill$ (8.1)

We can see this pattern in the bloodstream of the human body. After supplying tissue cells with oxygen, the bloodstream recirculates back to the lung. There, it is reoxygenated and circulates back again to supply oxygen. Likewise, in a tubular bioreactor, the low oxygen stream returns to an oxygenator. There, oxygen is transferred from a gas phase into the fluid before the medium recirculates into the bioreactor.

Thus, in a system where oxygen is supplied by the circulation of liquid, whether it is our body or a bioreactor, the oxygen content of the liquid must eventually be replenished by coming into contact with a gas phase. The human lung and the reactor oxygenator perform the same task.

For cell cultivation, two gas species are of particular interest: oxygen and carbon dioxide. The molar ratio of carbon dioxide produced to oxygen consumed is called the "respiratory quotient." Just as oxygen must be continuously supplied, carbon dioxide must be systematically removed. Carbon dioxide is produced by cells and must be continuously removed from the culture medium. Via either gas bubbles or the gas space in the flask, oxygen passes through the interface of the gas phase and the liquid phase. It is then transferred into the medium. Any carbon dioxide produced by the cells is excreted into the medium and then diffuses into the gas phase.

Mass transfer of any gas species from one phase to another can occur only when the two phases are not in equilibrium. When oxygen is in equilibrium between the two phases, the medium is already saturated with oxygen. In that case, there will not be any net transfer between the two phases.

In contrast, consider a liquid that is undersaturated with oxygen. When this liquid is put in contact with air containing 79% nitrogen and 21% oxygen, oxygen will diffuse into the liquid. Eventually, it will reach equilibrium in both phases. If the oxygen content in the air were to increase from 21% to 40%, the liquid phase would again be undersaturated with respect to oxygen. Oxygen would then continue to be transferred into the liquid phase until the dissolved oxygen concentration becomes equilibrated to the air (gas phase). If the oxygen level in the air is reversed back to 21%, oxygen will escape from the liquid into the gas phase, until equilibrium is reached.

Mass transfer between two phases is thus driven by concentration deviations from equilibrium. Such a deviation is often referred to as the "driving force for mass transfer." Quantifying the driving force, or the degree of deviation from equilibrium, is the first step toward calculating the rate of mass transfer.

8.3 Oxygen and Carbon Dioxide Concentration in Medium – Henry's Law

The oxygen level in the gas phase is usually described as the mole percentage or mole fraction (y_{O2}). It can also be expressed as partial pressure (P_{O2}), that is, the mole fraction of oxygen multiplied by the total pressure. If the ambient pressure is 1 atm (or 760 mm Hg) and the oxygen mole fraction is 0.21 ($y_{O2} = 0.21$), then the partial pressure of oxygen in the air is 0.21 atm (or 159.6 mm Hg). The oxygen concentration in a liquid phase (or the "dissolved oxygen concentration") is often expressed in units of mmol/L. When pure water is saturated with oxygen under a pressure of 1 atm and 21% O_2 at 37 °C, its oxygen concentration is 0.22 mmol/L. Because this concentration is the equilibrium value with an oxygen partial pressure of 159.6 mm Hg, it is sometimes expressed as 159.6 mm Hg. In the medical profession, it is common to express dissolved oxygen concentration in terms of its gas phase partial pressure (e.g., mm Hg).

As shown, there are several common, standard ways of expressing oxygen concentration. Since oxygen is a critical nutrient for the living world, we must be familiar and flexible with these different expressions, as they vary by industry.

The rate of oxygen transference between a gas phase and a liquid phase depends on how far its concentration is from equilibrium. To know this, it is important to determine the component's saturation level. The solubility of oxygen in water is very low. For such gas–liquid pairs, the solubility of a gas, which is sparingly solute in the liquid phase, is described by Henry's Law. Henry's Law states that at equilibrium, the concentration of a gas in the liquid phase (i.e., its solubility) is proportional to the partial pressure of the gas (Eq. 8.1, Panel 8.2).

Various concentration descriptors are also used for gas and liquid. For the same composition of gas and liquid, the numerical value of Henry's Law constant depends on the concentration descriptors used. Two of the most commonly used descriptors are (1) partial pressure for the gas phase, and (2) the mole fraction for the liquid phase. For process calculations, we may find it convenient to express liquid phase concentrations in mmol/L. In the medical profession, the gas phase composition is often expressed in mm Hg, with 760 mm Hg $= 1$ atm.

Henry's Law constants for different gases can be found in many reference sources. Because concentrations of the gaseous solute in the two phases can be expressed in many different ways, Henry's Law constants are also available (and convertible) with different combinations of units. For example, $x_A = P_A/H_A$ for a mole fraction of gas A in the liquid phase and partial pressure, and $c_A = P_A/H_A$ for molar concentration and partial pressure. When applying Henry's Law constants for calculating solubility, pay special attention to the units used.

For an ideal solution, Henry's Law applies well. The total solute concentration of most culture media is well within the range that an ideal solution approximation can be applied. However, in a large bioreactor, the air pressure is usually greater than 1 atm. The hydrostatic pressure may also be large, making it necessary to correct for the partial pressure instead of using ambient pressure.

The solubility of carbon dioxide (CO_2) can similarly be calculated using Henry's Law. However, in an aqueous solution, CO_2 associates with water molecules to become H_2CO_3 and then dissociates into HCO_3^- and H^+. Thus, the solubility of CO_2 will be affected by the pH of the solution (see Example 8.1).

8.4 Oxygen Transfer through the Gas–Liquid Interface

8.4.1 A Film Model for Transfer across the Interface

When we put an unsaturated liquid and a gas phase in contact, the concentration of the gas species in the two phases will not immediately reach equilibrium. It will take time for the solute gas from the gas phase to diffuse into the liquid phase. One can imagine the interface as being a stagnant liquid film, segregated from the bulk liquid and in contact with a gas film on the gas side (Figure 8.2). The bulk liquid and the bulk gas thus are not in direct contact with each other. The oxygen molecules must first

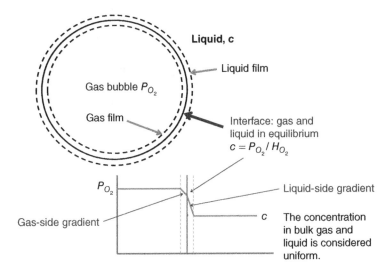

Figure 8.2 Oxygen transfer across a gas–liquid interface.

diffuse across a film, and then diffuse across a second film to reach the bulk of the other phase.

8.4.2 Concentration Driving Force for Interfacial Transfer

A concentration profile of oxygen across an interface of gas and liquid phases is depicted in Figure 8.3. In the figure, the scales of pressure (atm) and concentration (mM) are adjusted so that any pair of oxygen partial pressure in the gas phase and oxygen concentration in the liquid phase in equilibrium is connected by a horizontal line. For example,

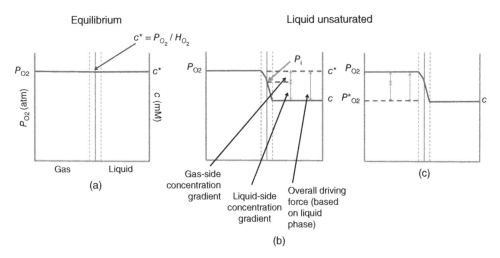

Figure 8.3 Oxygen concentration gradient across a gas–liquid interface and the driving force for oxygen transfer.

0.21 atm in the gas phase and its equilibrium concentration in water, 0.22 mM, will be at the same level on their corresponding axis (Figure 8.3a). At equilibrium, there is no net transfer of oxygen across the interface.

Next, consider a case that the oxygen concentration in the liquid phase (c) is below the saturation level (i.e., below the equilibrium level, where the gas phase partial pressure of oxygen is P_O) (Figure 8.3b). In the bulk liquid at some distance away from the interface, the concentration of oxygen is c. In the bulk gas phase, the oxygen fraction is y_{O2}, and its partial pressure is P_{O2}. The concentration at the interface where the two films meet is not known. One thus assumes that at this position, the two phases are in equilibrium. The concentration gradients exist in the two films, in the opposite sides of the interface.

The oxygen concentration in the liquid, if the liquid phase and gas phase were at an equilibrium, is denoted as c^*. c^* is thus described by Henry's Law ($c^* = P_{O2}/H_{O2}$). The magnitude that the liquid phase is away from the equilibrium is therefore ($c^* - c$). That magnitude is the driving force for oxygen to diffuse from the gas phase into the liquid phase. A larger driving force constitutes a faster rate of transfer. Because this driving force consists of two driving forces (or concentration gradients) from both liquid and gas films, it is referred to as overall driving force or overall concentration gradient.

The driving force can also be described in terms of the gas phase concentration. In this case, the oxygen partial pressure in the gas, if it were in equilibrium with the liquid (with an oxygen concentration designated by c), is denoted as P^*_{O2}. P^*_{O2} is thus the Henry's Law constant of oxygen, multiplied by the concentration c ($P^*_{O2} = H_{O2}c$). The actual oxygen partial pressure is P_{O2}. The driving force can also be described as $P_{O2} - P^*_{O2}$. Henry's Law constant relates the two different descriptions of the same driving force (Figure 8.3c).

Now, we consider the example of water being in contact with air. A bottle of water, which is at 50% of saturation with air being at 1 atm and at 37 °C, has a dissolved oxygen concentration of 0.11 mmol/L, or 79.8 mm Hg. If this bottle of water were in equilibrium with an N_2–O_2 mixture, then that mixture would have 79.8 mm Hg [or 10.5% (mol/mol)] oxygen and 89.5% nitrogen gas.

If this bottle of water is put in contact with air (with 21% O_2 and at 1 atm and 37 °C), it becomes unsaturated as the saturation concentration is 0.22 mM. It is thus 0.11 mmol/L away from saturation. In other words, each liter of water could hold 0.11 mmol more oxygen. One could also say the water is 79.8 mm Hg (159.6 − 79.8) below the point of oxygen saturation. This concentration difference from saturation is the driving force for oxygen transfer into water. One can then quantify the driving force using liquid phase concentrations ($c^* - c = 0.11$ mM) or gas phase concentrations ($P_{O2} - P^*_{O2} = 79.8$ mm Hg).

8.4.3 Mass Transfer Coefficient and Interfacial Area

In addition to the concentration difference between the two phases, the rate of oxygen transference depends on two other factors: (1) the area that is available for oxygen transfer, and (2) the mass transfer coefficient (Panel 8.3). The efficiency of mass transfer increases with increasing surface area available for transfer.

Panel 8.3 Oxygen Transfer through the Gas–Liquid Interface

- Three factors affecting Oxygen transfer rate (OTR):
  - Overall mass transfer coefficient (K_L or K_G)
  - Specific transfer area (i.e., the interfacial area)
  - Driving force (i.e., the gradient across the interface)

OTR can be expressed in terms of a liquid phase or gas phase equation:

$$OTR = K_L a \cdot (c^* - c) = K_L a \Delta c \tag{8.2}$$

$$OTR = K_G a \cdot (P_{O_2} - P_{O_2}^*) = K_G a \Delta P_{O_2} \tag{8.3}$$

$$K_L a = K_G a \cdot H_{O_2} \tag{8.4}$$

Consider the evaporation of water. In the evaporation process, water molecules are transferred from a liquid water surface into vapor form, in the air. If this water were contained in a cup, it would take longer to evaporate. By comparison, if that same amount of water is spread out in a large, flat dish, it will evaporate more quickly. Thus, using a large surface maximizes oxygen transfer.

For a biological example of this, we can again consider the human lung. When the lung is fully expanded, it yields an area of nearly 150 m², an area far larger than our entire body surface. The surface area for interfacial mass transfer is typically expressed as interfacial area per volume of liquid. Thus, it has a unit of inverse of length (L^{-1}, like cm^{-1}).

The other factor affecting the rate of oxygen transfer is the mass transfer coefficient. If a system is under vigorous mixing, the boundary layers at the interface will be thinner and the transfer rate will be faster. Consider, again, the water evaporation in a dish. This time, we swirl the dish and blow air across it. In this case, the mass transfer coefficient would be higher and the evaporation rate would be faster than when the dish is just laid to rest on a table. The mass transfer coefficient, as a quantification of the resistance for mass transfer, is affected both by the nature of the molecules that are being transferred and by the physical and chemical properties of the liquid.

The oxygen transfer rate (OTR) through an interface is calculated as the multiplicative product of the mass transfer coefficient, the specific transfer area, and the concentration driving force. We present two ways to describe the concentration driving force for oxygen transfer. First is in terms of liquid phase concentration ($c^* - c$) (Eq. 8.2 in Panel 8.3). Second is in terms of gas phase partial pressure ($P_{O2} - P^*_{O2}$) (Eq. 8.3 in Panel 8.3). Both quantities describe the magnitude that the oxygen concentration in the two phases is from the equilibrium.

A mass transfer coefficient (K_L or K_G) is used in conjunction with each form of driving force. K_L and K_G are called "overall mass transfer coefficients." A Henry's Law constant relates the two coefficients, like the two expressions of driving force (Eq. 8.4 in Panel 8.3). The mass transfer coefficient has the same units as velocity (L/t, like cm/sec), if we neglect the units related to converting different ways of describing the concentration. The overall mass transfer coefficient is a measure of conductivity, or the capacity to facilitate the transfer. Its inversion ($1/K_L$ or $1/K_G$) is a measure of the resistance.

The concentration difference $[(c^* - c)$ or $(P_{O2} - P^*_{O2})]$ is considered to be an overall driving force, to be associated with K_L or K_G. According to the two-film hypothesis, the overall transfer resistance $(1/K_L$ or $1/K_G)$ can be considered to consist of two components: a liquid-side resistance and a gas-side resistance. The concentration difference away from equilibrium in each side of the file is, thus, the concentration difference between the interface (which is denoted as c_i and P_i for the liquid-phase interface and gas-phase interphase, respectively) and the bulk, that is, $(c-c_i)$ and $(P-P_i)$, respectively (Eqs. 8.5 and 8.6 in Panel 8.4).

Panel 8.4 Overall Mass Transfer Coefficient and Gas/Liquid Phase Mass Transfer Coefficient

OTR can be expressed in terms of overall, gas-side, and liquid-side transfer coefficient and force:

$$OTR = K_L a(c^* - c) \tag{8.2}$$

$$OTR = k_l a(c_i - c) \tag{8.5}$$

$$OTR = k_g a(P_g - P_{gi}) \tag{8.6}$$

Assuming equilibrium at the gas–liquid interface:

$$Hc_i = P_{gi} \quad \text{thus,} \quad OTR = k_g a(Hc^* - Hc_i) \qquad c^* - c_i = \frac{OTR}{k_g H_a a}$$

From Eqs. 8.2, 8.5, and 8.6:

$$a(c^* - c) = \frac{OTR}{k_l} + \frac{OTR}{k_g H} = \frac{OTR}{K_L} \quad \therefore \frac{1}{K_L} = \frac{1}{k_l} + \frac{1}{k_g H} \tag{8.7}$$

Oxygen flux $= K_L(c^* - c) = k_l(c_i - c) = -k_g(P_{gi} - P_g)$

Overall resistance is the sum of resistances from liquid film and gas film at the gas–liquid interface.

Since there is no capacity to hold oxygen at the interface, the transfer rate through the liquid film and subsequently through the gas phase will be equal. P_i and c_i are considered to be in equilibrium. It can be shown that the overall resistance of transfer is the sum of the resistance in the liquid film and the resistance in the gas film (Eq. 8.7 in Panel 8.4).

8.5 Oxygen Transfer in Bioreactors

In a bioreactor for cell cultivation, oxygen is typically supplied by blowing air into the culture medium through a sparger. A sparger can be simply an air pipe with an open hole into the reactor. Another option would be a ring pipe with many open holes, sitting underneath the impeller. In small reactors, especially for mammalian cell culture, air is sometimes introduced into the medium through perforated, microporous stainless steel. In some small-scale instances of cell cultivation, especially those with a relatively low cell load in the reactor, oxygen may be supplied by recirculating the medium. However, because the medium has a very low oxygen-carrying capacity, the recirculation rate typically has to be high.

In bioreactors, air sparging is the most common means of supplying oxygen. Oxygen transfer has been a predominant issue in cell cultivation and the scale-up process. Since the late twentieth century, much of the effort to design effective bioreactors has focused on this issue.

Through mechanical agitation, the Rushton turbine and the large power input can increase oxygen transfer. The extensive agitation breaks up the air bubbles, to increase the interfacial area for transfer and to entrap air bubbles in the liquid. This increases the length of time the oxygen bubbles are held in the reactor, before they reach the top of the liquid.

Oxygen transfer in a stirred-tank bioreactor can be examined from two different aspects: (1) the supply of air into the reactor, and (2) the transfer of oxygen from gas bubbles into the liquid phase.

8.5.1 Material Balance on Oxygen in a Bioreactor

Air is typically supplied into the reactor as large bubbles. Rotating impellers further dispel these bubbles into swarms of smaller bubbles. As the bubbles rise up through the medium, oxygen transfers from the gas phase into the liquid phase (Figure 8.4). This rate of transfer is governed by the mass transfer principle. That is, it is dependent on local oxygen concentration gradients across the interface and the surface area, as well as the local mass transfer coefficient.

Figure 8.4 Oxygen supply through aeration in a bioreactor.

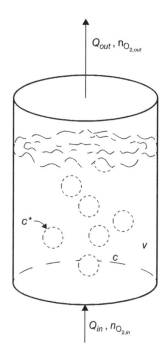

It takes some time for the bubbles to rise to the top of the culture fluid. Thus, the gas phase also occupies a finite amount of volume (V_g, Panel 8.5). Therefore, upon aeration, the total culture volume increases (to $V+V_g$), relative to an un-aerated culture (V), simply because of the entrapment of gas bubbles in the liquid. The total amount of interfacial surface area depends on the amount of gas in the liquid and the size of the bubble (Eqs. 8.8 and 8.9 in Panel 8.5).

Panel 8.5 Interfacial Area and Gas Holdup in Bioreactor

Surface area per volume of gas bubbles:

$$\text{Surface area of bubbles/volume of bubbles} = \frac{\pi d_p^2}{\pi d_p^3/6} = \frac{6}{d_p}$$

Total interfacial area: $V_g \left(\dfrac{6}{d_p} \right)$

$$\theta = \frac{V_g}{Q} \tag{8.8}$$

$$a = \frac{V_g}{V} \left(\frac{6}{d_p} \right) \tag{8.9}$$

V_g can be estimated by measuring the volume (or liquid height) in the reactor before (V) and after ($V+V_g$) aeration starts, but the average bubble diameter (d_p) is difficult to estimate.

As discussed in this chapter, material balances on oxygen can be performed on the oxygen being transferred from gas bubbles into the culture (Eq. 8.10 in Panel 8.6). The balance can also be performed on the gas entering and leaving the reactor. As the gas phase rises through the liquid phase, oxygen is transferred into the liquid phase. There, it is consumed by the cell. The gas exiting from the reactor has a lower oxygen content than the inlet air. The amount of oxygen carried out by the exit gas is thus less than that carried (by gas) into the reactor. The difference between the amount of oxygen supplied into the system and the amount carried out is the total amount that is transferred into the liquid phase (Eq. 8.11 in Panel 8.6).

Panel 8.6

Liquid phase balance:

$$OTR \cdot V = K_L a(c^* - c) \cdot V \tag{8.10}$$

Gas phase balance:

$$OTR \cdot V = (Q_{in} \cdot n_{O_2,in} - Q_{out} \cdot n_{O_2,out}) \tag{8.11}$$

$$OTR \cdot V = \frac{P}{RT}(Q_{in} \cdot Y_{O_2\ in} - Q_{out} \cdot Y_{O_2\ out}) \tag{8.12}$$

Gas phase transfer rate = liquid phase transfer rate:

$$OUR \cdot V = \frac{P}{RT}(Q_{in} \cdot Y_{O_2\ in} - Q_{out} \cdot Y_{O_2\ out}) = K_L a(c^* - c) \cdot V \tag{8.13}$$

Balance on oxygen transfer and consumption:

$$V\frac{dc}{dt} = OTR \cdot V - OUR \cdot V$$

$$= OTR \cdot V - q_{O2}x \cdot V \tag{8.14}$$

At a quasi–steady state:

$$OUR \cdot V = \frac{P}{RT}(Q_{in} \cdot Y_{O_2 \text{ in}} - Q_{out} \cdot Y_{O_2 \text{ out}}) \tag{8.15}$$

$$OUR = K_L a(c^* - c) \tag{8.16}$$

As the gas phase passes through the reactor, not only oxygen is transferred into the liquid phase. The CO_2 produced by cells through respiration and water vapor are also transferred into the gas phase. The exchange of oxygen, CO_2, and water is not necessarily in equal molar rate. Hence, the volumetric flow rate for the outlet stream may differ from that of the inlet stream. Assuming ideal gas behavior ($PV = nRT$), the molar flow rate of oxygen at its inlet and outlet is the total air flow rate, PQ/RT, multiplied by $y_{O2,in}$ and $y_{O2,out}$, respectively (Eq. 8.12 in Panel 8.6).

The OTRs calculated from the gas phase balance and from the liquid phase balance are equal (Eq. 8.13 in Panel 8.6). From the gas phase balance, it is evident that the airflow rate (Q) affects the difference in oxygen content between air intake and exhaust. With the same rates of cellular oxygen consumption, a slower flow rate yields a lower oxygen content upon exit. As will be discussed later in this chapter, the lower oxygen content will give a lower driving force for oxygen transfer in the liquid phase.

In the reactor, the rate of change in oxygen concentration is the balance between the OTR and oxygen consumption rate (Eq. 8.14 in Panel 8.6). In a bioreactor operation, the change of the oxygen consumption rate is relatively slow (usually on the order of many minutes), due to slow cell growth. Once a balance of supply and demand of oxygen is established, the change in the dissolved oxygen (dc/dt) and the exhaust gas composition is also relatively slow. One can assume that the system is at a quasi–steady state. Under such conditions, the OTRs measured through gas phase balance and through liquid phase balance are equal, and both should be equal in magnitude to the oxygen uptake rate (OUR) (Eqs. 8.15 and 8.16 in Panel 8.6).

8.5.2 Oxygen Transfer in a Stirred Tank

As described, the interfacial area is affected by gas holdup and the distribution of gas bubble diameter. Bubbles coalesce as they circulate in the medium; impeller agitation breaks the larger bubbles into smaller ones, increasing the interfacial area. The energy input from agitation also increases the mass transfer coefficient. The turbulence generated by agitation enhances the retention of bubbles in the liquid phase. Overall, the mixing mechanism and the level of agitation both have significant effects on the efficiency of oxygen transfer.

The magnitude of the local mass transfer coefficient and interfacial area varies significantly in different regions in the reactor. Near the impeller, especially at the tip of impeller blades, the bubbles are small. The oxygen transfer efficiency is high.

Most studies examine the reactor-wide average of mass transfer efficiency, rather than its distribution throughout the reactor. That is, generally speaking, the reactor-wide average value is used for the calculations of oxygen transfer capacity and for scale-up considerations.

The driving force for oxygen transference in a reactor may vary depending on the location in the reactor. For small reactors, we assume that both the liquid and gas phases are well mixed. This assumes that the gas composition measured at the outlet is the same as the concentration in the reactor. c^* at the air outlet thus should be used for driving force calculations. (*Note*: Not that in the air inlet!)

In a large reactor, there are many more factors affecting the driving force for oxygen transfer. The oxygen content in the exhaust gas tends to be lower in larger reactors versus smaller ones, even for the same process. This is because a smaller air flow rate for a given culture volume is often used, as will be discussed in Chapter 9. For very large reactors, the hydrostatic pressure in the bottom region of the reactor can have a significant effect on the value of c^*. This enhancing effect, however, diminishes toward the top of the reactor. If one assumes that the gas phase rises analogously to a tubular flow, the average driving force in the reactor is the logarithmic mean of the concentration gradient (Eq. 8.17 in Panel 8.7).

Panel 8.7

What Value of c^* to Use in a Bioreactor?

- In a small reactor, with the assumption of being well mixed, c^* is the oxygen concentration in the exit air.
- In a large reactor, use the logarithmic mean of the driving force at the air inlet and outlet.

$$(c^* - c)_{LM} = \frac{(c^*_{in} - c) - (c^*_{out} - c)}{\ln \dfrac{c^*_{in} - c}{c^*_{out} - c}} \tag{8.17}$$

Example 8.1   Balance of Oxygen Transfer and Consumption

A 500 mL cylindrical spinner flask is used to cultivate human induced pluripotent stem cells on microcarriers at 37 °C. In such a small-scale bioreactor, oxygen is transferred into the medium by diffusion from the air in the space above the medium. The diameter of the flask is 8 cm. The overall oxygen transfer coefficient (K_L) has been determined to be 5 cm/h. The specific oxygen consumption rate of the cell is 1×10^{-10} mmol/cell-h. For their optimal growth, the oxygen level should be maintained at 20% of saturation with air at 1 atm.

What cell concentration can be reached with this spinner flask?

Solution

First, we determine the oxygen concentration in the medium at saturation with air at 1 atm. The medium used is relatively dilute and can be assumed to behave like an ideal solution. Its oxygen solubility can be calculated using Henry's Law. Look up Henry's Law constant for oxygen in water.

Henry's Law constants for O_2 (atm/mole O_2/mole H_2O)

T, °C	5	20	25	30	35	37
$10^{-4} \times H$	2.91	4.01	4.38	4.75	5.07	5.18

$$c = \frac{P_{O_2}}{H}$$

$$= \frac{0.10 \text{ atm}}{5.18 \times 10^4 \text{ atm/mol } O_2/\text{mol } H_2O} \cdot \frac{55.5 \text{ mol } H_2O}{L \, H_2O}$$

$$= 0.22 \text{ mmol/L}$$

We next calculate the surface area available for transfer:

$$A = (4 \text{ cm})^2 \cdot \pi = 50.2 \text{ cm}^2$$

$$a = 50.2 \text{ cm}^2/500 \text{ cm}^3 = 0.1 \text{ cm}^{-1}$$

Dissolved oxygen concentration at saturation is 0.22 mM. At 20% of saturation, $C = 0.044$ mM, so $\Delta C = 0.18$ mmol/L.

Let the cell concentration be x:

$$OUR = q_{O_2} \cdot x$$

$$OTR = 5 \text{ cm/h} \cdot 0.1/\text{cm} \cdot 0.18 \text{ mmol/L} = 0.09 \text{ mmol/L} - h$$

Set $OUR = OTR$:

$$x = 0.09 \text{ mmol/L} - h \div 1 \times 10^{-10} \text{ mmol/cell-h}$$

$$= 9 \times 10^8 \text{ cell/L}$$

At a higher cell concentration, OUR will be greater than OTR, and dissolved oxygen will begin to decrease to below 20% of saturation.

8.6 Experimental Measurement of $K_L a$ and OUR

8.6.1 Determination of $K_L a$ in a Stirred-Tank Bioreactor

With extensive experimentation, oxygen transfer capacity of a "standard" reactor can be estimated from empirical correlations. Typically, such empirical correlations enable us to estimate oxygen transfer capacity of a reactor, using parameters related to air flow rate, agitation rate, and power input. More frequently, the oxygen transfer capacity of a reactor is determined experimentally for each particular reactor under different operating conditions. Ideally, the conditions under which the measurement is carried out should resemble cell cultivation, including fluid properties, cell types, the nature of the medium, and the effects of surface-active agents (for solubilizing hydrophobic nutrients or foam control). These factors often have a profound effect on interfacial mass transfer.

The direct measurement of oxygen transfer capacity thus yields realistic values that can differ significantly under different conditions. However, in experimental measurement, the interfacial area is very difficult to determine or to estimate. Most

Measurement of K_La
Dynamic Method in the Absence of Cells

Change gas composition (e.g., switch the gas from air to nitrogen after dissolved oxygen reaches a high level) to allow the dissolved oxygen concentration to change. Then measure the time profile of dissolved oxygen concentration. Plot dissolved oxygen concentration versus time.

$$\frac{dc}{dt} = K_La(c^* - c) \quad (8.18)$$

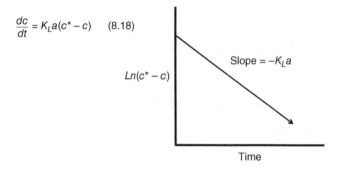

$Ln(c^* - c)$

Slope = $-K_La$

Time

Figure 8.5 Measurement of overall mass transfer coefficient (K_La) by degassing.

experimental measurements of mass transfer capacity of a bioreactor thus measure the combination of K_L and a and report the value as K_La. K_La is sometimes called the volumetric mass transfer coefficient.

To measure K_La, one can aerate a reactor that has a very low oxygen concentration under conditions similar to those for cell cultivation, but without the cell. The dissolved oxygen concentration will steadily increase until it reaches saturation. Without cells in the reactor, the change of dissolved oxygen in the medium is only caused by the transfer through the gas–liquid interface, as shown in Eq. 8.18 (Figure 8.5). The slope of the semilog plot of $c^* - c$ over time will be K_La. One can also perform the experiment by using nitrogen gas to strip oxygen from medium that is initially at a higher dissolved oxygen level (Figure 8.5). In this case, the slope in the semilog plot is $-K_La$. In most operations, the aeration rate changes over time, as does K_La. One can develop a correlation between K_La and air flow rate for use during cultivation.

8.6.2 Measurement of OUR and q_{O2}

Both OUR and the specific oxygen consumption rate (q_{O2}) can be measured in a flask by filling it with a cell suspension and not leaving any gas phase. In the absence of a gas phase, there is no oxygen supply in the measuring flask. The decrease in dissolved oxygen level is thus caused by cell consumption and can be measured with a dissolved oxygen probe. The slope of the oxygen concentration over time is the rate at which dissolved oxygen is consumed (i.e., the OUR). The specific oxygen consumption rate can then be obtained by dividing the OUR by the cell concentration in the cell suspension.

With an on-line gas phase oxygen sensor, such as a mass spectrometer, one may also measure the oxygen content in the inlet and exhaust gases. By taking the difference in the oxygen molar flow rate between the inlet and outlet, and with a known air flow rate, one can obtain a good estimate of oxygen consumption in the reactor and $K_L a$, as can be seen from Eqs. 8.15 and 8.16.

Example 8.2 Gas Phase Balance in an Aerated Bioreactor

A 10 L bioreactor is used to grow yeast cells for the production of recombinant human albumin. A quadrupole mass spectrometer is used to measure the composition of the intake and exhaust air. At a time point of the cultivation, the exhaust air has 19% oxygen, 2% carbon dioxide, and 79% nitrogen, on a moisture-free basis. The inlet air has 21% oxygen. The air flow rate is 20 L/min at 1 atm and 25 °C. The dissolved oxygen in the culture is 20% of saturation with air at 1 atm. The culture temperature is 30 °C.

What is the oxygen consumption rate of the culture? Estimate the volumetric oxygen transfer coefficient ($K_L a$).

Solution

The concentration of oxygen in the culture changes only slowly during cultivation. The transfer of oxygen into the culture medium and its uptake by cells can be assumed to be balanced, or at a quasi–steady state. The oxygen transfer rate into the culture is thus set to be equal to the oxygen uptake rate.

Since the nitrogen contents in the inlet and outlet air are both at 79%, the total molar air flow rate is thus the same at the inlet and outlet.

$$Q_{in} = Q_{out} = 20 \text{ L/min} = 1200 \text{ L/h}$$

The oxygen transfer rate is determined from the gas phase balance (see Eq. 8.12).

$$OTR \cdot V = \frac{1}{0.082 \cdot (273 + 25)} \cdot 1200 \cdot (0.21 - 0.19)$$
$$= 0.982 \text{ mol/h}$$
$$OTR = 98.2 \text{ mmol/L} - h$$

The oxygen consumption rate is thus also 98.2 mmol/h.

From Eq. 8.13, we see that the OTR calculated from the liquid phase balance and the value determined above should be the same:

$$OTR = K_L a(c^* - c) = 98.2 \text{ mmol/L} - h$$

At 30 °C, the saturation concentration of oxygen is 0.24 mmol/L. Assuming the small reactor is well mixed, c is thus 0.046 mmol/L.

But, what should c^* be? Should it be based on the equilibrium concentration at the air inlet? Or should it be based on air at the outlet? With the assumption that the reactor is well mixed, all the gas bubbles in the culture broth and the air at the outlet should have the same oxygen content. Therefore, c^* should be the equilibrium concentration based on the outlet gas composition:

$$c^* = (0.19/0.21)0.24 = 0.22 \text{ mmol/L}$$

From the liquid phase balance:

$$98.2 = K_L a(0.22 - 0.04)$$
$$K_L a = 545 \text{ h}^{-1}$$

8.7 Oxygen Transfer in Cell Immobilization Reactors

Cells are sometimes cultivated inside particles, for example yeast cells that are entrapped in agarose gel beads, or mammalian cells grown inside macroporous microcarriers. In these systems, oxygen not only needs to be supplied continuously to the culture medium, but also needs to diffuse from the liquid into the interior of the particle. As oxygen diffuses through the depths of the particle, cells begin to consume it. As a result, the concentration of oxygen decreases along the path of oxygen diffusion or along the depth of the particle.

The rate of oxygen decrease is dependent on how fast the cells consume oxygen and on the geometry of the particle they reside in. Further into the interior of the particle, oxygen may become depleted or become too low to support a high rate of growth or productivity. To avoid such limiting oxygen conditions, one can resort to using smaller particles or maintaining a higher overall oxygen concentration in the culture medium. The extent of oxygen transfer limitation can be estimated by comparing the mass diffusion and consumption rates. The theoretical calculation is presented as plots of two variables, often referred to as the effectiveness factor and Thiele's modulus, which are discussed in standard chemical reactor textbooks.

8.8 Concluding Remarks

The solubility of oxygen is very low in water. For cell cultivation, oxygen must be supplied continuously to meet cellular demand. The transfer rate of oxygen through the liquid–gas interface is affected by the mass transfer coefficient, the interfacial area, and the oxygen gradient between the medium and gas phases. To supply oxygen, the most direct and effective method is to bubble air through the bioreactor.

Further Reading

Cussler, EL (2009) Diffusion. Chapter 11, pp. 332–352. Cambridge University Press.

Garcia-Ochoa, F and Gomez, E (2008) Bioreactor scale-up and oxygen transfer rate in microbial processes: an overview. Biotechnol. Adv., **27**, 153–176.

Kirk, TV and Szita, N (2012) Oxygen transfer characteristics of miniaturized bioreactor systems. *Biotechnol. Bioeng.*, **110**, 1005–1019.

Nomenclature

A	cross-sectional area of reactor	L^2
a	specific area: interfacial surface area per reactor volume	L^2/L^3
c	concentration of a soluble gas or oxygen	$mole/L^3$ M/L^3

c^*	solubility or equilibrium concentration of a soluble gas or oxygen	mole/L^3 M/L^3
c_{in}^*, c_{out}^*	oxygen concentration in equilibrium with gas in the inlet and outlet	mole/L^3 M/L^3
c_o	oxygen concentration in the feed	mole/L^3 M/L^3
c_i	oxygen concentration at the interface of gas and liquid	mole/L^3 M/L^3
d_p	bubble diameter	L
F	volumetric medium flow rate	L^3/t
H	Henry's Law constant	(mole/L^3)/ (mole/L^3) or (mole/mole)/(mole/mole) or M/L^3/M/L^3
K_G	overall gas-side mass transfer coefficient	L/t
K_L	overall liquid-side mass transfer coefficient	L/t
k_g	mass transfer coefficient for gas side	L/t
k_l	mass transfer coefficient for liquid side	L/t
L	length of reactor	L
$n_{O_2,in}, n_{O_2,out}$	oxygen concentration in the gas at the inlet and outlet	mole/L^3
OTR	oxygen transfer rate	mole/L$^3 \cdot$ t
OUR	oxygen uptake rate	mole/L$^3 \cdot$ t
P	pressure	Force/L^2
P_A	partial pressure of A	Force/L^2
P_g, P_{gi}	partial pressure of a solute in the gas phase and at the interface	Force/L^2
P_i	partial pressure at the interface	Force/L^2
$P_{O_2}^*$	oxygen partial pressure in equilibrium with the liquid phase	Force/L^2
Q	volumetric gas flow rate	L^3/t
Q_{in}, Q_{out}	volumetric gas flow rate at the inlet and outlet	L^3/t
q	specific rate	mole/cell $\cdot$ t
q_{O2}	specific oxygen uptake rate	mole/cell $\cdot$ t
R	universal gas constant	
T	absolute temperature	K
V	volume of liquid	L^3
V_g	volume of gas in culture broth	L^3
x	cell concentration	cell/L^3
x_A	mole solute A/mole solution (concentration of A)	mole/mole
Y_{in}, Y_{out}	molar fraction of solute at the gas inlet and outlet	mole/mole
θ	average holding time of bubble in culture	t

Problems

A. Oxygen Uptake Rate and Oxygen Balance

A1 On-line measurement of cell concentration and the metabolic demand of glucose is not easily implemented because few sensors can directly measure cell or glucose concentrations. To circumvent this problem, many have computed the on-line determination of the oxygen uptake rate (OUR) by measuring the difference in oxygen partial pressures at the gas inlet and outlet.

In a 2-L culture, yeast glucose is fed continuously to maintain its concentration at a constant level. The yield coefficients for glucose ($Y_{x/glu}$) and oxygen ($Y_{x/o}$) are 0.4 g cell/g glucose and 1.0 g cell/g oxygen, respectively. The respiratory quotient (RQ) is 1.0. Air is supplied at 60 L/h. The dry gas composition (mole %) at a given time point during the exponential growth phase is shown in Table P.8.1.
a) What is the OUR?
b) What does the glucose-feeding rate (g glucose/h) need to be to support the cellular demand? The molar volume of an ideal gas under these operating conditions is 25 L/mol.

Table P.8.1 Gas composition at the inlet and outlet of a bioreactor.

	Air inlet	Air outlet
Nitrogen	78.9	78.9
Oxygen	21.1	17.8
Carbon dioxide	0	3.3

A2 The differentiation of stem cells into hematopoietic cells can be enhanced by culturing them under hypoxic (low-oxygen) conditions of 5% of saturation with air. The saturated concentration of oxygen in water equilibrated with air at 1 atm is 0.2 mM. The cultivation will be carried out in a stirred spinner flask with a diameter of 4 cm and a liquid height of 4 cm. The oxygen is supplied by surface aeration through a gas phase over the top of the liquid (in other words, there is no air bubbled in the medium).

Before the experiment, two tests were carried out. In the first test, stem cells at the cultivation concentration were placed in a sealed flask that was filled with medium and without a gas phase. The dissolved oxygen concentration was measured over time (data shown in Table P.8.2, test 1). In the second test, the oxygen concentration within medium in a reactor, without any cells, was first allowed to reach equilibrium with air at 1 atm. Then, the gas phase (50 mL in volume) in the flask was changed to nitrogen gas at a flow rate of 50 mL/min to strip off the oxygen from the medium. The high gas flow rate assures that the amount of oxygen in the gas phase is negligible. The oxygen in the medium gradually decreased over time, as shown in Table P.8.2, test 2. From the data given, estimate the percentage of oxygen

Table P.8.2 Time course data of oxygen concentration in two experiments.

Test 1 time (min)	Dissolved oxygen (% saturation)	Test 2 time (min)	Dissolved oxygen (% saturation)
0	90	0	100
6	80	5	82
12	70	10	67
18	60	15	55
24	50	20	45
30	40	25	37

in the gas phase that should be maintained to keep oxygen in the medium at the hypoxic condition of 5% of saturation.

A3 A 10-L stirred tank is used to cultivate mouse myeloma cells. Dry air is sparged from the bottom at 1.0 L/min to supply oxygen to the system. The gas composition at the outlet is 16% oxygen, 3.5% CO_2, 6.5% water vapor, and balanced nitrogen. The ambient pressure and temperature are 1 atm and 37 °C, respectively. The dissolved oxygen in the reactor is controlled at 20% of saturation with ambient air.

a) Assuming the reactor is at a steady state, what is the volumetric oxygen transfer coefficient $(K_L a)$?

b) The specific OUR of myeloma cells is 5×10^{-11} mmol/cell-h. What is the cell concentration?

c) What is the respiratory quotient (RQ) (i.e., mole CO_2 produced/mol O_2 consumed)? Assume that the mole fraction of CO_2 in the ambient air is negligible. The universal gas constant $R = 0.082$ L atm/mol-K.

A4 Bone marrow–derived stem cells are cultivated under hypoxic conditions to guide them to differentiate toward the hematopoietic lineage. The culture flask is placed in a 1-atm environment with 5% CO_2, 2% O_2, 4% H_2O, and balanced nitrogen. The cell density on the surface of the culture flask is 6×10^6 cells/cm^2. The cells consume oxygen at a rate of 3×10^{-10} mmol/cell-h. The medium covering the cells has a thickness of 5 mm. We can assume that the oxygen concentration at the interface is at an equilibrium between the medium and the gas phase. The diffusivity of oxygen in the medium is the same as that in water.

B. Oxygen Transfer in Reactors

B1 Gas phase balance using a mass spectrometer is a standard way of measuring the OUR and carbon dioxide evolution rate (CER). The differences in oxygen or carbon dioxide molar flow rates between inlet and outlet gas streams are taken as OUR and CER, respectively. The ratio of CER to OUR is taken as the on-line estimation of RQ.

    a) Write down the gas phase balance equations. What are the underlying assumptions in using gas balance as a real-time, on-line estimation of OUR and CER?

    b) The measurement of off-gas composition by mass spectrometry provides a reasonably accurate determination of OUR, but the estimation of CER is often prone to error, especially when sudden perturbations (such as a change in the air flow rate or the addition of a carbon source) stimulate metabolism. Examine Henry's Law constant and the kinetics of solubilization for O_2 and CO_2, and discuss why this is the case.

B2 A 40 m³ (working liquid volume) stirred tank is used for a batch culture of a recombinant bacterium. In the mid-exponential phase at an air flow rate of 30 m³/min, the dissolved oxygen is 15% of saturation with ambient air. The outlet gas composition is 16% O_2, 4.5% CO_2, and 5% H_2O. Calculate the OUR in mmole O_2/L-h. Estimate K_La for the transfer of oxygen. You can assume the system is at a quasi–steady state, and the air in the inlet is water-free.

B3 A 10-L stirred-tank bioreactor is used to grow *Bacillus thurigiensis* (BT) to produce BT insecticide. The air flow rate is 15 L/min at 1 atm, and the temperature is 37 °C. At the mid-exponential growth stage and a cell concentration of 15 g/L, the mass spectrometer measurement of the exit gas shows a composition of 19.2% O_2 and 1.8% CO_2, with the balance being N_2. The water vapor has been completely removed before air entry into the mass spectrometer. The dissolved oxygen concentration is 0.08 mM, which is equivalent to 40% of saturation with ambient air (oxygen content in the ambient air is 21%) at 1 atm. Calculate the volumetric oxygen transfer coefficient (K_La) and the specific oxygen consumption rate of the organism.

B4 A small bioreactor is used to cultivate a bacterium utilizing methane. The methane is fed together with air. The inlet composition of gas is 5% CH_4, 20% O_2, and 75% N_2. The dissolved O_2 in the culture is 0.05 mM. At equilibrium with 1 atm air (21% O_2 and 79% N_2), the dissolved oxygen level is 0.2 mM. For pure methane, the solubility in water is 1.4 mM. The stoichiometric ratio of oxygen consumption and methane consumption is 2 mole O_2 per mole of CH_4. The mass transfer coefficient (K_L) is inversely proportional to the square root of molecular weight. What is the concentration of CH_4 at the gas outlet? Assuming that both the gas and liquid phases in the reactor are well mixed, what is the concentration of dissolved methane in the reactor?

B5 A 50-L bubble column reactor is used to produce the biomass of a microorganism using methane (CH_4) as the carbon source. The yield coefficients are 1 g cells/g CH_4 and 0.8 g cells/g O_2. The aeration rate used is 1 vvm (1 L gas/L culture/min). The overall mass transfer coefficients (K_La) for both gas species are very similar and are estimated to be 1000 h^{-1}. The solubility of oxygen and methane (with pure gas at 1 atm) are 1.0 mM and 1.4 mM, respectively (e.g., at 1 atm of air, the saturated oxygen concentration is 0.21 mmol/L). Methane is to be blended with air before it is sparged into the reactor. You can consider the gas phase to be plug flow and the liquid phase to be well mixed in the reactor. For optimal cell growth, the dissolved oxygen and methane should be kept at 0.02 and 0.01 mM, respectively. For fire safety reasons, the exhaust gas should not contain more than 1% methane. The operating temperature is 25 °C. Determine the composition of the gas in the feed (i.e., the percent methane and percent air). The air has 21% O_2 and 79% N_2. To simplify the analysis, assume that RQ = 1.0 and that water vapor in the exit gas can be neglected.

C. Advanced Problems

C1 A lithotrophic anaerobic microorganism grows on hydrogen and carbon dioxide. Explain how it derives energy and makes biomass using carbon dioxide as the carbon source. In a culture, a gas mixture of nitrogen, hydrogen, and carbon dioxide is supplied through aeration into the liquid medium. The objective is to maintain the concentration of hydrogen and carbon dioxide (C_1 and C_2, respectively) at their optimal levels. The mass transfer coefficients for hydrogen and carbon dioxide are K_{L1} and K_{L2}, respectively. The yield coefficients are Y_1 and Y_2, and the solubility is C^*_1 and C^*_2 for pure hydrogen and carbon dioxide.

Set up equations that allow you to determine the ratio of hydrogen and carbon dioxide that should be used in the gas mixture. Define other symbols when needed; use 1 for hydrogen and 2 for carbon dioxide.

C2 Vinegar (acetic acid) can be produced by the aerobic oxidation of ethanol in a continuous process using *Acetobacter suboxydans*. The stoichiometry of this reaction is:

$$C_2H_5OH + O_2 \rightarrow CH_3COOH + H_2O$$

The oxygen required for cell growth is negligible compared to the oxygen required for acetic acid production. A 50,000-L tank has a diameter of 252 cm, and two 6-bladed turbine impellers, each measuring 30 cm in radius, are being operated at 150 rpm. Air is supplied at 5000 L/min at standard temperature and pressure (STP). The fermentation broth is Newtonian with a viscosity of 1 centipoise and a density of 1 g/cm³.

Describe the procedure that you will take to estimate the maximum productivity of acetic acid (g/L-h) that can be achieved if the oxygen transfer rate is the limiting factor.

Additional information: The dissolved oxygen is maintained at 16 mm Hg of oxygen. Neglect the hydrostatic head of the liquid on the partial pressure of oxygen.

Note: You will need to use an empirical correlation for gassed power and $K_L a$ for a stirred tank from a handbook.

C3 Insulin-secreting beta cells isolated from islets are immobilized in agarose for implantation into diabetic animals. The cell concentration in the agarose beads is 1×10^8/cm³. The specific oxygen consumption rate is 3×10^{-10} mmol/cell-h. The diffusivity of oxygen in agarose can be assumed to be 5×10^{-5} cm²/s. The dissolved oxygen concentration at the surface of the agarose beads is 0.2 mM. Estimate the largest diameter of the agarose beads one can use without inducing an oxygen diffusion limitation inside the bead.

C4 The diffusion coefficient of CO_2 in its dilute water solution is about 0.9 of that of O_2. K_L (the mass transfer coefficient) for CO_2 is thus about 0.9 of O_2. In a small bioreactor, the dissolved O_2 and CO_2 are kept at 0.05 mM and 5 mM, respectively. The (dry) air flow rate is 40 mmol/L-min at 25 °C. The outlet O_2 and CO_2 are 19% and 2%, respectively. The liquid phase can be considered to be at a quasi–steady state. The Henry's Law constants for O_2 and CO_2 are 0.9 and 0.04 [$H = P/c$ (P: atm; c: mmol/L)].

Are the gas phase and liquid phase balance results consistent with each other? Does the data support the hypothesis that the dissolution of CO_2 into the gas phase is rate-limiting? It was proposed to add carbonate anhydrase, which catalyzes the hydration reaction $CO_2(aq) + H_2O \rightleftarrows HCO_3^- + H^+$ in the culture to facilitate CO_2 removal. Does this proposal make sense for increasing the transfer of CO_2 into the gas phase?

C5 A 4000-L stirred-tank bioreactor is used to grow *Streptomyces* for the production of lincomycin. A mass spectrometer is used to measure the gas phase composition in the inlet and outlet. The compositions are given in Table P.8.3.

The airflow rate is 4000 L/min at STP. The fermenter is not perfectly mixed, and the dissolved oxygen concentration at the bottom of the fermenter is 0.017 mM, while at the top of the liquid it is 0.015 mM. The equilibrium relationship between dissolved oxygen and the partial pressure of oxygen is c (mmol/L) $= P$ (atm). In other words, at a partial pressure (P_{O2}) of 0.21 atm, the saturation-dissolved oxygen concentration is 0.21 mM. The pressure of air in both the inlet and outlet can be considered to be 1.0 atm. Determine the $K_L a$. The maximum specific growth of this organism is 0.3 h^{-1}, and the yield coefficient is 1 g cell/g O_2. If oxygen is the rate-limiting nutrient, what is the highest cell concentration that this fermenter can sustain?

The local oxygen transfer rate in an air-lift bioreactor varies with the vertical position. In the production of food-grade yeast, glucose is to be used as the substrate; the yield based on oxygen is 1 g cell/g O_2. The process is to be operated in a continuous mode with a dilution rate of 0.2 h^{-1}. To simplify our analysis, the respiratory quotient of yeast can be assumed to be 1.0. A production reactor is 30 m in liquid height and 5 m in diameter. The volumetric air flow rate is 0.6 vvm (volume air/volume liquid/min).

Table P.8.3 Gas composition at the inlet and outlet of a bioreactor.

	Inlet	Outlet
Nitrogen	79	77
Oxygen	21	17
Carbon dioxide	–	3.4
Water	–	2.6

C6 The relationship between the local overall mass transfer coefficient and gas superficial velocity is $K_L a = 0.008 + 0.63\ v_s$, where $K_L a$ is in s^{-1} and v_s is in m/s.

The equation is valid for V_s greater than 0.005 m/s. You can assume the gas phase behaves as plug flow, whereas the liquid phase is well mixed.

Estimate the output based on the oxygen transfer capacity of the bioreactor.

a) Identify the position within the reactor where the oxygen transfer rate is at its maximum. The gas holdup is 30%, and is a function of position too, but you can express your answer in ungassed liquid height instead of aerated liquid height. Assume the bubble rising velocity is constant.

C7 To maintain the freshness of fruit juice, the oxygen concentration inside of the juice box should be kept low. One idea is to use glucose oxidase, which uses glucose in the juice to produce gluconic acid and consume the oxygen that is in the juice as well as that which diffuses through the wall. The thickness of the wall is 0.2 cm. A bacterial glucose oxidase, with a K_m of 0.01 mM and 0.1 mM for oxygen and glucose, respectively, is to be immobilized on the inner wall of a cubic juice box. The glucose content of the juice is 30 mM. At saturation with air, the oxygen concentration in the juice is 0.2 mM, but it needs to be kept at no more than 0.002 mM. The permeability (which is equivalent to K_L in interfacial transfer) for oxygen is 10^{-5} cm/h. The design objective is to coat the box's inner wall with a layer of glucose oxidase at a thickness of 10^{-3} cm, so that all of the oxygen that diffuses through the wall is essentially consumed in the enzymatic reaction. To simplify the analysis, you can assume that, within the enzyme layer, the oxygen and glucose are well mixed and that their concentrations are uniform and the same as that in bulk juice.

What is the enzyme concentration that should be used in the coating layer? Express it with the proper units. You can express your answer in terms of $k_{cat}E_o$. Use D to represent diffusion coefficient of oxygen in the wall of the juice box.

C8 A special brand of vinegar is produced by a fermentation process using an obligate anaerobic microorganism. After fermentation, the viable cell concentration must be reduced from 10^8/mL to less than 1 survivor per liter. An easy way to kill the organism is by exposing it to oxygen. The cell death can be described by first-order kinetics with respect to the viable cell concentration. The first-order death rate constant is a function of oxygen, as shown here:

$$k = 0.1^*(c/0.21) \text{ min}^{-1} \text{ for } c < 0.21 \text{ mmol/L}$$
$$k = 0.1 \text{ min}^{-1} \text{ for } c \geq 0.21 \text{ mmol/L}$$

where c is the dissolved oxygen concentration. The Henry's Law constant for oxygen is $H = 0.21$ atm/0.21 mmol/L. The amount of oxygen required to kill the organism is so small that the killing process can be considered to consume no oxygen. The process for killing this microorganism involves bubbling air through the reactor to raise the dissolved oxygen from nil to saturation with air. The long-term presence of oxygen increases the frequency of spoilage. Therefore, after a holding period in which oxygen becomes saturated, nitrogen will replace air to strip off residual dissolved oxygen. The volumetric mass transfer coefficient ($K_L a$) is 5 h^{-1} under these operating conditions. To simplify the calculation, we can assume that when the level of dissolved oxygen increases to 95% of the saturation with air, the liquid can be considered to be saturated with oxygen. The killing process then enters the second (holding) stage in which the death rate constant is at saturation. Similarly, the stripping period can be considered complete when the dissolved oxygen decreases to 5% of the saturation with air.

a) Set up the equations necessary for determining the killing process.
b) Sketch the time profiles of the dissolved oxygen concentration and death rate constant. Try to be as quantitative as you can, and, qualitatively, the trend has to be accurate.
c) Determine the killing process.

9

Scale-Up of Bioreactors and Bioprocesses

9.1   Introduction

The cells of all living systems share a similar set of cellular machinery and biochemical reactions. Their size varies only over a range of tenfold, from 1 μm for small bacteria to a bit over 10 μm for eukaryotic cells. When we consider organisms, from a single bacterium to a multicellular organism such as plants and animals, the range of size becomes much larger.

In the natural evolution of living systems, scaling up is a persistent challenge. At the most basic level, cells perform their metabolic functions and other activities of "life" with elegant efficiency and little waste. In performing these same tasks, taller, bigger, and more complex organisms encounter additional challenges that simple bacteria do not face. For example, consider a cell that has increased tenfold in length scale. Can it still maintain the same metabolic rate per cell volume, without suffering limitations in the oxygen supply? How does a 30-m tree get adequate water and nutrients to its top? Can a tall conifer efficiently intake and process the material necessary to grow as fast as a short shrub? How does a massive elephant distribute oxygen throughout its body, using basically the same plug flow type of vascular system as a mouse?

Physical constraints can impose limitations. Taller plants and larger animals have adapted and evolved to perform the same tasks of "life" in different ways. As a result, no large creature is completely identical to its smaller counterpart.

Likewise, to scale up a given process in the lab or a plant, we must consider the same factors that affect the "scaling up" of organisms. For example, how do we distribute the optimal level of nutrients and oxygen to best reach their intended locations? Can we maintain the same physical configuration, as well as the operating conditions, in reactors of vastly different scales?

In this chapter, we discuss the specifics, challenges, and principles of process scale-up. We will often refer to examples in the natural world, of the adaptations and mechanisms that enable living things to grow in size.

Engineering Principles in Biotechnology, First Edition. Wei-Shou Hu.
© 2018 John Wiley & Sons Ltd. Published 2018 by John Wiley & Sons Ltd.
Companion Website: www.wiley.com/go/hu/engineering_fundamentals_of_biotechnology

9.2   General Considerations in Scale Translation

9.2.1   Process and Equipment Parameters Affected by Scale-Up

When scale changes, many physical parameters change. Different organisms evolve in vastly different ways. The vital physiological parameters critical for their fitness remain more constant; others change to accommodate the organism's increasing size.

For example, as an organism becomes larger, the surface area per body volume decreases. Hence, a larger organism cannot sustain the high metabolic activity of a smaller one. Some larger animals evolve to have longer legs to lift them off the ground and dissipate more heat when they are at high metabolic activity. Others crawl on the ground to better search for food, keeping their metabolism at relatively low levels. In becoming larger, each organism sharpens some critical functions while allowing other properties to change.

In the process of scale-up or scale-down, we face the same questions: which factors can stay constant, and which ones must change? We cannot keep everything constant. It is important to understand which physical and chemical parameters in the bioreactor will change with scale. Some changes can affect the physiology of cells and cause the productivity of product quality to change; others have a minimal effect on cell physiology. Then, we must consider the physiological variables that are important to process performance. Of these, which will be affected by those scale-sensitive physical and chemical parameters?

Given this information, we can better evaluate which physiological variables are most important and sensitive to scale. Then, we can focus on keeping each of these important parameters within its optimal range. For example, the oxygen transfer and CO_2 removal capacity of a reactor may be different at different scales. The supply of oxygen and the accumulation of CO_2 may have an effect on process performance; they may need to be controlled within a range. The metabolism of some cells is sensitive to the fluctuations of glucose. If glucose is continuously added to a culture, the necessary mixing time may be an important consideration in scaling up the bioreactor. Such sensitivity to glucose fluctuation is not universal, however. For other cells, glucose mixing time may not affect scale-up.

9.2.2   Scale Translation for Product Development and Process Troubleshooting

A typical processing development for a new product begins in small, laboratory-scale reactors. When a larger quantity of product is needed or the product is to be commercialized, the production process must be transferred to a pilot or manufacturing plant. Some product development will involve scaling up and the construction of a new production plant. In this case, we need to know the critical parameters that must be maintained in an optimal range. In the majority of cases, the production will be carried out in an established manufacturing facility. We must evaluate how a process will perform when the production reaches the manufacturing plant. This is a critical question that must be assessed early in process development.

At times, during the course of manufacturing, the process performance may deviate from the norm. The cause of the deviation must be found and corrected. For those cases, it is useful to have a scale-down model. We can translate the large-scale operation to a scale more suitable for experimental investigation. In both scaling up and down, it is important to understand which factors are affected by scale translation.

9.2.3   How Scale-Up Affects Process Variables, Equipment, and Cellular Physiology

As the scale changes, many aspects of the process itself will also change. In addition to the size of the bioreactor, other process changes might include: (1) the amount of materials to be processed, (2) the time period needed to process the material, (3) the preparation of the reactor, and (4) the method used to transfer the process stream. Depending on the context, all of these may change.

Changes in some of these factors may affect the cell's physiological variables that influence productivity. For example, the length of time involved to transfer cells from a seed culture to a large bioreactor will increase in large-scale operations. A longer transfer time may make oxygen starvation a concern. In the case that batch sterilization is used to process the medium, a longer duration of steam sterilization at the large scale may cause nutrients to degrade.

In large-scale operations, a greater amount of raw materials is used in each run. As a result, the manufacturing process goes through different lots of raw materials faster than at smaller scales. This could also pose a challenge in maintaining the supply chain and in ensuring quality control. Those issues are important for scaling up, but are also process and product specific.

9.2.4   Scale-Up of Equipment and Geometrical Similarity

In this chapter, we will discuss physical parameters in stirred-tank bioreactors, as affected by scale. We take this approach because the effect of these parameters on scale-up can be generalized as principles that will hold true in all processes. We will also focus on scaling up based on geometrical similarity, to better illustrate the effect of scale on different process parameters.

"Scaling up based on geometrical similarity" refers to maintaining the ratio of the main geometrical lengths constant in different scales. This means that we would keep the height and diameter, as well as the length of all internal parts (e.g., impeller and flow diverter), at the same proportion between the two scales. Consequently, different surface areas related to the reactor (including its cross-sectional area, area of the reactor wall, and area of the impeller) are increased by a factor of the geometrical proportionality to the second power. The volume of the reactor is increased to the third power.

In scaling up, when the reactor is increased by a factor of two in length, its related surface area increases by a factor of four; volume increases by eight. Thus, as a result of scaling up, the scale-related surface area per unit volume of equipment will decrease. In microbial fermentation, the decrease in surface area–to-volume ratio makes it more difficult to remove heat generated from metabolism and mechanical agitation. In mammalian cell processing, the generated metabolic heat is less of a concern. However, the process may still be sensitive to other variables related to scale change.

As the scale of a piece of equipment changes, the physical and mechanical parameters may not be constant. In general, it will not be possible to keep all key operating parameters constant between different scales. This may lead to changes in the chemical environment and affect cell physiology and productivity.

The practice of scaling up and scaling down is therefore not to keep scale-related parameters constant. Rather, it should define the operating range of scale-sensitive physical and mechanical parameters so that the cellular physiological states and critical process outcomes can be maintained within an acceptable range.

9.3 Mechanical Agitation

Since the stirred tank is the most widely used bioreactor, we will use a stirred-tank example to illustrate physical constraints on scale-up. In a stirred bioreactor, mechanical agitation keeps cells and microcarriers in suspension, provides mixing to create a more homogeneous chemical environment, and creates a flow pattern that increases the retention time of gas bubbles in the culture fluid, enhancing oxygen transfer. In the cases that cells are grown as aggregates, agitation also helps to reduce the formation of oversized particles.

In microbial fermentation, oxygen demand is rather high, often exceeding 150 mmole/L-h. To increase the efficiency of the oxygen supply, extensive agitation is used to break up air bubbles. In many fermentations of mycelial mold or actinomycete, extensive agitation is used to overcome the high viscosity of culture fluid and to reduce the mycelial pellet size to enhance oxygen transfer.

In cell culture processes, the oxygen demand is nearly two orders of magnitude lower than that in microbial fermentation. Adequate oxygen supply can usually be achieved with less intensive agitation, which also provides sufficient mixing and suspension of cells.

In a stirred tank, the flow patterns generated by different impellers are generally classified as one of two types: axial flow or radial flow (Figure 9.1). An axial-flow pattern refers to a primarily upward or downward flow, due to the pumping action of the impeller. In a radial-flow pattern, the liquid moves primarily outward toward the wall of the vessel.

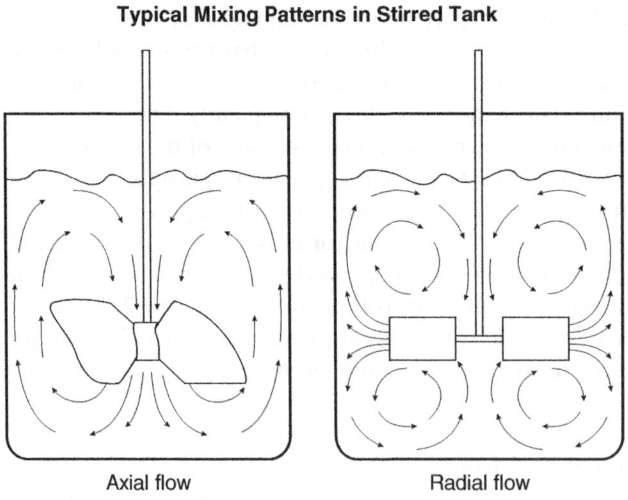

Typical Mixing Patterns in Stirred Tank

Axial flow   Radial flow

Figure 9.1 Two mixing patterns caused by two different classes of agitation mechanisms in a stirred tank.

Three Impellers Widely Used in Stirred Tanks

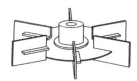

Rushton turbine

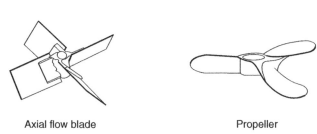

Axial flow blade   Propeller

Figure 9.2 Three types of impellers commonly used in stirred-tank reactors.

Rushton disk turbines, often used with multiple installations in large reactors, are the predominant type used in microbial fermentation (Figure 9.2).

In this design, the sparger is placed directly underneath the disk turbine. Gas bubbles from the sparger rise, hit the disk, and move outward beneath the disk. The blades, rotating at a fast speed, then break up the bubbles. In the region immediately surrounding the blade, a very high-energy dissipation is predicted by computer simulations. This would contribute to bubble breakup, but can also create a high shear zone, which can potentially damage cells. For the cultivation of larger plant and animal cells, impellers that generate axial flow are used; the shear fields generated by axial-flow patterns are lower than those generated with radial-flow patterns. These include propeller and pitch-blade type impellers. The impeller moves the liquid to create the lift for mixing and suspending solids. Sometimes in microbial fermentation, the propeller is used near the top of the liquid content to enhance gas bubble entrapment and to increase oxygen transfer. The mixing and shear field generated by the agitation of an impeller depend on the type of impeller, its diameter-to-tank ratio, and the rotation rate. For suspending large particles, such as the microcarriers used in mammalian cell culture, a larger impeller with a lower rotation speed is used.

9.4   Power Consumption and Mixing Characteristics

9.4.1   Power Consumption of Agitated Bioreactors

To design process equipment for a wide range of scale, we must be able to predict the relationship between the design variables and the scale. To develop this relationship, data are collected from different scales, under different conditions, and then plotted together. Using a wide range of experimental conditions, the resulting data will inevitably yield different correlations grouped by similar conditions.

For example, we can track the pump pressure required to deliver a given water flow in different pipes. Plotted by pressure drop and flow rate, the data will likely group together according to the size of the pipe used. A correlation may be observed within each group of data in the plot. These correlations may then be used for that particular scale or a range of scales. However, note that this correlation also depends on the specific experimental conditions. Outside of these specific conditions, we cannot depend on this correlation. Hence, we must be mindful of when these correlations are useful versus irrelevant.

To find a correlation that is applicable to different scales over a wide range of operating variables, the experimental data are often plotted in "dimensionless variables." These dimensionless variables are a combination of experimental variables. In such combinations, the units of dimensions from individual variables cancel each other out. The idea is that a correlation between dimensionless variables is not sensitive to scale. The dimensionless correlations obtained from experiments performed on different scales should hold on any scale. They should hold true, even in the range of variable values that have not been investigated.

Various relationships expressed in dimensionless numbers are fundamental to fluid mechanics, mass transfer, and heat transfer. The correlation between the friction factor (f) and Reynolds number (Re) is universally used in the design of almost everything involving fluid flow. This same correlation applies to the flow of city water in a pipe or transcontinental crude oil transport. The logarithmic plot of the friction factor and Re show two regions: at low Re, f decreases linearly until Re $= \sim 2000$. There is a short break, and then it continues at a relatively constant value in the high-Re region. The first region is recognized as the laminar- (or viscous-)flow region, and the constant tail is the turbulent-flow region.

A similar plot has been generated for power consumption in stirred-tank reactors. The Reynolds number is now denoted as Re$_I$ (impeller Reynolds number) to indicate that it is based on the length (diameter) of the impeller (Panel 9.1 and Figure 9.3). The dimensionless number for power consumption by the impeller is the power number, N_p.

Panel 9.1 Agitation of Newtonian Fluids

Impeller Reynolds number:

$$\frac{\text{Inertial force}}{\text{viscous force}} = \frac{mass \cdot acceleration}{(pressure)(area)} = \frac{(\rho D_I^3)(D_I N^2)}{\left(\mu \dfrac{ND_I}{D_I}\right) D_I^2}$$

$$R_{e_I} = \frac{\rho N D_I^2}{\mu} \tag{9.1}$$

Power number:
The representative velocity is approximated by impeller tip speed ND$_i$.

$$\frac{\text{External force}}{\text{inertial force}} = \frac{\dfrac{P}{ND_I}\dfrac{1}{D_I^3}}{\dfrac{\rho D_I^3}{D_I^3}N^2 D_I} \tag{9.2}$$

$$N_p = \frac{P}{N^3 D_I^5 \rho}$$

Nomenclature of a Stirred-Tank Layout

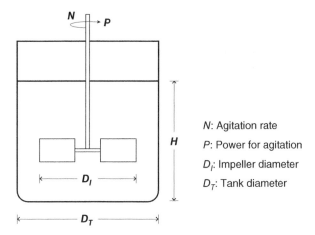

N: Agitation rate

P: Power for agitation

D_I: Impeller diameter

D_T: Tank diameter

Figure 9.3 Important physical parameters for the description of a stirred-tank bioreactor.

Correlations between N_p and Re_I have been generated for various types of impellers (Figure 9.4). They all exhibit behavior similar to the f versus Re plot for tube flow. In all these N_p versus Re_I plots, a linear decrease with increasing Re_I is followed by a constant value region. The linear decreasing region and the constant region represent the two correlations for viscous-flow and turbulent-flow regimes, respectively. In the turbulent regime, the power number is constant over a wide range of impeller Re_I's, but the value changes with different types of impellers.

In bioreactor operations, the flow is always in the turbulent regime. Viscous flow is encountered in a stirred tank only when a very viscous fluid, like glycerol, is used. Therefore, the power number for impellers is constant for a given type of impeller. In

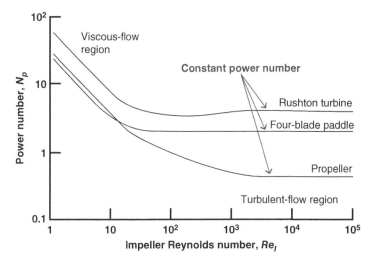

Figure 9.4 The correlation between a dimensionless power number (N_p) and impeller Reynolds number (Re_I) for different types of impellers in a stirred-tank reactor.

other words, the impeller power divided by $N^3D_I^5$ is constant, where N is the agitation rate (radian per unit time) and equal to $2\pi n$ [where n is the rotational speed (number of rotation per minute, or rpm)], and D_I is the impeller diameter.

From the value of the power number, we can calculate the power input (P) from the motor that is required to drive a given size of impeller at a specified rotation rate. Most stirred tanks larger than pilot plant scale have multiple impellers installed in them. Usually, the spacing between the impeller is approximately one impeller diameter distance apart. Under such conditions, the interference between two adjacent impellers is minimal. The power drawn by multiple impellers can be considered as the power calculated from the power number, multiplied by the number of impellers.

The power calculated from the power number is the amount of power required to drive the liquid content. During the operation with extensive aeration, there is substantial gas entrapment. As a result, the density of the fluid being pumped by the impeller is lower than the liquid alone, thus reducing the power consumption. The power calculated from the power number is often referred to as "ungassed power." Under aeration conditions, this is called "gassed power." Usually, an empirical correction is used for a particular type of reactor and process for use in making corrections of power consumption estimations. The extent of decrease of gassed power can be as high as 50% of ungassed power.

9.4.2 Other Dimensionless Numbers

Three other dimensionless numbers are frequently used to predict the performance of a stirred tank when the scale changes. Each of those three dimensionless numbers deals with one of the three important characteristics of bioreactor operations. These characteristics are: fluid velocity, volumetric flow rate, and mixing time.

First, we begin with fluid velocity. The maximum velocity in a mixing tank occurs at the tip of the impeller. This velocity can be represented by the multiplicative product of the rotation speed of the impeller times its radius, $\pi ND_I/2$ (*Note*: In this chapter, we will ignore 2 and π in the denotation and discussion of perimeter, area of circle, etc. The constant values such as 2 and π are cancelled out when comparing different scales. In the subsequent discussion, the tip velocity will be simply presented as ND_I to focus on the two scale-sensitive parameters, N and D_I.)

Next, we look at volumetric flow rate. The amount of fluid that the impeller can move (called "pumping," or Q_p) is directly dependent on its rotating velocity and the area of its blades. Since we are considering scale translation under geometrically similar conditions, we can use the length of impeller, instead of the impeller blade, to represent the length scale. The pumping, then, is the projected area (D_I^2) of the impeller multiplied by the velocity of its rotation (ND_I), which gives ND_I^3.

Last, we consider mixing time. The representative time scale (called "characteristic time") in a mixing tank is the inverse of rotation speed ($1/N$).

The dimensionless numbers for the three properties can be obtained by taking the representative velocity (v), liquid volumetric flow rate (Q_p), and mixing time (Θ), and dividing by their respective characteristic counterparts (e.g., ND_I, ND_I^3, and $1/N$). The plots of dimensionless velocity, pumping, and mixing time against Re_I all show profiles similar to the power number plot, with two distinct flow regimes: laminar (viscous) flow and turbulent flow. The values at high-Re_I turbulent regimes are relatively constant.

9.4.3 Correlation of Oxygen Transfer Coefficient

For a bioreactor, oxygen transfer capacity under various operating conditions is a very important parameter. Many factors affect the oxygen transfer capacity of bioreactors, such as the geometry, type, and spacing of impellers; the nature of fluid or fermentation broth; the aeration rate; and the impeller speed. Ideally, a dimensionless correlation should be developed to correlate the operational conditions and the oxygen transfer capacity of the reactor.

To develop empirical correlations, extensive experiments have been performed over a wide range of conditions. However, those correlations are not dimensionless. One thus needs to ensure that the correct units are used when applying those corrections. A typical correlation for a particular type of reactor includes the aeration rate or superficial velocity and the power input in the correlated variable (Panel 9.2). Correlations developed for different types of reactors have different values of exponents for the same variable.

Panel 9.2 Empirical Correlations for $K_L a$ in Mixing Tanks

$$K_L a = c(P_g/V)^\alpha V_s^\beta \qquad (9.3)$$

- Oxygen transfer is proportional to power input and superficial velocity.
- The correlation is not dimensionless (the value of c is dependent on units used).
- The value of power exponents (α, β) are less than 1.

9.5 Effect of Scale on Physical Behavior of Bioreactors

Using correlations based on dimensionless numbers, we can explore the effect of scale on different variables. We assume that equipment of different scales will remain geometrically similar, so the effect of the size of the reactor can be compared using a characteristic length, D. If the tank diameter increases tenfold, all the other reactor parts (tank height, impeller diameter, etc.) will increase by the same tenfold proportion in length.

In scaling up different processes, one needs to keep the most important variable(s) constant or within an acceptable range. The most commonly used criteria for scaling up are: (1) a constant $K_L a$, so that mass transfer can be maintained; (2) a constant impeller tip speed, to sustain a critical value of high shear velocity to break up agglomerating particles or pellets of mycelial cells; (3) a constant power input per volume (usually for less power-intensive processes, such as crystallization and blending); and (4) a constant mixing time.

Consider the case of scaling up by keeping the power input per reactor volume constant (Panel 9.3). Recall that the power number is constant in the turbulent region, and the power input (P) is proportional to $N^3 D^5$. The reactor volume is described by $\pi H D^2$. Because of geometrical similarity, we can represent H by D and ignore the constant π, which does not contribute to scale comparison. The reactor volume (V) is thus represented by D^3. In keeping P/V constant, $N^3 D^2$ is also constant in different scales. In scale-up, as D increases, the rotation speed must decrease by $1/D^{2/3}$. Larger reactors are operated at lower rotation speeds.

Panel 9.3 Scaling Up Geometrically Similarly by Keeping Power per Unit Volume Constant

---

$P/N^3 D_i^5 \rho$ is constant in a turbulent region. Density of water, ρ, is constant. Thus:

$$P = KN^3 D_i^5 \qquad (9.4)$$

The volume of a reactor can be expressed as the characteristic length raised to the third power (c is a constant):

$$V = \pi H D_T^2 = cD_i^3 \qquad (9.5)$$

The power per unit volume is constant:

$$P/V = KN^3 D_i^5 / cD_i^3 = \text{constant}$$

$$N^3 D_i^2 = \text{constant} \qquad (9.6)$$

All length can be represented by a characteristic length D.

Effect on Agitation Rate

From Eq. 9.6:

$$N^3 \propto D^{-2}$$

$$N \propto D^{-2/3} \qquad (9.7)$$

The agitation rate N decreases with increasing scale. When the diameter increases eight times, the agitation rate is reduced four times in the larger scale.

Effect on Impeller Tip Speed

Tip speed is described by N multiplied by D.
   From Eqs. 9.6 and 9.7:

$$ND \propto N^{-2}D^{-1} = (D^{-2/3})^{-2}D^{-1}$$

$$ND \propto D^{1/3} \qquad (9.8)$$

Tip speed increases with increasing scale, but only at 1/3 power of the length of scale.

Effect on Mixing Time

The dimensionless mixing time is θN. Its value is relatively constant in the higher number turbulent region.

$$\theta_m N = \text{constant}$$

$$\theta_m \propto N^{-1} \qquad (9.9)$$

From Eq. 9.7:

$$\theta_m \propto D^{2/3} \qquad (9.10)$$

Effect on Liquid Pumping

The capacity of liquid pumping (Q_p) can be described by the impeller tip speed, ND, multiplied by the area that it moves against the liquid, D^2.

$$Q_p \propto ND^3$$

From Eq. 9.7:

$$Q_p \propto (D^{-2/3})D^3$$

$$Q_p \propto D^{7/3} \tag{9.11}$$

Liquid-pumping capacity increases with scale. In scaling up, one examines the pumping capacity on a per volume basis (Q_p/V).

$$Q_p/V \propto D^{-2/3} \tag{9.12}$$

The pumping capacity per volume decreases with increasing scale.

By similar algebraic manipulation, one can also see that scaling up by a constant power per volume (N^3D^2) will lead to increasing the amount of total pumping (ND^3) with the scale. However, pumping per volume will decrease as the scale increases. For mixing time, the trend is an increase with scale.

Panel 9.4 compares the effects of scale-up by setting power per volume, agitation rate, or impeller tip speed at constant values. Cases considered include constant power per volume and constant agitation rate, constant pumping rate, and constant tip speed. Often, one chooses not to keep a single value constant, and not to scale with geometrical similarity.

Panel 9.4 Mechanical Agitation and Scale-Up under Different Scenarios

On scale-up geometrically $P/N^3D_i^5 = $ constant

Property	Small scale 100 liter	Plant scale 12,500 liter		
P	1.0	125	3125	25
P/volume	1.0	1.0	25	0.2
N	1.0	0.34	1.0	0.2
D	1.0	5	5	5
Q_p (pumping)	1.0	43	125	25
Q_p/volume	1.0	0.34	1.0	0.2
ND_i (tip speed)	1.0	1.7	5	1.0
D_i/D_T	0.33	0.33	0.33	0.33

Example 9.1   Power Consumption by Impeller in a Stirred-Tank Bioreactor
A laboratory 5 L stirred tank is equipped with a turbine impeller of 5 cm in diameter that operated at a variable speed from 10 to 200 rpm. For this type of impeller, the impeller power number is 5.5 if it is operated at a Reynolds number higher than 800. What is the difference in the power drawn by the impeller when it is operated at its lowest and highest speeds?

Solution
The relevant values for calculating the Reynolds number, Re_I, are:

Rotation speed:

$$N = 2\pi(10 \text{ min}^{-1}) = 1.05 \text{ s}^{-1}$$
$$N = 2\pi(200 \text{ min}^{-1}) = 21 \text{ s}^{-1}$$

Density: $\rho = 1000 \text{ kg/m}^3$ (use the properties of water)
Viscosity: $\mu = 0.001 \text{ kg/m-s}$
At 10 rpm:

$$Re_I = (1000 \text{ kg/m}^3)(1.05 \text{ s}^{-1})(0.05 \text{ m})^2/(0.001 \text{ kg/m-s}) = 2625$$

At 200 rpm: $Re_I = 52500$.

Both are higher than 800; thus, $N_p = 5.5$.
From the definition of power number N_p, the ratio of the actual power drawn at the speed is:

$$\frac{P_{10}}{P_{200}} = \frac{N_{10}^3}{N_{200}^3} = \left(\frac{10}{200}\right)^3 = \frac{1}{8000}$$

9.6   Mixing Time

In a batch stirred tank, in the absence of any chemical reaction, all materials in the reactor will eventually be uniformly distributed. When a reaction takes place in a reactor with inadequate mixing, the concentrations of reactant and catalyst (or the cell) may not be uniformly distributed. Whether mixing (or the uniformity of the reactant concentration) poses a concern in a reactor depends on (1) the relative magnitude of the reaction rate, and (2) the rate that the reactant can be transferred from a high-concentration region to a low-concentration region.

In bioreactors, the reaction rate is relatively low compared to convection caused by agitation, except when the fluid is highly viscous due to mycelial growth or the production of viscous product such as xanthan gum. In the vast majority of bioreactors, nutrients added at the beginning of the culture will eventually become uniformly distributed. Mixing problems may arise only for those components that are added continuously or intermittently during the cultivation.

9.6.1   Nutrient Enrichment Zone: Mixing Time versus Starvation Time

In bioprocesses, oxygen is almost always the first nutrient to be depleted. Due to its low solubility in the medium, it must be continuously supplied. If we take the fluid of a culture to fill up a sealed flask so that no gas phase is in contact with the fluid, the oxygen concentration in the culture fluid will decrease as it is being consumed by cells. At higher cell concentrations, the oxygen will be depleted faster. Continuous supply of oxygen is thus necessary to avoid its depletion.

Compared to oxygen, carbon sources are typically present at a much higher concentration. Their depletion time is much longer than that of oxygen. However, with a high consumption rate, even those nutrients can be depleted.

Supplying oxygen and other nutrients continuously or intermittently to a growing culture of cells is a routine operation. It should be noted, though, that this one aspect of the bioreactor is rather sensitive to scale.

When a nutrient solution is added to a stirred-tank bioreactor, it is typically added at a fixed position (e.g., through a feeding pipe entering from the top of the tank). The circulating liquid then carries the added nutrient to different regions in the reactor. Without mixing, the diffusion time for the nutrient to distribute would be too long to support the nutrient consumption in a bioreactor. If the circulation time is long, a concentration difference can develop between the feeding zone and a region farther away from the feeding zone.

Consider how a fluid element that is carrying cells moves around in the reactor. When it passes through the feeding zone, it acquires the nutrient. The fluid element and the cells then move away, while cells in the fluid element continue to consume the nutrient. When the fluid element returns to the feeding position, it again acquires the nutrient. It is important that the amount of nutrients acquired by the fluid at the feeding zone is sufficient for the cells before they return to the feeding zone again. Otherwise, starvation occurs. In other words, the circulation time should be shorter than the time-to-nutrient depletion.

Inadequate mixing may subject cells to cycles of feast and starvation. It may also affect nutrient utilization efficiency and alter a cell's metabolic state. This phenomenon is more frequently seen in large-scale, industrial bioreactors. To alleviate the adverse effects of nutrient fluctuation, multiple nutrient addition ports are sometimes installed in different regions of the reactor. In scaling up, mixing time and nutrient starvation time are thus important factors to consider.

9.6.2 Mixing Time

Earlier in this chapter, we discussed that the order of magnitude of the liquid pumping rate can be estimated by ND_I^3. The circulation time in a stirred tank thus will be in the order of magnitude of volume divided by the pumping rate (Panel 9.5). In an ideal well-mixed continuous stirred tank, it takes three holding times to replace 98% of the original content in the reactor (see Chapter 7). One can thus get an order-of-magnitude estimate of a mixing time, defined as the time that it takes for the content of the tank to reach 99% of the new steady-state value. That is often taken as four times of the circulation time (Panel 9.5). As was discussed in Section 9.3 on mechanical agitation, the mixing time will increase with the increasing size of the reactor.

In practice, the mixing time of a reactor is affected by: (1) the geometry of the reactor; (2) the spacing of the agitator; (3) the fluid properties; and (4) the mechanism of the nutrient addition port. When the mixing time is an important parameter for the process, it is measured under the operating conditions of the process.

Panel 9.5 Mixing Time

- Two important variables for mixing are approximated as a first-order process:

  Stirrer pump capacity $Q_P \propto ND_I{}^3$ Circulation time $\theta_c = \dfrac{V}{Q_P}$

- Mixing time:

  $$\theta_m \propto \frac{1}{N} \qquad\qquad (9.13)$$

- Definition: Time needed for tracer to reach 0.9 or 0.99 of final mixed concentration
- θ_m can be approximated by $4\theta_c$.

Measurement of Mixing Time

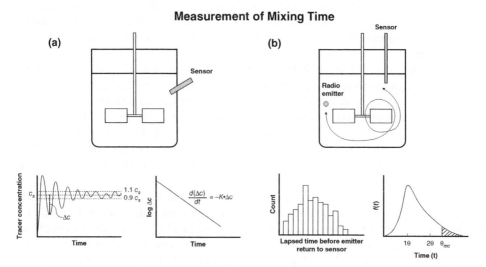

Figure 9.5 (a) Measurement of average (bulk) mixing time using a tracer. (b) Measurement of mixing time distribution in a stirred-tank reactor.

To measure the mixing time, we can inject a dye into the reactor. Then, we use a sensor fixed to a position in the reactor to record the dye concentration over time. The concentration will initially fluctuate, and then gradually settle to a steady value. The time needed for the concentration to settle within a narrow range of concentrations (such as 90 or 99% of its final homogeneous concentration) is considered the "mixing time" (Figure 9.5a). If one plots the concentration deviation from its final steady value, ΔC, the time profile can be approximated by first-order kinetics.

9.6.3 Mixing Time Distribution

The mixing time measurement described in Section 9.6.2 measures an averaged mixing time. If we follow a particular fluid element in a reactor, it will enter the nutrient feeding zone, exit it, and then come back in a repetitive cycle. However, the time interval between the element's consecutive returns to the feeding zone will not be uniform. Average mixing time is a useful measure of the characteristic of a reactor. In some cases, though, the distribution of mixing time profoundly affects cell physiology and reactor performance.

Imagine that we place a small ball that emits a radio signal in the fluid in the reactor (Figure 9.5b). The ball has the same density as the fluid and is carried around by the fluid motion. A sensor at a fixed position in the reactor would pick up the signal when the ball is close and record the time interval between consecutive returns. This time interval of return will be distributed over a range. Sometimes, the ball returns shortly after it moves away. At other times, it roams around the reactor for a while before returning to the sensor.

A histogram of the time-interval distribution can be converted to a mixing time distribution function. In general, the distribution follows a logarithmic-normal distribution. The mean or median of mixing time is a descriptor of mixing characteristics, but it does not present the complete picture of mixing. Given the same median mixing time, two reactors may still have a very different mixing time distribution.

Imagine that the location of the sensor is also the position of nutrient feeding. Those cells circulating with the particular fluid element receive nutrients only when the fluid returns to that position. When the circulation time is longer than the critical nutrient depletion time, cells may experience starvation, enter apoptosis, or suffer other irreparable damage. Thus, an exceedingly long circulation time can have a negative effect on the reactor performance.

In larger reactors, the mixing time increases. For processes that are sensitive to nutrient depletion, the mixing time distribution in the scaled-up reactor can be experimentally measured. This can help to ascertain whether the probability of an exceedingly long mixing time is acceptable.

9.7   Scaling Up and Oxygen Transfer

When scaling up a process, we aim to maintain the growth and production kinetics, as well as the productivity. To meet that goal, we normally provide all nutrients in quantities proportional to the scale to meet the metabolic needs of cells. It is relatively easy to supply medium components in proportion to cellular needs. To supply oxygen and remove CO_2 through gas aeration during scale-up, however, presents a challenge because of the constraints of physical factors.

9.7.1   Material Balance on Oxygen in Bioreactor

To supply oxygen to the growing cells in the bioreactor, air is blown through the culture fluid. This transfers oxygen from the gas bubbles into the culture fluid. The amount of oxygen in the exhaust gas from the reactor is less than that supplied into the reactor; the difference is the amount that has been transferred into the liquid phase.

As discussed in Chapter 8, material balances on oxygen can be performed on the gas entering and leaving the reactor (the gas phase balance). They can also be performed on oxygen being transferred from gas bubbles to liquid (the liquid phase balance) (Figure 9.6). The oxygen transfer rates calculated from the gas phase balance and from the liquid phase balance are equal at a steady state. The liquid oxygen transfer rate (OTR) is described by the overall mass transfer coefficient ($K_L a$) and the driving force (c^*-c).

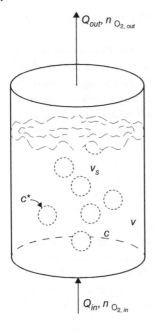

Figure 9.6 Oxygen transfer from gas phase in a stirred-tank bioreactor.

For small reactors, we assume that the liquid is well mixed. This assumes that the concentration measured at the outlet is the same as the concentration in the reactor. c^* at the air outlet should be used for the driving force calculation. For a large reactor, we assume that the gas phase behaves like a tubular reactor (plug flow); a logarithmic mean of the driving force is used. We assume that a quasi–steady state [i.e., the change in dissolved oxygen (dc/dt)] is very slow compared to the oxygen consumption and rate of transfer. Thus, the oxygen uptake rate (OUR) can be assumed to be the same as the OTR.

Consider a case where we scale up a reactor geometrically similarly, so that all physical length dimensions (e.g., diameter, height, etc.) increase proportionally. The reactor volume, V, thus increases proportionally with the tank diameter to the third power. The cross-sectional area of the tank, A, increases with diameter to the second power. To supply a sufficient amount of oxygen, we provide an air flow rate (Q) in proportion to the reactor volume. The superficial velocity (i.e., the amount of air passing through a cross section in the reactor) is described by the air flow rate divided by the cross-sectional area.

As the scale increases, the superficial velocity also increases proportionally with the scale factor. This will result in increased gas retention (or gas holdup) in the liquid phase. A very high air flow rate can cause flooding of the impeller, a situation in which the impeller moves air rather than liquid. Additionally, as superficial velocity increases, foaming becomes a more serious concern. To avoid those problems, in general, air flow per reactor volume decreases with increasing scale.

9.7.1.1 Aeration Rate and the Oxygen Transfer Driving Force

Now, we will examine the gas phase balance. In scaling up, if Q/V is kept constant and the OUR stays the same, the difference of oxygen levels at the inlet and outlet ($Y_{O2,in}$ and $Y_{O2,out}$, respectively) will also remain the same, as does the OTR. However, when scaling up, Q/V is likely to decrease, as discussed above. As can be seen in the gas phase balance, if OUR is to be kept unchanged, then $Y_{O2,out}$ has to be smaller to maintain the material balance.

Then, consider the liquid phase balance. The OTR is K_La multiplied by (c^*-c). Since the oxygen level at the outlet, $Y_{O2,out}$, is lower, c^* will also be lower. This leads to a lower driven force for oxygen transfer. The OTR cannot be kept at the same level as in the smaller scales with a lower c^* unless K_La is increased.

In any case, oxygen transfer will be affected in scale-up. It is not unusual to see the dissolved oxygen level in large-scale bioreactors be lower than in their smaller counterparts. For some processes where the oxygen consumption rate is lower, such as the mammalian cell culture process, oxygen can be used to enrich air and increase the driving force for oxygen transfer.

9.8   Other Process Parameters and Cell Physiology

So far, we have discussed the mechanical aspects of scaling up. The effects of scale-up on both mechanical agitation and aeration can be readily quantified. In addition to agitation and aeration, many other factors are affected by the scale of the operation. As the scale increases, the raw material process method may need to be modified because of equipment limitations or longer processing time.

In a large-scale operation, the equipment for the raw material processing is larger in size, which may alter the kinetics of dissolution or change the particle-size distribution. This may in turn affect the availability of nutrients to cells, and even possibly affect the efficiency of a cell's utilization of material. Large-scale operations also consume more raw materials, thus depleting a single lot of materials faster. This will cause more frequent changes among material lots, and such operations will be prone to raw material quality variations.

When scale increases, the fluid transfer and material-processing times also increase. For example, it takes longer to transfer the fluid from a seed reactor to the manufacturing reactor, possibly increasing the susceptibility to various cellular damage, such as cold ischemia or oxygen starvation. Some processes may involve temperature shifts during manufacturing. The heat transfer rate in a large-scale reactor is typically slower than in a small-scale one, resulting in a slow temperature change.

The physical parameters that change with scale may cause changes in the chemical environment. In large-scale operations, the gas flow rate per reactor volume is reduced, leading to not only reduced oxygen content, but also a higher CO_2 level in the exhaust gas. A higher CO_2 level in the gas phase is the consequence of a higher soluble CO_2 concentration in the fermentation broth. A high soluble CO_2 level may have a direct effect on cell physiology; it may also drive the pH down and require the addition of more base to neutralize the pH. Changes in physical parameters during bioreactor scaling up, therefore, may lead to changes in the chemical environment and the cell's physiological response.

When scaling up, the dissolved oxygen level in different scales is normally kept constant. This is typically accomplished by the control algorithm that adjusts the aeration rate of air, enrichment with oxygen, or (in small-scale operations) the agitation rate when the oxygen level in the reactor is below the set point. In large-scale operations, even though the dissolved oxygen is controlled at the same level as in the smaller scales, the control actions taken to maintain the same level will be different because of differences in K_La and gas flow rate. Such variations may not incur any physiological changes. But, in many cases, foaming becomes more pronounced in large-scale operations and thus requires the addition of antifoaming agents. The presence of a larger quantity of antifoaming agents in the culture fluid causes chemical, environmental, and possibly physiological changes.

Physical parameters, such as mechanical stress, may directly affect cell physiology. Cells may be sensitive to shear rate, or excessively strong energy dissipation near the gas discharge zone or the impeller tip. The average impeller pumping per reactor volume or velocity in larger scales is smaller than in smaller scales.

However, what affects cell metabolism or productivity may not be the average value, but the frequency in which cells encounter a high and damaging mechanical stress. This is akin to our discussion on the mixing time distribution: what is most critical may not be the average mixing time, but the frequency of occurrence of a very long recirculation time that causes nutrient starvation and triggers apoptosis. In large reactors, even though the average shear rate is lower, frequent exposure to high mechanical stress may be more damaging to cells.

The most critical factor in scaling up is the physiological outcome. The scaling up of physical parameters, such as mechanical stress distribution, may directly impact a cell's well-being, either by causing structural damage or by inducing a cell's mechanosensing-based response.

In addition, physical parameters may cause the chemical environment to indirectly change at different scales. For example, the change in the air flow rate for a given reactor volume after scaling up may decrease CO_2 stripping from the reactor. The resulting accumulation of CO_2 may affect pH and cause an increased osmotic pressure in the medium. This, in turn, may affect cells' metabolic activity. In scaling up and scaling down, a good understanding of those direct and indirect effects will facilitate the design of experimental conditions. Using this information, we can create a scale-down model suitable for investigation of large-scale processes in laboratory conditions.

9.9   Concluding Remarks

In scale translation, we cannot keep all physical parameters constant. By keeping one set of parameters constant between two scales, some other factors are bound to become different. Therefore, we must define the critical range of various scale-sensitive variables and choose scale-up criteria to ensure that the operation is within the optimal regions.

One may choose not to scale up geometrically similarly. Most large-scale stirred-tank reactors have a larger height-to-diameter ratio compared to small-scale ones. Nevertheless, the physical constraints of scaling up are the same regardless of whether one scales up geometrically similarly or not. In scaling up, the gas flow rate is also likely to change in its proportion to the reactor volume. This causes the mass transfer characteristics to be different for different scales.

Given that multiple physical parameters of the bioreactor cannot be easily manipulated or controlled in the desired range, one may resort to selecting cells that are less sensitive to those parameters. Understanding scale-sensitive parameters, as well as a sound knowledge of estimating the range of those critical parameters, will greatly help in the scale translation of cell culture processes.

In process development involving scale translation, we should aim to reproduce or predict the conditions of physical constraints, as well as the resulting chemical environment. We should also keep in mind that although the chemical environment in a stirred tank can be kept relatively homogeneous with a sufficient degree of mixing, cells within a bioreactor are continually exposed to a varied physical environment. They move from a high-shear zone near the impeller to a low-shear zone. Their exposure to the environment with varying degrees of mechanical energy dissipation follows a distribution, and the distribution changes with changing scale. Perhaps in constructing a scale-up or scale-down model, one should employ a reactor with varying agitation patterns that reproduce the distribution of the identified critical parameter(s).

It may not be possible to replicate all physical and chemical parameters on drastically different scales. Ultimately, one should identify the critical physical parameters and aim to control them within an acceptable range, to ensure a high level of cellular physiological performance at all scales.

Further Reading

Junker, BH (2004) Scale-up methodologies for *Escherichia coli* and yeast fermentation process. J. Biosci. Bioeng., **97**, 347–364.

Nienow, AW, Nordkvist, M and Boulton, CA (2011) Scale-down/scale-up studies leading to improved commercial beer fermentation. Biotechnol. J., **6**(8), 911–925.

Serrano-Carreon, L, Galindo, E, Rocha-Valadez, JA, Holguin-Salas, A and Corkidi, G (2015) Hydrodynamics, fungal physiology and morphology. Adv. Biochem. Eng. Biotechnol., **149**, 55–90.

Nomenclature

a	specific area, interfacial surface area per reactor volume	L^2/L^3
C	a constant	
c	concentration of oxygen	$mole/L^3$
c^*	concentration of oxygen in equilibrium with gas phase	$mole/L^3$
c_s	steady-state concentration	$mole/L^3$
D	characteristic length of reactor	L
D_I	impeller diameter	L
D_T	diameter of stirred-tank reactor	L
H	height of stirred-tank reactor	L
K_L	overall mass transfer coefficient	L/t
N	agitation rate	t^{-1}
$n_{O_2,in}, n_{O_2,out}$	oxygen concentration in the gas at inlet and outlet	$mole/L^3$
N_p	power number	
ND_I	impeller tip speed	L/t

OUR	oxygen uptake rate	mole/L^3·t
P	power for agitation	Force·L/t
P_g	gassed power for agitation	Force·L/t
Q_p	liquid pumping capacity in stirred tank	L^3/t
Q_{in}, Q_{out}	volumetric gas flow rate into and out of reactor	L^3/t
Re_I	impeller Reynolds number	
t	time	t
V	volume of reactor	L^3
v_s	superficial gas velocity	L/t
ρ	density of fluid	M/L^3
μ	viscosity	M/L·t
θ_c	circulation time	t
θ_m	mixing time	t
θ_{mc}	critical mixing time	t
θN	dimensionless mixing time	

Problems

A. General Concepts

A1 Metabolic heat generated during microbial growth needs to be removed by using cooling water in industrial bioprocesses. Discuss why heat removal can be a problem in scaling up. *Note*: Typically, cooling is accomplished by using a water jacket around the wall of the reactor and by using water-circulating cooling coils inside the reactor.

A2 The mean mixing time in a bioreactor has been estimated to be inversely proportional to the agitation rate when the bioreactor is scaled up geometrically. If the power input per unit volume is to be kept constant, what will happen to the mean mixing time when scaling up? If the mean mixing time is to be kept constant in scaling up, what will happen to the power input per unit volume and maximal velocity (or tip speed)?

A3 The "pumping" of liquid in a stirred-tank bioreactor can be approximated by the area moved by the impeller blade's rotational movement and the rotation speed. How will you represent pumping using *N* and D_i? How will pumping be affected by scaling up if the power per unit volume is kept constant when scaling up geometrically similarly?

A4 A laboratory bubble column bioreactor is being used to grow yeast cells at 30 °C. The fluid flow caused by the bubbling of air is the only means of mixing provided. The cells consume oxygen at 5 mmol/g-cell-h. Air is being supplied at 2 L air/L-min to maintain dissolved oxygen at 20% of saturation with air, or 0.04 mmol/L. The exhaust air contains 18% oxygen and 3% CO_2 after removing moisture. The reactor is to be scaled up by increasing its height three times while keeping the diameter of the reactor constant. The air will be supplied at the same rate as in the original scale. The rationale is that the increased height will increase the holding time of the air bubbles in the reactor, and thus increase the oxygen transfer efficiency for a given air supplied to the reactor. Calculate the new exhaust air composition on a dry basis (i.e., after the moisture is removed). Will the dissolved oxygen concentration be the same if the cell concentration is maintained at the same level as in the original bioreactor? Assume that the cells can sustain the same specific oxygen consumption rate until the dissolved oxygen is reduced to 0.02 mM and that, below that rate, the growth rate (and specific oxygen consumption rate) decreases linearly until both the growth rate and oxygen consumption cease at oxygen depletion. You can also assume that $K_L a$ is the same at the same air superficial velocity.

B. Mechanical Agitation

B1 In scaling up a bioprocess using the fungus *Penicillium acremonium* (a mold) from 10 L to 1000 L, the power input per unit reactor volume is to be kept constant to ensure proper dispersion of mycelia. The reactor is scaled up in a geometrically similar way. What will be the ratio of the agitation rate and the impeller tip speed between the small- and large-scale bioreactors? How will pumping be affected by scaling up?

B2 An anaerobic biofuel production process is carried out in a 10,000-L stirred tank with a diameter of 2 m. In order to ensure proper mixing, agitation will be provided using a marine propeller at an agitation rate of 100 rpm. Assume the impeller diameter is half of the tank diameter. *Note*: With a marine impeller, when the impeller Reynolds number is greater than 3×10^3, the power number has a value of 1.0. What is the power of the motor that is required to accomplish this? You can assume the medium has the properties of water.

B3 In non-Newtonian mycelial fermentations, there is a substantial resistance to oxygen transfer between the bulk liquid and the mycelial clumps. The impeller tip speed was shown empirically to be proportional to the liquid-to-clump mass transfer coefficient. In scaling up the process, the agitator power per unit volume as well as the ratio for impeller diameter to tank diameter will be maintained constant. Two options of scaling up will be explored: (a) using a tall stirred tank of the same diameter but with three agitators ($n = 3$); and (b) using a shorter fermenter with two agitators ($n = 2$) in a wider stirred tank. If you are to maximize the mass transfer to the clump, which approach will you take? How do the two options differ?

B4 A large fermenter equipped with a motor is being refitted for a new process. An item being investigated is the diameter of the impeller. The number of agitators on the shaft is to be kept equal. If you are to maximize impeller tip speed, should you choose a smaller or larger impeller? How do they differ?

B5 A stirred-tank bioreactor is 12 m in height and 3 m in diameter. Two Rushton turbines, with a diameter of 0.45 of the tank diameter, are installed. The agitation rate is 100 rpm. The working volume reaches 70% of the tank height. You are designing a scaled-down version using a 200-L reactor that is geometrically similar. What should be the operating conditions if you are to keep the power-per-unit volume constant or to keep the tip speed constant?

B6 A bubble column bioreactor with no mechanical agitation used for an insect cell culture is being scaled up geometrically similarly. It is proposed to keep the gas superficial velocity constant when scaling up. Under the operating conditions, the bubble size is relatively constant; so are the mass transfer coefficient and the bubble-rising velocity. One can thus consider the bubbles to be freely rising at their terminal rising velocity, which is affected by only bubble size and the properties of the surrounding fluid (which is the same in both small and large scales). Since the height of the reactor increases with scale, the traveling time of each bubble also increases. Using the information provided, develop a correlation between the overall mass transfer coefficient and the scale. Will the oxygen transfer rate follow the same relationship you developed for the mass transfer coefficient? Discuss two separate conditions: (a) the amount of oxygen that is transferred is small, so the partial pressure of oxygen in the inlet and outlet does not change appreciably; and (b) the opposite of the first condition, where the oxygen partial pressure decreases very profoundly. You can assume the liquid phase in the reactor is well mixed.

C. Oxygen Transfer

C1 In a small, 10-L stirred-tank bioreactor, air is supplied at 10 L/min. The air inlet and outlet oxygen concentration is 21 and 19%, respectively. In scaling up to 1000 L, the superficial velocity is kept constant. Assuming the oxygen demand of the culture remains constant, what is the outlet oxygen concentration? If the respiratory quotient is one, compare the CO_2 concentration at the outlet of the gas stream between the small and large scales. You can neglect the evaporation of water and the effect of hydrostatic pressure.

C2 A 100-L bioreactor is used to grow recombinant *E. coli* with air being supplied at 100 L/min. The dissolved oxygen level is maintained at 0.1 mM. The solubility of oxygen is 0.2 mM. The inlet air has 21% oxygen (79% nitrogen), while the outlet on a dry basis (after removing the moisture carried from the reactor) is 19% O_2, 2% CO_2, and 79% nitrogen. In the scale-up process, a reactor 125 times larger is constructed in a geometrically similar way. The superficial velocity of air flow can only be increased 2.5-fold above that of the 100-L scale.
a) What is the K_{La} in the 100-L scale?
b) If one can keep K_{La} constant in the 12,500-L scale, can you keep the same oxygen uptake rate as in the 100-L scale? If yes, what is the dissolved oxygen concentration?

Most likely, the K_L cannot be kept constant because the power input per unit volume needs to be reduced in the large scale. If the agitation power per unit volume in the 12,500-L reactor is 50% of that in the 100-L reactor and the air–liquid interfacial area per unit volume (a) is proportional to the air superficial velocity, can the oxygen transfer rate be sustained? What is the new dissolved oxygen level?

C3 A bubble column bioreactor is used to grow insect cells. Air bubbles, introduced at the bottom of the reactor, rise up and exit at the top. This type of reactor is scaled up by keeping the superficial velocity constant. In such cases, the K_L (mass transfer coefficient) is also constant. A 10-L culture with an airflow rate of 5 L/min maintains the dissolved oxygen level at 0.15 mM. The air in the inlet has 79% N_2 and 21% O_2. The outlet air has a composition of 79% N_2, 20% O_2, and 1% CO_2. The molar volume of air is 25 L/mol.

The same cell concentration and oxygen consumption are seen in a scaled-up, 270-L reactor. The two reactors are geometrically similar. You can assume the bubble size in the two reactors is the same.

a) What is the oxygen uptake rate (OUR; mmol/L-min) in the 10-L reactor?

b) What is the exit gas composition in the 270-L reactor if the OUR is maintained constant?

c) If K_L is the same in both scales, what is the dissolved oxygen level in the 270-L reactor?

D. Advanced Problems

D1 A 1-m^3 reactor is to be scaled up 125-fold in a geometrically similar manner. The 1-m^3 fermenter operates at 120 rpm with an air flow rate of 1 m^3/min. The oxygen partial pressure at the air inlet and outlet is 0.21 and 0.17 atm at 25 °C, respectively. The flooding air flow rate is 3 m^3/min, so the air flow rate must be maintained below that. Both fermenters are to be operated basically in the turbulent regime with an impeller Reynolds number greater than 10^4. The power number can be assumed to be a constant value of 6.5. In both cases, the gassed power can be assumed to be 0.6 of ungassed power. The gassed power input per unit liquid volume will be held constant in the scaling up.

Correlation for $K_L a$:

$$K_L a = A(P_g/V)^{0.5} V_s^{0.5}$$

where A is a constant; and P_g is the gassed power. The flooding superficial velocity is the same in both reactors. However, in the large reactor, the gas flow rate will be in the range of 0.5 to 0.7 of flooding gas velocity, as opposed to 0.333 in the 1-m^3 fermenter. The dissolved oxygen concentration in the large tank is to be maintained at the same level as in the small tank, at 0.03 atm of P_{O2} (or 0.03 mM).

a) What is the operating agitation rate in the large fermenter?

b) If $K_L a$ of the large fermenter is to be kept at the same value as that of the small fermenter, what is the gas flow rate? What is the oxygen transfer rate?

c) If the same oxygen transfer rate is to be accomplished in the two scales, what are the values for $K_L a$ and the gas flow?

To simplify the analysis, neglect the hydrostatic pressure effect.

10

Cell Culture Bioprocesses and Biomanufacturing

10.1 Cells in Culture

Near the beginning of the twentieth century, scientists began to isolate and culture cells from animal tissues. Within a few decades, a variety of cells had been isolated and cultured in laboratories. In the early periods, the only tissue-derived cells that could be cultured were anchorage-dependent. This means that they had to attach to a surface in order to grow. These cells also underwent senescence and died after a certain number of generations in culture.

By the middle of the twentieth century, the first human cell line, HeLa, was successfully isolated from a cervical tumor and cultured *in vitro*. This cell line, having been isolated from a cancerous tissue, could be cultured in the laboratory indefinitely. But the cell line was not normal in its growth regulation, chromosome number, and morphology. This was in stark contrast to the normal diploid fibroblastic cell strains, WI38 and MRC5, which were isolated from human lung tissue a decade later. These fibroblastic cells could be passaged in culture for tens of generations, but eventually would undergo senescence. They also had characteristic morphology and behavior. When cultivated in a dish, they spread and showed an elongated shape.

Neighboring WI38 and MRC5 cells also aligned with each other, displaying a characteristic "social" pattern. These cells also exhibited contact inhibition, meaning that their growth stopped once cell coverage of the surface reached confluence. Such characteristics are typically lost in cell lines of cancer origin.

Around the time that the WI38 and MRC5 cell lines were isolated, a mouse fibroblastic cell line, 3T3, was also established. 3T3 cells are morphologically normal and appear to exhibit contact inhibition, but can be passaged and cultured continuously without undergoing senescence. The chromosomal makeup of these cells, however, is not normal. A normal mouse cell has 40 chromosomes (2n); 3T3 cells are aneuploid and have an abnormal number of chromosomes. Lines like WI38 and MRC5, which are normal diploid and have a limited lifespan, are often called "cell strains." In contrast, 3T3 and HeLa can be cultured forever but are not diploid. They are called "cell lines."

Early efforts to develop cell culture were driven by the need to find a replacement for animals and animal tissues. At the time, these tissues were used to grow viruses for use as vaccines. After their successful isolation, normal human diploid fibroblast cells became the major production vehicle for many viral vaccines. Until the arrival of protein biologics in the last two decades of the twentieth century, vaccines were the main products produced in animal cells.

Engineering Principles in Biotechnology, First Edition. Wei-Shou Hu.
© 2018 John Wiley & Sons Ltd. Published 2018 by John Wiley & Sons Ltd.
Companion Website: www.wiley.com/go/hu/engineering_fundamentals_of_biotechnology

In the past decade, there has been continuous growth in mammalian cell culture bioprocessing. This is primarily driven by the expansion of therapeutic protein production in the pharmaceutical industry. Over the past ten years, both the range and quantity of products have significantly increased. Also fueling this growth are the increasing numbers of candidate biologic drugs under development. These drugs have potential as a more effective treatment for many diseases.

10.2   Cell Culture Products

10.2.1   Vaccines

For over two centuries, scientists have used viruses as vaccines. In the eighteenth century, it was observed that dairy maids who had regular contact with cowpox seemed less susceptible to its deadlier cousin, smallpox. This began a long road of scientific research, curiosity, and experimentation about how to build immunity in individuals. Today, widespread immunization is one of the most important healthcare accomplishments of the twentieth century.

Early vaccine production relied on using animals, either by direct infection of the animals or by *in vitro* infection of animal tissues. Virus production in cell culture processes was a major step forward, enhancing process robustness and product quality consistency.

Cell culture processes are now used to produce a large number of viral vaccines (Table 10.1). The total number of cell lines used for human vaccine production is relatively small (Table 10.2). Most are human normal diploid cells, although a dog cell line (MDCK) and a monkey cell line (Vero) have also been used.

Some viral vaccines are made of viruses that have been killed or inactivated after product isolation. Others are live-attenuated viruses that have been adapted to become nonvirulent. The inactivation is often accomplished by a long incubation of the virus with formalin. This renders the virus incapable of infection, while still retaining its immunogenicity.

Table 10.1 Major viral vaccines produced using tissues or animal cells.

Targeted infection or disease	Vehicle of production	Type of vaccine
Poliomyelitis	Human diploid cell line, monkey kidney	Live attenuated, inactivated
Measles	Chicken embryo	Live attenuated
Smallpox (vaccinia)	Chorioallantois, tissue cultures	Vaccinia
Yellow fever	Tissue cultures and eggs	Live attenuated
Influenza	Cell culture (MDCK, Vero)	Attenuated
Rabies	Duck embryo or human diploid cells	Inactivated
Japanese B encephalitis	Mouse brain (formalinized), cell culture	Inactivated

Table 10.2 Major cell strains and lines for human biologics production.

Human vaccines	
Primary cells	Green monkey kidney cells (no longer used)
	Chicken embryo cells
Cell strains	MRC5 (human lung fibroblast)
Cell line	Vero (monkey kidney epithelial cells), MDCK (dog kidney cells)
Recombinant proteins	
Species cell line derived from	
Human	HEK 293
Mouse	C-127, NSO, hybridoma cells, SP2/0
Chinese hamster	CHO
Syrian hamster	BHK

Attenuated viruses elicit a host immune response by actively replicating in the patient. This allows us to use a lower dose for immunization. However, live-attenuated viruses also carry a low risk of reverting to their wild type and causing an active infection in the patient.

With recombinant DNA technology, virus surface antigens may be produced using engineered host cells. The protein antigen is then used to vaccinate, without using the actual virus particle itself. It is also possible to employ a recombinant virus by inserting the gene coding for the surface antigen of the target virus into the genome of the carrier virus. The carrier virus can then be used as the vaccine without posing the risk of infection by the target virus.

10.2.2 Therapeutic Proteins

Today, virtually all protein biologics are produced using recombinant DNA technology by cloning the protein gene into mammalian or microbial cells. However, well before the arrival of recombinant DNA technology, many protein biologics had been derived from tissue, blood, and even cell culture. Examples include insulin, urokinase, Factor VIII, and interferon.

The first generation of recombinant DNA therapeutic proteins, including human growth hormone and insulin, were produced in *Escherichia coli*. The second wave of human therapeutic proteins, including tissue plasminogen activation (tPA), erythropoietin (EPO), and Factor VIII, required extensive posttranslational modifications (such as multiple disulfide bonds and complex glycosylation) not attainable in microbial hosts.

Cell lines derived from animals can perform those posttranslational modifications. However, the glycans produced by cells of different species have different structures. The need to produce proteins that bear glycans closely resembling those on human

Table 10.3 Some therapeutic proteins produced in mammalian cells.

Trade name	Type	Therapeutic use	Host cell
Cerezyme	β-glucocerebrosidase	Gaucher's disease	CHO
Activase	Tissue plasminogen activator	Acute myocardial infraction	CHO
Epogen/Procrit	EPO	Anemia	CHO
Aranesp	EPO (engineered)	Anemia	CHO
Follistim/Gonal-F	Follicle-stimulating hormone	Infertility	CHO
Benefix	Factor IX	Hemophilla A	CHO
Enbrel	TNF receptor fusion	Rheumatoid arthritis	CHO

glycoproteins largely necessitated the use of mammalian cells (instead of insect or chicken cells) for the production of therapeutic proteins (Table 10.3).

Early protein therapeutics were human-native molecules but were produced in insufficient quantities or in a mutated form under some disease states. Many subsequent products were antibodies (Figure 10.1 and Table 10.4). Some of these antibodies were targeted against antigens that are expressed under diseased conditions. For example, an antibody against the epidermal growth factor (EGF) receptor, HER2, is now used to treat some forms of breast cancer that overexpress it.

Some early therapeutic antibodies were acquired by immunizing laboratory animals. They still retained part of the amino acid sequence of the immunized species in their structure, earning the name of "chimeric" antibodies. Later generations of antibody molecules were all humanized, meaning that even the amino acid sequence in the variable region had been converted to human sequences.

Many therapeutic proteins are fusion proteins, with the fused domains being of human origin. A prominent example is the fusion molecule containing the Fc fragment of immunoglobulin G (IgG) and the tumor necrosis factor-α (TNFα) binding fragment of the TNF-α receptor. This molecule was developed by then Immunex (now Amgen) for inhibiting TNF-α and suppressing its inflammatory effect.

Recently, more nonnatural proteins are being explored for treating various diseases. Bispecific antibodies use antigen binding sites from two different antibody molecules to simultaneously target multiple antigen molecules. For example, bispecific T-cell engager (BiTE) antibodies bring the T-cells of the host's immune system to the target cancer cells for close-range killing of cancer cells. Antibodies can also be tagged with drugs or cytotoxic agents [called antibody–drug conjugates (ADCs)] to deliver cytotoxic agents to the specific diseased cells.

10.2.3 Biosimilars

Two decades after their introduction, the patents for many of the mammalian, cell-based therapeutics have expired or soon will be expiring. This has opened the gateway for the production and distribution of "generic" protein biologics at a lower cost, which will make them available to those regions of the world where the price of a patent-protected drug has been prohibitively high.

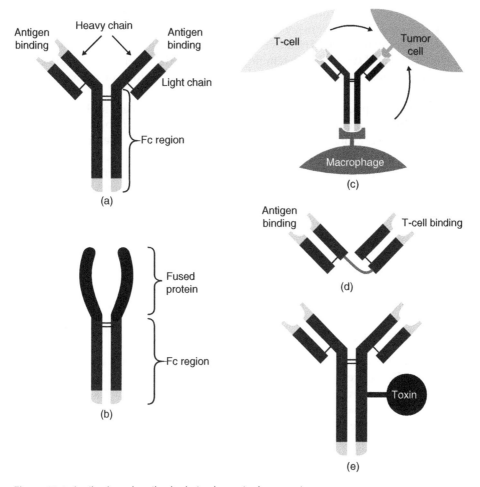

Figure 10.1 Antibody and antibody-derived protein therapeutics.

Table 10.4 Recombinant antibodies produced in mammalian cells.

Trade name	Antibody type	Therapeutic use	Host cells
Zenapax (Daclizumab)	Humanized, anti-α-subunit T-cell interleukin-2 (IL2) receptor	Prevention of acute kidney transplant rejection	NS0 (mouse myeloma cells)
Remicade	Anti–tumor necrosis factor (TNF)	Active Crohn's disease	SP2/0 (mouse myeloma cells)
Herceptin	Anti-HER2	Metastatic breast cancer	CHO
Humira	Anti-TNF	Rheumatoid arthritis	CHO
Xolair	Humanized anti-IgE (immunoglobulin E)	Moderate to severe asthma	CHO
Avastin	Anti–vascular endothelial growth factor (VEGF)	Metastatic colorectal cancer and lung cancer	CHO

The generic versions of protein biologics are also called "biosimilars" or "follow-on biologics." Biosimilars differ from generic drugs, which are small molecules of synthetic chemicals or natural products. Traditional drugs, like penicillin and statins, have a very well-defined chemical structure. Protein therapeutics, on the other hand, are too complex to be completely characterized by their chemical composition or primary amino acid sequences.

Like their name implies, biosimilars may not be exactly equivalent to their patented counterparts. In fact, since the biological and clinical equivalency of a biosimilar cannot be determined merely by its structural similarity to its patented therapeutic, biosimilars need to be individually tested in clinical trials (albeit on a smaller scale than what is required for the initial approval of the original innovative protein therapeutic). This is different from "generic" drugs made of small molecules, for which clinical trials are not required to receive the approval from regulatory agencies. Although biosimilars face a higher technical and regulatory barrier to entry than generic drugs, the potentially expanded reach of protein therapeutics has triggered rapid growth in this industry, in many regions of the world.

10.3   Cellular Properties Critical to Biologics Production

10.3.1   Protein Secretion

Approximately 30% of all cellular proteins in a secretory cell, such as a hepatocyte (liver cell) or a B-cell, are destined for organelles, membranes, and/or secretion. These proteins must first be processed through the endoplasmic reticulum (ER) before being delivered to their destinations. Most of those proteins are translated in the cytosol but are translocated into the ER as nascent protein molecules. Recombinant DNA protein molecules follow the same path. They get translated from mRNA in cytosol and pass through the ER membrane into the ER lumen while being synthesized.

10.3.1.1   Folding in the Endoplasmic Reticulum
Proteins destined for secretion have a leader sequence at their amino terminus that serves as a signal peptide (see Figure 7.11). After translation is initiated in the cytosol, signal recognition particles (SRPs) bind to the signal peptide of the growing nascent protein, causing translation to pause. The SRPs, along with the nascent protein (only the beginning segment of the amino terminus), then dock to a receptor on the ER membrane. The nascent polypeptide is then transferred to a translocon on the ER membrane. Translation elongation then resumes, and the elongating polypeptide passes through the channel of the translocon into the ER lumen.

Folding of the polypeptide starts immediately upon its translocation into the ER lumen, and the signal peptide on the elongating polypeptide is cleaved. A class of ER chaperones and other proteins that facilitate protein folding then assist in the folding of nascent protein molecules to prevent aggregation. Their actions require cellular energy, in the form of adenosine triphosphate (ATP). An important ER luminal chaperone, BiP (also known as GRP78), is also a component of the translocon complex. In addition to BiP, other major ER luminal chaperones include calnexin, calreticulin, and protein disulfide isomerase (PDI).

Glycosylation starts while the protein is still being translated and folded in the ER (see Figure 7.11). A preassembled glycan core, anchored on the interior side of ER membrane, is translocated to the asparagine residue of the glycan binding site on the protein. The presence of glycans increases protein solubility, prevents aggregation, and enhances protein stability. Each glycan core has three glucose molecules at its terminus. As the protein is being folded, the glucose molecules on the glycan are also being trimmed. The absence of the glucose on the glycan is an indication that the folding has been completed. Folded protein molecules are then transported to other organelles, while unfolded ones are recycled for further folding. Unfoldable ones are eventually exported for degradation.

Protein molecules that have completed the folding process are exported from the ER in membrane vesicles. These vesicles bud off from the ER membrane, translocate to their destinations, and then fuse with the membrane of the target organelle or cytoplasmic membrane to release the cargo protein. Membrane trafficking is the main form of molecular transfer from the ER to different organelles.

10.3.1.2 Membrane Vesicle Translocation and Golgi Apparatus

Secretory proteins in the vesicles are taken from the ER to the cis-Golgi. Both the cis-Golgi and the trans-Golgi comprise an array of vesicles and tubules. They are located on the opposite sides of the medial-Golgi. The medial-Golgi, typically containing three to seven stacks of cisternae, is the main site of glycan elongation for glycoproteins.

After a protein molecule is translated, a finite amount of time passes before it is completely folded and glycosylated. With an average protein of about 350 amino acids in length, for example, the translation takes only tens of seconds to be translated. However, the time required for a protein molecule to be secreted after its synthesis can range from 30 minutes to a few hours. The α1-protease inhibitor is among the fastest-secreted proteins, with a half-life of about 28 minutes. Transferrin, on the other hand, takes around two hours to be secreted. In general, a protein molecule spends more time in the ER than in the Golgi apparatus before it is secreted out of the cell.

Even for the same protein, the secretion time is not uniform for all molecules. Some molecules are secreted faster than others. This suggests that the progression of protein molecules through the secretory pathway does not resemble plug flow; a significant degree of mixing occurs in the secretion process. Of course, it is also possible that some protein molecules may exit from different sections of Golgi apparatus and are secreted without completing the entire journey of transversing through Golgi apparatus. The entire secretory pathway is not likely to be like that of a giant continuous flow reactor [continuous stirred-tank reactor (CSTR)]. Its behavior is probably closer to that of a few CSTRs in series, because the amount of time that molecules of the same protein spend in the ER and the Golgi apparatus is not uniform. Also, the duration that each molecule spends to get their glycan extended will be different.

Eventually, the cargo at the trans-Golgi network (TGN) is transported in membrane vesicles to its final destination, be it the plasma membrane (as is the case for secreted proteins) or other organelles. If the contents of the early compartment were translocated through vesicle trafficking (called forward, or anterograde, trafficking), the latter compartment would grow while the early compartments would shrink. For this reason, vesicle trafficking relies on retrograde transfer in addition to anterograde transfer. The ER, Golgi, lysosomes, and endosomes are all part of the secretory network. They communicate through the dynamic trafficking of membrane vesicles.

10.3.2 Glycosylation

The vast majority of recombinant therapeutic proteins are glycoproteins. The extent of glycosylation and the structure of the glycans on those glycoproteins may have a profound effect on their activities and circulatory half-life. Glycans are classified as O-linked or N-linked glycans. O-glycans attach to the polypeptide through the −OH group of serine or threonine. N-glycans link to proteins through the amide group of asparagine. For N-linked glycan, the asparagine is in an Asn-X-Thr/Ser recognition sequence (where X indicates no specificity).

N- and O-glycans attached to proteins are structurally heterogeneous. The glycans attaching to the same attachment site of different glycoprotein molecules are often structurally different. Such heterogeneity is called "microheterogeneity."

Multiple glycosylation sites are often present on a protein molecule, and not all glycan attachment sites on a protein molecule may be occupied. Different protein molecules may have different combinations of occupied and free sites; such differences in the occupancy on different attachment sites is called "macroheterogeneity."

The glycan structure affects the half-life of proteins in blood. For instance, the presence of carbohydrates delays clearance of recombinant interferon from blood, and a higher sialic acid content can increase the circulation half-life for erythropoietin.

Glycans on a glycoprotein may also affect its biological activities. Many therapeutic antibody IgG molecules facilitate the killing of target cells through antibody-dependent cellular cytotoxicity (ADCC). The ADCC activity of an antibody is affected by its glycan structure: a molecule without a fucose on its mannose core has 50-fold higher ADCC activity than one with a fucose (Figure 10.2).

10.3.3 Protein Secretion and Glycan Heterogeneity

The first part of N-glycan synthesis involves the assembly of a high-mannose backbone, which is first synthesized on the exterior side of the ER membrane as a lipid-bound

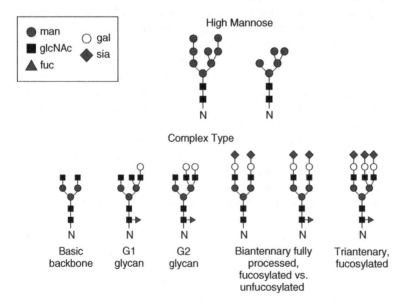

Figure 10.2 Examples of N-glycans.

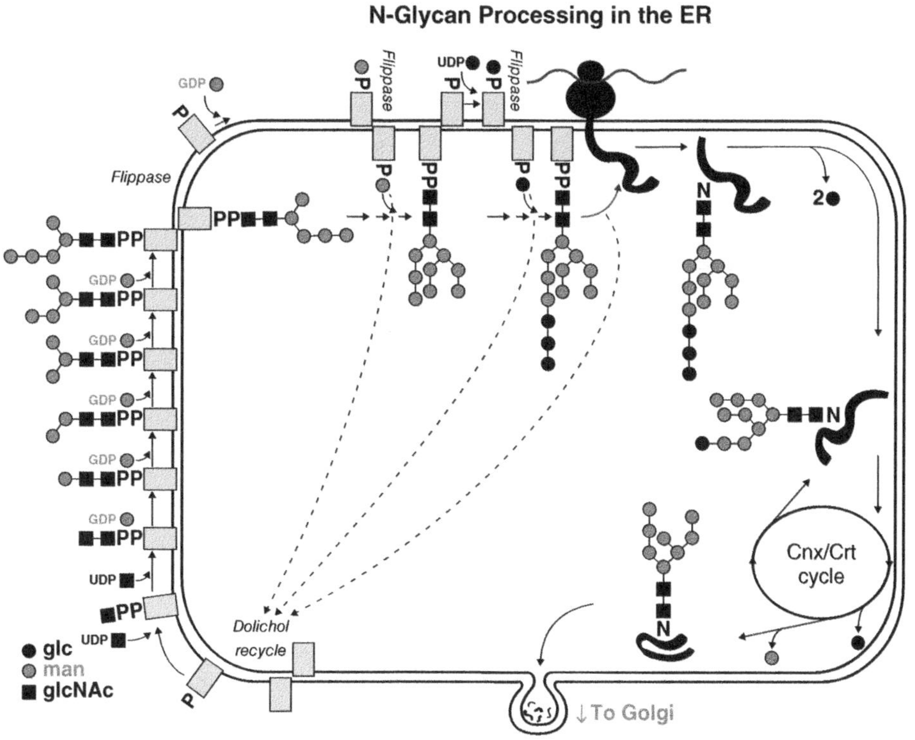

N-Glycan Processing in the ER

Figure 10.3 Protein N-glycosylation in the endoplasmic reticulum.

oligosaccharide (Figure 10.3). The glycan backbone synthesized outside is then flipped over to the interior of the ER. Inside the ER, the backbone acquires an additional four mannose and three glucose molecules to become a mature core. The mature core is then transferred to its binding site (Asn-X-Thr/Ser) on a nascent protein molecule.

Inside the Golgi, mannose is trimmed from the N-glycan core (down to three) before extension takes place (Figure 10.4). Three main carbohydrate molecules constitute most of the extended glycan: N-acetyl glucosamine, galactose, and sialic acid. They are synthesized outside the Golgi, derivatized to become nucleotide sugar, then transported into the Golgi through specific antiporters that transfer a nucleotide sugar in and simultaneously a spent nucleotide monophosphate out.

A number of glycosyltransferases participate in extending the glycan by adding different nucleotide conjugated monosaccharaides to the growing core glycan through different glycosidic bonds. Intermediate glycans on the protein have more than one sugar that can be extended, and each sugar may have more than one hydroxyl group that is available for extension. Each growing glycan, thus, has multiple available reaction paths for extension. The extension reaction does not take place on all of the available reaction sites. In some cases, the reactions of adding those sugars to different branches of the glycan may occur in different sequential orders, but lead to the same product. In other cases, the addition of a particular glycosidic linkage in one location may hinder the reaction of other locations; thus, the reaction itself leads to different glycan structures. The extending glycans can grow into different numbers of branches, creating structures such as biantennary and triantennary glycans.

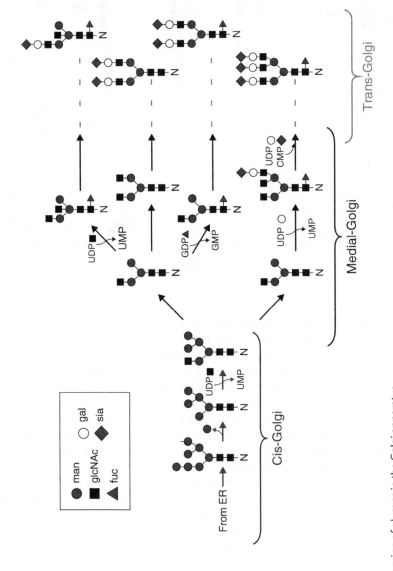

Figure 10.4 Extension of glycans in the Golgi apparatus.

The diversity of glycans is potentially enhanced by the distribution of the time each molecule spends in the Golgi apparatus for glycan extension. The extension reactions take time to carry out. As protein molecules are further modified, the accessibility of the glycan for further extension also varies. In fact, for many proteins, including IgG, the glycans are mostly, not completely, extended.

It is thus not surprising that a very large number of glycan structures can be formed in the N-glycosylation pathway. The web of glycan extension reactions forms a complex network that, when drawn out graphically, indeed resembles a network of diverging and converging paths leading to a number of different fully extended N-glycan structures.

In cell culture processing for the manufacturing of therapeutic proteins, it is a challenge to ensure that the pattern of glycans in the product is consistent across different manufacturing sites. Also, it must stay within the specified quality range over time, even if the supply of raw materials changes.

10.4 Nutritional Requirements

10.4.1 Chemical Environment *In Vivo* and in Culture

Microorganisms grow in environments that may change over a wide range of temperature, pH, and other chemical conditions. In contrast, multicellular organisms developed to maintain a narrow range of chemical and physical conditions for their organs, tissues, and individual cells, even if the conditions of its habitat fluctuate. Thus, cells in our body do not tolerate a wide range of environmental perturbations. In fact, mammalian cells are much more fragile than microorganisms, thriving in a much narrower range of chemical and physical conditions.

Mammalian cells are relatively sensitive to the osmotic pressure of their surroundings. The general range they experience is 270–320 mM (sometimes expressed as mOsm, emphasizing that the total molarity accounts for all dissociated species, e.g., 10 mM NaCl accounts for 20 mM of osmolarity). The optimal pH for growth is usually 7.0–7.4, while the optimal temperature is a rather narrow range of 36.5–37.5 °C. Cells also are sensitive to the relative concentrations of ions. In our body, cells maintain a membrane potential of about −50 mV and maintain a concentration gradient for a number of ions across the cytoplasmic membrane. For example, the concentration of Na^+ is about 15 times greater outside of the cell, while the concentration of K^+ is about 15 times greater inside of the cell (Table 10.5).

It is worth noting that any bioprocess medium is designed to meet the requirement of growth, as well as that of production. For instance, an optimal medium contains a higher concentration of Na^+ than K^+. In the course of bioprocess, the period of growth and that for production are often segregated. The production period commences only after a long period of growth to expand the cell population.

The medium used for growth and production may be different. In the growth period, the chemical environment is kept in the optimal range for growth. It is not unusual that, in the production period, a medium with a higher osmolarity and lower temperature and pH is used, if that favors the product formation. In some cases, nonoptimal conditions are preferentially employed because the cellular stress response increases cellular productivity of some recombinant proteins.

Table 10.5 Approximate concentrations of nutrients in cellular and culture environment.

	Interstitial (mM)	Intracellular (mM)	Typical medium (mM)
Na^+	140	14	140–150
K^+	4.0	140	3–5
Ca^{2+}	1.2	0.01	0.3–1.8
Mg^{2+}	0.7	20	0.4–0.8
Cl^{--}	108	4	100–130
HCO_3^-	28.3	10	14–45
$HPO_4^{3-}, H_2PO_4^{2-}$	2	11	1–5
SO_4^{3-}	0.5	1	0.4–0.8
Amino acids	2	8	Total 10–15
Lactate	1.2	1.5	0
Glucose	56		5–25
Protein	2	4	
Total chemical species (mmole/L)	301.8	302.2	300–340
Corrected osmolar activity (mM)	281.3	281.3	280–320

10.4.2 Types of Media

10.4.2.1 Basal Medium and Supplements

Cell culture medium consists of basal medium and growth supplements. The basal medium is the nutrient mixture consisting of sugar (typically glucose), amino acids, vitamins, various salts, and so on. It provides a nutritional source for deriving energy and making new cell mass and product. It also provides balanced salt concentrations and osmolarity to allow for optimal cell growth (Panel 10.1 and Table 10.5). In designing a process and selecting a medium for a particular cell line, one has to keep stoichiometric balance in mind. Depending on the amount of cells to be reached, the medium must supply sufficient amounts of all components, including inorganic materials (phosphate, potassium, magnesium, etc.), in the medium.

Panel 10.1

Basal Medium

- Sugar
- Amino acids
- Fatty acids, lipid precursors
- Vitamins, nucleosides
- Bulk salts, trace elements
- pH buffer

Supplements

- Serum, hydrolysates
- Growth factors
- Carrier proteins

Most cultured cells require additional supplements, including growth factors, phospholipids, soy hydrolysate, serum, and so on. These supplements promote cell growth by providing constituent components for specific signaling pathways, or they may supply special nutritional needs (such as delivering cholesterol). In stem cell culture, supplements may direct cellular differentiation or maintain cells at a particular differentiation state.

10.4.2.2 Complex Medium, Defined Medium

Traditional cell culture medium contains up to 15% animal serum, in addition to its basal medium. Such a medium, containing some materials of largely undefined chemical composition, is called a complex medium. Many supplements commonly used in industrial processes (e.g., plant hydrolysates and soy phospholipids) also fall into this category.

A chemically defined medium consists of only those components whose chemical composition is entirely known and characterized. A chemically defined cell culture medium typically includes glucose, amino acids, salts, nucleotides, and vitamins. Lipids or phospholipids obtained from natural sources may not be considered as "chemically defined." Modern industrial cell culture processes have largely eliminated serum and employ medium free of animal components, to minimize the risk of animal viruses or prior contamination. The trend is moving toward using chemically defined medium for better process robustness and easier product recovery.

10.5   Cell Line Development

10.5.1   Host Cells and Transfection

Many commonly used host cells for the generation of industrial production cell lines were derived from rodents. For example, Chinese hamster ovary (CHO) cells were originally from hamster, and NS0 cells were from mouse myeloma cells.

A transgene (a gene of interest, i.e., the product gene) coding for the product protein is typically introduced into the host cell using a plasmid vector (Figure 10.5 and Panel 10.2). The transgene is driven by a strong promoter and is followed by a polyadenylation signal for proper mRNA formation. In addition to the transgene, the plasmid carries a gene that confers a selectable trait, such as antibiotic resistance. After transfection, a selective pressure can be applied to enrich for those cells that have internalized the plasmid.

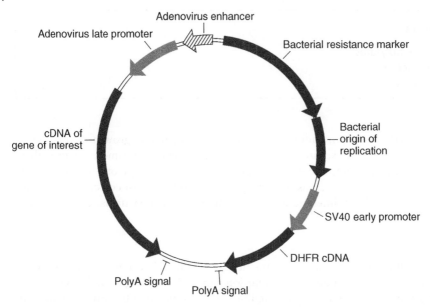

Figure 10.5 A vector for the expression of a transgene using a DHFR amplification system.

Panel 10.2 Elements in a Vector for Transgene Expression

- Promoter
- Coding sequence of gene of interest
- Poly adenylation signal
- Selectable marker
- Other elements of plasmid (cloning site, origin of replication in bacterial host)

The plasmid also consists of elements that are necessary for it to be propagated in bacteria, for generating many copies of itself for transfection into the host cell. The plasmid does not replicate in mammalian cells and would otherwise be lost as cells multiply. By applying selective pressure over an extended period, the surviving cells are likely to have the plasmid integrated into a chromosome and be stable transfectants (Figure 10.6).

10.5.2 Amplification

To achieve high productivity, the producing cell must have a high level of the mRNA for the product protein. In general, the amount of mRNA for a gene accumulated in a cell increases with the promoter strength that drives the transcription and the number of copies of the gene. To increase the transcript level, the number of transgenes in a cell is often "amplified" to tens or even hundreds of copies. This is done by including an amplification marker near the transgene in the plasmid. After the plasmid is integrated into the chromosome, the transfected cells are subjected to a high concentration of an inhibitor to the amplification marker. The high concentration of inhibitor kills the vast majority of cells, sparing only those that have multiple copies of the resistance genes.

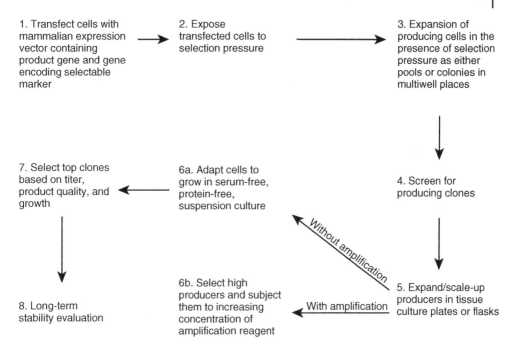

Figure 10.6 Schematic of cell line development for the production of recombinant proteins.

A commonly used amplification system is the dihydrofolate reductase (DHFR) system (Figure 10.6). The enzyme DHFR is required for the synthesis of nucleoside hypoxanthine and is inhibited by methotrexate (MTX). By dosing cells with a very high concentration of MTX, only cells with multiple copies of *DHFR* will survive. As the amplification marker multiplies, the adjacent integrated transgene(s) will also be co-amplified, thus giving rise to high levels of transgene transcription and translation. Through this process, some cells develop a very high capacity for product synthesis.

Typically, single-cell cloning is performed on the population of surviving cells after selection and amplification (at the least, after amplification). The surviving cells make up a very heterogeneous population. Single-cell cloning, with the entire population started from a single cell (thus, a clone), generates a more homogeneous population. This is typically done by sorting a single cell into each culture well using flow cytometry.

Cloning can also be performed by dispensing cells (approximately 0.2 cells per well) into multiwelled culture plates. At an average of 0.2 cell/well, the Poisson distribution predicts that the chance of a well having two cells is low. Therefore, one can assume that all cells arising from a given well all originated from the same cell and are a clonal population. Without single-cell cloning, the population of the producing cells is heterogeneous and poses a high risk of a population drift, which occurs when a subpopulation outgrows other cells in the original population over time. This may lead to changes in productivity or even product quality, and it is a risk that must be minimized if not eliminated.

After single-cell cloning, the productivity of each clone is then assessed. Those with high productivity are isolated. To establish the producing cell line, the product quality is further assessed in two respects: (1) protein sequence (lest a mutation in the product gene may have occurred in the course of cell line development), and (2) structure and glycosylation pattern. The cell line is also evaluated for its long-term stability. To generate a sufficient amount of cells for manufacturing over many years, the single cell that was first isolated must undergo a large number of cell divisions. Long-term cultivation is thus carried out to ensure that the probability of major alteration in the cell's production characteristics is stable. This occurs before the selected producing cell line is expanded for the establishment of a cell bank for manufacturing.

10.6 Bioreactors

The oldest means of producing therapeutic products from mammalian cells is to use the animals themselves, or their tissues. For example, the cowpox virus vaccines that were used against smallpox (mentioned at the beginning of this chapter) were produced in cows 200 years ago, by Edward Jenner. Also, many viral vaccines and antiserums developed in the first half of the twentieth century used animals or animal tissues as production vehicles. The use of animal tissues (e.g., chicken eggs or rodent brains) for virus production is still practiced today for some viruses. However, those are exceptional cases. Cultured cells are the standard vehicle for the production of biologics, both proteins and viruses. The quantity of products produced each year varies widely, from hundreds of kilograms of proteins produced in stirred-tank reactors of tens of cubic meters to merely a thousand liters of a virus-infected culture medium. Only a few different types of reactors are used for those processes. They are described further in this section.

10.6.1 Roller Bottles

Roller bottles are cylindrical, screw-capped bottles with a total volume ranging from 1 to 1.5 L. They are suitable for a culture volume of 0.1–0.3 L (Figure 10.7). The bottle is

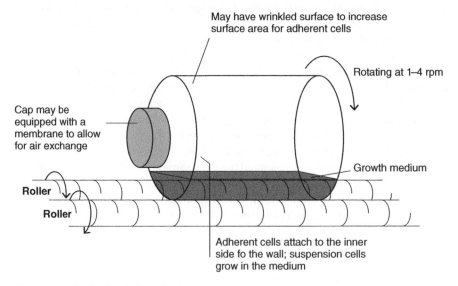

Figure 10.7 A roller bottle for cell culture.

placed on two rotating rods. The rotational motion of the rods makes the bottle rotate around its own long axis at 1–4 rpm.

Following the invention of these bottles in the 1930s, they replaced culture dishes and flasks in the industrial cultivation of adherent cells. Roller bottles are still widely used for the production of viral vaccines. Their use is labor-intensive and prone to microbial contamination, however, because of extensive manual operations in manipulating cell inoculation, medium exchange, and cell harvest. In some manufacturing facilities, a fully automated, robotic roller bottle handling system is used.

10.6.2 Stirred-Tank Bioreactors for Suspension Cells

For large-scale production of biologics, stirred-tank bioreactors are typically the reactor of choice (Figure 10.8). They provide a homogeneous environment for the entire culture. Online environmental monitoring and control can also be implemented. Oxygen transfer is generally not a limiting factor, even at high cell densities. Oxygen demand in a typical cell culture bioreactor is at least 20 times lower than in typical microbial fermentation on a per-culture volume basis. Although mammalian cells are generally more susceptible to mechanical stress caused by the agitation of impellers, the adverse effects of mechanical agitation are not significant with most agitation designs and typical power input.

Most cell culture bioreactors use pitch blade impellers to provide more pumping power for mixing. Most industrial cell culture operations for the production of rDNA proteins employ stirred-tank bioreactors, operated in fed-batch mode.

Figure 10.8 A stirred-tank bioreactor for cell cultivation.

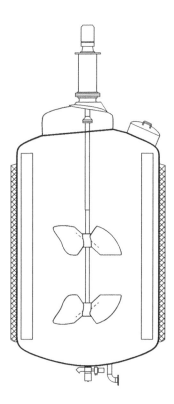

10.6.3 Stirred-Tank Bioreactor with Microcarrier Cell Support

Many cell lines used in the production of biologics can proliferate in suspension without attaching to a surface. Most normal diploid cell strains, primary cells isolated from tissues, and stem cells, conversely, are anchorage dependent. For large-scale operations with these cells in a stirred-tank bioreactor, microcarriers are used to provide an adherent surface. Conventional microcarriers are 150–300 μm beads that support cell attachment and growth on their exterior surface (Figure 10.9). The carriers may be made of solid (such as polystyrene) or microporous materials (such as cross-linked dextran). Macroporous microcarriers that have large pores to allow cells to grow in their interior are also used in many cases, especially when cells in the interior of the carriers are not to be recovered.

Microcarriers can be made of many different materials, including dextran, gelatin, polystyrene, glass, and cellulose. The density of microcarriers is usually in the range of 1.02–1.05 g/cm^3, and the diameter is often 150–250 μm. In this range, the terminal settling velocity of the microcarrier is slow enough that a very gentle agitation condition can be used to keep the microcarrier in suspension, minimizing mechanical stress on cells.

In general, microcarriers made of dextran, cellulose, or gelatin have a wettable surface. They are sometimes coated with collagen or other adhesion molecules to enhance cell adhesion and spreading. Macroporous microcarriers contain large internal pores for

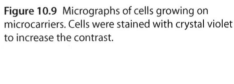

Figure 10.9 Micrographs of cells growing on microcarriers. Cells were stained with crystal violet to increase the contrast.

Human fibroblast FS4 on cytodex 1
microcarriers

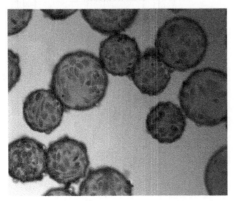

CHO on cytodex 3 microcarriers

cells to populate inside. Since most macroporous microcarriers have a larger diameter (500 μm–2 mm) than conventional microcarriers, cells in the interior of microcarriers are more likely to be subject to oxygen limitations, due to the long diffusional distance.

10.6.4 Disposable Systems

Over the past decade, there has been an emerging trend to employ disposable bioreactors for industrial cell cultivation. The design of those disposable, or single-use, bioreactors ranges from large containers with various external mixing motions (shaking, rocking, or rotating) to stirred plastic tanks with a simple impeller. Some disposal vessels have the ability to monitor pH and dissolved oxygen on-line, with scales up to 2000 L.

There are many benefits to using disposable bioreactors. They incur a much smaller capital investment than building a conventional production plant with stainless-steel vessels and auxiliary facilities. The lead time necessary to construct a manufacturing plant is also reduced by opting for using disposable systems. With a ready-to-use system, the personnel demand is greatly reduced. Most importantly, to meet regulatory standards in between production runs, standard reaction vessels and equipment must be thoroughly cleaned and validated. Those tasks are very labor-intensive. With single-use bioreactors, much of that laborious effort can be eliminated.

Constructed from plastic materials, a disposable system cannot sustain the mechanical stress incurred in a large-scale stirred tank made of stainless steel. Thus, the volume of a single-use reactor is limited and often not sufficiently large for the manufacturing of many products. To increase throughput, disposable reactors can be operated in a continuous mode and at much higher cell concentrations. Engineers are increasingly exploring the concept of continuous operations in single-use biomanufacturing facilities. Continuous processing may also have the advantage of steady-state operation, which could generate product of more consistent quality.

10.7 Cell Retention and Continuous Processes

10.7.1 Continuous Culture and Steady State

Mammalian cells in culture convert most of the glucose they consume into lactate. The accumulation of lactate in culture impedes growth. Thus, the cell and product concentrations achieved in a batch culture or a simple continuous culture are rather low.

If a continuous process is used, the lactate and other metabolites that accumulated in the medium must be removed by medium replenishment. However, the flow rate used in a simple continuous culture is constrained by the cell growth rate, in that the rate of cell removal in the fluid flow must not be higher than the rate of cell growth. To allow for both a high flow rate and a high cell density in a continuous culture, a cell-recycling system can be used. By recovering cells from the effluent flow and returning them to the reactor, we can operate a continuous culture beyond the limitations of the dilution rate (flow rate divided by the bioreactor volume). With a higher cell concentration in the reactor, the overall throughput of the reactor is then also higher.

Cell recycling relies on a cell separation device that can be installed internally in the bioreactor or external to the reactor. This device allows for the withdrawal of an effluent stream from the reactor, and the stream has a substantially lower cell concentration than what is in the reactor. This is accomplished through the use of a centrifuge, a settling tank, or a membrane filtration device.

Virtually all continuous cell culture processes used in biomanufacturing employ cell recycling. The operation is frequently referred to as perfusion. Until recently, perfusion cultures were mostly seen in the production of labile products such as Factor VIII. With increased equipment reliability and the demonstration of success, there has been increasing interest to explore the use of continuous culture in biomanufacturing.

10.8   Cell Culture Manufacturing – Productivity and Product Quality

10.8.1   Process and Product Quality

The manufacturing process for protein therapeutics is similar to that of traditional biochemicals, such as antibiotics and *E. coli*–based recombinant proteins. A typical process entails the propagation of a couple of seed expansion reactor cultures before reaching the production reactor (see Figure 1.5). The process cycle also tends to be longer. Many cell culture manufacturing processes are operated in fed-batch modes that last 10–15 days. Some are operated as continuous perfusion processes and last 2–6 months.

The recovery process of cell culture products is simpler than that for bacterial-based recombinant proteins (see Figure 13.8). The vast majority of processes now employ a medium with a relatively low concentration of proteins, to ease the purification operation. With high product concentrations in the range of 5–10 g/L, the product molecule should be the predominant protein in the medium at the end of the cell culture process. The product isolation and purification process is substantially simpler than separating intracellular protein products.

After years of research efforts on cell lines, media, and process development, the productivity of cell culture processes has increased by over two orders of magnitude. This is quite a leap from their infancy a quarter century ago. One area that still can be improved is our ability to control the quality of production, notably the glycosylation profile of the product.

In addition to the N-glycosylation discussed earlier in this chapter, other posttranslational modifications on the product, such as O-glycosylation and phosphorylation, are also important product quality attributes. A cell's ability to perform posttranslational modifications varies with species, tissue origin, and even cell clone. Furthermore, the supply of precursors and cofactors may change with the chemical environment in the culture and with the cell's metabolic state. Thus, the pattern of posttranslational modifications on the product may differ even with subtle variations in culture conditions.

Most glycoproteins in blood circulation have heterogeneous glycans on their molecules. Variations of glycan structure on the protein molecule are not unexpected, but are a reflection of the reaction kinetics of their synthesis. However, cells under different states of differentiation or disease, and in different tissues, also express different glycans, thus underlining the potential biological significance of glycan variation. The glycan structure on some therapeutic proteins is known to affect the protein's biological activities. Thus, confining glycan distribution to an acceptable range is important for the quality control of the end product. The prediction and control of glycan patterns in cell culture bioprocesses remain a challenge.

10.8.2 Product Life Cycle

Over the past decade, increased productivity has allowed for more flexibility in selecting a process. Except when the product is labile, fed-batch cultures in large, stainless-steel bioreactors have been the norm. However, times are changing. Previous standard practices are continually revised, improved, supplemented, and even replaced with more efficient solutions. Today, it is becoming more common to use disposable reactors. Continuous manufacturing is also being explored. The higher productivity achieved in manufacturing plants has also fueled deep interest in how to improve the efficiency of process and product development.

Intellectual property rights have been critical to the growth of the pharmaceutical industry. As we will discuss here, biologics research is time and capital-intensive, and highly competitive. There is no guarantee that the drug will ever reach the market; every candidate drug is subject to regulatory approval. Pushing any candidate drug through this process requires years of effort, resources, and capital. Without intellectual property rights, there would be little incentive for any company or shareholders to pursue such a high-risk endeavor. Thus, the initial discovery of a compound's biological effects is only the very first step in a long journey toward clinical applications. Figure 10.10 describes an estimated timeline for the development of a biologic drug.

Upon discovery, the candidate drug is first evaluated in animals and *in vitro* systems. Scientists will check and confirm its biological effects and/or possible toxicity. At the same time, cell lines are developed for the drug's possible entry into clinical trials. If there is a decision to pursue clinical trials, the next few steps begin: process development for cell cultivation and purification, and product characterization.

Within a few months, the materials for clinical trial are made and the Phase I clinical trial begins. Phase I tests the safety of the candidate drug in a small number of patients. Phase II tests the drug's efficacy and further evaluates its safety. For Phase III trials, a large number of patients are enrolled. This larger study helps to evaluate the statistical significance of the drug's efficacy and examine its possible adverse effects.

As shown, clinical trials and approval by the regulatory agency are a long process. However, it should be noted that the manufacturing process starts before the drug is even approved. First, materials are prepared for the late-stage (Stage III) clinical trials. Later in the process (but still pending approval for market), inventory is manufactured and stockpiled. This way, the drug can reach the patients and the market as soon as the drug is approved.

The manufacturing process for any drug is strictly regulated. Producers must follow Good Manufacturing Practice (GMP) in manufacturing operations. This also applies to the documentation of materials (e.g., raw materials and supplies), procedures, and the final product. Even pending official approval, producers adhere to GMP as early as the Phase II trials.

The patent for a drug has a limited effective duration; it provides only a limited period of exclusivity for its holder. A few years before the patent expires, competitors begin process development to create the biosimilar version of the biologics. They will charac-terize the protein drug using chemical and biochemical means, tailoring their cell lines

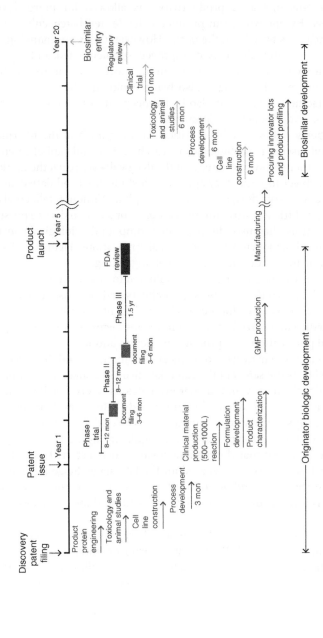

Figure 10.10 Estimated timeline for the development of originator biologics and biosimilars.

and process to develop a protein product nearly identical to the originator drug. Since the drug has been used for years, they need to conduct only a small-scale clinical trial. Biosimilars reach the market as a lower cost alternative to the original.

10.8.3 Product Manufacturing

10.8.3.1 Platform Process

The regulatory process can take years. By the time a new drug is approved, the patent has already lost a few years of its lifespan. Thus, to recoup the resources already invested, it is critical that the drug reaches the market sooner than later. As a result, there is always high demand for fast, efficient process development for new drugs.

In the past decade, process research and development have taken a platform approach. A standard platform process is established for each step of development, from cell line construction and medium formulation to cell culture process and product recovery. The platform process may not be the best for every cell line and product, but it does yield adequate process performance overall. Using this approach requires less process development time for new products. A platform process is used for all new candidate products in the early stages of their drug feasibility evaluation. More effort to optimize the process is invested only for those candidates that pass early evaluation.

The continual push for higher efficiency has also spurred changes in process research and development. Increasingly, experiments are statistically designed to test multiple variables over a wider range for each variable, simultaneously. Biotechnologists are analyzing larger and larger sets of time course data, in order to extract physiological and process information for process optimization.

10.8.3.2 Manufacturing

As discussed, manufacturing of the product begins even before the drug is approved. If the manufacturing is to be performed in a new plant, the decision to construct that plant must be made far in advance (e.g., by the time Phase II trials are midway). At that time, it is not yet known whether the drug will be approved. It is also not yet known which disease indications can be treated by the drug. Therefore, it is difficult to estimate the market size.

As an example timeline, it could take three years to go from first plant designs to final regulatory approval. There are several expensive steps along the way, such as designing the plant, selecting a site, raising capital, constructing the facility, and training workers. However, final regulatory approval of the drug is not guaranteed. Hence, the success of this whole process hinges on the last step. For a manufacturing plant of biologics, the capital investment is typically upward of USD $200 million. Because we can never be certain of the final regulatory outcome for a new drug candidate, its market size, and the required manufacturing capacity, constructing a new plant is a particularly high-risk investment.

Companies and investors are looking for ways to mitigate this significant financial risk; this demand for alternatives has spawned new trends in biomanufacturing. Over the past decade, more contract manufacturing organizations (CMOs) have sprung up in various parts of the world. These firms specialize in cell culture and recovery process, and produce materials for clients. By serving many clients, CMOs can use their own manufacturing facility efficiently. In this way, CMOs can help to meet production needs, while clients avoid the financial risk of constructing a new plant for a drug that may never be approved.

Producers are always looking for alternative ways to meet manufacturing needs with a smaller capital investment. Aside from CMOs, another option is to use smaller, disposable bioreactors and other auxiliary facilities. For example, instead of using a stainless-steel bioreactor of $10\,m^3$ or larger, plastic disposable bioreactors of up to 2000 L scale can be used. There are some drawbacks to these options, however. Because this type of reactor is constructed with plastic materials, it cannot withstand high levels of mechanical stress, such as those caused by agitation. Also, its maximum size is limited. To meet the needs of the production capacity using smaller reactors, the productivity per reactor volume must be enhanced. This particular need has fueled effort in process intensification, to drive the cell and product concentrations higher. It has also pushed the development of continuous processes, to increase the throughput of the plant. In those plants, even some downstream process equipment is for single use.

The cost and specifics of the plant itself can change as well. In a plant using a disposable bioreactor, the process becomes more modular. Less time is required for the construction of the manufacturing plant; the cost of construction is also much less. This has resulted in another trend: distributing the manufacturing capacity of biologics around the world. Hence, many pharmaceutical organizations have manufacturing plants in several locations around the world. This also enables companies to take advantage of certain regional advantages and technical and economic strengths.

10.9 Concluding Remarks

Cell culture processes are the most important bioprocesses in the pharmaceutical industry. In this chapter, we have discussed the important steps for taking cells from a laboratory to a manufacturing process. The evolution of manufacturing technology for cell culture biologics – from large, fixed, stainless-steel tanks to small, flexible, disposable bioreactors – highlights the continual need to reevaluate and improve previous norms. The increasing global distribution of manufacturing capacity reflects the interconnection between organizations and biotechnology professionals alike. However, multiple production sites can be a challenge to manage and maintain. Consistent and reliable product quality is critical to bioprocesses for biologics production. As manufacturing is distributed across multiple sites and multiple organizations, controlling product consistency will become another key challenge in bioprocess science research.

Further Reading

Geering, B and Fussenegger, M (2015) Synthetic immunology: modulating the human immune system. Trends Biotechnol., 33, 65–79.

Hauser, H (2015) Cell line development. In: Animal cell culture, ed. Al-Rubeai, M, pp. 1–26. Springer.

Kim, BJ, Diao, J and Shuler, ML (2012) Mini-scale bioprocessing systems for highly parallel animal cell culture. Biotechnol. Prog., 28, 595–607.

Zhou, W and Kantardjieff, A (2013) Mammalian cell cultures for biologics manufacturing. Adv. Biochem. Biotechnol., 139, 1–10.

Problems

A1 Why would you produce recombinant therapeutic proteins in mammalian cells rather than in *E. coli*?

A2 In the dihydrofolate reductase (DHFR)–methotrexate amplification system, DHFR catalyzes the conversion of folate to tetrahydrofolate (FH4). FH4 is required for the biosynthesis of glycine from serine, for the biosynthesis of thymidine monophosphate from deoxyuridine monophosphate, and for purine biosynthesis. Chinese hamster ovary (CHO) cells that are deficient in DHFR were isolated after mutagenesis and selected using high levels of radioactive (3^H-labeled) deoxyuridine. This is possible because cells with functional DHFR convert [3^H] deoxyuridine to [3^H] dTMP, which is further incorporated into DNA and causes cell death; DHFR-deficient cells do not undertake this conversion and can survive.

DHFR-deficient cells require the addition of thymidine, glycine, and hypoxanthine in their growth media. DHFR-deficient cells also do not grow in the absence of added nucleosides unless they acquire a functional *DHFR* gene. Look up the pathway involving DHFR in a standard biochemistry textbook, and point to the reactions that allow this selection system to work and why the cell requires certain nutrients for growth.

A3 The primary advantage of the DHFR system has been its use for the selection of cells in which the *DHFR* gene has been amplified. Methotrexate (MTX), a folate analogue, binds DHFR and inhibits its activity, thereby leading to cell death. When cells are selected for growth in MTX, the surviving population contains increased levels of *DHFR*, which results in an amplification of DHFR enzymatic activity. Upon further increasing the concentration of MTX, the only surviving cells are those that have amplified *DHFR* significantly, often several thousand copies of *DHFR*.

In CHO DHFR-deficient cells, *DHFR* can be used as a selection marker in the co-transformation of heterologous genes. In this process, *DHFR* is coamplified and coexpressed with the heterologous gene. One can enrich for transformed cells by selecting cells in increasing concentrations of MTX. In this way, DHFR-deficient CHO cells can be a useful host.

In some cases, mutant DHFR (instead of wild-type DHFR) is used in the plasmid construct as the amplification marker. Explain why mutant enzyme is used and the special property that the mutant enzyme may confer.

A4 A typical transfection process uses 5 μg of plasmid (size 15 kb) to transfect 10^6 of DHFR-deficient cells in a 3-mL suspension. The transfection efficiency is 80% (i.e., 80% of cells receive enough copies of the plasmid that makes the presence of plasmids in the cell detectable) by the calcium phosphate precipitation method. For cells to develop MTX resistance, the plasmid DNA needs to enter the nucleus, be transcribed, be processed into mRNA, and be exported into the cytoplasm for translation.

After transfection, cells are plated under two different conditions: (1) neither MTX nor nucleotides, and (2) with 20 nM MTX but without nucleotides. After two days, the two conditions yielded 5000 and 20 surviving cells, respectively. We can assume that since we're using the calcium phosphate precipitation method, only 1% of the particles containing plasmids are taken up by cells and the uptake of the phosphate particle is random. The number of particles taken up by cells follows a Poisson distribution; only 0.1% of plasmids taken up by the cells enter the nucleus, and only 1% of those will be transcribed and translated into protein. The rest is degraded.

a) A wild-type cell has two copies of *DHFR*. If one had plated the transfected cell in a nucleotide-deficient medium without MTX, how many survivors may one get?

b) From the survival rate in the absence of MTX, estimate the number of plasmids necessary for a cell to take up in order to survive the selection.

c) Give an order-of-magnitude estimate of the number of plasmids that have been taken up by those 20 surviving cells in 20 nM MTX.

d) Is it possible that those 20 survivors from wild-type cells have survived because of other reasons? If yes, how?

e) How many plasmids in the nucleus are needed for the cell to survive MTX selection?

Consider the selection of cell line development using a DHFR–MTX system: after the first transfection of the plasmid, among cells that survived, only a small fraction will have the *DHFR* gene integrated into a chromosome (e.g., be stably transfected), while those cells in which plasmids fail to integrate will eventually lose plasmids and die. Write down a formula including a probability term for every step involved to allow for an order-of-magnitude estimation (or even experimental determination) of the frequency of the resulting stably transfected cells. This can provide a quantitative estimate of how many cells you should start with in order to have a desired number of clones that can be productively evaluated.

A5 Mammalian cells in culture convert a large portion of the glucose that they have taken up to lactate. The pathway to form lactate is:

$$\text{Pyruvate} + \text{NADH} \rightarrow \text{Lactate} + \text{NAD}^+$$

Lactate is then excreted into the medium by a transporter that exports lactate along with a proton to maintain charge neutrality.

Propose a strategy to minimize lactate formation by metabolic engineering. You just need to state your overall strategy with an enzyme or target name. You do not need to describe what kind of plasmid or selection will be used. Specifying the pathway or name of the target and the aim of the genetic manipulation will be sufficient.

A6 The plasma cells in our body produce antibody at a rate of 2000 molecules/s. How many picograms of antibody can it produce per day? Assume that each IgG heavy-chain and light-chain mRNA molecule can be used to initiate translation at a rate of 0.5/s, and elongate at a rate of 10 amino acids/s. The heavy-chain and light-chain transcripts are 1000 bp and 500 bp, respectively. Estimate the copy number of mRNAs for the heavy chain and light chain, and the transcription rate of heavy-chain and light-chain genes. You can assume the half-life of IgG mRNAs is 2 h. Note that the plasma cell has only one copy of functional heavy-chain and light-chain gene because of allele inactivation to prevent one cell from producing two different kinds of antibody.

A7 Continue with the preceding problem. Using DHFR amplification, producing cells with the productivity of the plasma cell has been obtained. It has 10 copies each of IgG heavy-chain and light-chain genes. Estimate the transcription rate of those genes if every copy of the transgene amplified in the genome is active. Discuss why it take multiple copies of the product gene to have the same secretion rate that the plasma cell can achieve with a single copy of the gene.

A8 The process of protein secretion entails first synthesizing and folding of the protein in the endoplasmic reticulum (ER), and then further processing in the Golgi apparatus. The Golgi apparatus is further segregated into different subcompartments. Those cellular compartments that the protein molecules pass through resemble a number of bioreactors in series. Propose alternative reaction models for the glycosylation of protein molecules in the Golgi based on the flow characteristic in the Golgi compartments (i.e., consider that the protein molecules transit through like in a plug flow reactor or in a continuous stirred-tank reactor).

A9 Antibody molecules constitute a very large portion of mammalian cell culture products. Most of them are "humanized" antibodies. Draw a schematic of an antibody (IgG) molecule, and explain the difference between a chimeric antibody and a humanized antibody.

A10 Explain single-cell cloning. What is it, and how do you ensure that the cells grown in a well or in a colony originated from a single cell (i.e., a clone)?

A11 You are planning to create a cell line that will be expanded to support the production of a product in a 10,000-L bioreactor with a final cell concentration of 10^{10} cells/L at 30 runs per year for 20 years. How many cell doublings would the cell have gone through, from its original parent cell after cell cloning until the production is over? Compare that number to the number of doublings you can estimate for the growth of a single cell all the way up to the total number of cells in a human (about 2–4×10^{12}).

A12 A cell line was cultivated on microcarriers to produce a plasminogen activator product. Because direct sparging causes serious foaming in a high-protein medium, oxygen is supplied into the culture medium by diffusion from the headspace to the liquid. There is no bubbling aeration into the liquid. In a laboratory-scale bioreactor, the oxygen transfer rate is measured by a degassing method from the cell-free growth medium. The dissolved oxygen is first allowed to reach a saturation level of 0.21 mM before nitrogen gas is introduced into

Table P.10.1 Dissolved oxygen
level during a degassing period of
$K_L a$ measurement.

Time (min)	Dissolved oxygen level (% saturation)
0	100
5	80.4
10	64.6
20	41.8
30	27.0
40	17.4

the headspace to strip the dissolved oxygen from the medium. The flow rate of the nitrogen gas is sufficiently high so that the oxygen concentration on the gas side can be assumed to be zero. A dissolved oxygen profile during the degassing period is shown in Table P.10.1. Both the diameter of the vessel and the height of liquid are 10 cm.

In a separate experiment, the oxygen uptake rate of CHO cells was measured in a 30-mL vessel filled with a suspension of cell-laden microcarriers. Since there is no headspace in the vessel, no oxygen can be transferred into the liquid during the period of measurement. The rate of decrease of dissolved oxygen concentration was determined at two different cell concentrations. In both measurements, the initial dissolved oxygen concentration was 0.18 mM or 90% of saturation with air at 1 atm. At 2×10^6 cell/mL, the dissolved oxygen decreased to 0.12 mM in 40 min; while at 5×10^6 cell/mL, it decreased to 0.03 mM. Since the doubling time for cells is 18 h, the cell concentration can be considered to be constant during the oxygen uptake measurement. To sustain the maximum productivity of plasminogen activator, it is necessary to maintain a dissolved oxygen concentration above 15% of saturation.

a) What is the maximum cell concentration achievable in this laboratory vessel?
b) Practically, the maximum cell concentration achievable in a stirred vessel is about 6×10^6 cells/mL because of the limitation on the microcarrier concentration that can be used. If the vessel is to be scaled up in a geometrically similar way, what is the upper limit of vessel size (assuming the height-to-diameter ratio remains the same) that can still sustain this cell concentration? Assume that air at 1 atm is to be used in the headspace.

A13 It is suggested that perfluorocarbon be used to increase the oxygen transfer capacity in a cell culture bioreactor, thus avoiding direct aeration. Perfluorocarbon is very hydrophobic and is not miscible with water. It has a high solubility of oxygen (about 10 times of that of water) and can possibly be used as plasma substitute for blood transfusion. It can be applied in emulsion form and is usually used as an "oxygen carrier" to carry oxygen from the lung to the tissue in a patient being treated. Design a stirred tank that allows perfluorocarbon to be used for oxygenation. Do you envision any problem with this proposal?

A14 A major difference between cultivating mammalian cells and bacteria in a bioreactor is the aeration (gas-sparging) rate. A high aeration rate causes extensive mechanical force–related cell death. Often, oxygen-enriched air is used to reduce the air flow rate in a large-scale bioreactor. In a $10\,m^3$ bioreactor, the target cell concentration is 2×10^{10} cells/L with a specific oxygen consumption rate of 2×10^{-10} mmol/cell-h. The dissolved oxygen level is to be maintained at 20% of saturation with ambient air (21% oxygen). The airflow rate cannot exceed $1\,m^3$/min. The mass transfer coefficient (K_L) is 40 cm/h. The interfacial area (a) of the gas bubble can be estimated from the specific surface area of the bubble and the bubble rising velocity ($\sim$1 cm/s) (see Chapter 8).

One potential problem of using oxygen-enriched air at a reduced air flow rate is the accumulation of CO_2. Because the Respiratory Quotient (RQ) for cells is very close to 1.0, for every mole of O_2 consumed, 1 mole of CO_2 is produced. Unless CO_2 is removed, it will accumulate and quickly exceed the tolerable concentration of 40 mM. The mass transfer coefficient for CO_2 is about 90% of that of oxygen. The CO_2 concentration in the inlet gas can be considered to be nil. Set up equations of gas balance for determining the gas flow rate and oxygen levels in the inlet gas that are sufficient for oxygen transfer while still maintaining CO_2 below the tolerance limit. The Henry's Law constants for O_2 and CO_2 (37 °C) are 0.9 and 0.04 for $C = P/H$ (P in atm, C in mM).

A15 The production cell line used in biomanufacturing must be shown to have originated from a single cell after the introduction of the transgene and any subsequent genetic manipulation. In other words, the cell line must be started as a clone. Arising from a clone is sometimes referred to as "clonality." Often, evidence must be provided to the regulatory agency to show a greater than 99% confidence of clonality. Single-cell cloning is often practiced by placing cells with a known concentration and volume into each of a 96- and 384-well plate, so that the probability of having two cells in a well is acceptably low. It has been recommended to perform two rounds of single-cell cloning to achieve a 99% confidence of clonality. Explain this argument: why are two rounds of cloning more assuring? Look up "Poisson probability table." If you are to perform two rounds of single-cell cloning (i.e., perform single-cell cloning, then after the cell grows to a colony of around 1000 cells, distribute them into wells for single-cell cloning again) with the same cell concentration per well, what should the number of cells per well be so that a 99% confidence level can be achieved?

A16 Inoculating cells into a microcarrier culture is very different from plating cells onto a Petri dish. On a Petri dish, cells can move around the surface to distribute themselves to the area that has empty surface for growth, whereas in a microcarrier culture, once attached to a microcarrier bead, cells cannot relocate from one microcarrier to another. Hence, microcarrier beads that failed to receive any cell during cell inoculation will remain empty throughout the cultivation period.

The cell attachment to microcarriers during inoculation can be considered to be completely random events. The distribution of cells on microcarriers can thus be approximately by a Poisson distribution. Consider a suspension of microcarrier beads of 200 µm at a concentration of 20,000 microcarrier beads/mL.

What concentration of cells should be inoculated into the microcarrier suspension so that the fraction of beads with fewer than 3 cells/bead is lower than 2% of all microcarriers? Plot the histogram of cells per bead.

On a microcarrier bead, the cell attached to the bead will grow until the entire surface is covered by cells (called "reaching confluence"). The microcarrier used can support 80 cells per bead when reaching confluence. If the doubling time of the cell is 12 h, how long will it take for beads initially with 3, 6, and 8 cells to reach confluence? Discuss the effect of cell distribution on microcarriers on the growth kinetics.

11

Introduction to Stem Cell Bioprocesses

11.1 Introduction to Stem Cells

Stem cells are cells that have the potential to differentiate into more specialized cell types. They are capable of reproducing more stem cells with the same ability to self-renew and differentiate. Stem cells reside in various tissues of our bodies, and act to replenish differentiated cells lost to natural turnover or tissue impairment (Panel 11.1). Hematopoietic stem cells (HSCs) reside in bone marrow where they can divide and differentiate into various types of blood cells. Replenishing our blood cells is an important mechanism, since most of them have a very limited lifespan. HSCs can reconstitute the blood of a patient and also a patient's entire blood-forming system after a bone marrow transplant. Normally quiescent, stem cells in the liver can be triggered upon tissue damage to proliferate and differentiate to hepatocytes.

Panel 11.1 Core Properties of Stem Cells

- Self-renewal: extended cell division without loss of potency
- Differentiation to specific cell type(s)
- *In vivo* **functional reconstitution** of a tissue

Stem cells that reside in different tissues typically maintain their presence at a very low level, undergoing cell division only as needed. As they proliferate, some of their progenies become differentiated cells. Others remain as stem cells to maintain their baseline levels. This ability to self-renew and differentiate is a core feature of all stem cells. From serial transplant experiments, we know that stem cells from tissues like the hematopoietic system and the liver pass on to many generations of offspring, suggesting that they have very high, or perhaps unlimited, self-renewal capability *in vivo*.

The decision whether a stem cell stays in a quiescent state, undergoes self-renewing proliferation, or embarks on differentiation depends on its internal "program" (or regulatory circuits) and environmental cues. In the past 2 decades, major advances in stem cell science have provided a better understanding of these control processes and have made the isolation or generation of stem cells possible *in vitro*. These cells can then be cultured to expand and/or differentiate into different specialized types of cells. Stem cells may also be transplanted into a patient to reconstitute tissue and restore impaired functions. Stem cell technology holds the promise of dawning a new era of regenerative medicine.

Engineering Principles in Biotechnology, First Edition. Wei-Shou Hu.
© 2018 John Wiley & Sons Ltd. Published 2018 by John Wiley & Sons Ltd.
Companion Website: www.wiley.com/go/hu/engineering_fundamentals_of_biotechnology

11.2   Types of Stem Cells

A variety of stem cells have been identified in tissues, and many have been successfully cultured. Depending on their tissue of origin and state of development (i.e., embryo or adult), each stem cell's differentiation potential is different. Such potential is evaluated by the different germ layer lineages they can differentiate to. For instance, in an early embryonic developmental stage, cells can differentiate into all three germ layers: ectoderm, mesoderm, and endoderm. Ectoderm subsequently develops into cells of the nervous system; mesoderm gives rise to muscle, bone, and cartilage; and endoderm becomes gut tissue. Stem cells that can differentiate into lineages of all three germ layers are pluripotent. Multipotency refers to the capacity of cells to differentiate into lineages of more than one, but not all three germ layers, whereas unipotent stem cells can renew and differentiate to only one cell type (Panel 11.2).

Panel 11.2 Stem Cell Potency

- **Totipotent:** Can differentiate to all cell types of the adult, including placenta
- **Pluripotent:** Can differentiate to all cell types in three germ layers
- **Multipotent:** Can differentiate to a limited number of cell types in more than one germ layer but not all three
- **Unipotent:** Can only differentiate to one specific terminal cell type but has self-renewal capability

---

Embryonic stem cells (ESCs) are derived from the inner cell mass of an early developing mammalian embryo. These cells have the potential to differentiate into all three germ layers and are thus pluripotent. Adult stem cells that are derived from postnatal organisms often have a more limited differentiation capacity, and are mostly confined within the lineage they are derived from.

11.2.1   Adult Stem Cells

Adult stem cells are classified according to their tissue of isolation. They are believed to be responsible for tissue homeostasis, or repair/regeneration following an injury, and they have been found throughout bone marrow, digestive tract, heart, kidney, and brain tissues. In the intestine, stem cells reside in the crypt of the epithelial lining. Stem cells are present at the base along with Paneth cells (Figure 11.1). Paneth cells, possibly with other surrounding cells, provide the signaling cues for the proliferation of intestinal stem cells. As they proliferate, they also move outward and differentiate into adsorptive, enteroendocrine, or other types of epithelial cells.

These adult stem cells have tremendous capacity for self-renewal. In the course of differentiation, progenitor cells may arise in a middle stage of differentiation. These progenitor cells are also capable of proliferation and can continue to differentiate into multiple lineages. However, although (Panel 11.3) progenitor cells differ from stem cells in that they have a limited proliferation capacity, the distinction between stem cells and progenitor cells can be, at times, less than definitive.

Stem Cells in Intestinal Epithelium—Example of Adult Stem Cells

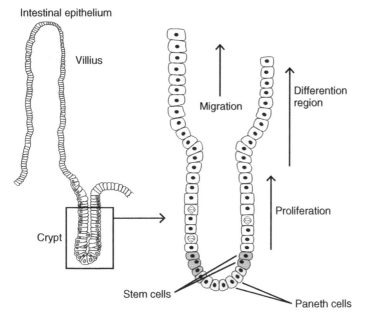

Figure 11.1 Intestinal epithelium stem cells.

Panel 11.3 Adult Stem Cells

- Derived from adult tissue (bone marrow, brain, muscle, etc.)
- Limited differentiation potential capacity
  - Most are restricted to the cell type of either the tissue of origin or the germ layer of origin.
- May have "plasticity, which is" the ability to break lineage restriction and become more potent (differentiate to cells of other germ layers)

11.2.1.1 Hematopoietic Stem Cells

HSCs are the stem cells that generate circulating blood cells (Figure 11.2). Our blood consists of cells of two major lineages. The myeloid lineage includes red blood cells (erythrocytes) that carry oxygen and platelets that help clot blood, as well as neutrophils, basophils, eosinophils (which together are called granulocytes), and macrophages to help fend off infections from microbes. The lymphoid lineage includes B-lymphocytes (which produce antibodies) and T-lymphocytes (which play many key roles in directly killing invaders and facilitating immune responses).

The lifespan of many of those blood cells is rather short. In an average adult, nearly 10^{11} blood cells are generated in bone marrow each day. The occurrence of HSCs in bone marrow is estimated to be only 1 in 10^3 to 10^4 in a mouse. HSCs are able to differentiate into a variety of specialized hematopoietic cells, such as lymphocytes, natural killer cells, and megakaryocytes. Clearly, the hematopoietic system has a tremendous regenerative and differentiation capacity.

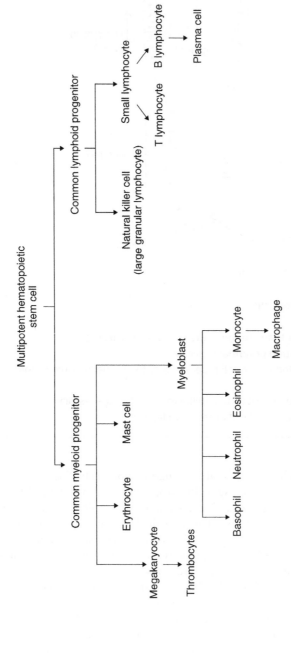

Figure 11.2 Development of hematopoietic stem cells into blood cells of different lineages.

Beside hematopoietic stem cells, many progenitor cells, which have a more limited differentiation and proliferation capacity, are also present in bone marrow. For example, lymphoid progenitor cells can proliferate but can only generate lymphocytes as the mature cell type. In a bone marrow transplant, HSCs are transplanted together with those progenitors and other cells. In some cases, HSCs are enriched before transplant by passing the cell mixture through an adsorption column, which selectively retains $CD34^+$-expressing hematopoietic lineage cells. Bone marrow–based HSC transplant has been practiced for over 5 decades; it has been highly successful for reconstituting hematopoietic systems in patients with hematopoietic cancers, or patients with other cancers who are coping with severely damaged hematopoietic systems resulting from systemic chemotherapy.

Paradoxically, although HSCs have been used widely in stem cell therapy, their *in vitro* culture and expansion have been elusive. HSCs have been isolated based on the unique set of proteins expressed on their surface, which allows for labeling with fluorescent antibodies and FACS (fluorescence-activated cell sorting) isolation. The purified HSC can be used for transplant. However, the use of purified HSCs instead of a mixed bone marrow population is limited due to the high cost of purification.

11.2.1.2 Mesenchymal Stem Cells

In addition to HSCs, mesenchymal stem cells (MSCs) are isolated from the bone marrow. These cells adhere to plastic surfaces *in vitro*, have a fibroblast-like morphology, and express a set of defined cell surface markers. They can be directed to differentiate to osteroblastic, adipocytic, myogenic, and chondrocytic lineages by simply altering their culturing conditions. For example, exposure to a cocktail of dexamethasone, ascorbic acid, and beta-glycerophosphate guides MSCs to differentiate toward the osteogenic lineage, while treatment with transforming growth factor beta-1 (TGFβ1) drives them to a chondrocytic fate. They can be derived from adults and thus are easier to acquire and establish than embryonic stem cells. Although MSCs were first isolated from bone marrow stroma, they have since been isolated from umbilical cord blood and many different tissues, including blood, adipose, and muscle tissue.

MSCs have been used in clinical trials for bone and cartilage regeneration, myocardium repair, and so on. They also secrete cytokines and have the capability of immune modulation. In clinical trials, they have been tested for treatment of graft-versus-host disease (GVHD).

11.2.1.3 Neuronal Stem Cells

Our body's nervous system has an effective repair mechanism to counter tissue damage caused by inflammation and degeneration. Much of the preventive actions against this irreversible nervous tissue damage are attributed to the response of somatic neural stem/precursor cells (NPCs). Multiple NPCs, such as neural stem cells, neuronal progenitor cells, and glial progenitor cells, have been isolated from the adult nervous system. However, the native repair system is incapable of repairing damaged cytoarchitecture of a tissue.

The central nervous system (CNS) consists of two major cell types: neural cells, which transmit electrical and chemical signals to carry information, and glial cells, which provide structural support and protection, and maintain homeostasis. An area of future regenerative application using NPCs is in replacing impaired neural cells or glial cells.

Transplanted NPCs may also contribute to a "bystander effect," in that they do not directly participate in repair, but they sense inflammatory signals in damaged tissue and secrete signals to facilitate the native, local cells to undergo the different repair mechanisms. It should be noted, as in other tissues, that the exploration of nervous system regeneration is not limited to using NPCs or other nervous system–derived cells. Other stem cells, including bone marrow, mesenchymal, and embryonic stem cells, are also being explored to regenerate nervous tissues.

11.2.2 Embryonic Stem Cells

ESCs were first isolated from mouse embryos in 1981. But it was not until more than a decade later that human ESCs were isolated. In early mammalian embryogenesis, the fertilized egg develops into a blastocyst, which consists of an outer layer (the trophectoderm) surrounding a group of cells (the inner cell mass, or ICM) (Figure 11.3). The trophectoderm will ultimately develop into the placenta, while the inner cell mass will develop into all tissues comprised within the embryo, as well as some tissues outside the embryo needed for embryo development.

When culturing excised ICMs on a layer of mouse embryonic fibroblasts, some cells grew out and form colonies (Figure 11.3). These cells can be passaged in culture and are capable of differentiating into the different cell types representing multiple tissues. When cultured ESCs are transplanted into developing blastocysts in an immune-deficient mouse, they contribute to cells from tissues derived from all three germ layers. Upon subcutaneous injection into syngeneic mice, cultured ESCs also form teratocarcinomas, wherein the tumor displays cell markers from all three germ-layer lineages.

Mouse ESCs (mESCs) and human ESCs (hESCs) are truly pluripotent; they have a virtually unlimited capacity for self-renewal. Unlike fibroblasts or cells isolated from other tissues, ESCs do not undergo senescence. Various protocols have been developed to guide them to differentiate into liver cells, pancreas cells, nervous cells, and others. However, it is important to recognize a distinction between ESCs and other adult stem cells. Many adult stem cells are resident stem cells *in vivo*. They may be quiescent most of the time, but they are present for a long duration. Upon isolation and provided with the appropriate chemical cues in the culture medium, adult stem cells develop the capability to proliferate *in vitro* while still retain their stem cell state. In contrast, ESCs are isolated from a blastocyst state, at which ICM cells are proliferating and undergoing further differentiation into other cell types. It is possible that, *in vivo*, the ESC state is transient, whereas other adult stem cells have a more stable state of existence in our body. Through the culture conditions, the transient state of pluripotency is made unending in the *in vitro* culture of ESCs as long as the appropriate culture conditions are maintained.

11.2.3 Induced Pluripotent Stem Cells and Reprogramming

It has recently been shown that the potency of stem cells and somatic cells is subject to change through a process called "cellular reprogramming." In 2006, Shinya Yamanaka made a landmark discovery in the generation of pluripotent stem cells. By transfecting somatic cells (in Yamanaka's case, fibroblasts) with a combination of four genes expressing exogenous transcription factors [*Oct4*, *Sox2*, *Klf4*, and *c-Myc* (OSKM)], while simultaneously using culture conditions that are used for growing ESCs, pluripotent

Figure 11.3 Derivation of embryonic stem cells from inner cell mass on a feeder layer of mouse embryonic fibroblasts, and morphology of the derived embryonic stem cells on the feeder layer.

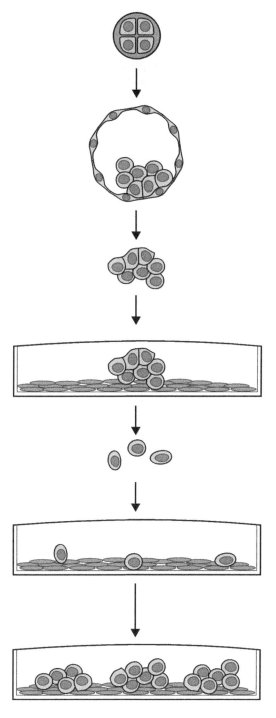

stem cells arose from the transfected cells at a low but finite frequency (about 10^{-3} to 10^{-4}) (Figure 11.4). The introduction and expression of OSKM genes initiate extensive cellular reprograming that remodels the epigenome of the somatic cells. Thus, in a small fraction of cells, this reprogramming allows the expression of genes that confer pluripotency and the suppression of others that lead to differentiation.

Phenotypic changes are attributed to changes in the methylation patterns of specific segments of DNA, chemical modifications of histone proteins, and altered interactions between histones and chromosomes. These changes do not incur DNA sequence changes, and are considered to be epigenetic alterations. Such alterations change the accessibility of DNA to transcription factors and other DNA-processing proteins. Eventually the endogenous pluripotency genes become expressed in a small fraction of completely reprogrammed cells to give rise to induced pluripotent stem cells (iPSCs).

iPSCs are morphologically similar to ESCs, display highly similar gene expression profiles, and differentiate *in vivo*, as well as *in vitro*, into all three germ layers. iPSCs have been derived from a number of different tissues and species. A variety of reprogramming methods have been introduced, including a somewhat different combination of genes for cells of different tissue origins, and the use of small-molecule factors. The successful derivation of iPSCs has made possible to derive pluripotent stem cells from somatic cells of different individuals, and may be used for autologous stem cell transplant, which would incur minimal risk of GVHD.

11.3   Differentiation of Stem Cells

In their native environment, often called the "stem cell niche," stem cells differentiate into their target cell types in response to the signaling molecules and surface substrates added to their microenvironment by surrounding cells. For example, under hypoxic conditions, cells in the kidney produce erythropoietin, which is transported to the bone marrow by circulating blood and binds to the receptor on the surface of bone marrow progenitor cells to stimulate their proliferation and differentiation into red blood cells.

Cultured stem cells can be similarly directed to differentiate toward a particular cell fate; this feat typically is accomplished by changing the culture environment to conditions that mimic the known *in vivo* differentiation cues. This may entail plating cells at a permissive range of density on surfaces coated with extracellular matrix molecules, in the presence of specific growth factors or cytokines, or inhibitors of specific signaling pathways.

Differentiation protocols typically involve stepwise changes of culture conditions to guide the cells' development through different stages and along the desired lineage (Figure 11.5). For example, the differentiation of ESCs toward the hepatic lineage involves plating ESCs on a Matrigel-coated surface in the presence of the growth factors Activin and WNT3A for 6 days to perpetuate their conversion to endoderm. This is followed by replacing the growth-factor cocktail with a combination of FGF2 and BMP4 for an additional 4 days, and then FGF1, FGF4, and FGF8 for the next 4 days until the cells begin to exhibit hepatocyte progenitor markers. By the fourth stage, cells in culture media supplemented with HGF and follistatin begin to express more mature hepatic markers, such as albumin and cytochrome p450 isozymes.

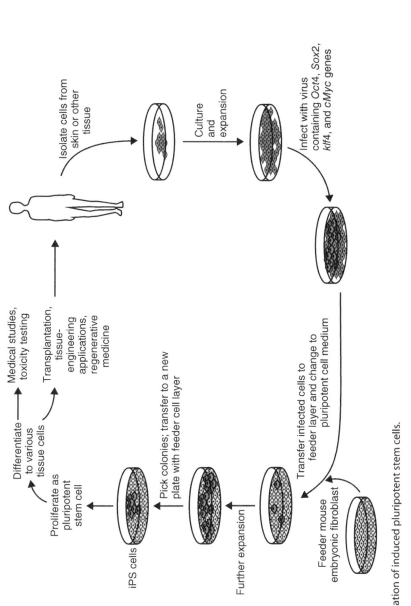

Figure 11.4 Derivation of induced pluripotent stem cells.

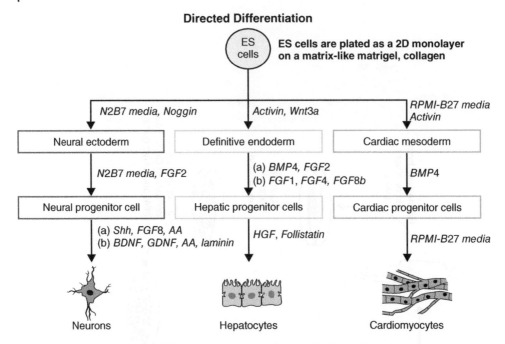

Figure 11.5 Directed *in vitro* differentiation of stem cells toward different lineages.

Induction toward the cardiac mesoderm involves fewer factors. In this case, activin treatment for 1 day is followed by treatment with BMP4 for 4 days, after which the cells are cultured normally for about 2 weeks and yield a population of cardiomyocytes.

11.4 Kinetic Description of Stem Cell Differentiation

In general, the cultivation of stem cells involves two aspects: (1) the expansion, or renewal, of the stem cell population, maintaining the cell at its potency state; and (2) the directed differentiation of the stem cell to the target differentiated state. A quantitative description of the population must include the distribution of the population among the different differentiation states. The propensity of a population to undergo expansion, maintenance, and differentiation is dependent on the environmental cues that the population is subject to. A kinetic description of stem cells entails the expression of the specific rate of cell proliferation (μ) and differentiation ($\emptyset$). These specific rates are dependent on chemical cues like growth factors and signaling molecules. Each subpopulation of a particular differentiation state has its characteristic proliferation and differentiation potential. The potential to differentiate to a particular cell type is possessed only by cells that are at a preceding differentiation state. For example, pluripotent cells do not directly differentiate to the hepatic state without reaching the endodermal state first.

We will now illustrate the process of developing a quantitative description of stem cell differentiation. As an example, we consider the differentiation of pluripotent stem cells to the hepatic lineage with an intermediate state of endodermal cells.

A quantitative description describing a population of pluripotent stem cells differentiating toward the hepatic lineage will have descriptors of the concentrations of pluripotent, endodermal, and hepatic cells. They are denoted as n_p, n_e, and n_h, respectively (Eqs. 11.1, 11.2, and 11.3). By convention, the descriptor of cell concentration for a stem cell culture is based on cell number (n) instead of cell biomass (x). The different differentiation states of a cell population are often characterized by their function, morphology, and gene expression pattern. Increasingly, the transcript and protein level of specific marker genes are used to classify cells into different differentiation states because they can be readily measured using quantitative PCR or flow cytometry.

The three differentiation states – pluripotent, endodermal, and hepatic states – are characterized by high-level expression of transcription factor Oct4, transcription factor Foxa2, and albumin, respectively. The expression of each of those three markers at the specific state is virtually mutually exclusive; only one marker, but not the other two, is expressed at a given state. Cells at the three states reside at three different regions in a plot of the three marker genes (Figure 11.6). The equation describing the dynamics of the three subpopulations can be written as the balance between growth and differentiation terms (Eqs. 11.1, 11.2, and 11.3 in Panel 11.4). The differentiation process is considered as irreversible.

In the development of multicellular organisms, cell growth and differentiation are regulated by various cues. Such cues could include growth factors, signaling molecules, and, in some cases, surface substrates. This is in contrast to the microbial world, where

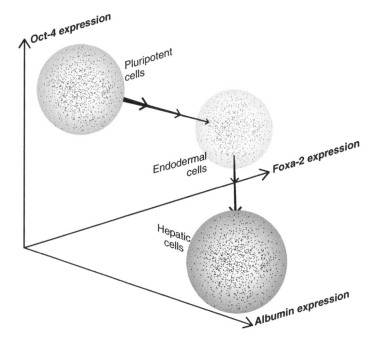

Figure 11.6 A 3D space of marker gene expression (transcript or protein) levels depicting cells at a pluripotent state, endodermal state, and hepatic state. Each dot represents a cell with the corresponding levels of marker gene expression. The three envelopes represent regions of the pluripotent, endodermal, and hepatic states.

the growth rate depends on nutrient levels, as described by the Monod model. As in the Monod model, an empirical relationship between the growth or differentiation rate and the concentration of signaling molecules may be established. Typically, cellular differentiation does not respond to increasing levels of the regulatory signal molecule in saturation kinetics, as seen in the Monod model. The differentiation rate often rises quickly after reaching a threshold concentration of the cue, followed by a rapid decline. An example of an expression for a differentiation rate is shown in Panel 11.4.

Panel 11.4

Balance on pluripotent cell:

$$\frac{dn_p}{dt} = \mu_p n_p - \Phi_{p \to e} \cdot n_p \tag{11.1}$$

Balance on endodermal cell:

$$\frac{dn_e}{dt} = \mu_e \cdot n_e + \Phi_{p \to e} \cdot n_p - \Phi_{e \to h} \cdot n_e \tag{11.2}$$

Balance on hepatic cell:

$$\frac{dn_h}{dt} = \mu_h \cdot n_h + \Phi_{e \to h} \cdot n_e \tag{11.3}$$

During differentiation, μ's may be negligible as cell proliferation is limited. The differentiation of a cell type to the next differentiation stage may depend on the level of a "signaling cue" (C_A, C_B), which may have an optimal range beyond which it becomes inhibitory.

$$\Phi_{p \to e} = \frac{\Phi_{p \to e, max} C_A^{\,n}}{K_1 + C_A + K_2 C_A^{\,m}} \tag{11.4}$$

$$\Phi_{p \to h} = \frac{\Phi_{p \to h, max} C_A^{\,n'}}{K_1' + C_B + K_2' C_B^{\,m'}} \tag{11.5}$$

Such a model treats the population as consisting of three discrete subpopulations. The evolvement of one subpopulation to another downstream in the differentiation path is a step function, crossing from one state to the next state instantaneously without a transient subpopulation. In reality, the differentiation from one state to another takes time as the expression of the marker genes for one state subsides while those for the next state gradually emerge (Figure 11.6). Although this can potentially be dealt with by introducing more subpopulations in transit from one differentiation state to another, there may be a better approach. We could employ a multiscale model that describes the kinetics of the expression of intracellular markers. Then, we can define the state of the cell according to the expression level of those markers.

The differentiation state of cells can thus be represented by (e_o, e_f, and e_a), where e_o, e_f, and e_a are the transcript levels of Oct4, Foxa2, and albumin, respectively (Eq. 11.6 in Panel 11.5). In the pluripotent state, e_o is above a threshold value, while e_f and e_a are at the background level (Eq. 11.7 in Panel 11.5). Similarly, in the endodermal and hepatic states, e_f or e_a are above a threshold value, while the other two marker genes are at low levels (Eqs. 11.8 and 11.9 in Panel 11.5). The intracellular expression level of the marker genes is the balance of the expression rate as regulated by the differentiation cues and the degradation.

Panel 11.5 Intracellular Variables

The differentiation state of a cell population is characterized by the expression levels of Oct4, foxa2, and albumin (e_o, e_f, e_a).

$$n = n(e_o, e_f, e_a) \tag{11.6}$$

Concentration of the pluripotent, endodermal, and hepatic cells:

$$n_p : n(e_o \geq e_{o,p}, \, e_f^-, \, e_a^-) \tag{11.7}$$

$$n_e : n(e_o^-, \, e_f \geq e_{f,e}, \, e_a^-) \tag{11.8}$$

$$n_h : n(e_o^-, \, e_f^-, \, e_a > e_{a,h}) \tag{11.9}$$

where $e_{o,p}$, $e_{f,e}$, and $e_{a,h}$ are the threshold expression levels of Oct4, foxa2, and albumin genes for pluripotent, endodermal, and hepatic cells, respectively.

Cells of a differentiation state express the marker gene above a threshold, but do not express the marker of the other differentiation states (as indicated by "-").

The expression level of those genes is a function of the "signaling cues":

$$\frac{de_{i,j}}{dt} = \eta_{i,j} \prod_k \left(\frac{C_k^{\beta_i}}{K_{i_1} + C_k + K_{i_2} C_k^{\alpha_i}} \right) - \delta_{i,j} e_{i,j} \tag{11.10}$$

where $i = o, f, a$; $j = p, e, a$; and $k =$ different sigaling cues.
$\eta_{o,p}$ $\eta_{f,e}$ and $\eta_{a,h}$ are positive.
$\eta_{o,e}$ $\eta_{o,h}$ $\eta_{f,p}$ $\eta_{f,h}$ $\eta_{a,p}$ and $\eta_{a,e}$ are all 0.

Empirical equations (Eqs. 11.7, 11.8, 11.9, and 11.10) that describe the expression rate as a function of external cues and assume the combined effect of different factors is multiplicative may be used. Alternatively, one may seek to develop a mechanistic model that links an external signal to the signaling pathway and transcriptional regulation. However, such a model would be rather complex. Often, we do not have a complete picture of the path from external cue to gene expression control in order to devise a mechanistic model.

11.5 Stem Cell Technology

Stem cells hold great promise for regenerative medicine (Panel 11.6). A new and exciting avenue of tissue regeneration is to stimulate endogenous stem cells to initiate their intrinsic repair mechanisms. Except for such wholly *in vivo* modulation of stem cells, most stem cell applications require *in vitro* culturing of stem cells for cell expansion, to extend their availability to multiple doses for the same patient or to a larger number of patients to benefit from economy of scale.

In some cases, undifferentiated stem cells may be transplanted into patients, where they will hone in on their own natural niche and reconstitute tissue. Stem cells may also be directed to a particular differentiated lineage before transplantation. The cells used for transplantation may be allogeneic or autologous (Panel 11.7). Autologous transplant has been practiced using bone marrow stem cells for over 5 decades. The discovery of iPSCs has significantly raised the prospect of autologous stem cell therapy.

Panel 11.6 Stem Cells Applications

1) Therapeutic cell transplantation
   - Cells delivered to relevant tissues to enable regeneration of function
2) *In vitro* models for drug toxicity screening
   - Stem cells (iPS) from a wide spectrum of populations differentiated into hepato- or cardiomyocytes for assessment of drug toxicity
3) Extracorporeal applications
   - Bioartificial liver devices
4) Tissue engineering
   - Biomimetic scaffolds

Panel 11.7 Cell Transplantation for Restoration of Function

- Autologous transplantation involves using one's own cells
  - Low risk of immune rejection
  - Time-consuming processing
- Allogeneic transplantation involves using cells from donors
  - High risk of immune rejection
  - Cells can be banked and readily available
- Examples
  - Bone marrow, umbilical cord blood cells transplant to cure diseases like leukemia
  - Cardiac stem cells, generated from the heart tissue of patients, help reduce tissue scarring after heart attack
  - ES cell-derived retinal pigment epithelial cells resulted in restoring eyesight in blind women after transplant

---

In addition to therapeutic applications in regenerative medicine, stem cells may also be used for *in vitro* studies. They can be directed to differentiate to various lineages and may be combined to form tissue analogues with the unique activities and functions of different human tissues. Because of ethical issues, some tissue samples of certain genetic or racial backgrounds may not be attainable for studies. Since iPSCs can be derived from adult somatic cells, these tissue analogues can be constructed using iPSCs isolated from patients of different racial and genetic backgrounds and greatly increase the diversity of the cell source. These tissue analogues constitute a repertoire of various human physiological models that can be used for studying homeostatic or disease states and for the discovery of treatment for dysfunctions. This will advance our understanding of genetic diversity and greatly enhance our capability to develop medicines for various genetic diseases.

11.6   Engineering in Cultivation of Stem Cells

Most microbial productions of biochemicals are carried out under submerged conditions in stirred-tank bioreactors, as are most processes for the production of protein therapeutics using recombinant mammalian cells. Although many cells that are used in biomanufacturing started out as adherent on solid surfaces, almost all industrial cell lines are adapted to grow in suspension before the process of scaling up is initiated. Conversely, all stem cells are adherent and require a surface to attach, spread, and grow. Typically, they are cultivated on a flat surface coated with suitable extracellular matrices in a culture flask.

Some, especially embryonic stem cells, are grown on a monolayer of mouse fibroblastic cells that have been previously irradiated or chemically treated. This renders the fibroblastic cells incapable of proliferation but still metabolically active. Since these "feeder" cells are typically of rodent origin, their presence in the production of any product destined for human therapeutic use is highly undesirable. Furthermore, their use is a barrier to scaling up. Most industrial-scale productions of stem cells will likely be converted to feeder-free cultures by using specially developed medium. Nevertheless, surface attachment is still a requirement for stem cell cultivation.

Therapeutic applications of stem cells may be autologous or allogeneic. Autologous applications require relatively small quantities of cells. For allogeneic applications, a large quantity of cells needs to be produced. This makes a culture flask–based process too labor intensive, prone to microbial contamination, and uneconomical. To cultivate anchorage-dependent stem cells in a stirred tank, suitable surfaces must be provided. One way of achieving this is through the use of microcarriers. Microcarriers are polymeric (e.g., cross-linked dextran or polystyrene) beads that are 150–200 μm in diameter and have a chemically modified surface to facilitate cell adhesion. The chemical modification may be treatment with extracellular matrix proteins, such as collagen or adhesion peptides, or derivatization with charged moieties.

Microcarriers make it possible to have a large surface area for cell growth in a stirred-tank bioreactor. However, the surface they provide is highly segregated, unlike that in tissue culture flasks. A flask provides a contiguous surface that allows cells to move, migrate, and even adjust local distribution. In contrast, in a microcarrier culture, beads of 3×10^{-4} cm^2 surface area each are segregated from one another. After initial cell attachment, cells do not migrate from one surface to another, even if one surface becomes crowded while others are still relatively sparsely populated.

A microcarrier culture is started by mixing cells and suspended microcarriers in a reactor. Cell attachment to microcarriers is a random process; the number of cells attached to each microcarrier is not uniform but follows a distribution. Stem cells, like other adherent cells, show a growth rate dependence on cell density. At the start of the culture, a low cell density may retard cell growth. Also, upon approaching confluence, the growth rate will slow down and eventually cease. Thus, depending on the distribution of the number of cells per microcarrier, one may observe uneven cell growth on microcarriers. The segregation of the growth surface in a microcarrier culture needs to be considered when evaluating stem cell growth kinetics and in process design.

Stem cell culture processes are distinct from conventional bioprocesses in product recovery (Figure 11.7). The recovery of conventional products, including proteins and biochemicals, typically involves first the removal of cells from the product stream, followed by the isolation and purification of products through a few sequential steps of chromatographic separations and/or extraction, perhaps in conjunction with membrane operations. The purified products are then packaged to their final form using

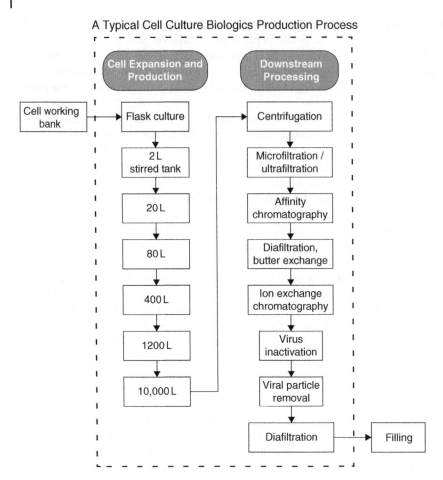

Figure 11.7 A typical manufacturing process for the production of therapeutic proteins.

crystallization, precipitation, and so on. Products that are going to be administered in a sterile form for injection are subjected to virus inactivation and removal, if necessary, and are sterilized by membrane filtration before final finishing.

For stem cell bioprocesses, living cells are the product (Figure 11.8). The live cell product cannot be passed through a sterilization step. In order to retain a high viability rate, the procedure of isolating, purifying, and concentrating the cells must be relatively simple and induce little to no mechanical and chemical stresses. Furthermore, the number of unit operations must be kept small.

Stem cell differentiation often results in a heterogeneous population containing not only cells of the desired state of differentiation, but also other cells. The product cell population will likely need to be separated from the other cells. In hESC therapy, even a contamination of a small number of undifferentiated ESCs will increase the risk of tumor formation in the patient. Thus, product cell isolation must be highly selective and robust.

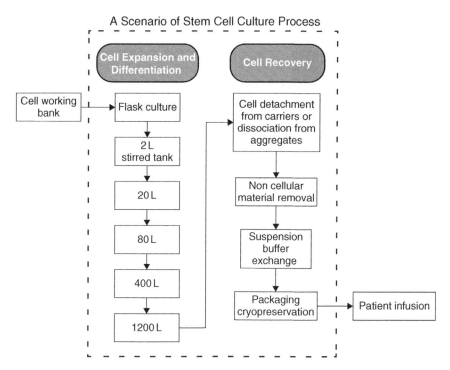

Figure 11.8 A possible manufacturing process for the production of stem cells for regenerative medicine.

11.7 Concluding Remarks

Translating stem cell applications into biomanufacturing processes will pose many challenges for bioengineering. The translation of work with stem cells, from discovery in the laboratory to future clinical applications, will also require the implementation of stem cell cultivation as a process technology. The fields of engineering and medicine are now interdependent, and we must partner effectively to realize our shared goals. To realize the full potential for stem cell–based regenerative medicine, bioengineers must boldly address these challenges.

Further Reading

Diogo, MM, da Silva, CL and Cabral, JMS (2012) Separation technologies for stem cell bioprocessing. Biotechnol. Bioeng., **109**, 2699–2709.

Lanza, R and Atala, A (2014) Essential of stem cell biology, 3rd. ed. Academic Press, San Diego, CA.

Tandon, N, Marolt, D, Cimetta, E and Vunjak-Novakovic, G (2013) Bioreactor engineering of stem cell environments. Biotechnol. Adv., **31**, 1020–1031.

Nomenclature

C_A, C_B	concentration of signaling cue	mole/L^3
e_o, e_f, e_a	expression level of gene *Oct4*, *foxa2*, and albumin	mole/L^3
$e_{o,p}$	threshold expression level of *Oct4*; above that, becomes pluripotent state	mole/L^3
$e_{f,e}$	threshold expression level of *foxa2*; above that, becomes endodermal state	mole/L^3
$e_{a,h}$	threshold expression of albumin; above that, becomes hepatic state	mole/L^3
subscript p, e, h	denote pluripotent, endodermal, and hepatic cells	
$K, K_1, K_2,$ $K_1', K_2', m,$ n, m', n'	kinetic parameters for differentiating a gene expression	
n_p	concentration of pluripotent cells	cell/L^3
n_e	concentration of endodermal cells	cell/L^3
n_h	concentration of hepatic cells	cell/L^3
$\eta_{i,j}$	basal rate of expression of gene i in cells of differentiation state j	1/L$^3\cdot$t
μ_j	specific growth rate of cell in differentiation state j	t^{-1}
$\Phi_{e \to h}$	specific differentiation rate from endodermal to hepatic state	t^{-1}
$\Phi_{p \to e}$	specific differentiation rate from pluripotent to endodermal state	t^{-1}
δ_{ij}	the degradation rate constant of the state j	t^{-1}

Problems

A1 Using no more than five sentences, briefly describe what an iPSC is. Which of the following will be more likely to enable autologous stem cell–based therapy: adult stem cells, embryonic stem cells, or iPSCs?

A2 Pluripotent stem cells, when injected into immune-deficient mice, may cause the formation of teratomas. In general, upon injection of about 10^4 pluripotent stem cells into an immune-deficient mouse, one tumor can be found. Tumor formation is thus a risk factor that must be safeguarded in any stem cell transplantation. Consider the case of using a stem cell–derived liver cell to treat liver disease. It is estimated that 5% of the cell mass of a liver will be needed to treat an adult patient. An adult liver has about 10^{11} cells. In the course of directed differentiation of pluripotent stem cell to liver cells, the transcript level of pluripotent genes decreases over time, while the liver-specific genes become highly expressed. In a pure population of pluripotent stem cells, a pluripotent gene (*Oct4*) is expressed at a relative level of 10^3 copies per cell. After the pluripotent stem cell is subjected to differentiation, the *Oct4* transcript decays with a half-life of 1 day. After a 20-day differentiation period, the resulting liver cell, however, has an undetectable *Oct4* transcript level (at a sensitivity of 10^{-2} per cell). With this data, estimate the probability of a pluripotent cell persisting after hepatic differentiation and the tumor-forming risk involved in giving a patient a sufficient number of liver cells derived from pluripotent cells. What else can you do to: (a) get a better estimate of the risk involved, and (b) further reduce the risk?

A3 The reprogramming of differentiated cells to become iPSCs is attributed to epigenetic events caused by the expression of exogenous transcription factors introduced into the cell and the chemical environment used to cultivate them. Explain the epigenetic changes involved in the process of pluripotency induction.

A4 Why are iPSCs likely to make a major impact in drug discovery and in the study of diseases of genetic abnormality?

A5 In this problem, we use a recombinant process as an analogy to design a hypothetic stem cell culture process. A typical recombinant immunoglobulin G (IgG) production process employs a 10,000-L reactor as its production reactor. In a 12 to 14-day process, Chinese hamster ovary (CHO) cells are grown to a concentration of 10^{10} cells/L and produce about 5 g/L of IgG. After purification, about 70% of the IgG produced is recovered as product. In the clinic, the dose for each patient is about 500 mg. Assume a stem cell culture process requires a similar economy of scale to be financially viable (i.e., the same number of doses is to be produced in a manufacturing batch). What will a liver cell transplantation process look like? Consider the case that pluripotent stem cells will first be expanded, and then differentiated into liver cells or liver precursor cells before transplantation. Assume that 10^{10} cells are to be used for each patient. Develop a plan that includes the selection of reactors for different stages of cell expansion and differentiation.

A6 Pluripotent stem cells from mice can be directed to differentiate to become hepatic cells. The differentiation occurs in stages by changing the growth factors and signaling molecules in the medium initially to endoderm induction medium, then on day 4 to hepatic induction medium and day 12 to hepatic maturation medium. During the differentiation, the differentiation state of the cell is determined by the expression of marker genes by flow cytometry, and the cell composition (percentage of the total initial population) at different times is shown in Table P.11.1. During differentiation, cells do not proliferate and a small fraction of cells die; thus, the total at a given time may not be 100. Use the data to estimate the specific rate of differentiation from one stage to the subsequent stage.

Table P.11.1 Percentage of cells.

Day	Pluripotent	Endodermal	Early hepatic	Late hepatic
0	100	0	0	0
4	40	60	0	0
8	5	51	35	0
12	3	26	61	1
16	0	3	64	20
20	0	0	22	65

12

Synthetic Biotechnology: From Metabolic Engineering to Synthetic Microbes

12.1 Introduction

For centuries, humans have harnessed the metabolic and biosynthetic capacities of microbes in order to create desirable products and better quality of life. Through microbial strain improvement, humans have (sometimes unknowingly) selected or enriched microbes to produce better products. Only much later, scientific investigation began to reveal the mechanisms underlying this process. Centuries ago, people just chose to cultivate more of the results they preferred. For example, by repeatedly and serially selecting *koji* (*Aspergillus oryzae*) to make better soy sauce or saké, humans cultivated favorable microbial strains. Concerted efforts to improve industrial microbial strains are as old as classical biotechnology. The industrialization of microbial biochemicals and enzymes in the twentieth century is only possible because of the extensive strain improvements used to enhance production.

The primary goal of strain improvement is to increase a product's concentration in culture by altering a native organism (or "producing cell"). With recombinant DNA technology, direct genetic intervention became the method of choice to enhance productivity. Producing cells are engineered to eliminate a byproduct, to use a more economical substrate, or to employ a more efficient synthetic route. We mutate or eliminate endogenous enzymes, as well as introduce foreign proteins, to direct cells to produce a new product. Taking metabolic engineering one step further, scientists can now transplant biosynthetic pathways from one organism to another. The expanded synthetic capacity of microbial cells has caused great excitement in "synthetic biology."

Like many new areas of scientific exploration, synthetic biology includes a wide variety of research activities. One area of focus is the engineering of microorganisms to enhance or add new biosynthetic capabilities. Another is the construction of an artificial regulatory circuit that can turn transcription and/or translation "on" and/or "off" in a prescribed way. Synthetic biology can also include the synthesis of a genome or the reconstitution of a microorganism in the laboratory.

Ultimately, synthetic biology may lead to a new, combined technological skill set that is the combination of all the above. For example, one could use a synthetic organism with a synthetic regulatory circuit to control the biosynthesis of a new compound. Synthetic biology began with the humble goal of improving industrial microorganisms. It shares common threads with metabolic engineering. It is now engineering the genomes of microbes and using artificial genetic elements to extend the synthetic capacity of industrial microorganisms even further.

Engineering Principles in Biotechnology, First Edition. Wei-Shou Hu.
© 2018 John Wiley & Sons Ltd. Published 2018 by John Wiley & Sons Ltd.
Companion Website: www.wiley.com/go/hu/engineering_fundamentals_of_biotechnology

12.2 Generalized Pathways for Biochemical Production

Microbes have evolved to never waste resources. They make only enough cellular components or metabolites to ensure their own well-being. To make a native microbe into an industrial producer of a metabolite, we must overcome its natural tendency to avoid wasteful metabolism. Before discussing the strategies for retooling a microbe's biosynthetic capacity, we will start by examining the general material flow used to synthesize a biochemical product (Figure 12.1).

Cells take up glucose, ammonium, or other carbon and nitrogen sources through their transporters. They metabolize them to generate more biomass and to synthesize products. The metabolic reactions are grouped into pathways (like glycolysis, fatty acid oxidation, etc.). All of the different pathways, together, form a large reaction network. In eukaryotic cells, the reactions and their reaction products are partitioned into different organelles (Figure 12.1). For example, citrate is formed in the tricarboxylic acid (TCA) cycle, occurring in the mitochondrial matrix. Citrate is then exported into the cytoplasm, where it can be converted to acetyl CoA (and oxaloacetate) for use in the synthesis of lipids, polyketide antibiotics, and terpenoids. The latter two are major classes of secondary metabolites.

In plants, terpenoid synthesis takes place across three different compartments (the cytosol, mitochondrion, and peroxisome plastid). By compartmentalizing specialized biochemical reactions into discrete compartments, eukaryotic cells are able to localize special reactions or compounds that might not be compatible with one another in the cytosol, or that may even be harmful to the cell. In each organelle, the reactions form a separate subnetwork that communicates with cytosolic reactions through transporters in the organelle membrane.

Many industrial biochemicals are intermediates for the cellular pathways that constitute a cell's primary metabolism (represented by "O" in Figure 12.1). For example, lysine and tryptophan are products of amino acid biosynthesis. Citrate is an intermediate of the TCA cycle. The metabolites with a more complex structure, including most secondary metabolites (represented by "P"), are synthesized by drawing precursors from primary metabolism reactions. The synthesis of secondary metabolites may also be compartmentalized into multiple organelles. Thus, the boundaries between primary metabolism and secondary metabolism may overlap. In industrial manufacturing, a precursor may need to be supplied in the medium to increase productivity.

Finally, the product is excreted out of cells through a transporter. However, before or after secretion, a product may be degraded or converted to a side product, "Q." Some products, notably the energy-storage compounds of microbes (such as polyhydroxybutyrate), accumulate intracellularly and will halt production when their concentration reaches the cell's holding capacity.

Microbes minimize resource waste by adjusting the flux of a pathway, according to their environment and cellular needs. Often, the first enzyme after a divergent branching point plays a regulatory role in controlling the flux. Control can be exerted on the enzyme's activity (feedback inhibition or activation), as well as on the synthesis of new enzymes at the transcriptional level (repression or induction). The control of enzymatic activity is often modulated through the binding of a metabolite to the enzyme, in a process called allosteric regulation. Therein, when a metabolite accumulates to a high level, it reduces the activity of its producing enzyme and prevents further

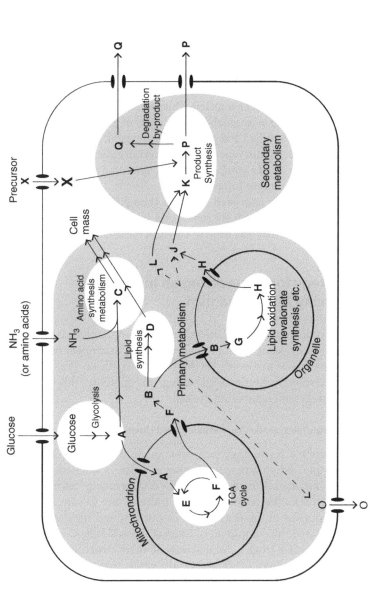

Figure 12.1 A general material flow in the production of biochemical products. Carbon and nitrogen sources, sometimes also precursors, are imported through transporters. Multiple biochemical pathways and/or multiple compartments (in eukaryotic cells) may be involved in the synthesis. Side products may also be formed. The product is secreted through a transporter.

metabolite accumulation. An enzyme may not be needed all of the time; it is synthesized (i.e., its gene is transcribed and translated) only when the reaction it catalyzes is needed.

To this point, our discussion has focused mainly on primary metabolites, or products that are produced to support cell growth. Secondary metabolites, on the other hand, are produced only after cell growth ceases. The genes for their synthesis are simply not expressed during cell growth and are only expressed (or "turned on") when the growth stage is over.

The synthetic capacity of a product in a microorganism is thus controlled at multiple points in its pathway: from the transport of substrates and precursors, to their participation in reaction pathways, to product excretion. This use of multiple layers of control minimizes wasteful production. To increase the productivity of the microbe, then, these constriction points need to be identified and intentionally relaxed. This is done through a concerted dismantling of the regulatory mechanism. There are always multiple control points involved in the biosynthesis of a compound. Thus, once one constriction point is removed, another one will become rate-limiting.

To succeed, a synthetic biology approach requires thoughtful design and a well-planned blueprint for the final, desired microbe. This is in stark contrast with the empirical, incremental approach that was used in classical industrial microbiology.

12.3   General Strategy for Engineering an Industrial, Biochemical-Producing Microorganism

To create an industrial producing microorganism, one can either enhance the productivity of a native producer or construct the necessary synthetic machinery in a new host microbe. Either way, a first step is to identify the biosynthetic pathway that leads to the desired product and understand its bottleneck(s) to high productivity. This information helps us to develop a strategy for engineering the cell.

12.3.1   Genomics, Metabolomics, Deducing Pathway, and Unveiling Regulation

Advances in DNA sequencing technology have now made the genome sequence of a microbe readily accessible. When a new producing microbe is isolated, its genome sequence can be quickly obtained. Complementary RNA sequencing can help to identify the mRNAs expressed (i.e., the transcriptome). It allows one to gain information on the protein coding sequence of the enzymes, transporters, and regulators that are involved in the biosynthesis of the compound of interest. Possible allosteric regulation on the enzymes can be learned from a comparison of their structural similarity to orthologues that have been studied. Orthologues are genes that are from the same ancestor gene but have evolved separately in different organisms. The pathway can then be verified by measuring the reaction intermediates or metabolites. Finally, by studying the accumulation of a pathway's biosynthetic intermediates, its potential rate-limiting steps can be elucidated.

To reach our desired result, we must decide on the most effective strategy. We can either: (1) enhance the productivity of a pathway in its native host, or (2) reconstruct the

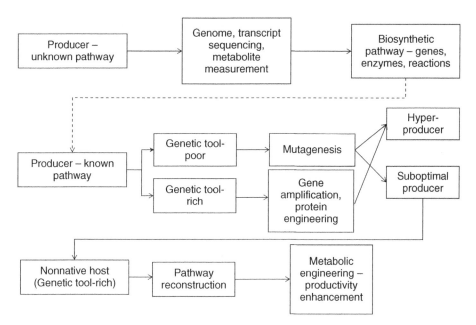

Figure 12.2 A general strategy of engineering a microorganism for the production of a biochemical.

biosynthetic pathway in a new host. Toward that end, we consider the tools available for genetic modification of the native host (Figure 12.2). If the genetic toolbox of the native host is underdeveloped, it would be difficult to use random mutagenesis to isolate high producers. In that case, it may be easier to transplant the pathway into a new host. On the other hand, if the native producer is especially suited for the overproduction of the product (like the *Clostridium* family is for solvent production), then one may opt to work within the native host.

12.3.2 Introducing Genetic Alterations

To isolate better strains of producers, we take advantage of the genetic heterogeneity within a producing cell population. Because DNA replication is not error-free, mutations are introduced into the genome at a low frequency, eventually leading to a somewhat heterogeneous population. For a given pathway, then, a small fraction of the mutated population may harbor genetic variations that affect the activity of the synthetic enzyme, reduce the effect of feedback inhibition, or affect the pathway's productivity through other ways. By plating a diluted cell suspension on nutrient agar plates, cells can be spread far apart from each other on the surface of the agar, and eventually form colonies that each arose from a single cell. The cells in a colony are, thus, a pure population and identical in nature. Colonies can then be screened to find one or a few that have the desired properties (e.g., higher productivity). By repeating this selective isolation of natural mutants multiple times, one can isolate a higher producer.

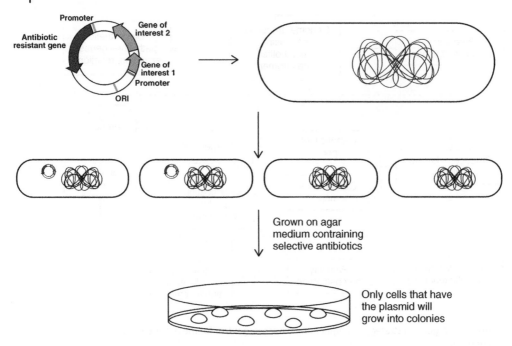

Figure 12.3 Transformation of a microorganism with plasmids. The plasmid carries a selective marker gene that is expressed to produce a protein that allows those cells that received the plasmid to have a growth advantage under selective conditions. The plasmid also contains the gene(s) that enable or enhance those cells to produce the product.

To further accelerate the process, mutations can be introduced through the use of chemical mutagens, x-rays, or UV irradiation. This increases the frequency of genetic alteration in a random manner. Most of the mutations are unrelated to the trait that one seeks to change; some can even be detrimental, retarding cell growth or suppressing productivity. Typically, only a fraction of cells will receive the intended genetic alteration. To generate high-producing strains, we must effectively enrich and identify the cells that have the correct genetic changes and acquire the desired traits.

In recent years, it is more common that specific (as opposed to random) genetic modifications are imparted by introducing a plasmid that carries a desired gene(s) into the producing cell (Figure 12.3). When expressed as the protein product, the introduced gene will impart a new activity to affect product synthesis.

Plasmids are small, circular DNA elements that can replicate autonomously in the host cell and pass along the genes they carry. In a bacterial cell, a plasmid may be present in multiple copies, ranging from a few to several hundred. Many microbial species can take up exogenous DNA molecules at sufficiently high frequencies (called transformation). However, only some of the cells in a population are ultimately transformed. Thus, the plasmids used for recombinant DNA work usually contain a selective gene. This gene encodes for a protein that makes the host cell resistant to an antibiotic (e.g., a protein that can pump the antibiotic out of the cell or a protein that can degrade the antibiotic). After transformation, cells are then placed into a posttransformation medium. This contains a lethal concentration of the antibiotic, which selects for transformed cells and kills all untransformed cells.

12.3.3 Isolating Superior Producers

To identify and isolate cells that have a desired trait, we screen a large population of cells. We can find those that have the target phenotype, or impose selective pressure. This selective pressure kills or suppresses those that do not possess the desired trait, so that the desired mutants can be enriched.

12.3.3.1 Screening of Mutants with the Desired Phenotype

To screen clones, we largely rely on growth assays and product assays. We can readily assess cell growth by the size of the colony. Ideally, the product level of each clone is quantified directly in the agar plate on which cells grow. Some products can be detected when they react with an indicator reagent; this forms a color or fluorescent-reaction product. The amount of product produced is shown in the size of the color zone surrounding each colony. Some secreted products can be assayed by measuring growth, or lack of growth. These bioassays, thus, involve plating the microbe to be screened over the indicator organism (Figure 12.4).

A lysine producer can cross-feed lysine into the agar and allow a lysine auxotroph (the indicator organism) to grow. The extent of growth of the autotroph surrounding the colony indicates the amount of lysine produced. Similarly, antibiotic generated by an antibiotic producer inhibits the growth of a microorganism that is sensitive to the antibiotic. The size of the growth inhibition zone, then, would be an indicator of the antibiotic concentration produced.

It should be noted that as photon-sensing techniques, image analysis, and other microanalytical platforms become more widely available, integrated robotic systems are now commonly used to automate the screening process. Indeed, such robot-based, high-throughput screening is a key component in industrial biochemical discovery.

12.3.3.2 Selection of Mutants with the Target Trait

Selection of mutants with the target trait largely falls into two categories: positive selection and negative selection. In positive selection, only cells with the target trait will grow. In negative selection, only cells without the target trait will grow.

Positive selection is frequently used in the transformation of bacterium using a plasmid vector, as described in this chapter. This has been used in the strain improvement

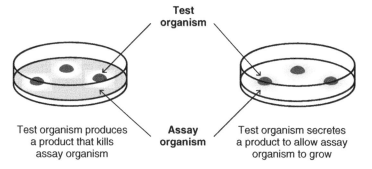

Figure 12.4 Bioassays for screening microorganisms that produce a product that suppresses or enables the growth of an assay organism. The former uses an assay organism that is sensitive to the inhibition of the product. The latter uses an auxotroph that is dependent on the metabolite for growth.

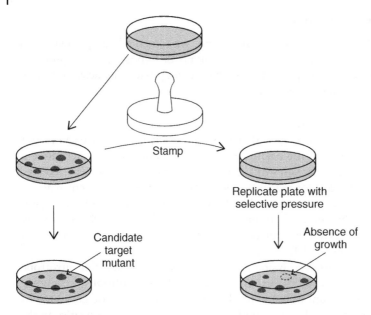

Figure 12.5 A replica plate method to detect conditional mutants whose growth is suppressed only under some conditions.

of antibiotic-producing cells. A high producer must also have a mechanism to tolerate a high concentration of its own antibiotic. After mutagenesis, mutants that survive in high levels of the antibiotic are often also higher producers.

An example of negative selection is the isolation of temperature-sensitive yeast cells. Normally, yeast cells can grow at temperatures ranging from 23 to 36 °C. To isolate mutants that grow at the permissive temperature of 23 °C, but not at the nonpermissive temperature of 36 °C, cells are plated on agar plates to grow into colonies. Replica plates are then created using a stamp. The stamp picks up some cells from each colony and transfers them to another clean plate. After incubating for a period, the new plate will have an identical spatial localization of the colonies as the original plate (Figure 12.5). On each, replica plate colonies of corresponding locations originate from the same initial cell. By placing the replica plates in two different temperatures, one can compare cell growth of the two replicas and identify mutants that have temperature sensitivity.

Mutants with altered feedback regulation may also be enriched by selection. Lysine can suppress its own synthesis by inhibiting aspartate kinase, the first enzyme in the lysine biosynthesis pathway. A common strategy is to use chemical analogues to select mutants whose aspartate kinase is not subjected to lysine inhibition. These analogues are structurally similar to lysine, but are not metabolized by enzymes that use lysine as a substrate, nor are they incorporated into proteins by translation. In the presence of a high concentration of a lysine analogue, aspartate kinase becomes inhibited. This halts the biosynthesis of lysine and other amino acids in the aspartate family pathway. Without the supply of those amino acids, the wild-type cells cannot grow. In contrast, mutants that have deregulated feedback control of aspartate kinase will no longer be subjected to inhibition by the analogue. Thus, they will be able to synthesize lysine and grow. Such analogue-based selection of deregulated mutants has created many important industrial producers of lysine and other amino acids for over half a century.

12.3.4 Mechanisms of Enhancing the Biosynthetic Machinery

For over five decades, industrial microorganisms came from screening natural isolates. Once a producer of a biochemical or a bioactive compound was found, a strain improvement program (including random mutagenesis, screening, and selection) was then established to isolate strains with high productivity. In the past two decades, we have developed better mechanistic understanding of the genetic changes required to make industrial microorganisms better producers. With our increased ability to engineer the genetic makeup of microorganisms, these mechanistic perspectives have transformed into strategies for the rational design of industrial microorganisms. These strategies will be described in this section (Panel 12.1).

Panel 12.1 Strategies to Engineer Biochemical Production Machinery

Enhancing the biosynthetic machinery
    Relaxing the constriction point in the pathway
        Augmenting the rate of controlling reactions
        Relaxing allosteric regulation
    Channeling precursor supply
    Eliminating product diversion
    Enhancing product transport
    Rerouting the pathway to increase the efficiency
    Transplanting heterologous genes/pathways to reprogram biosynthesis
Engineering host cells – beyond the pathway
    Altering substrate utilization
    Manipulating time dynamics of production
    Increasing product tolerance
        Engineering product transporter
        Engineering product binding protein
        Adaptation – altering membrane composition
        Engineering stress response

12.3.4.1 Relaxing the Constriction Points in the Pathway

Microbes economize their resources with multiple levels of control. First, different catabolic pathways contribute different weights in energy generation to meet a cell's demand. Likewise, different anabolic pathways synthesize different cellular materials that are needed at various proportions. The capacity of different pathways is optimized to meet cellular needs. Under normal growth conditions, most pathways are not running at their maximal capacity. Then, when the need surges, the pathway can handle the increase.

However, pathways are not "overdesigned" to become wasteful. For example, the pentose phosphate pathway (PPP) and fatty acid synthesis pathway operate at a flux that is only a fraction of glycolysis. The capacity of enzymatic reactions is determined by the enzyme concentration and k_{cat}. Enzymes in different pathways have their characteristic $k_{cat}E_o$, or r_{max}. Regulating the capacity of different pathways and enzymes is the cell's first line of control over synthetic flux. This level of capacity control is often exerted at the gene expression level, through a concerted control of transcription induction and repression.

Also, different pathways compete for shared and co-substrates, like ATP, NADH, and NADPH. Many pathways have divergent points from which a shared intermediate is divided into two reaction routes. For example, in glycolysis, glucose-6-phosphate, glycerate-3-phosphate, and pyruvate are all diverted to synthesize ribulose-5-phosphate, serine, and alanine, respectively. The competitive strength, or the magnitude of the flux of a pathway, is determined by the capacity of each competing enzyme and its affinity for the shared substrate or co-substrate. The competitive strength can be described as r_{max}/K_m (or $k_{cat}E_o/K_m$) (see Eq. 4.14 in Chapter 4). When two enzymes of two pathways are competing for a shared substrate, the one that has a higher $k_{cat}E_o/K_m$ will have a higher flux.

12.3.4.1.1 Augmenting Rate-Limiting Enzymes or Even the Pathway A common approach to increase the flux of a pathway is to increase the competitive strength of the enzyme that is rate-limiting. In many cases, this ends up being the enzyme governing the first irreversible reaction in the pathway. The effect on enzyme strength is achieved by either (1) increasing the expression level of the enzyme, or (2) engineering it to decrease its K_m for its substrate. After relaxing the rate-limiting point, other points further downstream in the pathway may then become flux constricting. For this reason, it is often beneficial to amplify the genes of multiple enzymes to elicit larger increases in productivity. It is even possible to increase the genes of an entire pathway to increase the capacity of synthesizing the product. Indeed, some high-producing penicillin strains that were obtained through random mutagenesis were found to have multiple copies of a gene cluster from the penicillin biosynthetic pathway.

12.3.4.1.2 Removing Feedback Regulation Flux through a pathway is controlled by either feedback regulation exerted by the product or a reaction intermediate on the enzyme of one or more key reaction(s) (Figure 12.6). The elimination of the feedback inhibition and repression removes a constraint that limits the product from accumulating at a high level. One may use an orthologous gene that is not subject to feedback inhibition and then express it using a constitutive promoter. This overrides both feedback inhibition and repression. Alternatively, one may change the structural element in the enzyme that transmits feedback inhibition, which is typically the binding site of the feedback inhibitor. For example, one could mutate the inhibitor-binding site to reduce or eliminate substrate binding and enzyme inhibition. This is akin to increasing the inhibition constant, $K_{I,E}$, in Eq. 4.50 in Chapter 4. If feedback regulation is exerted through the transcription of the enzyme (i.e., repression), the repressor protein can be modified to change the binding site of the inhibitory ligand (i.e., the metabolite responsible for feedback repression). One may also modify the DNA binding site of the repressor so that the regulator site on the DNA is always at an "on" state.

12.3.4.2 Channeling Precursor Supply

Glycolysis, the TCA cycle, and other intermediate metabolism pathways supply the precursors for synthesizing various biochemicals. For instance, the glycerol backbone

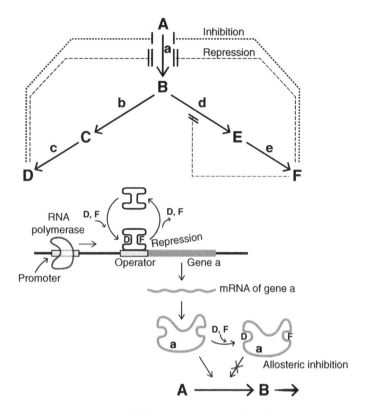

Figure 12.6 Feedback inhibition and repression. The binding of the metabolites (D, F) to the repressor protein (I) and the enzyme (a) causes the repression and allosteric inhibition, respectively. Protein engineering to impair their binding can alleviate feedback regulation.

of glyceride and microbial biodiesel is derived from dihydroxyl-acetone-3-phophophate (DHAP) in glycolysis. Pyruvate, oxaloacetate (OAA), and α–ketoglutarate give rise to many amino acids. Penicillin is derived from lysine, valine, and cysteine, while isoprenoids and alkaloids require acetyl CoA for their synthesis.

A large increase in the product flux is possible only if the supply of precursors is also augmented. For example, lysine biosynthesis relies on the supply of aspartic acid, which is derived from OAA in the TCA cycle. OAA can also be synthesized by the carboxylation of pyruvate. A flux analysis would quickly show that OAA generated through the TCA cycle must be retained in the TCA cycle to keep it running, and does not generate an overall net increase of oxaloacetate. For high productivity of lysine, OAA is supplied by the carboxylation of pyruvate. In the metabolic engineering of *Corynebacterium glutamicum* for lysine production, the pyruvate carboxylase is amplified to supply the precursor (aspartate).

In the quest for increased productivity, the flux of precursor supply must be increased. The strategy of gene/pathway amplification and deregulated feedback inhibition (as described here) for the biosynthetic enzyme is thus also applied to precursor pathways.

12.3.4.3 Eliminating Product Diversion

Sometimes, the product of interest is not the final compound in a biosynthetic pathway, but is further metabolized or degraded. In other cases, the product biosynthesis pathway may have a reaction branch that leads to a byproduct. These situations will reduce the product yield and possibly increase the complexity of product recovery. To increase the reaction yield, one can mutate or knock out the key competing enzyme leading to the byproduct.

12.3.4.4 Enhancing Product Transport

The vast majority of industrial biochemicals is excreted and accumulates in the culture broth. Very few metabolites can freely diffuse through the lipid bilayer membrane; they require a transporter for their export. As biosynthetic capacity increases in strain development, there must be a corresponding increase in product export. In industrial processes, some biochemical products accumulate in the medium to very high levels. The excretion of those products at such high extracellular levels is not possible with facilitated diffusion unless the intracellular level is higher than the extracellular level; this gives a favorable chemical potential gradient across the membrane.

A very high intracellular accumulation of some metabolites (such as lysine, succinate, and glycerol) would cause a significant change in the intracellular pH, osmolality, or solvation proprieties. Cells that produce products at such high levels must possess a mechanism to counter the change caused by their high-level accumulation. Otherwise, they must employ an active transport mechanism to keep the intercellular level low, while using energy to pump the product out.

Compared with our knowledge of biosynthetic pathways, our current understanding of the transporters for biochemicals is much less. Many hyperproducers of amino acids, organic acids, and secondary metabolites generated through mutagenesis have been found to have enhanced transport systems for their products. Transporters have increasingly become targets of metabolic engineering, to enhance the productivity of biochemicals.

12.3.4.5 Rerouting Pathways

In the microbial world, different routes may be used to carry out the same chemical conversion. A microbe may have only one of those routes, or may have multiple routes and use different ones depending on growth conditions. These routes usually have different numbers of reaction steps or energy efficiencies. Each provides a fitness advantage in a changing natural environment. For the purposes of biochemical production in a controlled environment, one route may have a higher productivity or yield.

For example, while glutamate dehydrogenase is used to take up ammonium with a high-energy efficiency of 1 mole of NADPH per mole of ammonium, its affinity for ammonium is only moderate and works well only in high-ammonium concentrations. The glutamine synthetase (GS)–glutamate synthase (GOGAT) system, on the other hand, has a higher affinity for ammonium. It is capable of taking up ammonium even at low concentrations, but at a higher energy cost of 1 mole of ATP and 1 NADHP per mole of ammonium (Figure 12.7). For biomass generation or production of amino acids, it is beneficial to engineer an organism that uses the GOGAT system to employ the glutamate dehydrogenase system.

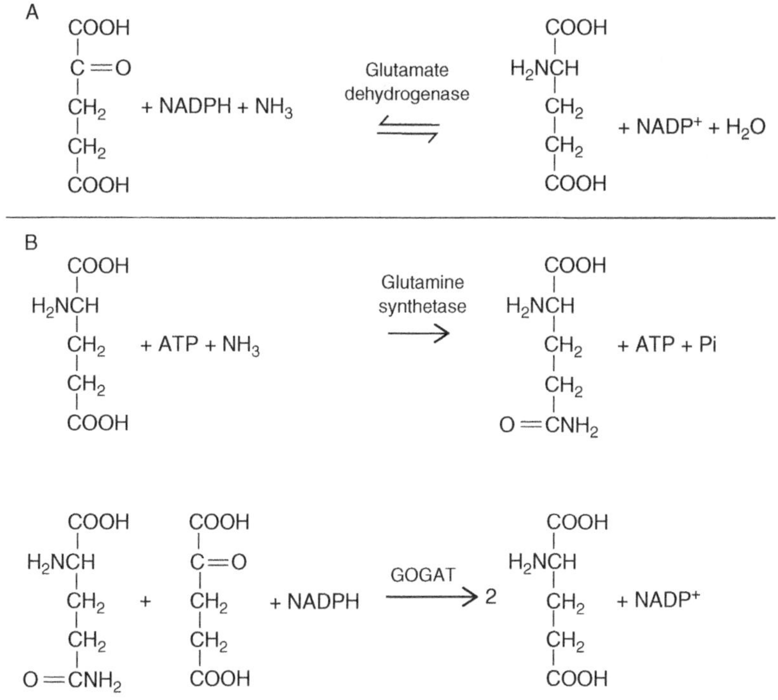

A

$$\underset{\begin{array}{c}\text{COOH}\\|\\\text{C}=\text{O}\\|\\\text{CH}_2\\|\\\text{CH}_2\\|\\\text{COOH}\end{array}}{\quad} + \text{NADPH} + \text{NH}_3 \quad \underset{\text{Glutamate}\atop\text{dehydrogenase}}{\rightleftharpoons} \quad \underset{\begin{array}{c}\text{COOH}\\|\\\text{H}_2\text{NCH}\\|\\\text{CH}_2\\|\\\text{CH}_2\\|\\\text{COOH}\end{array}}{\quad} + \text{NADP}^+ + \text{H}_2\text{O}$$

B

$$\underset{\begin{array}{c}\text{COOH}\\|\\\text{H}_2\text{NCH}\\|\\\text{CH}_2\\|\\\text{CH}_2\\|\\\text{COOH}\end{array}}{\quad} + \text{ATP} + \text{NH}_3 \quad \underset{\text{Glutamine}\atop\text{synthetase}}{\longrightarrow} \quad \underset{\begin{array}{c}\text{COOH}\\|\\\text{H}_2\text{NCH}\\|\\\text{CH}_2\\|\\\text{CH}_2\\|\\\text{O}=\text{CNH}_2\end{array}}{\quad} + \text{ATP} + \text{Pi}$$

$$\underset{\begin{array}{c}\text{COOH}\\|\\\text{H}_2\text{NCH}\\|\\\text{CH}_2\\|\\\text{CH}_2\\|\\\text{O}=\text{CNH}_2\end{array}}{\quad} + \underset{\begin{array}{c}\text{COOH}\\|\\\text{C}=\text{O}\\|\\\text{CH}_2\\|\\\text{CH}_2\\|\\\text{COOH}\end{array}}{\quad} + \text{NADPH} \quad \underset{\text{GOGAT}}{\longrightarrow} 2 \quad \underset{\begin{array}{c}\text{COOH}\\|\\\text{H}_2\text{NCH}\\|\\\text{CH}_2\\|\\\text{CH}_2\\|\\\text{COOH}\end{array}}{\quad} + \text{NADP}^+$$

Figure 12.7 Two pathways for nitrogen assimilation into amino acids are (A) the glutamate dehydrogenase pathway, and (B) the glutamine synthetase–GOGAT pathway.

For the uptake of glucose, *E. coli* and many other microbes use the phosphotransferase system (PTS), which uses one phosphoenol pyruvate (PEP) to phosphorylate glucose, to generate pyruvate, and to import glucose as glucose 6-phosphate (Figure 12.8). For products that use PEP (or an intermediate upstream of PEP) as a precursor, the PTS diverts 50% of carbons from glucose to pyruvate. By knocking out the PTS system and replacing it with a non-PTS system (for instance, by using enzymes from a different organism), one can improve the yield based on carbon.

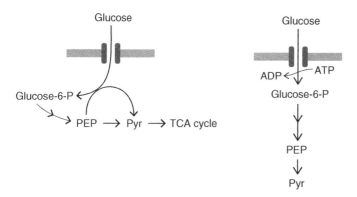

Figure 12.8 Two different glucose uptake systems in bacteria.

12.3.5   Engineering Host Cells – Beyond the Pathway

A hyperproducing organism of an industrial biochemical typically possesses many preferred and useful traits. High production capacity is just one such feature. The organism must also grow well in an economical medium, and have favorable production kinetics in order to rapidly produce the peak amount of product. Additionally, it must sustain high productivity even after it has generated a high concentration of the product. A hyperproducer is typically obtained through extensive screening of a large number of mutants. If the relationship between the genotype and the phenotype is known, one can also metabolically engineer the organism to attain the desired phenotype.

12.3.5.1   Altering Substrate Utilization

A microorganism's ability to use a substrate depends on whether it has the required transporter and metabolic enzymes. Some microbes can use a wide range of compounds as their carbon and nitrogen sources; others can only use a limited number. Conversely, some compounds, like glucose, can serve as the carbon source for most microorganisms. Others, such as xylose, are only used by a small number of microbes. In many cases, the producing organism may not have the capacity to use an economical carbon or other nutrient source. Thus, one can introduce the genes that are needed to generate a transporter system for the uptake of the new substrate. This way, the enzyme(s) needed to convert the substrate to an intermediate that can enter a metabolic pathway are already present in the cell.

Researchers have engineered a yeast cell to acquire the transporter and isomerization enzyme(s) that convert xylose to ribulose-5-phosphate or xylulose-5-phosphate (Figure 12.9). The two compounds can enter the PPP and then enter glycolysis. Xylose constitutes about 40% of sugars derived from cellulosic waste. The introduction of a xylose utilization pathway into yeast, and the use of a yeast cell's endogenous metabolic capability, would allow it to convert xylose to ethanol.

12.3.5.2   Manipulating the Time Dynamics of Production

The classical model of primary and secondary metabolism describes that primary metabolites are produced while cells are actively growing (trophophase). In this model, secondary metabolites are produced only in idiophase (a slow-growth or stationary phase) (Figure 12.10). In a manufacturing process, the volumetric productivity is highest

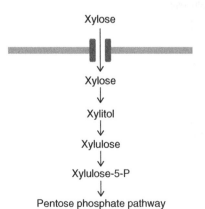

Figure 12.9 The introduction of a xylose transporter system and a pathway leading to its utilization through the pentose phosphate pathway.

Classical Processes

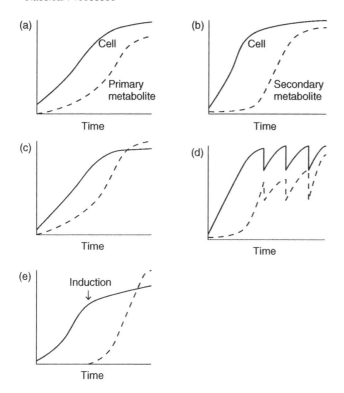

Figure 12.10 Time dynamics of biochemical metabolite production. Classical processes for the production of (a) a primary metabolite and (b) a secondary metabolite. Industrial fed-batch cultures for the production of (c) primary and (d) secondary metabolites, versus (e) the industrial production of metabolites using an inducible system.

(1) when cell concentration is also high, and (2) when cells are at a highly active state for a long period after reaching their peak cell concentration. For primary metabolites and secondary metabolites alike, fed-batch cultures have become common practice. In a fed-batch culture, concentrated substrate is added continuously or intermittently after the culture is started to prolong the production. In the production of many secondary metabolites, often, a portion of the reactor content (and the product) is harvested after a period of cultivation and replenished with fresh medium to allow for perpetual cell growth and product accumulation. The withdrawal of culture broth and replenishment of medium are repeated a few times in each process run.

Through metabolic engineering, we can control the timing of product formation. Key enzymes of the product biosynthesis pathway can be turned on after a rapid period of growth and a high cell concentration have been reached. The production is then continued until the product accumulates to high levels (Figure 12.10). The method of "switching on" of production depends on the promoter used: (1) an inducible promoter that responds to the addition of an inducer, (2) one that responds to a new substrate (such as methanol), or (3) a promoter that is turned on upon the depletion of a nutrient (such as phosphate). As the "synthetic circuit" becomes more mature, we can expect more applications for those circuits, to control the time profile of product synthesis.

12.3.5.3 Increasing Product Tolerance

Some products are growth inhibitory even to the cells that produce them. For example, the anticancer drug mitomycin C can crosslink DNA, and is toxic even to its producer. Cells producing such toxic compounds often equip themselves with a defensive mechanism, such as an export system or a binding protein, to alleviate the toxic effect.

Even a biochemical product that is not growth inhibitory can cause extreme stress to the cell. When it accumulates to the high level needed for industrial manufacturing, it can retard growth and productivity. Ethanol, organic solvents, and organic acids are notable examples. Thus, conferring producing cells with greater tolerance to the products they produce has long been recognized as one way to increase productivity. Depending on the nature of the product, tolerance can be developed through different mechanisms. A few such mechanisms are described in the remainder of this subsection (and see Panel 12.1).

12.3.5.3.1 Product Transporter Efficient excretion of the product reduces its intracellular accumulation and possibly lessens its growth-inhibitory effect. The transport rate of a facilitated diffusion transporter is affected by the difference between intracellular and extracellular concentrations. As the product accumulates to a high level in culture, its excretion rate reduces. As a result, the intracellular level may reach an inhibitory level. The overexpression of its transporter may enhance the excretion of the product molecule. As a result, a higher tolerance of the product is achieved.

In other cases, active (i.e., efflux) transporters may be used. The identification of an efflux transporter that can transport a given product can be a challenge. It is often identified by examining the gene expression profile of mutants that are tolerant to a high concentration of the product. Once the transporter is identified, the gene can be cloned and used to increase product tolerance.

12.3.5.3.2 Product Binding Protein Antibiotics that are toxic to cells are often bound to binding proteins in the intracellular environment of the producing cell. By engineering the producing cells to overexpress these binding proteins, we may achieve tolerance to these compounds.

12.3.5.3.3 Stress Responses Intracellular accumulation of metabolites may cause high osmolality, extreme pH, and/or the denaturation of cellular proteins. Under stressful conditions, cells produce stress-responsive proteins to cope. Industrial high producers of ethanol have been reported to express higher levels of stress response proteins, simply to maintain the proper folding of their cellular proteins. Also, the lipid composition of the cellular membrane in high-ethanol producers may also change, in order to maintain integrity under high-solvent-contact conditions. Exploitation of a cell's stress response system may lead to increased productivity.

Example 12.1 Metabolic Engineering of Lysine Biosynthesis

Lysine is an essential amino acid for mammals. It is supplemented in animal feed to give a balanced diet. Lysine is produced by a fermentation process using a bacterium, *Corynebacterium glutamicum*, with glucose and ammonium as the carbon and nitrogen sources, respectively. Glucose is converted to α-ketoglutarate, which provides the

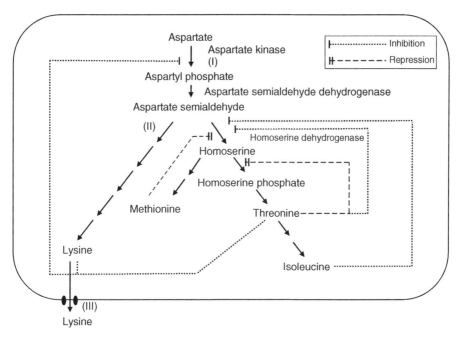

Figure 12.11 Regulation of flux distribution in aspartate family amino acids biosynthesis.

carbon skeleton for ammonium assimilation to glutamate. Glutamate then transfers its amino group to oxaloacetate (also derived from glucose) in the TCA cycle to form aspartic acid. Aspartic acid can then be converted through the aspartate family pathway to form lysine.

The biosynthetic pathway of lysine also leads to the synthesis of methionine, threonine, and isoleucine. The pathway is subjected to feedback inhibition and feedback repression (Figure 12.11). These feedback regulations on the aspartate pathway minimize wasteful overproduction of those amino acids and allow cells to better use their resources. Threonine and lysine together inhibit aspartate kinase, the first enzyme in the pathway. Threonine and methionine repress the first enzyme in the branch pathway for their own biosynthesis. Isoleucine inhibits the first enzyme in its own biosynthetic branch.

Metabolic Engineering

For the production of L-lysine as an industrial biochemical, these regulations must be overcome. This would allow the biosynthetic pathway to remain active even after the product accumulates to high levels. Furthermore, restricting the flux toward the branch from aspartate semialdehyde to methionine, threonine, and isoleucine will channel more of the carbon flow to lysine. However, complete shutoff of the branch toward methionine, threonine, and isoleucine makes cells become autotrophs, thus necessitating the exogenous supply of these amino acids. Additionally, the feedback regulation of lysine should be removed to allow for high-level accumulation.

Through classical mutagenesis and repeated selection and screening, *Corynebacterium glutamicum* was engineered to be a very high lysine producer. A metabolic engineering approach can generate a hyperproducer at a much faster pace. The aspartate family pathway in *Corynebacterium* is not as tightly regulated as it is in *E. coli*. To introduce a deregulated pathway, one would replace the first enzyme pivotal in its flux regulation with an engineered enzyme that is no longer inhibited by feedback regulation from lysine and threonine (I). The gene for this enzyme can then be amplified many times within the producer. Instead of shutting down the flux into the isoleucine branch, the first enzyme in the competing (and the desired) branch leading to lysine is amplified (II). This channels the flux to lysine production without making the mutant, an auxotroph of isoleucine. Subsequently, a transporter that "pumps" the lysine against a concentration gradient using ATP could be cloned in to enhance the export of the product (III). The final product concentration from this engineered mutant growing in an industrial bioreactor can reach 120 g/L, which would not be attainable without such an active transport system.

12.4 Pathway Synthesis

Advances in high-throughput DNA sequencing have significantly expanded the repository of microbial genome sequences. This work has provided a rich gene repertoire of enzymes. We can then mine this collection, to identify candidate enzymes with a particular catalytic activity. Candidates are then screened for their ability to enhance the performance of existing pathways and to create new ones.

Our opportunities to explore new pathways are growing rapidly, thanks to the decreasing cost of DNA synthesis and our increasing ability to assemble large DNA segments. Biochemicals are increasingly being produced in nonnative host cells. Even chemicals that have not been previously synthesized in any known organisms can be synthesized using a *de novo* synthetic pathway.

12.4.1 Host Cells: Native Hosts versus Archetypical Hosts

In selecting a host cell for biochemical production, two main factors must be considered: (1) the genetic tools available to genetically engineer the host cell, and (2) the versatility of the host cell to overproduce the target molecule. Since *E. coli* and the yeast, *Saccharomyces cerevisiae*, are by far the bacterium and the eukaryotic cell most amenable for genetic manipulation, they are the natural choice for host cells. *E. coli*, in particular, is commonly used in the conceptual demonstration of the construction of a new pathway. *E. coli* is extremely versatile and can adapt to a wide range of growing conditions. It can tolerate a wide variety of compounds (even to very high concentrations), including those that are produced only in plants. However, *E. coli* is not the best host cell for all biochemicals. For example, its preference for neutral pH for growth makes it less attractive for the production of organic acids.

Based on our knowledge of the industrial microorganisms already studied, we know that some genera of microorganisms are more likely to become hyperproducers of certain classes of biochemicals. Those native hosts have specific genetic characteristics that make them suitable for the production of specific groups of compounds (Panel 12.2).

Panel 12.2 Host Cells Used for Biochemical Production

Microorganism	Biochemical product	
	Native product	Metabolically engineered product
Escherichia coli		Ethanol, succinic acid, amino acids, 1,3-propanediol
Corynebactrium spp.	Phenylalanine, glutamic acid, lysine, and other amino acids	Amino acids, organic acids, pantothenic acid
Lactobaccillus spp.	Lactic acid	Flavors
Clostridium spp.	Acetone, butanol	Acetone, butanol (enhanced production)
Saccharymyces cerevisiae	Ethanol	Xylitol, lactic acid
Aspergillus niger	Citric acid	Citric acid, itaconic acid, oxalic acid

For example, the amino acid synthesis pathways in *Corynebacterium* are not subject to stringent feedback regulation. This makes them more amenable to the overproduction of amino acids. The genome *Streptomycetes* often includes a large set of genes related to secondary metabolism. This group of microorganisms has an inherent capability for antibiotic biosynthesis. As another example, *Aspergillus* spp. are known for the production of organic acids; likewise, *Clostridia* for solvents. Platform host cells of each of these genera can be derived and used for further engineering and to produce specialized biochemicals.

12.4.2 Expressing Heterologous Enzymes to Produce a Nonnative Product

Microbes have been engineered to overproduce products that are not native to them. This is done through the introduction of heterologous genes (meaning from a different organism) or even an entire pathway. Notable examples include the production of ethanol in *E. coli*, the synthesis of lactate in yeast, and the accumulation of intracellular polyhydroxyalkanoate (PHA) in *E. coli*. All three compounds are naturally produced by microorganisms other than *E. coli*.

Microbes have also been engineered to produce products that were not made in microbes before. For instance, *E. coli* has been engineered to synthesize isobutanol, a potential biofuel. It can also synthesize 1,3-propanediol, a chemical used in the production of many polymeric materials, and traditionally made from chemical synthesis (Figure 12.12). These products are derived from reaction intermediates of primary metabolism. The addition of a relatively small number of heterologous genes, some that confirm modified enzyme substrate specificity, allowed the creation of these new pathways in a host.

More complicated pathway engineering can involve the transplant of a natural pathway in plants, sponges, and so on into microbes like *E. coli* or yeast. One notable example is the construction of pathways for the synthesis of the malaria drug, artemisinin, in yeast. Artemisinin was originally produced in the medicinal plant, *Artemisia*. Another

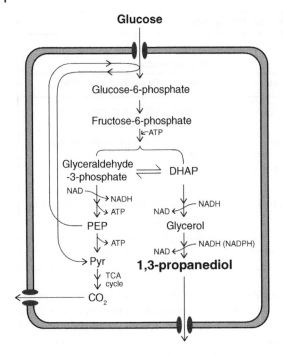

Glucose

Figure 12.12 The engineering of a glycolysis pathway for the synthesis of 1,3-propanediol.

example is the synthesis of carotenoids in *E. coli*. Successful exploitation of a microbe's synthetic capacity is further illustrated by the synthesis of methyl-valerolactones in *E. coli*. This compound can be used for the synthesis of a rubbery material that has many possible applications in plastic, food, and even the medical industry. It can now be made economically through a biological route (Figure 12.13).

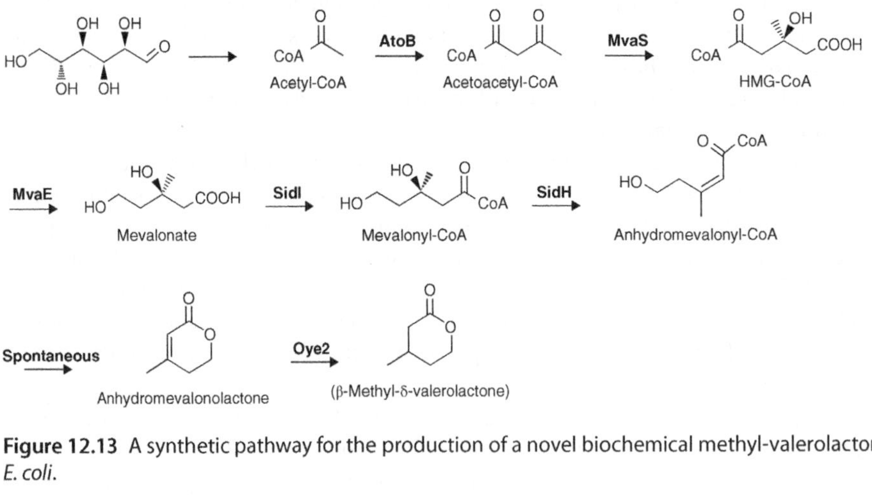

Figure 12.13 A synthetic pathway for the production of a novel biochemical methyl-valerolactone in *E. coli*.

12.4.3 Activating a Silent Pathway in a Native Host

Producers of secondary metabolites, such as members of *Actinomycetes*, often harbor multiple pathways and have a large synthetic potential for those natural products. However, many of these pathways are not active. Their products are not detected or not known, because the pathways are either (1) dormant, (2) not transcribed under tested culture conditions, or (3) devoid of promoters in their genes. Some of these pathways may confer novel secondary metabolites that possess biological activities with therapeutic potential.

Different strategies have been applied to activate dormant pathways. The genes of the entire putative pathway can be cloned into a new host organism. Alternatively, a promoter(s) may be inserted upstream of the genes, to activate their transcription in the native host organism. Activating those silent pathways will allow for an exploration of their utility and expand the natural product repertoire.

12.5 Stoichiometric and Kinetic Considerations in Pathway Engineering

To maximize the substrate yield of a synthetic pathway, one has to consider both material flux and energetic cost. Some products are metabolites of energy metabolism and generate ATP during its production (e.g., ethanol, lactate, and glutamic acid). Many others are at a more reduced state than their substrate. For example, both 1,3-propanediol and glycerol are more reduced than glucose. Their biosynthesis requires the supply of energy, by converting a portion of the substrate to generate ATP, NADH, or NADPH.

In determining the theoretical yield, the energetic cost must not be neglected (see Chapter 4). With the increasing repertoire of enzymes from a vast number of species, one can design different possible combinations of reactions and optimize a pathway to give a high theoretical conversion yield of the product. It should be noted that the reactions in a pathway must be balanced to give rise to a steady-state flux. Such balance hinges on the nature of the enzymes making up the pathway, specifically the kinetic properties (K_m, r_{max}, inhibitory constant, etc.).

In principle, one can modulate the expression level of the enzymes to give a balanced reaction flux. Unfortunately, such kinetic data are scarcely available except for the major pathways in model organisms, like *E. coli* and *S. cerevisiae*. Nevertheless, order-of-magnitude estimates may be performed to evaluate the kinetic behavior of the reaction network, based on the chemical nature of the reaction involved in an enzymatic reaction.

Example 12.2   Engineering of Glycolysis for 1,3-Propanediol Production
E. coli has been metabolically engineered to produce 1,3-propanediol (13PD). To evaluate the feasibility of 13PD synthesis, heterologous enzymes were first introduced to take DHAP to glycerol (GLY), and then to 13PD. Subsequently, a second generation of metabolic engineering was performed by replacing the glucose uptake system with one that uses ATP instead of PEP. Later, the co-substrate for the reduction of DHAP to 13PD

was switched from NADH to NADPH. Calculate the maximum yield from glucose for the first-generation and the second-generation process.

Solution

The pathway leading from glucose to 13PD is shown in Figure 12.14. Each flux is given a number. Note that all reactions in a straight path without any converging or diverging flux are treated as a single flux. For example, fructose-6-phosphate to DHAP and glyceraldehyde-3-phophate is given the flux J_3, even though two enzymatic reactions are involved. Also, note that for glucose uptake and glycerol conversion to 13PD, the fluxes (J_1, J_9) for the first-generation process are labeled as (J_{1a}, J_{9a}). In the second- and third-generation processes, they are labeled as (J_{1b}, J_{9a}) and (J_{1b}, J_{9b}), respectively. The stoichiometric equations for reactions and fluxes 7, 11, and 12 are shown in Figure 12.14. For all the other reactions, the stoichiometric coefficient for all species involved in the reaction is 1. In this system, the cell takes up glucose, and secretes 13PD and CO_2. These two reactions are given an input/output flux q_{glc}, q_{CO2}, and q_p, respectively. We also

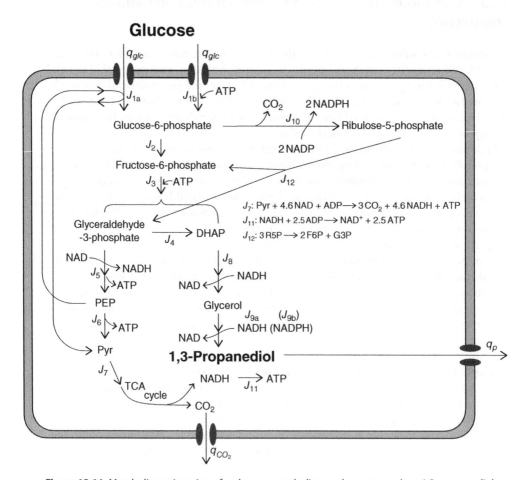

Figure 12.14 Metabolic engineering of a glucose metabolism pathway to produce 1,3-propanediol.

give the ATP generation rate a flux q_{ATP}. In this solution, we will set q_{glc} to 100 and find the maximal q_p possible. In reaction 11, we assigned a value of 2.5 ATP per mole of NADH oxidized. The conversion yield obtained will vary somewhat with the value used.

First-Generation Process

For the first-generation process, the PPP (J_{10}, J_{12}) is not involved. The material balance equations (excluding PPP) can be set up and a stoichiometric matrix can be devised as shown here:

Mass balances:

Glucose: $q_{glc} - J_1 = 0 \rightarrow J_{1a} = q_{glc} = 100$

G6P: $\quad J_{1a} - J_2 = 0$

F6P: $\quad J_2 - J_3 = 0$

DHAP: $J_3 + J_4 - J_8 = 0$

GAP: $\quad J_3 - J_4 - J_5 = 0$

PEP: $\quad -J_{1a} + J_5 - J_6 = 0$

PYR: $\quad J_{1a} + J_6 - J_7 = 0$

CO_2: $\quad 3J_7 - q_{CO_2} = 0 \rightarrow q_{CO_2} = 3J_7$

GLY: $\quad J_8 - J_{9a} = 0$

1,3PD: $J_{9a} - q_p = 0$

ATP: $\quad -J_3 + J_5 + J_6 + J_7 + 2.5J_{11} - q_{ATP} = 0$

NAD: $\quad -J_5 - 4.6J_7 + J_8 + J_{9a} + J_{11} = 0$

Stoichiometric matrix:

Pathway A	GLC transport	GLC-G6P (A)	G6P-F6P	F6P-DHAP/GAP	GAP-DHAP	GAP-PEP	PEP-PYR	PYR-CO$_2$	DHAP-GLY	GLY-PD (A)	NADH-NAD	ATP export	13PD transport
	q_{glc}	J_{1a}	J_2	J_3	J_4	J_5	J_6	J_7	J_8	J_{9a}	J_{11}	q_{ATP}	q_P
GLC	1	−1	0	0	0	0	0	0	0	0	0	0	0
G6P	0	1	−1	0	0	0	0	0	0	0	0	0	0
F6P	0	0	1	−1	0	0	0	0	0	0	0	0	0
DHAP	0	0	0	1	1	0	0	0	−1	0	0	0	0
GAP	0	0	0	1	−1	−1	0	0	0	0	0	0	0
PEP	0	−1	0	0	0	1	−1	0	0	0	0	0	0
PYR	0	1	0	0	0	0	1	−1	0	0	0	0	0
GLY	0	0	0	0	0	0	0	0	1	−1	0	0	0
13PD	0	0	0	0	0	0	0	0	0	1	0	0	−1
ATP	0	0	0	−1	0	1	1	1	0	0	2.5	−1	0
NAD	0	0	0	0	0	−1	0	−4.6	1	1	1	0	0

The maximum yield that can be achieved in this case is when minimal carbon from the glucose input is wasted to become CO_2, or when J_7 is minimal. Since $J_7 = J_{1a} + J_6$, the smallest J_7 possible is when $J_6 = 0$. We thus set $J_6 = 0$ and $J_7 = J_{1a} = 100$. Also, by looking at the pathway, we can see that $J_{1a} = J_2 = J_3$; $J_8 = J_{9s} = q_{ATP}$. The stoichiometric matrix reduces to:

Pathway A	GLC transport	GLC-DHAP/GAP	GAP-DHAP	GAP-PEP	PYR-CO$_2$	DHAP-13PD transport	NADH-NAD	ATP export
	q_{glc}	J_{1a}	J_4	J_5	J_7	J_8	J_{11}	q_{ATP}
GLC	1	−1	0	0	0	0	0	0
DHAP	0	1	1	0	0	−1	0	0
GAP	0	1	−1	−1	0	0	0	0
PEP	0	−1	0	1	0	0	0	0
PYR	0	1	0	0	−1	0	0	0
ATP	0	−1	0	1	1	0	2.5	−1
NAD	0	0	0	−1	−4.6	2	1	0

Solving for the system of equations, we obtain $J_5 = J_7 = J_{1a} = J_{9a} = 100$, $J_4 = 0$, and so on. The resulting flux of the solution is shown as a vector here:

$$X = \begin{bmatrix} q_{GLC} \\ J_{1b} \\ J_2 \\ J_3 \\ J_4 \\ J_5 \\ J_6 \\ J_7 \\ J_8 \\ J_{9a} \\ J_{11} \\ q_{ATP} \\ q_P \end{bmatrix} = \begin{bmatrix} 100 \\ 100 \\ 100 \\ 100 \\ 0 \\ 100 \\ 0 \\ 100 \\ 100 \\ 100 \\ 360 \\ 1000 \\ 100 \end{bmatrix}$$

100 moles of glucose are converted to 100 moles of 13PD. The conversion yield is (3 mole C/mole 13PD)/(6 mole C/mole glucose) × 100/100 = 0.5 mole C 13PD/mole C glucose. 50% of glucose carbons are converted to 13PD. On a mass basis, the theoretical maximum of conversion is 92/180 = 0.51 g/g. A major constraint that reduces the yield is that for every mole of glucose taken up, one mole of PEP is also consumed in glucose transport.

Second-Generation Process

In the second-generation process, glucose is transported by an ATP-dependent system. The material balance equations can be set up as:

Glucose:	$q_{glc} - J_1 = 0 \rightarrow J_{1b} = q_{glc} = 100$
G6P:	$J_{1b} - J_2 = 0$
F6P:	$J_2 - J_3 = 0$
DHAP:	$J_3 + J_4 - J_8 = 0$
GAP:	$J_3 - J_4 - J_5 = 0$
PEP:	$J_5 - J_6 = 0$
PYR:	$J_6 - J_7 = 0$
CO_2:	$3J_7 - q_{CO_2} = 0 \rightarrow q_{CO_2} = 3J_7$
GLY:	$J_8 - J_{9a} = 0$
1,3PD:	$J_{9a} - q_P = 0$
ATP:	$-J_{1a} - J_3 + J_5 + J_6 + J_7 + 2.5J_{11} - q_{ATP} = 0$
NAD:	$-J_5 - 4.6J_7 + J_8 + J_{9a} + J_{11} = 0$

Many segments of the pathway do not have branched reactions. Those nodes [(glucose, G6P, F6P), (PEP, PYR, CO_2), and (glycerol, 13PD)] and their fluxes can be combined ($J_{1a} = J_2 = J_3$; $J_8 = J_{9a} = q_P$; $J_5 = J_6 = J_7$). We obtain a reduced stoichiometric matrix:

	Pathway A (ATP-dependent HK) (reduced)	GLC transport	GLC- DHAP/GAP	GAP-DHAP	GAP-CO_2	DHAP-13PD	NADH-NAD	ATP Export
		q_{glc}	J_{1a}	J_4	J_5	q_P	J_{11}	q_{ATP}
GLC		1	−1	−0	0	0	0	0
DHAP		0	1	1	0	−1	0	0
GAP		0	1	−1	−1	0	0	0
ATP		0	−2	0	3	0	2.5	−1
NAD		0	0	0	−5.6	2	1	0

$$J_{1a} = q_{glc} = 100$$
$$J_5 = 100 - J_4$$
$$q_P = 100 + J_4 = 200 - J_5$$
$$5.6J_5 = 2q_P + J_{11} = 400 - 2J_5 + J_{11}$$
$$7.6J_5 - J_{11} = 400$$
$$3J_5 + 2.5J_{11} - q_{ATP} = 200$$
$$J_5 = \frac{1}{3} \times (200 + q_{ATP} - 2.5J_{11}) \tag{E2.1}$$

q_P is maximized if J_5 is minimized. This happens when $q_{ATP} = 0$. Thus, J_5 equals:

$$\frac{1}{3} \times (200 - 2.5 J_{11}) \tag{E2.2}$$

Solving the system of the two equations E2.1 and E2.2 gives $J_5 = 54.55$ and $q_P = 145.45$, and the resulting flux of the solution is shown as a vector here:

$$X = \begin{bmatrix} q_{GLC} \\ J_{1b} \\ J_2 \\ J_3 \\ J_4 \\ J_5 \\ J_6 \\ J_7 \\ J_8 \\ J_{9a} \\ J_{11} \\ q_{ATP} \\ q_P \end{bmatrix} = \begin{bmatrix} 100 \\ 100 \\ 100 \\ 100 \\ 45.45 \\ 54.55 \\ 54.55 \\ 54.55 \\ 145.45 \\ 145.45 \\ 14.55 \\ 0 \\ 145.45 \end{bmatrix}$$

The carbon yield is $\frac{q_P}{q_{glc}} \times \frac{3C}{6C} = 72.73\%$.

Third-Generation Process

For the third-generation process in which glucose transport uses an ATP-dependent system and the co-substrate for the reduction of glycerol to 13PD is switched to NADPH, the stoichiometric reactions will include PPP. Note that for the molecular transformation segment of PPP, the reactions are combined into one stoichiometric equation 12 (flux J_{12}). It can be shown that the five reactions in that segment will yield stoichiometric equation 12 with R5P, F6P, and G3P as the reactants or products. The material balance equations can be set up as:

Glucose:	$q_{glc} - J_1 = 0 \rightarrow J_{1b} = q_{glc} = 100$
G6P:	$J_{1b} - J_2 - J_{10} = 0$
F6P:	$J_2 - J_3 + 2J_{12} = 0$
DHAP:	$J_3 + J_4 - J_8 = 0$
GAP:	$J_3 - J_4 - J_5 + J_{12} = 0$
R5P:	$J_{10} - 3J_{12} = 0$
PEP:	$J_5 - J_6 = 0$
PYR:	$J_6 - J_7 = 0$
CO$_2$:	$3J_7 + J_{10} - q_{CO_2} = 0 \rightarrow q_{CO_2} = 3J_7 + J_{10}$
Glycerol:	$J_8 - J_{9b} = 0$
1,3PD:	$J_{9b} - q_P = 0$
ATP:	$-J_{1b} - J_3 + J_5 + J_6 + J_7 + 2.5J_{11} - q_{ATP} = 0$
NAD:	$-J_5 - 4.6J_7 + J_8 + J_{11} = 0$
NADP:	$J_{9b} - 2J_{10} = 0$

The stoichiometric matrix is shown here:

Pathway B	GLC transport	GLC-G6P (B)	G6P-F6P	F6P-DHAP/GAP	GAP-DHAP	GAP-PEP	PEP-PYR	PYR-CO$_2$	DHAP-GLY	GLY-PD (B)	G6P-R5P	NADH-NAD	R5P-GAP/F6P	ATP export	13PD transport
	q_{glc}	J_{1b}	J_2	J_3	J_4	J_5	J_6	J_7	J_8	J_{9b}	J_{10}	J_{11}	J_{12}	q_{ATP}	q_P
GLC	1	−1	0	0	0	0	0	0	0	0	0	0	0	0	0
G6P	0	1	−1	0	0	0	0	0	0	0	−1	0	0	0	0
F6P	0	0	1	−1	0	0	0	0	0	0	0	0	2	0	0
DHAP	0	0	0	1	1	0	0	0	−1	0	0	0	0	0	0
GAP	0	0	0	1	−1	−1	0	0	0	0	0	0	1	0	0
R5P	0	0	0	0	0	0	0	0	0	0	1	0	−3	0	0
PEP	0	0	0	0	0	1	−1	0	0	0	0	0	0	0	0
PYR	0	0	0	0	0	0	1	−1	0	0	0	0	0	0	0
GLY	0	0	0	0	0	0	0	0	1	−1	0	0	0	0	0
13PD	0	0	0	0	0	0	0	0	0	1	0	0	0	0	−1
ATP	0	−1	0	−1	0	1	1	1	0	0	0	2.5	0	−1	0
NAD	0	0	0	0	0	−1	0	−4.6	1	0	0	1	0	0	0
NADP	0	0	0	0	0	0	0	0	0	1	−2	0	0	0	0

We now have a system of 13 linear equations with 14 unknowns (i.e., 15 − 1 because q_{glc} is known). Since there are fewer equations than unknowns in the system, it is an underdetermined system with its degree of freedom being 1 (14 − 13). We will treat this as an optimization problem and find the condition that the conversion yield to 13PD is maximum in the solution space. We will search for the optimal values (i.e., maxima or minima) of an *objective function* that is subject to *equality* and/or *inequality constraints*. In our system, the objective function is the 13PD production flux (q_P). It is constrained by a system of stoichiometric equations and bound values of fluxes. The optimization problem for the third-generation process can be written as:

Maximize $c^T \cdot X$
Subject to $A \cdot X = 0$
And $LB \leq X \leq UB$

where c is the objective coefficient vector (1×15); A is the stoichiometric matrix (13×15); X is the flux vector (15×1); and LB and UB are vectors of lower and upper bounds of fluxes, respectively (15×1).

Now we will use a linear optimization algorithm in MATLAB (i.e., interior-point-legacy linear programming) to find the maximum conversion yield from glucose to 13PD. First, we define the upper and lower bounds for each flux.

The upper and lower bounds are set for each flux. Fluxes of irreversible processes (i.e., $J_1, J_3, J_6, J_7, J_{10}, q_{ATP}$, and q_P) have their lower bounds of 0 for their nonnegative nature.

The upper bound of J_7 is set at 100 since 100 is the value of the first-generation process; reducing it will increase carbon flow to 13PD. The upper bound of 13PD production flux (q_p) is set at 200, the maximum possible by carbon balance. The bounds of other fluxes can be set at an arbitrarily wide range (e.g., from $-100/100$ to $-1000/1000$). This is to ensure that the system is not overly constricted to result in a nonoptimal solution or no solution. The objective coefficient and bound vectors of the system are shown here:

$$c^T = \begin{bmatrix} 0 \\ 0 \\ 0 \\ 0 \\ 0 \\ 0 \\ 0 \\ 0 \\ 0 \\ 0 \\ 0 \\ 0 \\ 0 \\ 0 \\ 0 \\ 1 \end{bmatrix}; LB = \begin{bmatrix} 100 \\ 0 \\ -100 \\ 0 \\ -100 \\ -200 \\ 0 \\ 0 \\ -200 \\ -200 \\ 0 \\ -1000 \\ -100 \\ 0 \\ 0 \end{bmatrix}; UB = \begin{bmatrix} 100 \\ 100 \\ 100 \\ 100 \\ 100 \\ 200 \\ 200 \\ 200 \\ 200 \\ 200 \\ 100 \\ 1000 \\ 100 \\ 1000 \\ 200 \end{bmatrix}$$

The algorithm searches for a solution with the objective function of maximizing q_p. The resulting flux of the solution is shown as a vector here:

$$X = \begin{bmatrix} q_{GLC} \\ J_{1b} \\ J_2 \\ J_3 \\ J_4 \\ J_5 \\ J_6 \\ J_7 \\ J_8 \\ J_{9b} \\ J_{10} \\ J_{11} \\ J_{12} \\ q_{ATP} \\ q_P \end{bmatrix} = \begin{bmatrix} 100 \\ 100 \\ 27.82 \\ 75.94 \\ 68.42 \\ 31.58 \\ 31.58 \\ 31.58 \\ 144.36 \\ 144.36 \\ 72.18 \\ 32.48 \\ 24.06 \\ 0 \\ 144.36 \end{bmatrix}$$

The calculated yield is $\frac{q_P}{q_{glc}} \times \frac{3C}{6C} = 72.18\%$ carbon.

After solving the flux balance equations, it is a good practice to plot the flux data on the metabolic chart for visualization. Figure 12.15 shows the molar fluxes of the third-generation process. It can be seen that some flux toward the TCA cycle is necessary to supply ATP, even though the use of NADPH has reduced the need of NADH generation.

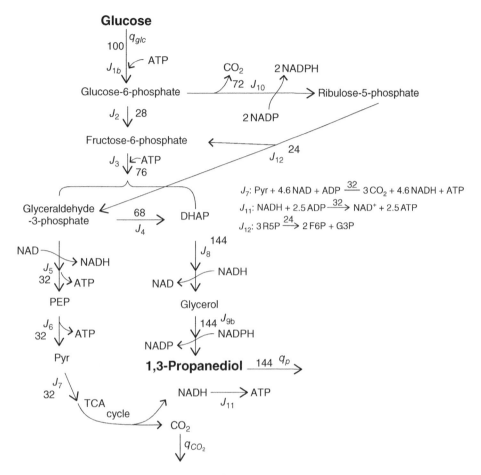

Figure 12.15 Metabolic fluxes in the metabolically engineered 1,3-propanediol pathway when the conversion yield is maximum.

12.6 Synthetic Biology

The effort to engineer metabolic pathways is turning increasingly toward synthetic methods. Looking forward, scientists have also begun to explore synthetic approaches for constructing synthetic control circuits. They are even contemplating the construction of a purely synthetic microorganism. We will discuss this further in the remainder of this chapter.

12.6.1 Synthetic (Cell-Free) Biochemical Reaction System

To synthesize a cell from scratch, one would need the soluble cytoplasmic contents of a cell. This would include the machinery for transcription, translation, DNA replication, energy generation, a cell membrane, and a genome.

Before we can "create" a synthetic cell, scientists have explored a cell-free synthesis system. Cell-free systems have long been used for translating proteins from transcripts.

This type of system includes the plasmid that is used to synthesize the mRNA transcript of the protein product, the amino acids, tRNAs, and the translation machinery (including ribosomes, rRNAs, amino acids, and ATP). It includes almost every element of a cell except for a cell membrane and a genome, and cannot replicate to produce more of itself.

Because each peptide bond requires at least 2 ATP (for "charging" the tRNA with an amino acid), ATP is in high demand. A cell-free protein synthesis system is supplied with an energy source (glucose), just like a microbial culture. The energy pathway components include glycolysis, the TCA cycle, electron transfer, and oxidative phosphorylation enzymes. Also included are other proteins needed to convert glucose to CO_2, in order to recharge ATP from ADP/AMP. Nucleotides for RNA synthesis are also supplied. All of the components are derived from *E. coli* whole-cell lysate. Using such a system, a target protein can be produced in the cell-free bioreactor and accumulate to high levels. The system is attractive for the production of proteins that are toxic to the cell.

The cell-free synthetic system is not limited to producing proteins. Using transcripts for multiple enzymes, the possibility of using a cell-free system for biochemical production is also being explored (Figure 12.16). The cell-free system gives us an opportunity to design a complex biochemical system by reconstituting major cellular components. It can perform all biochemical aspects of a cell, except for DNA replication.

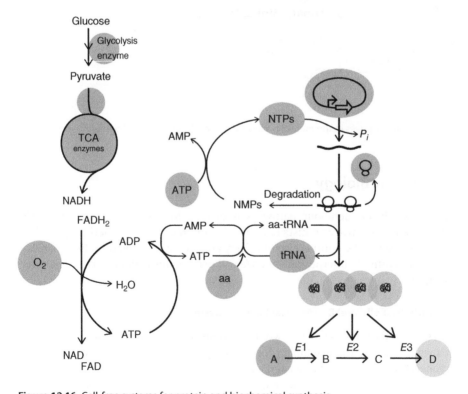

Figure 12.16 Cell-free systems for protein and biochemical synthesis.

12.6.2 Synthetic Circuits

12.6.2.1 Artificial Genetic Circuits

In the microbial world, there is a very large collection of regulatory DNA/RNA and protein elements that control gene expression and allow the cell to respond to changing conditions. Such DNA/RNA elements would include promoters and operators; protein elements would include repressors and transcription factors.

Only a few of these elements are well characterized. Using a synthetic biology approach, a small number of well-characterized regulatory elements can be cleverly combined to create various control circuits. This can exert negative and positive controls (repression and induction, respectively) on gene expression. These circuits can give rise to a wide range of dynamic behaviors.

Important to this approach is the BioBricks project. The core of the project is a collection of different DNA elements (or "bricks") that conform to certain restriction-enzyme assembly standards (http://partsregistry.org). All DNA elements have been constructed following set rules. They have specific restriction sites at their ends so that they can be efficiently stitched together in the right order. Each DNA element is a part of the larger assembly. The bricks are divided into categories of parts (e.g., promoters, terminators, ribosome biding sites, plasmid backbones, and primers) that constitute a full translational unit (i.e., protein) (Figure 12.17).

Individual parts of the BioBrick can be assembled into circuits. That circuit can then be combined with an "expression unit" to become a device to deliver an output (like protein synthesis). This is the response to an input(s) of an external inducer or repressor molecule(s).

A gene can be made to periodically oscillate with varying amplitude and frequency. Using a pair of inducers/repressors, a Boolean-type (such as "A and B" or "A not B") gate can be created. For example, the gene is turned on only when inducers A and B are both present, or when only A is present but not B. These devices can function as sensors (measurement), inverters, protein generators, and more. A collection of circuits and devices can be assembled into a system, just like a production plant. The system can potentially be used in an artificial cell or a reduced cell for the manufacturing of products.

12.6.2.2 Synthetic Signaling Pathway

Microbial cells respond to signals produced by their own population or even by other species within their ecosystem. A signal may elicit coordinated actions among members of a population. For example, in quorum sensing, cells use signaling molecules to gauge their own density. This is to synchronize their behavior when cell density reaches a particular level. Such quorum sensing is seen in biofilm formation and in the onset of antibiotic biosynthesis. As the cell concentration increases, a high signaling molecule concentration triggers biofilm formation. In antibiotic biosynthesis, the accumulation of a signaling molecule triggers cells to synchronously produce the antibiotic.

Signaling pathways play crucial roles in regulating cell growth, differentiation, metabolism, and immune response. A signaling system typically consists of a signaling molecule(s), a surface receptor(s) on the membrane, binding- and signal-transmitting proteins that trigger cellular response (such as phosphorylation), and an actuator that

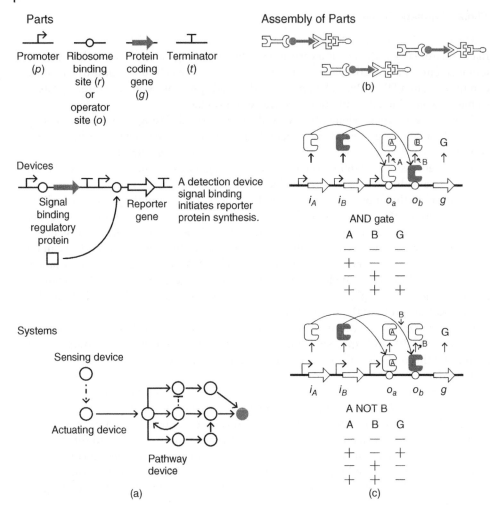

Figure 12.17 Concept of BioBricks and *in vitro* assembly of DNA fragments. (a) DNA parts, devices, and systems. (b) Lego-like assembly of DNA fragments *in vitro*. (c) Logic gates formed from BioBricks.

initiates specific endpoint events, like the initiation of transcription. A control circuit that regulates cellular behavior may include one or more such signaling pathways (Figure 12.18).

A signaling control system can be used to elicit communal responses, such as synchronizing the population to respond to an intrapopulation or interpopulation cue. It may even allow for communal manufacturing, in which different members serve complementary roles in production. For instance, one subpopulation may produce an intermediate that another subpopulation develops into a product. The synthesis occurs only when one of the subpopulations reaches a threshold density. This may allow for each subpopulation to operate under its own intracellular conditions that are optimal for its own task. This is similar to a modern chemical plant with multiple reactors for reactions that require different conditions.

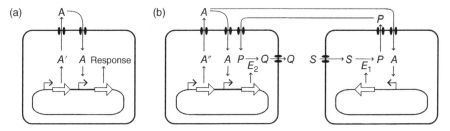

Figure 12.18 Synthetic circuit coupled signaling. (a) Intrapopulation signaling. The signal may be processed as it is secreted. When it accumulates to high levels, the signal triggers the expression of genes that cause the population density-dependent response, such as biofilm formation and antibiotic biosynthesis. (b) Interpopulation (community) signaling. An example of cooperative chemical synthesis from raw material *S* to product *P*.

12.6.3 Synthetic Organisms

One long-term goal of synthetic biotechnology is to create a self-replicating bioreactor. This would entail wrapping a cytoplasmic membrane around a synthetic biochemical reaction system, along with a synthetic genetic and signaling control system. The system would autoreplicate, like a living cell.

Some have envisioned a "top-down" approach. Starting from an existing cellular system, non-essential genome regions would be stripped off. The DNA assembly that carries a new synthetic capability would be inserted. This approach seeks to build a "new" cell using an old cell body. In contrast, another possible approach is "bottom-up." This starts from scratch and seeks to create an artificial cell with the desired, synthetic potential. For the top-down approach, we must first seek insight into the genes that are essential for cell survival. Then, we can develop methods for eliminating the non-essential genes from the genome of the target cell. To use a bottom-up approach, we must understand how a cell can be created from a solution of biological molecules.

12.6.3.1 Minimum Genome and Reduced Genome

The size of an *E. coli* genome varies among different strains, ranging from 3.6 Mbp [3.6×10^6 base pairs (bp)] to 4.6 Mbp. Over 50 different strains of *E. coli* have been sequenced. The number of genes in the *E. coli* genome is estimated to be about 4000–5500, but only 20% of the genome sequences are present in all strains that have been sequenced so far. Many genes in the genome are thus not "essential" for cell survival in the laboratory, or are functionally substitutable by other genes. This set constitutes a "reduced genome," in which genes that are not essential for cell growth in a controlled environment are deleted.

Research regarding which genes are absolutely necessary for a self-replicating microorganism (excluding viruses) has been performed on parasitic microorganisms in the genus *Mycoplasma*, particularly the species *Mycoplasma genitalium*. Mycoplasma cells lack a cell wall and many biosynthetic pathways for synthesizing building blocks. They have a very small genome compared to other bacteria. By inactivating the genes one by one, a subset of 381 genes was identified as being essential for growth.

Choosing an appropriate method for identifying these "essential" genes highly depends on the cell host and culture conditions used. The outcome may also be affected by the method of gene inactivation, since some methods may still allow some protein domains to remain partially functional. It is not surprising, then, that the minimal gene sets identified by different laboratories are not entirely identical. For instance, the estimated number (from different investigators) of essential genes in *M. genitalium* has ranged from 270 to 381. These essential genes are known to be involved in energy metabolism, transport, protein synthesis, the structure of the cell envelope, and nucleic acid metabolism. Interestingly, a large number (~40) are not annotated, and their functions are not well documented (Panel 12.3).

Panel 12.3 Functional Groups in Minimum Gene Sets of Parasitic Microorganism Genes That Are Essential for Proliferation in a Stress-Free Environment Containing All Nutrients

Function	Approximate number of essential genes
Protein synthesis	10–90
Transport and binding	35
Cell envelope	35
Energy metabolism	25
DNA metabolism	20
Protein processing	20
Nucleotides	15
Transcription	10
Others	30
Unknown/hypothetical	80

Similar efforts have also been undertaken in *E. coli* and yeast cells. By systematically knocking out genes in *E. coli*, it was determined that out of about 4000 total genes, only 1617 are essential for growth in rich medium. Such studies may not identify all essential genes, as a gene may not be essential as long as a compensatory gene is present, or a gene may be essential only under some culture conditions. Subsequently, double and triple knockouts of almost all combinations of gene pairs have also been carried out.

12.6.3.2 Chemical Synthesis of a Genome

For the generation of a synthetic life form, viruses offer the advantage of having rather small genomes. Importantly, a virus acquires its membrane or envelope in the host cells that it infects. Therefore, to produce the "synthetic" virus, one needs only to synthesize the nucleic acid genome for transfection into the host cell. Chemical synthesis of a viral genome, followed by the generation of infectious virus, has been carried out in laboratories. This includes the poliovirus, whose genome is made of an RNA molecule. After chemical synthesis of the DNA counterpart of the poliovirus genome, this nucleic acid was transcribed *in vitro* to make infectious viral RNA, which was then used to make infectious viral particles within human cells.

The chemical synthesis of bacterial genomes and the reconstitution of a bacterial cell are far more complicated than making a "synthetic" virus. Even a small bacterial genome of *Mycoplasma* (genome size ~800 kbp) was too large for chemical synthesis or polymerase chain reaction (PCR) amplification as a single molecule. One bottleneck in the synthesis of large DNA segments (too large to be amplified by PCR) is the need to insert the DNA fragment into a vector and clone it into bacteria. By growing a large quantity of the bacteria, one can then isolate the DNA fragment for further manipulation. This step is labor-intensive and costly.

Another challenge is to maintain different DNA fragments in the correct order when stitching them together. One brilliant way of designing the fragment is to engineer its ends like Lego toys, so that they can only be connected to matching parts. This approach has allowed chemically synthesized DNA segments to be connected within large pieces before they are cloned into bacteria for amplification. Using this approach, scientists assembled a *Mycoplasma* genome by connecting chemically synthesized ~24 kbp segments into sections of <100 kbp in the correct order. This was followed by cloning into a bacterial artificial chromosome (BAC), a vector that can incorporate large chunks (Mbp) of DNA. This process allowed those segments to be amplified. It also provided sufficient quantities for the final assembly of the complete bacterial genome into a yeast artificial chromosome (YAC) (Figure 12.19). Further growth of these transformed yeast cells

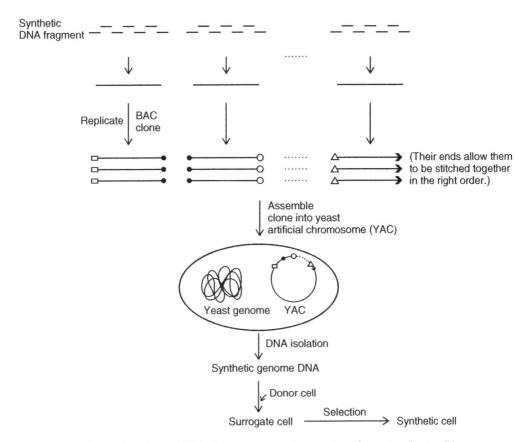

Figure 12.19 Chemical synthesis of DNA of the genome, and generation of a semisynthetic cell by replacing the genome of a surrogate donor.

then allowed a sufficient quantity of the synthetic bacterial genomes to be purified and circularized.

12.6.3.3 Surrogate Cells from a Synthetic Genome

The chemically synthesized genome of a bacterium, like its natural counterparts, has all of the information needed to make every cellular component. However, it still requires other cellular machinery to transcribe, translate, and enclose its progenies in cytoplasmic membranes. This has been accomplished by transplanting the synthetic genome into a surrogate (recipient) *Mycoplasma* cell. The recipient cell is then able to make cellular components from both genomes. After cell division, the daughter cell may contain either the host cell genome or the synthetic genome. With appropriate selective pressure, the cell carrying the original recipient cell genome can be eliminated, and those carrying the synthetic genome can be isolated. After a few more generations, all the cellular materials are derived from the synthetic genome.

12.7 Concluding Remarks

In the past century, we have already exploited the metabolic capabilities of many microorganisms. Current research owes much to this previous, exploratory work. First, scientists increased the metabolic capabilities of industrial microorganisms by biochemical means. Later, scientists began to use targeted genetic manipulation. By the turn of the century, we began to move pathways between organisms, and to mix enzymes and pathways from different species into chosen host cells.

There have been rapid advances in genomic sciences and technology, from DNA sequencing to chemical synthesis of DNA. This progress has dramatically narrowed the gap between making a blueprint for a pathway and producing product molecules in the host cell. We have even been able to create new cells with a synthetic genome using surrogate cells.

Scientists now envision a designer cell with a synthetic genome that specializes in synthesizing a novel synthetic product. These designer cells are capable of replication and production only in a bioreactor environment. They may be tailored to execute tasks that cannot be undertaken by ordinary microbes, in specialized environments. Synthetic biology may be the start of a completely new era of biotechnology.

Further Reading

Dietrich, JA, Fortman, JL, Juminaga, D and Keasling JD (2011) Microbial production of plant-derived pharmaceutical natural products through metabolic engineering: artemisinin and beyond. In: *Biocatalysis for green chemistry and chemical process development*. Eds. Tao, J and Kazlauskas, RJ. John Wiley & Sons.

Kazlauskas, RJ and Kim, BG (2011) Biotechnology tools for green synthesis: enzymes, metabolic pathways, and their improvement by engineering. In: *Biocatalysis for green chemistry and chemical process development*. Eds. Tao, J and Kazlauskas, RJ. John Wiley & Sons.

Lee, JW, Na, D, Park, JM, Lee, J, Choi, S and Lee, SY (2012) Systems metabolic engineering of microorganisms for natural and non-natural chemicals. *Nature Chem. Biol.*, **8**, 536–546.

Minty, JJ, Singer, MC, Scholz, SA, Bae, CH, Ahn, JH, Foster, C, Liao, JC and Lin, N (2013) Proc. *Natl. Acad. Sci.*, **110**, 14592–15497.

Nakamura, CE and Whited, GM (2002) Metabolic engineering for the microbial production of 1,3-propanediol. *Curr. Op. Biotechnol.*, **14**, 454–459.

O'Brien, EJ, Monk, JM and Palsson, BO (2015) Using genome-scale models to predict biological capabilities. *Cell*, **161**, 971–987.

Swartz, J (2012) Transforming biochemical engineering with cell-free biology. *AICHE J.*, **58**, 5–13.

Wendisch, V (2014) Microbial production of amino acids and derived chemicals: synthetic biology approaches to strain development. *Curr. Opin. Biotechnol.*, **30**, 51–58.

Problems

A. Synthetic Pathways

A1 Citric acid is a major industrial chemical. It is biologically produced using yeast (*Saccharomyces cerevisiae*) or molds (*Aspergillus*). Propose a metabolic engineering strategy to generate an overproducer of citrate. You only need to discuss the biosynthetic pathway. There are other aspects that should be considered, including transport and acidity in the cell, but your answer should just focus on engineering the pathway.

A2 A metabolic reaction that generates product F is shown here:

$$A \xrightarrow{E_1} B \xrightarrow{E_2} C \underset{E_3}{\xrightarrow{(C+D)\to F}} F \xrightarrow{E_4} G$$

$$\uparrow \quad \downarrow$$
$$D$$

C and D together are used to make F, where D is supplied externally. F is secreted out of the cell, but is also further converted to G. E represents enzymes: $E1$ is inhibited by F, and $E3$ is induced by D. You are asked to use metabolic engineering to make the cell overproduce F. A and D can freely enter the cell (i.e., their transport rate can be assumed to be very high). Describe your metabolic engineering strategy.

A3 In a microbe, compound A is the substrate of two enzymes, E_1 and E_2, to produce the products C and D, respectively. Under the condition of 0.2 mM A, C and D are both produced at 1 mmol/L · h under normal culture conditions. The K_m of A for E_1 and E_2 equals 0.1 mM and 0.2 mM, respectively. You want to produce four times more C than D using metabolic engineering, simply by amplifying one of the two enzymes. The supply rate of A is unchanged. Which enzyme do you choose, and by how much should you amplify it?

A4 A microorganism uses the aspartate family amino acid biosynthesis pathway to synthesize lysine and isoleucine. The pathway is the same as shown in Figure 12.11 except that the only feedback regulation controlling the synthesis is the inhibition of homoserine dehydrogenase by isoleucine. The (K_m, r_{max}) for Enzyme II (as shown in Figure 12.17) and homoserine dehydrogenase (HD) at the aspartate semialdehyde branch are (0.1 mM, 5 mM/h) and (0.3 mM, 1 mM/h), respectively, when there is no isoleucine present. The organism now produces lysine at 3.75 mM/h, but virtually secretes no isoleucine. All the leucine synthesized is used for its own growth, which is estimated to be 0.1 mM/h. The intracellular concentration of aspartate semialdehyde is measured and determined to be 0.3 mM.

The flux upstream is now deregulated, and the r_{max} for aspartate kinase is increased by 30 times. You are asked to apply contemporary metabolic engineering techniques to increase the lysine production by 15-fold, to take advantage of the increased supply of aspartyl phosphate. The idea is to keep the intracellular level of isoleucine and its synthetic flux constant, while increasing the flux of lysine synthesis. How would you do it?

A5 Under certain conditions, yeast cells can produce glycerol as a way of regenerating NAD for glycolysis. They can convert dihydroxyacetone phosphate to glycerol-3-phosphate. The phosphate is then removed to become glycerol. The enzymes involved in these reactions are glycerol-3-phosphate dehydrogenase and glycerol-3-phosphate phosphatase, respectively.

a) You are asked to metabolically engineer the organism so that it can convert most of its glucose to glycerol, instead of ethanol and CO_2, under anaerobic conditions. Propose a strategy that allows you to achieve this. You can amplify, knock down (i.e., suppress expression), or knock out (i.e., eliminate) genes using genetic engineering.

b) Using your approach, how much glycerol can be produced from each mole of glucose? Are any other by-products (such as CO_2) produced? Show your reasoning or calculations.

A6 The first enzyme in the amino acid biosynthetic pathway of the aspartate family is aspartokinase. It is a controlling step in lysine production, and it is both inhibited and repressed by lysine. For lysine production using a recombinant cell, what can you do to overcome this bottleneck to generate a hyperproducer? Propose a metabolic engineering strategy to boost productivity.

A7 A microorganism has been engineered to overproduce gluconate from glucose. Recall that phosphogluconate is the reaction product of G6P (glucose

6-phosphate) oxidation (catalyzed by the enzyme G6P dehydrogenase). By cloning phosphogluconate phosphatase into the organism, a portion of the phosphogluconate is diverted to gluconate instead of going to the subsequent reaction of phosphogluconate dehydrogenase to form ribulose-5-phosphate (see the pathway in Figure 3.5). In a culture of this microorganism, G6P is consumed at a rate of 100 mM/h in the PPP, while gluconate is produced at 80 mM/h. The r_{max} and K_m of G6P dehydrogenase, phosphogluconate phosphatase, and phosphogluconate dehydrogenase have been measured to be (400 mM/h, 0.3 mM), (300 mM/h, 0.2 mM), and (500 mM/h, 0.05 mM), respectively. With this set of enzyme data, how fast is gluconate being produced?

Now the organism is to be further engineered so that 80% of G6P goes to gluconate, instead of downstream to ribulose-5-phosphate. You want to further amplify phosphogluconate phosphatase. How will you accomplish this? You can assume the NADP level does not affect the reaction rates.

A8 A pathway that converts A to B, and then to products C and D, is shown. C exerts an inhibitory effect on the formation of B. C is excreted to an extracellular environment by transporter T_4.

$$A \xrightarrow[r_1]{E_1} B \xrightarrow[r_2]{E_2} C \xrightarrow[r_4]{T_4} C_e$$
$$E_3 \downarrow r_3$$
$$D$$

The kinetic expressions of the reactions involved are:

$$r_1 = \frac{6.25\,A}{0.5 + A + 2C}$$
$$r_2 = \frac{22\,B}{1 + B}$$
$$r_3 = \frac{B}{0.1 + B}$$
$$r_4 = \frac{10C}{2 + C}$$

All the quantities of concentration have units of mM, and rates have units of mmol/h.

a) Under normal growth conditions, the culture produces 2 mmol/h of C in the culture. The amount of D produced is completely consumed by biosynthesis. What is the supply rate of A?

b) In an effort to increase the productivity of C, the cell has been metabolically engineered to change the enzymes E_1 and E_2. As a result, the kinetics of E_1 and E_2 are:

$$r_1 = \frac{6.25A}{0.1 + A}$$

$$r_2 = \frac{22B}{0.1 + B}$$

Why are those changes made? What is the new production rate of C and D, assuming the concentration of A remains the same as in part (a)?

c) In order to make the process economical, the production rate of D must be maintained at the original level as in (a), and the production rate of C has to be 100 times of D. Assume that the supply rate of A is not limiting. How can this be accomplished? Be specific and quantitative in your answer. When this is accomplished, what is the intracellular concentration of C? *Note:* There may be more than one way to accomplish this.

A9 Yeast cells can produce glycerol as a way of regenerating NAD^+ for glycolysis when ethanol fermentation is blocked. They convert dihydroxyacetone phosphate (DHAP) to glycerol-3-phosphate. The phosphate is then removed to yield glycerol. The enzymes involved are glycerol-3-phosphate dehydrogenase and glycerol-3-phosphate phosphatase, respectively.

A producing organism has been engineered such that the pathway from pyruvate to ethanol has been deleted.

a) It is now proposed to use an engineered enzyme that uses NADPH as a cofactor instead of the original enzyme, which uses NADH to reduce DHAP. Explain why this is better and where the NADPH will come from.

b) With this new enzyme, is it possible to channel most of the glucose to glycerol? If not, what additional step will be needed to channel the maximal amount of glucose (excluding those that generate ATP for cell growth through generation of pyruvate) to glycerol?

c) Draw the flux from glucose to indicate the flux distribution that converts all of the glucose to glycerol.

B. Synthetic Genome and Organism

B1 Mycoplasma, the parasitic microbe, has been used to identify an "essential gene set." These genes have been categorized into about a dozen functional classes, some of which have more genes than others, and are considered to be "key" functional classes. List four such pivotal functional classes, and give your reasons for why you think each class you chose is important.

B2 What is the approximate size of an *E. coli* genome? What is the size of a mammalian genome? How many times is a typical mammalian genome greater than a bacterial genome (10, 100, or 1000)? Compared to a bacterium, does a mammal have as many times of genes as its genome size difference (i.e., if a mammal's genome is 100 times larger than a bacterium's genome, does it also have 100 times more genes)?

B3 Gibson assembly is a method of assembling multiple fragments of DNA in a single reaction. Its invention in 2009 has made major impacts in synthetic biology. Use sentences or a sketch to explain Gibson assembly.

B4 If you can synthesize a minimum bacterium cell that is to be grown only in bioreactors (and therefore does not need to evolve and survive in nature) for the production of an artificial and toxic protein, what are the essential functional classes (e.g., translation and protein synthesis) that must be included in the cell's genome?

13

Process Engineering of Bioproduct Recovery

13.1   Introduction

Biological products can be derived from a wide variety of sources, including plant and animal tissues, human blood, and the production streams of bioreactors. The technology used to separate biological products has evolved over time to match changes in the sources of origin. For instance, distillation was long used to produce ethanol and other solvents, while sedimentation and centrifugation were used to isolate starch and other solids.

For many decades, precipitation was the predominant method for isolating blood proteins. Beginning in the 1950s, the large-scale biochemical manufacturing of antibiotics, statins, and amino acids prompted the development of new biochemical separation methods, such as liquid–liquid extraction, adsorption, continuous filtration, and crystallization. These unit operations can be scaled up to a throughput of thousands of liters per hour. Increased demand for enzymes further drove the development of modern chromatographic materials for the separation of proteins. However, these tools largely remained in the laboratory and in small-scale manufacturing, until the emergence of recombinant proteins in the last quarter century.

The most valuable protein products are now therapeutics. These proteins are primarily administered via injection, and require a much higher purity and more stringent quality standards than those require for traditional biochemicals and industrial enzymes. The much higher commercial value of these therapeutic products further drove the development of biomanufacturing technology. Researchers quickly realized the challenges inherent in isolating recombinant proteins from *E. coli* and from cell culture medium. A period of innovation in bioseparation technologies followed the initial commercialization of recombinant proteins, including many new chromatographic materials, new designs for chromatographic columns, and new membranes for microfiltration and ultrafiltration were introduced to the market. These advances in bioseparation played a major role in the success of modern biotechnology.

In spite of all of the technological advances that have occurred, the fundamentals of the unit operations involved in bioseparation remain the same. This chapter will give a brief overview on the practice of bioseparation. The principles of various unit operations will be highlighted here. Chromatographic separations, which form the core of modern bioseparation, will be discussed in the next chapter.

Engineering Principles in Biotechnology, First Edition. Wei-Shou Hu.
© 2018 John Wiley & Sons Ltd. Published 2018 by John Wiley & Sons Ltd.
Companion Website: www.wiley.com/go/hu/engineering_fundamentals_of_biotechnology

13.2 Characteristics of Biochemical Products

Many biotechnological products are isolated from plant or animal tissues. Examples include proteases, coloring agents, and the antimalaria drug, artemisinin. Although this list remains long, there has been a steady trend to produce plant- and animal-derived products using recombinant DNA technology. In this chapter, we will focus on recovering products from the process stream of the production bioreactor.

Reflecting the diversity of the media used in upstream production, the starting materials for bioseparation can either be rather simple or highly complex. For instance, the medium used to produce therapeutic proteins is relatively simple. Its composition is better defined because of the high value and the demand of the high purity of the product. Using a better defined medium eases downstream-removal of impurities and the high value of the product afforded it to employ more expensive medium. In contrast, the cost of goods for community biochemicals is highly sensitive to the raw materials used. Their media often contain waste products from other processes. For example, molasses (the residual materials of refining sucrose from sugarcane) and corn steep liquor (residual fluid from cornstarch processing) are often used in microbial fermentation. A fermentation process that uses cruder and less pure components will also have a larger amount of residual solids and other impurities at the end of the bioprocess. Paradoxically, many processes that employ complex starting materials and produce commodity chemicals also must keep their cost of recovery process low while still achieve a very high recovery yield in order to be economically competitive.

The product concentration achieved in most bioprocesses is relatively low, resulting in low concentrations of starting material for bioseparation (Table 13.1). Except for a few cases (including glutamic acid and lysine fermentation, and the production of some solvents like ethanol and 1,3-propandiol), the product concentration from bioprocessing rarely reaches into the molar range. For protein products, the concentration on a molar basis is even lower. Even with a highly productive bioprocess and a high antibody product concentration of 10 g/L, the product stream will still consist of 99% water and only 1% of product. Therefore, in general, the first step in product recovery is to increase the product concentration to reduce the process volume.

Table 13.1 Concentrations of product in the bioreactor.

	Molecular weight	Concentration	Approximate price
Ethanol	46	70–100 g/L	$1/kg
Amino acids	110 (Average)	20–120 g/L	$10/kg
Vitamin B12	1355	2–5 g/L	$300/kg
Cellulase	~300 K	15 g/L	
Penicillin		30 g/L	$100/kg
Recombinant antibody	180 K	1–5 g/L	$1000/g
Recombinant interferon		25 g/L	
tPA		500 mg/L	$400/mg
Polysaccharide (xanthan gum)		30 g/L	

The vast majority of biochemical and protein products produced in bioprocesses are excreted out of a cell (i.e., extracellular). Thus, the first step in the separation of these "extracellular" products is the removal of residual cells. Conversely, some products are "intracellular" and accumulate inside a cell. For example, long-chain fatty acids and lipids sometimes accumulate in a cell, and will even form droplets or otherwise visible structures. Similarly, while recombinant proteins produced in mammalian cells or fungi are often engineered to be secreted, those produced in recombinant *E. coli* are retained within the cell (Figure 13.1). When driven by an inducible promoter, the production of recombinant proteins is typically only induced after a high cell density is reached. The synthesized protein molecules accumulate intracellularly and form aggregates, called inclusion bodies, which are often visible using an electron microscope. Another way of producing recombinant protein in *E. coli* is to engineer the cell to secrete the protein molecules into the periplasm, or the space between the inner cell membrane and the cell wall.

The two different methods of protein synthesis are contrasted by very different recovery processes. With periplasmic synthesis, a gentle cell lysis breaks the cell wall and is sufficient to release biologically active protein molecules into solution. This process, however, generates product in a relatively dilute stream that also contains many other cellular materials. Synthesizing the protein within inclusion bodies, on the other hand, allows for the rapid isolation of a relatively pure form of the product. In this process, cells are lysed and the inclusion bodies can be isolated using centrifugation, from which the relatively pure protein product can be further purified.

The quality of a conventional product molecule is assessed by its chemical purity. For biotechnology products, however, purity alone is often insufficient as the quality index – protein-folding status is also important. The most important distinction between the two production methods described here is in the folding status of the protein molecule. Since the periplasmic space is more oxidative than the intracellular environment, human proteins produced in *E. coli* can form disulfide bonds and fold correctly. In contrast, protein molecules synthesized intracellularly and isolated from inclusion bodies are generally misfolded and inactive. Thus, upon recovery, these proteins need to be refolded to become biochemically active. In addition to the correct folding of protein, the quality of the product can be assessed by many other means, depending on the application. These other assessments may include:

1) *Biochemical/biological activity*: Enzymes and many other proteins are valued for the specific reaction they catalyze or the specific activity they confer. The price of the product, thus, is dependent more on its activity than its mass. For example, baker's yeast is valued for its ability to ferment when mixed into dough. For some applications, the product even does not need to be pure in its chemical composition as long as its activity is pure, or free of contaminating activities. For example, a restriction enzyme preparation that cuts DNA at a specific sequence may not need to be a highly purified enzyme, but it must not contain other DNA-cutting or DNA-modifying enzymes.

2) *Contaminant*: In addition to meeting purity requirements, a product may also need to be free of certain contaminating agents or molecules. The restriction enzyme discussed here is just one classical example. A similar example is therapeutic

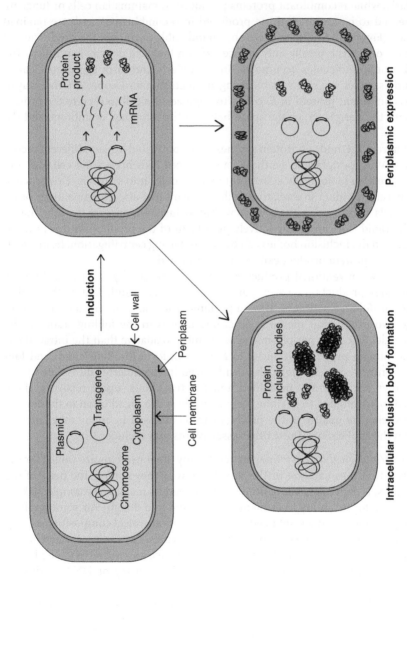

Figure 13.1 Synthesis of recombinant protein in *Escherichia coli* as intracellular inclusion bodies and periplasmic soluble molecules.

protein produced in nonhuman mammalian cells (such as Chinese hamster ovary cells). Since nonhuman protein contaminants may be immunogenic or cause other adverse effects in human recipients, therapeutic proteins produced in nonhuman mammalian cells may only carry a very small amount of host cell proteins. Likewise, blood products that are derived from pooled human plasma must be free of known viruses, even though those products are often mixtures of proteins rather than purified molecules.

3) *Structural consistency and integrity*: Chemical structural integrity has always been an important index of product quality, especially for labile compounds. For complex biologics (especially when the product consists of a virus, a cell, or its components), chemical integrity is difficult to quantify. Even then, the quantification of compositional characteristics is necessary to assess product consistency and to detect structural or compositional variants. For example, during the purification of protein therapeutics, some chemical modifications of the protein will occur at low frequency; disulfide bonds may rearrange and/or some amino acids may get oxidized. With ever-stronger instruments of detection, such as mass spectrometry, one can see the "quality" change in much greater detail than before. It is not surprising, that product integrity is becoming an even greater emphasis for modern bioseparation processes.

13.3   General Strategy of Bioproduct Recovery

13.3.1   Properties Used in Bioseparation

To separate two objects, whether visible particles or invisible molecules, there must be some difference in their chemical or physical properties (Table 13.2). Size and density are the two properties most often used in separation. Differences in size allow for separation using depth filtration and tangential-flow microfiltration or ultrafiltration, while sedimentation and centrifugation utilize both size and density differences, which cause particles to move at different velocities under a given gravitational or centrifugal field. Molecules, large and small, are also separated by charge. In the case of electrophoresis, the force field used to capitalize on mobility differences among different molecules is the electric field. Liquid–liquid extraction (both solvent extraction and aqueous two-phase extraction) takes advantage of the partition of solutes into two immiscible phases due to their propensity to "solubilize" in each phase.

Chromatography separates solutes similarly, except that this process distributes a solute through both a liquid phase and a solid (or adsorbent) phase. While the partitioning of a solute between two liquid phases (i.e., their concentration ratio at equilibrium) is often relatively constant over a wide range of concentrations and described by a single partition coefficient, the distribution of a solute between the liquid phase and the adsorbent in chromatography is affected by its concentration in a nonlinear fashion. The relationship between the concentrations in the two phases at equilibrium is called an "isotherm." A chromatographic process uses the difference of isotherms between two solutes to separate and purify them.

Table 13.2 Properties used in different bioseparation methods.

Unit operation	Properties used in separation	Application
Filtration	Size	Solid removal
Centrifugation	Size, density	Solid removal
Microfiltration	Size	Solid removal
Extraction		
Solvent extraction	Partition	Isolation
Aqueous two-phase extraction	Partition	Isolation
Absorption		
Ion exchange chromatography	Charge	Purification, isolation
Affinity chromatography	Molecular interaction	Purification, isolation
Hydrophobic interaction chromatography	Protein–ligand interaction	Purification, isolation
(Transition) metal ion chromatography	Sequence-specific tag–metal interaction	Purification, isolation
Elution chromatography (liquid chromatography and HPLC)		
Gel permeation chromatography	Size and shape of molecules	Purification
Reverse-phase chromatography	Size, molecular interaction	Purification
Chromatofocusing	Charge, mobility	Purification
Displacement chromatography	Molecular interactions	Purification
Electrophoresis	Charge, mobility	Purification
Ultrafiltration	Size	Purification, isolation
Reverse osmosis	Size, molecular diffusivity	Purification, isolation
Precipitation	Solubility	Isolation
Crystallization	Solubility, molecular interactions	Purification, polishing

HPLC = High-performance liquid chromatography.

Many properties of solute molecules and their environment can be utilized to alter the behavior of the "adsorption isotherm." For instance, ion exchange relies on varying the affinity of solute molecules to the adsorbent under different pH or salt concentrations. Gel permeation chromatography takes advantage of different pore penetration depths caused by size and topological differences of macromolecules. Hydrophobic interaction chromatography seeks to amplify the differences in the hydrophobic binding affinity of proteins under different solvent conditions. Affinity chromatography utilizes specific molecular interactions between the ligand and the solute.

Most biological molecules are soluble in aqueous environments to a varying degree. The processes of precipitation and crystallization use the "solubility property" of solutes. By changing the temperature, pH, or composition of the solution, a solute is selectively "pushed" into a region of oversaturation, thus allowing a product or a contaminant to enter a different (solid) phase and be separated.

13.3.2 Stages in Bioseparation

A typical separation process consists of a number of unit operations. Each operation takes advantage of a different property to separate the solute and the contaminant. A typical bioseparation process involves four different stages: (1) cell and solid removal, (2) product isolation, (3) product purification and (4) product polishing. Depending on the nature of separation different unit operations are suitable for use in different stages.

13.3.2.1 Cell and Solid Removal

The core of a bioprocess is the cell, which, besides synthesizing the product, can also cause its own destruction. Once the process is over and the aeration (i.e., the supply of oxygen) stops, cells begin to deteriorate. Chilling to a lower temperature slows, but does not prevent, the loss of viability and cell lysis. The release of cellular enzymes not only complicates the recovery of the product, but also may degrade or modify the product. Many process media contain solid components in addition to cells. Except for some liquid–liquid extractors that are capable of handling solids, most types of purification equipment and chromatographic columns are designed for clarified liquid and not solid-containing fluid. Thus, the first stage of bioseparation is typically to remove cells and other solids from the process stream. The unit operations frequently used at this point are depth filtration, microfiltration, and centrifugation.

13.3.2.2 Product Isolation (Capture) and Volume Reduction

In most cases, the product concentration in the process stream from bioreactors is relatively low. After the removal of cells and solids, the next task is to selectively increase the concentration of the product and reduce the volume of the process stream. This allows the unit operations downstream to use smaller-sized equipment and reduce the processing time.

For an intracellular product, the reduction of the process stream is achieved by harvesting and breaking up cells to release their intracellular product into a smaller volume of fluid. To release protein inclusion bodies, one may mechanically break the cell using a homogenizer or a stainless-steel bead mill. Alternatively, one may use an enzyme- or chemical-based method, such as lysozyme and highly concentrated guanidine chloride, to disrupt the cell and to release and solubilize the product. To release periplasmic proteins, milder osmotic shock may be applied. Many intracellular lipid products can be released by direct solvent extraction. With all of these approaches, the product is invariably released into a process stream with a smaller volume.

For extracellular products that can be favorably partitioned into a second phase (typically an organic phase, although an aqueous-liquid phase is also possible), liquid–liquid extraction can be used to transfer the product into a smaller volume of solvent. A second extraction is then employed to transfer the product solute back to a fresh aqueous phase. This requires the manipulation of the chemical environment to alter the partition behavior of the solute. First, the partition of the solute favors the solvent phase. After the transfer of most solutes to the solvent phase, the chemical environment is then altered to favor partition to the aqueous phase in the second extraction. These extraction steps can achieve both volume reduction and product enrichment simultaneously.

Volume reduction and product capture are also accomplished through the use of adsorption chromatography. Normally, a clarified stream is used to avoid clogging the packed-bed column and prolong the life of the column, and a membrane filtration step is typically placed before adsorption. This is particularly important when an expensive adsorbent for protein isolation is used. In this step, a process stream with a large volume is passed through the column, and the product is adsorbed. The column is then "washed" to remove the entrapped contaminants, and the product is then eluted into a smaller volume of elution buffer for subsequent operations.

Passing a process stream through a packed-bed column is a slow process, especially when a compressible chromatographic medium (i.e., adsorbent) is used and the flow rate needs to be very slow, such as in the case of protein adsorption. It is often necessary to concentrate the process stream before feeding it into the adsorption column. This can be accomplished by using an ultrafiltration membrane device. By choosing a membrane with an appropriate molecular weight cutoff (MWCO), the product remains in the retentate while the smaller solutes and the fluid flow through the pore of the membrane.

13.3.2.3 Product Purification

After the product is captured in a manageable volume, the next stage is to further purify it and remove contaminants. Adsorption chromatography and elution chromatography are effective and frequently used at this stage for both small-molecular-weight compounds and large proteins. Adsorption chromatography that has a high selectivity can accomplish product isolation and purification in a single step. Different chromatographic steps are typically employed in tandem. Different steps rely on different properties of the solute to achieve a high degree of purification. For example, cation exchange, anion exchange, and hydrophobic interaction chromatography can be combined into the same purification operation for improved results.

Electrophoresis is effective for separating molecules that are different in size and charge. It is an important tool that has been used for three decades to separate DNA molecules of different sizes, and has been used extensively in DNA sequencing. However, the process releases a large amount of heat and is not used at a process scale for protein isolation.

Precipitation is used to purify both small-molecular-weight compounds and proteins. To purify a product protein from a mixture, precipitation is carried out using a number of steps consisting of different temperature, pH, and solvent compositions that alternatively precipitate out contaminating proteins or product. The process is called fractional precipitation; however, precipitation alone does not generate a very high-purity protein product. It is used in the isolation of different components of blood proteins, since the presence of a small amount of other proteins does not pose any risk to the patient. For many industrial proteins that do not require a very high degree of purity, precipitation is a very effective method. Conversely, for recombinant therapeutic proteins produced in bioreactors, this method will need to be followed by other steps, such as chromatography, to obtain the high degree of purity required.

Crystallization is a more effective means of purification than precipitation. Crystals are inherently pure, barring some small amount of inclusions. Crystallization is used as a final purification step in the manufacturing of many biochemicals, including amino acids and sucrose.

13.3.2.4 Product Polishing

The product generated in the purification is typically a bulk product that still needs to be converted to the final form of product. For example, crystals harvested in purification are dried and size selected to become the final product for packaging. Sometimes, purified products are re-dissolved and crystallized into a particular crystal structure and size to meet the market need. For example, glutamic acid is converted to sodium glutamate crystals. Polishing is the final stage of bioseparation operations. Common unit operations for polishing are drying, lyophilization (freeze drying), spray drying, crystallization, and buffer replacement using diafiltration. The solution used to carry the product in the final purification step is often not the solution required for the final bulk product. For instance, in many cases, a product needs to be kept in a stabilizer for subsequent drying or for a long shelf life. Products that ultimately need to be injected into a patient must be free of pyrogens (compounds that cause fever, even when present at a very small amount), and the water used to solubilize the product must be of water for injection (WFI) grade, obtained by following stringent regulatory guidelines.

13.4 Unit Operations in Bioseparation

13.4.1 Filtration

Filtration and centrifugation are the two main methods used to remove cells and other solids from the process stream. Traditional filtration, in which a filter cloth is used to collect solid particles as the stream flows perpendicularly to the cloth, is called dead-end filtration. On the surface of the filter, the solid particles form a filter cake that serves to collect more particles (Figure 13.2). It has long been established that the velocity of the liquid flowing through the filter cake is proportional to the pressure drop across the bed, but inversely proportional to the thickness of the filter cake (Eq. 13.1 in Panel 13.1). The inverse of the proportionality constant in the equation, R, is called the cake resistance. As more fluid passes through and a thicker cake is formed, the fluid velocity will decrease unless higher pressure is applied.

A simplified analysis is to assume that the flow across the filter cake is in the viscous-flow regime, the particles are uniform spheres, and the drag force exerted by the spherical particles on the fluid can be approximated by the drag force of fluid flow through a collection of cylindrical conduits in the bed that, together, has the same void fraction and specific surface area (surface area per unit cake mass) as the cake. This analysis leads to a relationship known as the Kozeny equation (Eq. 13.2 in Panel 13.1). This equation shows that the pressure drop across the filter bed increases

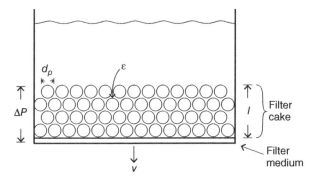

Figure 13.2 Idealized filtration process, with the filter cake formed by uniform incompressible spheres.

with decreasing diameter of the particle to the second power, but increases with the void fraction.

Panel 13.1

Darcy's Law:

$$v = K\frac{\Delta P}{\mu l} = \frac{\Delta P}{R\mu l} \tag{13.1}$$

Kozeny equation:

$$v = Z\frac{\varepsilon^3 d_p^2 \Delta P}{(1 - \varepsilon)^2 \cdot \mu l} \tag{13.2}$$

The Kozeny relationship was derived for incompressible solids. Cell biomass, proteins, and many biological materials are compressible. At a very high drop in pressure, these materials collapse into a gel, rendering the cake impermeable. Sometimes, filter aids are used and result in better filtration characteristics. In such cases, the cell particles adsorb to the porous filter-aid particles while the fluid flows through the interstitial space of the filter-aid particles.

13.4.2 Centrifugation

Centrifugation and sedimentation take advantage of the density difference between the solid particle and the fluid that triggers the particle to settle in a gravitational or centrifugal force field. At a low particle Reynolds number region, the terminal settling velocity of the particle can be derived from Stokes' law, which describes the drag force exerted by the fluid on a settling small particle. By balancing the drag force and the buoyancy force of the particle, one can obtain the terminal settling velocity of the particle (Eq. 13.3 in Panel 13.2). In a batch centrifugation, a particle is collected when it reaches the position of collection (typically, the bottom of a centrifuge tube) (Figure 13.3).

Panel 13.2

The terminal velocity of a particle with a low Reynolds number in a gravitational field:

$$v_g = \frac{d_p^2 \cdot g(\rho_s - \rho_l)}{18\mu} \tag{13.3}$$

The settling time for a vertical distance H:

$$t_\theta = \frac{H}{v_g} \tag{13.4}$$

The volume of the settler:

$$V = H \cdot L \cdot W \tag{13.5}$$

The maximum holding time for capturing a settling particle:

$$t_\theta = \frac{V}{F} \tag{13.6}$$

The maximum throughput:

$$F = \frac{V}{t_\theta} = \frac{H \cdot L \cdot W}{\left(\dfrac{H}{v_g}\right)} = (L \cdot W \cdot v_g) = v_g \cdot A \qquad (13.7)$$

Increasing the settling area by increasing the number of settling channels:

$$A_N = N \cdot A \qquad (13.8)$$

The holding time with a multiple-channel settler:

$$t_\theta = \frac{H/N}{v_g} \qquad (13.9)$$

The throughput:

$$F = \frac{V}{t_\theta} = v_g \cdot A \cdot N \qquad (13.10)$$

The acceleration in a centrifuge with a rotational radius r and angular velocity w:

$$a = r\omega^2 \qquad (13.11)$$

The terminal settling velocity:

$$v_a = v_g \frac{a}{g} \qquad (13.12)$$

The throughput in a centrifuge:

$$F = V_g \cdot \Sigma \qquad (13.13)$$

In continuous sedimentation, the direction of fluid flow and the direction of the force field that causes the settling are not parallel. The fluid flow carries a particle toward the outlet, while the force field drives the particle toward the collection plate. A particle entering the settling tank is thus moving toward the outlet, and toward the collection plate in the gravitational field. At the surface of the collection plate, the fluid velocity is zero under viscous flow conditions. If a particle reaches the surface of the collection plate before the end of the plate, it will not be carried by fluid to the exit but will settle, and is then considered to be "collected."

The minimal fluid holding time for the collection of particles that enters at the top of liquid in the inlet in the continuous settler is thus the particle-settling time (i.e., the settling distance, H, divided by the particle's terminal settling velocity, v_g, (Eq. 13.4)). The maximal capacity (or the throughput) of the settling device, in terms of volumetric flow rate, is thus the volume of the settler, V, divided by the minimal fluid holding time (Eq. 13.6). The capacity of a settler can thus be expressed as the terminal settling velocity of the particle multiplied by a surface area, A, (Eq. 13.7).

Imagine that the settling tank is now modified by adding $N-1$ plates, laid in parallel to the bottom settling plate to create N channels for fluid flow (Figure 13.4). If these N plates are equally spaced, the settling distance for the particle is now reduced to l/N, as is the settling time. As a result, the throughput is increased by N times. It can be seen that the capacity is proportional to the settling surface area (Eq. 13.10).

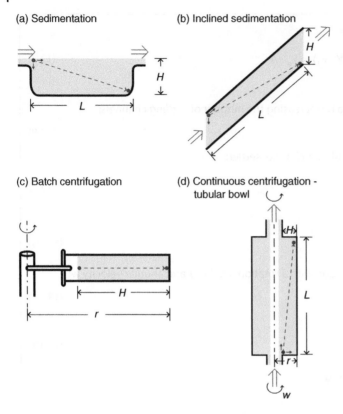

(a) Sedimentation

(b) Inclined sedimentation

(c) Batch centrifugation

(d) Continuous centrifugation - tubular bowl

Figure 13.3 Sedimentation and centrifugation in batch mode and continuous mode the dash line represents the trajectory of a particle settling to the "particle capturing surface".

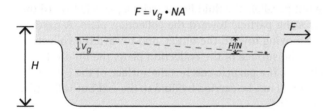

$$F = v_g \cdot NA$$

Figure 13.4 Continuous sedimentation and the effect of settling surface area.

In centrifugation, the settling velocity is increased by centrifugal acceleration, a, which is the radius of rotation, r, multiplied by the square of angular velocity, ω, (Eq. 13.11). The acceleration, v_a, is quantified by the ratio of centrifugal acceleration to gravitational acceleration, g, which is often referred to as g, multiplied by the terminal settling velocity (Eq. 13.12). The expression for the capacity of a centrifuge, F, is similarly represented by two components, the particle-settling velocity and a surface-area term (expressed as Σ, or sigma factor) that includes the relative acceleration g (Eq. 13.13). In selecting a centrifuge or a settler, one thus determines the terminal settling velocity of the particle in the process fluid, and then finds the value of the Σ factor for the required throughput.

13.4.3 Liquid–Liquid Extraction

Liquid–liquid extraction exploits the differential solubility of the product solute and the contaminating solute(s) in an immiscible, second liquid phase (usually an organic solvent or a solvent-rich phase). If the product solute is highly soluble in the second phase, but not the contaminating compounds, the extraction of the product into a smaller volume in the second phase achieves volume reduction and some degree of purification simultaneously. The effectiveness of liquid–liquid extraction is dependent on the partition coefficient, q, of the product (Eq. 13.14 in Panel 13.3) and the selectivity with respect to the contaminating solute, α, (Eq. 13.15).

Panel 13.3

Equilibrium is described by a linear isotherm:

$$q = Ky \tag{13.14}$$

The selectivity is:

$$\alpha = \frac{K_1}{K_2} \tag{13.15}$$

Material balance on the solute for batch extraction:

$$V_f y_f = V_f y_e + W_s q_e \tag{13.16}$$

Operating line:

$$q = -\frac{V_f}{W_s}(y - y_f) = \frac{q_e}{y_e - y_f}(y - y_f) \tag{13.17}$$

Extraction may be carried out in a batch extractor, which is typically a mixing tank that disperses the organic phase into the continuous aqueous phase to provide a large interfacial area for the rapid transfer of a solute into the organic phase. After a period of mass transfer, the two phases reach equilibrium. The content of the extractor is then allowed to separate into light and heavy phases for product recovery. Typically, the solute is extracted back into an aqueous buffer for further product isolation and purification.

The performance of an extractor can be analyzed using an equilibrium plot (Figure 13.5). In general, the equilibrium curve is linear until the concentration of the solute becomes very high. Along the equilibrium line, the solute concentration in the two phases is proportional, as described by the slope K. From the material balance for the batch operation (Eq. 13.16 in Panel 13.3), one can draw a line for the batch extraction to describe the material balance equation from the feed condition to the final equilibrium conditions (Eq. 13.17). The yield of the operation is thus $[(y_f - y_e) / y_f]$.

Most extraction operations are carried out in countercurrent continuous extractors, as this can give a higher degree of purity and recovery yield. The extractor is divided into a number of stages (Figure 13.6). Some mechanical (e.g., agitation) or structural (e.g., materials to promote more turbulent flow) mechanisms are provided in each stage to facilitate the two phases to reach equilibrium before the heavy and light streams exit at the opposite ends of the stage. Upon exiting, the light and heavy phases move to the next stage, as described by $n-1$ and $n+1$, respectively. On a plot of concentration, the equilibrium between the two phases is described by a line (Figure 13.7).

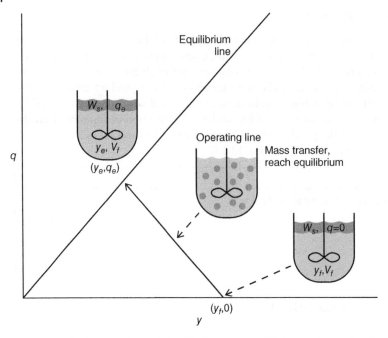

Figure 13.5 Batch extraction: relationship between equilibrium and operating (mass balance) lines.

Continuous Counter-current Operation

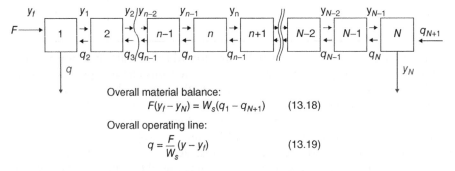

Overall material balance:
$$F(y_f - y_N) = W_s(q_1 - q_{N+1}) \quad (13.18)$$

Overall operating line:
$$q = \frac{F}{W_s}(y - y_f) \quad (13.19)$$

Figure 13.6 Depiction of countercurrent continuous extraction.

We examine a case in which a fresh light-phase solvent (no solute, $q = 0$) at a volumetric flow rate of W_s is flown in at stage N, and also a feed containing the product solute at y_f and at a volumetric flow rate of F enters at stage 1. The stream exiting stage n is given a subscript n (where y_n and q_n are solute concentrations in the heavy and light phases exiting stage n, respectively). We assume that the streams leaving stage n are in equilibrium (i.e., $q_n = Ky_n$). The overall material balance of the solute dictates that the slope of the material balance line is F/W_s (Eqs. 13.18 and 13.19). We want to determine how many stages are needed if the extraction is to achieve a 90% recovery; in other words, the heavy stream exiting stage 1 shall have $y_1 = 0.1y_f$. The material balance line (called the operating line) thus intercepts the $q = 0$ axis at $0.1y_f$, and extends with a slope of F/W_s until it intercepts the $y = y_f$ line. This ending point of the operating line is (y_f, q_1), according to the material balance. Since q_1 is the concentration leaving stage 1, it is in equilibrium with y_1. By extending a horizontal line of $q = q_1$, the intercept with the equilibrium line is (y_1, q_1).

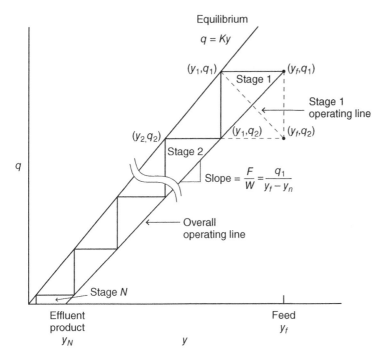

Figure 13.7 Stage-wise material balance in countercurrent continuous extraction.

Next, we perform material balance on stage 1. It can be shown that at y_1 on the operating line, q is q_2. The three points (y_f, q_1), (y_1, q_2), and (y_1, q_1) thus define stage 1; the first two points are concentrations at the two ends of stage 1, and the third is the concentrations of the exiting streams and is at an equilibrium. The procedure of drawing the horizontal and vertical lines can then be repeated to obtain the concentrations for subsequent stages, as shown in Figure 13.7.

If we increase the number of stages, but maintain the flow rates of the feed and the solvent, the operating line will shift to the left and its intercept with the $y = y_f$ line will move upward while its slope remains unchanged. Thus, the concentration of the product solute (q_1) recovered by the light stream is higher, and the remained unrecovered solute (y_N) will be lower.

13.4.4 Liquid Chromatography

Liquid chromatography and liquid-liquid extraction have very similar operating principles: both rely on the distribution of solute molecules over two phases and utilize the differences in the distribution behavior of the product solute and contaminating solutes to achieve their separation. However, in liquid–liquid extraction, both phases can be easily moved around and the equilibrium relationship is linear, while in liquid chromatography, the solid phase (adsorbent) is stationary, the isotherm is rather nonlinear, and the mass transfer process into the solid phase is rate-limiting.

Liquid chromatography is practiced in two ways, either as adsorption or as elution chromatography. Adsorption is used when both the affinity and the selectivity of the product solute are very high. Therein, a large volume of the feed is fed into the chromatographic column to allow the product solute to adsorb, while most contaminants flow through the column. After the adsorption (feeding) step, the column is washed with a buffer to remove entrapped contaminant. Afterward, the adsorbed product solute is eluted.

In elution chromatography, a small volume (usually only a small fraction of the column volume) of feed is added to the top of the column, followed closely by an elution buffer. The product solute and other impurities then move down the column with the elution buffer. Because of the differences in their affinity to the adsorbent, the product solute and impurities move at different speeds. By making the column sufficiently long, the product and impurities become separated by some distance. The portion of eluent that contains the product at a high purity is then collected at the end of the column. Liquid chromatography will be discussed in detail in Chapter 14.

13.4.5 Membrane Filtration

Membranes are used in three different types of recovery processes: (1) microfiltration uses membranes with pore sizes of 0.2–2 μm to retain microbial or animal cells. (2) Ultrafiltration employs submicron-pore membranes with different MWCOs to keep molecules larger than the MWCO in the retentate and let other solutes pass through with the permeate; both ultrafiltration and microfiltration are used extensively in bioseparation. (3) Reverse osmosis uses membranes with pores of nanometer sizes to allow water molecules to pass through while rejecting various solutes. Reverse osmosis is largely used in the preparation of pure water, or in the concentration of juices, salts, and so on. Finally, (4) nanofiltration, with the pore size in the range of 1 nm, is used in virus removal in recombinant protein manufacturing.

Most membrane filtration is operated in a tangential flow pattern. Therein, the fluid flows in a direction perpendicular to the membrane. While it flows across the length of the membrane, the pressure difference across the membrane (or transmembrane pressure, ΔP_{TM}) pushes some fluid through the pores to become permeate (or filtrate), while the rest of the fluid exits from the lumen side of the membrane downstream (Figure 13.8). The exit stream (retentate) is then returned to a reservoir, from which the stream is recirculated back to the membrane filter. After each pass, the recirculating stream loses some of its volume. In microfiltration, particles larger than the pore size are retained. In ultrafiltration, molecules or particles that are smaller than the MWCO pass into the permeate.

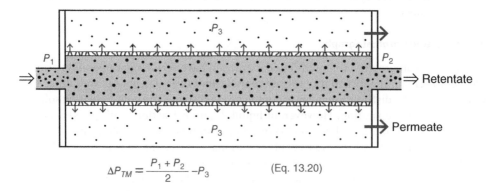

$$\Delta P_{TM} = \frac{P_1 + P_2}{2} - P_3 \qquad \text{(Eq. 13.20)}$$

Figure 13.8 Tangential-flow membrane filtration.

The filtration membrane is often configured into small tube-like fibers (hollow fibers) and packed into a casing. Sometimes, the device is organized like multiple parallel sheets of membranes, yielding more filtration surface area per unit volume. Because the diameter of the fiber or the clearance between parallel membranes is small, a large head pressure is needed to generate a fast flow rate. To drive the filtrate through the membrane, a significant drop in pressure is also necessary. The flux across the membrane is thus dependent on the transmembrane pressure drop that is often calculated by averaging the pressure drop across the membrane at the two distal ends (Eq. 13.20 and Figure 13.8).

In general, the flux across the membrane is proportional to the pressure drop over a range. At a high drop in transmembrane pressure and a high flux (even below the membrane-bursting pressure, i.e., the pressure at which the membrane mechanically breaks down), the molecules at a high flux that are rejected accumulate near the wall of the membrane. Those molecules then diffuse from the region of higher concentration, near the wall, back to the bulk liquid and are carried out by the bulk flow. This phenomenon is called concentration polarization. The high concentration of particles near the membrane may cause membrane fouling and reduces flux. Operating at a very high flux is not necessarily optimal as it may induce flux reduction. An overall better performance is attained using a more moderate pressure drop and flux.

13.4.6 Precipitation and Crystallization

Both precipitation and crystallization achieve separation by taking advantage of the solubility limit of the product solute (Figure 13.9). In order for a phase change (i.e., from a solution to a solid) to take place, the concentration of the solute is increased to exceed its solubility by changing the solvent or salt concentration, or by decreasing the temperature. Under oversaturated conditions, molecules begin to form either amorphous aggregates or ordered structures. As the molecules agglomerate, they become visible particles that grow in size and quantity, and ultimately become precipitates or crystals. Small-molecular-weight compounds have a higher propensity to form structurally ordered crystals that are of high purity. Crystallization is thus a better purification operation than precipitation. However, proteins and other macromolecules are difficult to crystallize, especially on an industrial scale.

The solubility behavior of many small-molecular-weight compounds, product solutes, and contaminants alike can be characterized easily. From their solubility profile, one can devise the precise conditions that favor crystallization of the product solute, while minimizing the precipitation of the contaminants. The precipitation or crystallization is thus an effective means for the purification of many small-molecular-weight products, such as amino acids and organic acids.

Precipitation is also used in the industrial separation of proteins, especially for products that do not require a very high degree of purity, such as industrial enzymes or human albumin purified from plasma for transfusion. To achieve a higher degree of purification, the precipitation process is often carried out in steps. The solubility behavior of a protein changes with pH, salt, and solvent concentrations and temperature (Figure 13.9). Adjusting the pH to the isoelectric point (pI) of a protein causes it to have no net charge and a low solubility, thus making it easy to precipitate. In the first few steps, pH, temperature, and other chemical aspects of the environment are adjusted to

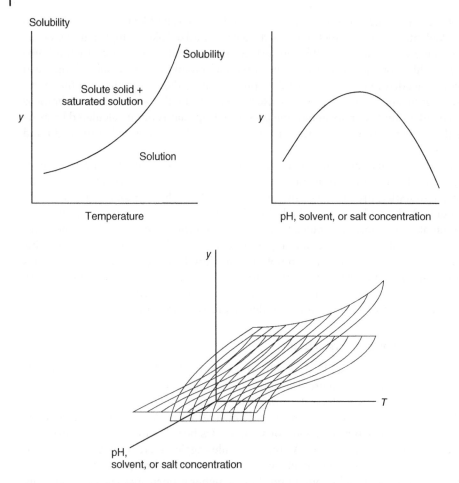

Figure 13.9 Solubility of a solute at different temperature or different pH, solvent, and salt concentration. Combining differential solubilities of two proteins (or solutes) in different temperatures and chemical conditions to achieve fractional precipitation.

create conditions that cause groups of contaminant proteins to precipitate out while keeping the product in solution. At the end, the product protein is then precipitated from the final solution. Such a process is called fractional precipitation.

13.5 Examples of Industrial Bioseparation Processes

13.5.1 Recombinant Antibody IgG

A typical process flowchart for the manufacturing of therapeutic antibody immunoglobulin G (IgG) is shown in Figure 13.10. The production cells are typically grown up in a large quantity before they are aliquoted into a number of small ampules and stored in liquid nitrogen as a master cell bank. To create a working cell bank, one of those ampules is thawed and cells are expanded to prepare more ampules. For each manufacturing run, a frozen ampule from the working cell bank is taken out to initiate the inoculation train, starting from a tissue culture flask and gradually expanding the culture volume to reach the production reactor. After production, the IgG product accumulated in the supernatant of the culture is recovered.

The cell suspension from the reactor is chilled in a holding tank to prevent cell lysis, and centrifugation is used to remove the cells. The bioreactor is then subjected to a cleaning-in-place (CIP, i.e., automated reactor cleaning using a combination of high temperatures and caustic solutions) operation, to ready it for the next manufacturing run.

To prepare for subsequent chromatographic steps and membrane processes, the particulate matters in the supernatant must be removed by microfiltration. In this operation, the permeate is harvested while the retentate (which contains particulates) is discarded. A microfiltration operation cannot process the entire volume to completion, but leaves a residual volume of retentate at the end of the operation that is too small to fill the recirculation loop. This residual volume still contains valuable product, so a small volume of buffer is added to the residual process stream to carry the remaining product into the permeate.

After the first stage of cell removal, the process stream is concentrated using ultrafiltration. For some very high-productivity processes, this step may not be necessary, as the product concentration may already be approaching a level at which IgG aggregates. This ultrafiltration step also serves as buffer exchange, replacing the culture medium that is rich in nutrients and prone to microbial contamination with a buffer saline that provides the optimal conditions for the next step of affinity chromatography.

The retentate from ultrafiltration is then fed into a Protein A affinity chromatography column. Protein A, a surface protein of *Staphylococcus aureus* that has a very high affinity to the Fc region of IgG, provides an extremely high selectivity for IgG molecules. This chromatography step not only captures the product at a very high yield, but also gives a very high degree of purification. After washing with a buffer to remove residual contaminants, the adsorbed IgG is eluted with a small volume of low-pH buffer. The recovered IgG is thus highly concentrated and pure.

The host cell used in the production, often the Chinese hamster ovary cell, is non-human. Mammalian genomes are known to contain endogenous retroviruses that may form virus particles. To reduce the risk of contamination by those viruses, the product is kept at a low pH for a few hours. Sometimes, instead of low-pH inactivation, another means of virus inactivation is used. After virus inactivation, a nanofiltration step is used to remove the remaining, inactive virus particles. Because protein aggregation sometimes occurs, a gel permeation chromatography step may also be employed to remove aggregated IgG (not shown).

Host cells may also shed cellular proteins and DNA fragments, which go on to contaminate the product. To reduce the level of those contaminants, anion exchange chromatography is used. In order to load the anion exchange column, the IgG solution must have an appropriate salt concentration and pH. Of note, a diafiltration step is often needed before ion exchange to change the chemical buffer of the IgG solution. After purification, the product solution is filter sterilized using a microfiltration membrane. Then, the buffer is exchanged to a formulation solution based on WFI.

Table 13.3 depicts a simulated product profile from a production process that employs a chemically defined medium. Note that the starting material is relatively "clean," with only a small amount of contaminating proteins. It should be noted that a very high

Example of Recombinant Antibody (IgG) Manufacturing Process

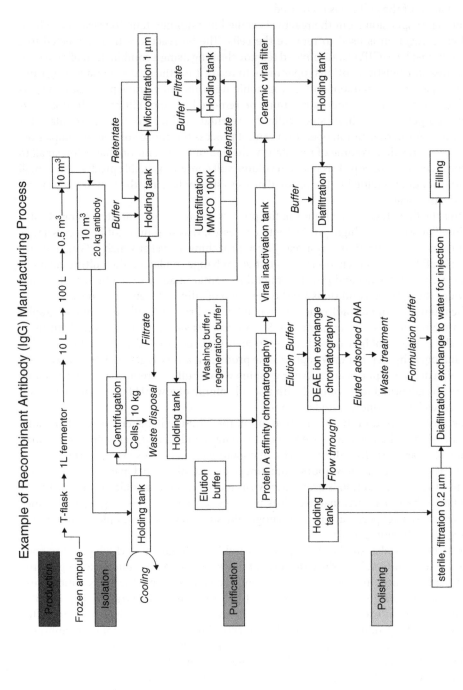

Figure 13.10 Flowchart of recombinant antibody production.

Table 13.3 Simulated concentration profile in IgG bioseparation.

Step	Volume of output stream (L)	Antibody in output stream (g)	Total protein in output stream (g)	Specific antibody (g antibody/ g total protein)	Total DNA (ng)	Specific DNA (ng/mg protein)
Cell culture broth	1000	1500	1600	0.93	–	ND
Centrifugation	950	1450	1500	0.97	–	ND
Microfiltration	1200	1420	1470	0.97	–	ND
Ultrafiltration	60	1420	1450	0.98	286	0.2
Affinity chromatography	18	1380	1390	0.99	90	0.1
Diafiltration	20	1380	1390	0.99	–	ND
Gel permeation chromatography	80	1200	1200	>0.99	90	0.2
Ion exchange	100	1150	1150	>0.99	90	0.2
Diafiltration	20	1150	1150	>0.99	9.5	0.01

ND = Not determined.

degree of product purity is required in this case. Note also that after microfiltration, a small increase in the process volume is seen because of the addition of fresh buffer solution to the retentate toward the end of filtration, in order to carry more product into the permeate and increase the recovery yield. Through affinity chromatography, the total volume decreases and the purity increases, thus reflecting its dual roles in product isolation and purification.

13.5.2 Penicillin

Penicillin G is an important precursor for making many semisynthetic β-lactam antibiotics. Its production process is very similar to that of antibodies, except that the scale is substantially larger and the medium is much more complex. It does not command the same price as biologics, and cost-of-goods is a major factor in making process decisions. The medium components are of lower purity and include less expensive by-products of other processes, such as corn steep liquor. A manufacturing seed train typically starts from spore stock, with a gradual expansion of culture volume until the production scale is reached (Figure 13.11).

At the end of production, the biomass is separated using a rotatory drum filter for continuous filtration (Figure 13.12). The drum has a filter cloth wrapped around its frame, and the filter rotates along with the frame through three sections: filtration, washing, and drying. The lower section is the filtration section. As the rotating filter cloth reaches the filtration section, it also reaches into the reservoir that contains fermentation broth. A vacuum is then applied to suck the broth through the filter and into the drum, and to deposit a cake on the filter medium. The cake thus gets thicker as the cloth rotates from the front to the back of the drum. After the filtration section, the filter cloth carries the cake into the washing section, where water is spread on the cake and vacuum is applied to suck the water though the cake. The washed cake is then carried to the drying section,

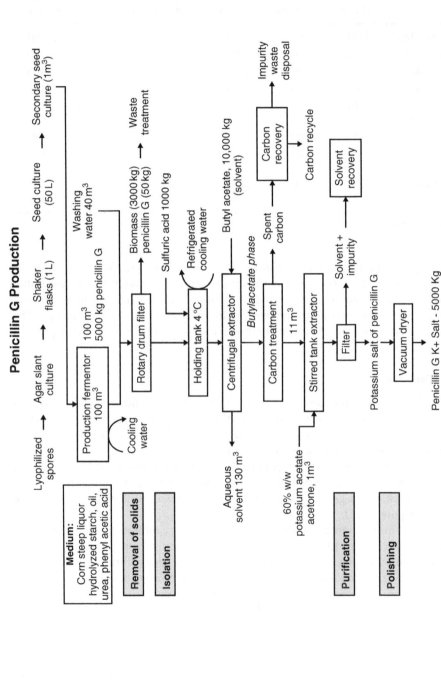

Figure 13.11 Flowchart of penicillin G production.

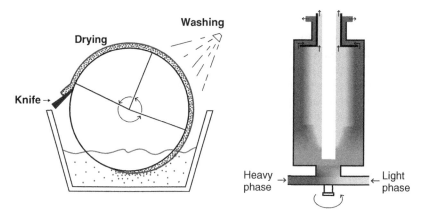

Figure 13.12 Continuous rotary drum filter and continuous centrifugal extractor.

where the air passing through the cake by vacuum suction dries it. Upon reaching the front of the drum, the cake is peeled off by a knife blade.

The penicillin G molecule has a carboxylic group that dissociates at neutral pH to carry a negative charge. At a low pH, it is undissociated and carries no charge. When it carries no charge, its solubility is high in an organic solvent but is low in an aqueous solution. After filtration, the pH of the filtrate is adjusted to a low value to allow the uncharged penicillin molecules to partition into butyl acetate. However, penicillin G is pH sensitive (that is why it cannot be taken orally) and is degraded rapidly at a low pH. The extraction process is thus carried out using a centrifugal extractor to allow for rapid extraction (Figure 13.12). In a modern manufacturing process, the biomass filtration step is bypassed. Rather, the whole broth is fed into the extractor. The biomass is then discarded with the extracted broth.

The heavy (aqueous) stream and light (solvent) stream both enter the extractor and rapidly mix together. In a centrifugal field, the heavy phase moves outward while the organic phase moves inward, and it becomes two streams again at the exit. Penicillin is then extracted from the aqueous phase into the organic phase. The volume of the organic phase is much smaller than that of the aqueous phase, so the penicillin becomes concentrated. Most of the contaminating solutes in the fermentation broth are not soluble in the organic solvent and are left behind. Subsequently, penicillin is extracted back into an aqueous phase with a neutral pH. The spent solvent is then recovered by distillation for reuse. In this extraction step, penicillin G is isolated into a smaller volume and a high degree of purification is also accomplished.

An adsorption step of active carbon is often used to clean up the solution. Active carbon adsorbs impurities and colored materials. Afterward, filtration is used to remove active-carbon solids. Since penicillin G is used mostly as a raw material to make other more valuable antibiotics, it does not need to be at a high degree of purity. After it becomes precipitated as potassium salt and recovered by filtration, the dried penicillin is packaged as a bulk product.

13.5.3 Monosodium Glutamate

Monosodium glutamate is a commodity food additive. Its commercial value is only in the range of a few dollars per kilogram, so its fermentation cost has to be low and its recovery yield must be high in order to be economically competitive. A key to achieving a high recovery yield is to reduce the number of unit operation steps, or at least reduce the amount of unrecovered product that is discarded instead of being recycled into the process stream. Even if the yield in each single step is seemingly high, after successive steps, the yield becomes low. For example, a combination of five recovery steps with a recovery yield of 95% in each step gives an overall yield of only 77% ($=0.95^5$). To increase the yield of recovery, one has to reduce the number of steps, or recycle the "waste" stream and discard only a small amount of waste stream that contains glutamate.

After fermentation, the broth is heated to evaporate water (Figure 13.13). By lowering the pH and decreasing the temperature to decrease solubility, glutamic acid begins to crystallize out of the oversaturated fluid. After the crystals are separated by centrifugation, the supernatant (which is made up of the mother liquor, cells, and other lighter solids) is discarded. This is the only major stream that allows glutamic acid to be discarded.

Two subsequent crystallization steps are carried out in neutral pH to change the crystal structure and to increase its purity. In both steps, the mother liquor is not discarded, but is used to dissolve the crystals that are harvested after the first and second crystallization steps. In this first stage of product recovery, glutamic acid is recovered at a high yield and its purity is increased drastically. The glutamic acid crystal still has color and is not in the form of sodium salt. In the second stage, the crystal is re-dissolved and active carbon adsorption is used to remove the color. Subsequently, cation exchange chromatography is used to elute glutamic acid as monosodium salt. After another crystallization and being dried, the crystals in the target size range are sieved as the product.

13.5.4 Cohn Fractionation

Cohn fractionation is a historically important method for separating proteins in blood plasma, and is used to isolate albumin. This process generated the albumin that was used in World War II for the blood transfusions of many soldiers, and has since been the foundation for many variations of blood fractionation processes used in industry today. This method employs relatively simple solvents (ethanol and acetate) that are medically safe. As can be seen in Figure 13.14, this process utilizes the different solubilities of different proteins under varying ranges of temperature, pH, and hydrophobicity, to serially precipitate out different products. Each fraction that is separated is a mixture of blood proteins with different enriched components. Fraction V is the main product, albumin; however, albumin is also precipitated out in other fractions. The recovery yield is thus low compared to modern processes.

13.6 Concluding Remarks

Biotechnological products encompass a wide range of compounds of vastly different chemical and physical properties, quality and purity requirements, and commercial values. Most of them are synthesized only at relatively low concentrations in the production process. Their recovery process is largely divided into four stages: solid

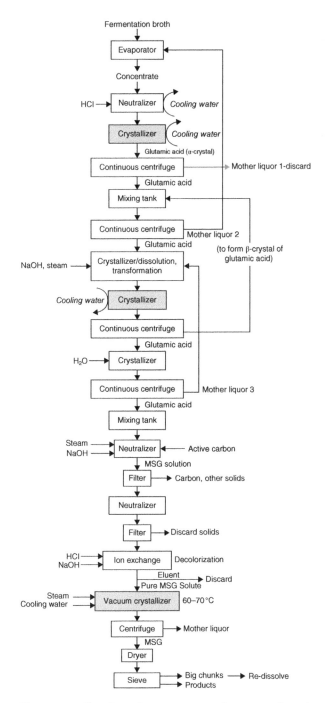

Figure 13.13 Flowchart of recovery process for monosodium glutamate production.

Cohn Fractionation of Blood Proteins

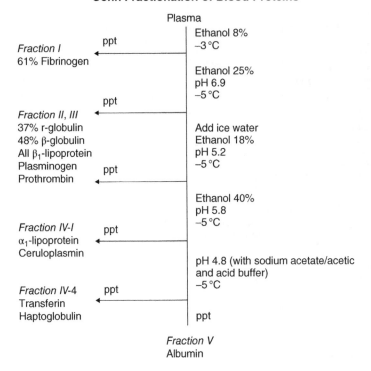

Figure 13.14 Flowchart of Cohn fractionation for blood protein production.

removal, product isolation, product purification, and polishing. Various unit operations are employed to accomplish the tasks needed for those four stages. Unit operations taking advantage of different chemical and physical properties to separate the product from contaminating solutes are combined to achieve a high degree of recovery yield. Despite the diversity of the biotechnology products, the common principles of bioseparation processes can serve as a guide in devising a recovery process of new products.

Further Reading

Belter, P, Cussler, EL & Hu, WS 1988, *Bioseparations: downstream processing for biotechnology*, Wiley, New York.

Cramer, SM & Holstein, MA 2011, 'Downstream bioprocessing: recent advances and future promise', *Current Opinion in Chemical Engineering*, 1, 27–37. Available from: http://dx.doi.org/10.1016/j.coche.2011.08.008. [19 July 2016].

Harrison, RG, Todd, P, Rudge, SR and Petrides, DP 2015, *Bioseparations science and engineering: topics in chemical engineering*, 2nd edn, Oxford University Press, New York.

Ladisch, MR 2001, *Bioseparations engineering: principles, practice, and economics*, Wiley-Interscience, New York.

Nomenclature

A	area of the settler	L^2
A_N	area of multiplate settler	L^2
a	acceleration	L/t^2
d_p	diameter of the particle	L
F	volumetric flow rate	L^3/t
g	gravitational acceleration	L/t^2
H	height of the settler	L
K	partition coefficient	$(\text{mole}/L^3)/(\text{mole}/L^3)$, $(M/L^3)/(M/L^3)$
L	length of the settler	L
l	thickness of filter cake	L
N	number of settling channels	
P	pressure	Force/L^2
ΔP	pressure drop across the bed	Force/L^2
q	solute concentration in organic phase	mole/L^3, M/L^3
R	filtration resistance	
r	radius of rotation	L
t	time	t
t_θ	holding time of the liquid stream in the settler	t
V	volume of the settler	L^3
V_f	volume of aqueous phase	L^3
v_a	particle-settling velocity with an acceleration of a	L/t
v_g	particle-settling velocity under gravity	L/t
W	width of the settler	L
W_s	volume or volumetric flow rate of solvent phase	L^3, L^3/t
y	solute concentration in aqueous phase	mole/L^3, M/L^3
y_e	solute concentration in aqueous phase at equilibrium	mole/L^3, M/L^3
y_f	initial solute concentration in aqueous solution	mole/L^3, M/L^3
Z	a constant	
μ	viscosity	$m/L{\cdot}t$
ε	voidage	
ρ_l	density of the liquid	m/L^3
ρ_s	density of the particle	m/L^3
ω	angular velocity	$1/t$
Σ	sigma factor in centrifuge design	L^2
α	selectivity	

Problems

A1 A coronavirus is produced in a microcarrier culture of Vero cells (a green monkey kidney cell line) for use as a vaccine. The microcarriers are beads of 200 μm diameter that are suspended in a bioreactor and provide surfaces for cells to attach and grow. After cells grow to reach a high concentration, seed virus is added to the culture to infect the cells and replicate the virus. After a period of time, cells are lysed and the virus is secreted into culture medium.

To recover the virus, microcarriers are first removed by filtering through a 50 μm screen, and the remaining supernatant is used in the recovery. The diameter of the virus particles is 150 nm. You are given the following process equipment. How will you arrange their sequence of use? Explain your decision. You can use any equipment multiple times if you wish.

a) Microfiltration membrane with 0.1 μm pores
b) Gel permeation elution chromatography column (retaining globular proteins with a molecular weight less than 300,000)
c) Ultrafiltration membrane with a molecular weight cutoff (MWCO) of 200,000 Daltons (pore size of 10 nm)
d) Cation exchange chromatography column.

A2 Protein X (molecular weight 100,000) is produced in yeast as an excreted product. It is purified by a series of unit operations. The concentration of Protein X is expressed in units/L. The isoelectric point of Protein X is 6.5. The feed stream (i.e., the effluent stream from the bioreactor) has the composition shown in Table P.13.1.

The unit operations used include (not in the operating sequence):

a) Ultrafiltration with a MWCO of 40,000
b) Diafiltration with a MWCO of 4000
c) Microfiltration with a pore size of 1.0 μm
d) Cationic ion exchanger with gradient elution
e) Affinity adsorption column with antibody against Protein X with maximum adsorption pk_a at 8.5.

The composition of the solution after each of these operations, but not in the right order, is listed in Table P.13.2.

Table P.13.1 Composition of a feed stream into a product recovery process.

	Per liter
Cells	1 g
Protein X	1000 units
Total protein (in solution)	2.0 g
Organic medium components (amino acids, etc.)	3.0 g
Inorganic salts	10 g
pH	7.1

Note: The feed volume is 1000 L.

Table P.13.2 The composition of a process stream after different unit operation steps.

Solution	1	2	3	4	5
Cells (g/L)	0	0	0	0	0
Protein X (units/L)	500	9800	8×10^5	9×10^6	1×10^6
Total protein (g/L)	1.0	18.5	500	560	625
Organic medium components (g/L)	1.5	1.2	0	0	0
Inorganic salts (including buffer) (g/L)	8	5	5	5	5
Volume collected (L)	1800	90	1	1.3	0.8
pH	7.0	7.0	4.0	6.0	6.0

The final bulk product is a pure Protein X solution at pH 7.4. Match the unit operation with the composition of the streams. Write your solution in a flowchart format.

Explain your arrangement and the purpose of each step. Do you need any additional steps to meet the final product specification?

A3 A virus surface antigen with a molecular weight of 100,000 Dalton was produced by recombinant mouse cells in culture. The culture fluid at the end of the production stage contained 10^{10} cells/L and 500 mg/L of product protein. The rest of the fluid consisted of approximately 900 mg/L of other proteins that were medium components or cell debris. A majority of the other proteins have molecular weight greater than 50,000 Dalton.

Since this product is for human use in an injectable form, purity requirements are very stringent. The purification unit operations used are shown below (not in the sequence used):

a) Microfiltration (pore size: 0.2 μm)

b) Cation exchange adsorption column (pH step change elution using acetate buffer)

c) Gel permeation chromatography (molecular weight range: 50,000 to 150,000 Dalton)

d) Ultrafiltration (MWCO: 10,000 Dalton)

e) Anion exchange adsorption column (NaCl gradient elution)

f) Diafiltration (this is a form of ultrafiltration, except that the retentate is recycled while a new buffer solution is continuously added; it is used frequently to exchange the salts in the solution rapidly)

g) Silica gel DNA adsorption column (DNA is adsorbed with a high selectivity).

The relevant purity data after each step of the recovery process are shown in Table P.13.3, including the specific activity (expressed as mg product/mg total proteins), the concentration of product in the fluid collected after each recovery step, and the concentration of contaminating DNA.

You are asked to reconstruct the unit operation used in the recovery based on the purity data shown in the table. For each step there may be multiple possible unit operations that can give rise to the same outcome. Give a short explanation of your chosen answer for each step.

Table P.13.3 Relevant product purity data after different unit operation steps.

Step	Product concentration (μg protein/cm³)	Purity (μg product/μg total protein)	DNA content (pg/mg)	Possible methods used
1	500	0.36	ND	
2	5000	0.35	ND	
3	9000	0.50	ND	
4	9000	0.50	ND	
5	8000	0.95	2	
6	1000	>0.99	0.1	
7	800	>0.99	0.1<	
8	3000	>0.99	0.1<	

ND = Not determined.

A4 A DNA polymerase that is stable at 90 °C is cloned into *E. coli* for production. The *E. coli* is not thermophilic; virtually all the native proteins are destroyed upon exposure to 60 °C for 30 min. The DNA polymerase is to be used in a polymerase chain reaction (PCR) to make DNA products for gene chip construction. The polymerase thus does not need to be very pure except that it should not have DNA/RNA contamination or any other DNA processing enzyme contamination. Propose a bioseparation scheme for this product.

A5 What is affinity chromatography? What does a Protein A column separate?

A6 You are asked to design a process to purify plasmids from *E. coli* cells. The plasmids are closed circular supercoiled DNA, much smaller than the bacterial genome, but much larger than proteins. This plasmid's molecular weight is 3,000,000 Dalton. Being a DNA molecule, it is also highly negatively charged because of the phosphate group. You are given a centrifuge to harvest cells, and a protocol of osmotic shock to break up cells and to release intracellular materials, including the plasmid. You are also given:
a) A cation exchanger column with a $-CH_2-SO_3^-$ functional group as its ligand
b) An anion exchanger column with $(CH_2-CH_2)_3-N^+$ on its ligand
c) A gel permeation column that separates molecules in the molecular-weight range of 100,000 to 2000 (the solutes outside this range do not get absorbed)
d) An ultrafiltration unit (you can use it multiple times).
How are you going to organize them (from centrifuge, osmotic shock process) into a process? Explain your rationale.

A7 You are asked to develop a separation process for purifying IgG molecules produced by recombinant animal cells. You are given the following equipment.

Propose a recovery process. You can use some equipment more than once. You do not have to use them all:

a) Protein A affinity chromatography column
b) Cation exchange chromatography column
c) Ultrafiltration membrane
d) Centrifuge
e) Microfiltration hollow-fiber system
f) Gel permeation chromatography column
g) Vacuum crystallization reactor.

14

Chromatographic Operations in Bioseparation

14.1 Introduction

Column chromatography is used extensively in the purification of biological products. It is a relatively cost-effective method to separate individual compounds from many others, even those with rather similar chemical or physical properties. As such, it is used to produce many fine chemicals and pharmaceuticals with a high degree of purity. Other unit operations commonly used for bioproduct recovery, such as membrane filtration and liquid–liquid extraction, cannot deliver the high selectivity of chromatography. Chromatographic separation is also suitable for large-scale operations, such as those used to produce pharmaceutical products.

The basic operation of liquid chromatography involves passing a solution through a column that is packed with a chromatographic medium, or adsorbent, made of highly porous and spherical beads. The adsorbent is often referred to as the "stationary phase." The solution that carries the solute through the column is called the "mobile phase."

The success of chromatographic separation depends upon the binding characteristics of the solutes in relation to the adsorbent. The strength of the binding, or "affinity," of the solute and the adsorbent depends on the chemical and physical properties of the solute, the solvent, and the adsorbent. The adsorbent is often made of a matrix material that is the material and mechanical core of the stationary phase. To provide greater degrees of selectivity, a matrix material may be chemically modified to incorporate specific ligand molecules or chemical functional groups. Depending on the type of ligand attached, the same matrix material can be used to make adsorbents for cation exchange, anion exchange, or other types of chromatography.

Depending on the operating mode, liquid chromatography can be run as an adsorption process or as elution chromatography. An adsorption process typically involves three stages: adsorption, washing, and elution. During adsorption, a feed solution containing a product solute is continuously passed through the column. The product, having a higher affinity to the adsorbent, binds to the column while the contaminating solutes pass through. Feeding solution flows continuously over the column until it becomes "saturated" with the solute (i.e., the adsorbent can no longer bind any more solute). In most cases, it takes a large volume of the feeding solution (many times the volume of the adsorbent in the column) to saturate the column. After saturation, the column is then washed with a buffer to remove residual contaminating molecules. Then, the product solute is released from the adsorbent by passing an elution buffer through the column.

Engineering Principles in Biotechnology, First Edition. Wei-Shou Hu.
© 2018 John Wiley & Sons Ltd. Published 2018 by John Wiley & Sons Ltd.
Companion Website: www.wiley.com/go/hu/engineering_fundamentals_of_biotechnology

During elution, the product emerges at the end of the column as an enriched or even purified stream. In some applications, the reverse occurs. For example, contaminating solutes (usually minute components in the mixture) can be adsorbed on a column, while the desired product solutes pass through and can be collected in a pure form. After the product solute is eluted, a regeneration buffer is passed through the column to revert it to the original state, ready to receive the feeding solution again.

In elution chromatography, a small amount of feed is added to the top of the column. Then, an elution buffer is used to "push" the solutes downward through the column. Solutes that have different binding affinities to the adsorbent matrix will move at individual speeds through the column and emerge at the end of the column at different times. Given the right operating conditions and timing, a product solute can be isolated in a fairly pure fraction.

While both adsorption and elution chromatography separate solutes by exploiting their differences in binding affinity, they have very different outcomes. The adsorption process takes in a very large amount of the feed and elutes the purified product in a smaller volume, thus concentrating the product. Elution chromatography, on the other hand, starts with a small volume of feed and uses a large volume of elution buffer, thus diluting the product solute (Figure 14.1).

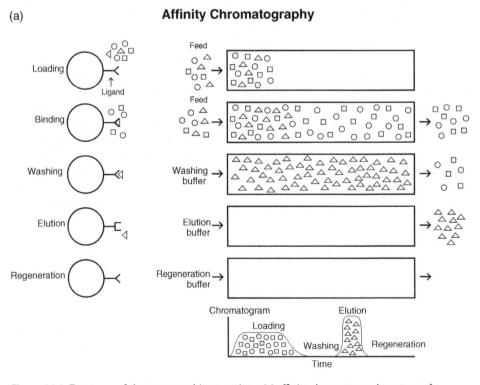

Figure 14.1 Two types of chromatographic operations: (a) affinity chromatography, a type of adsorption; and (b) elution chromatography.

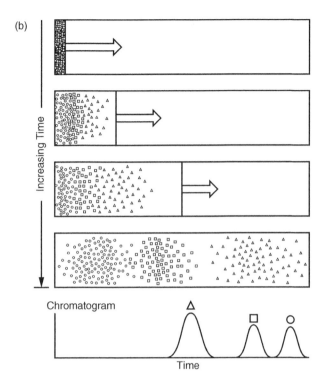

(b)

Increasing Time

Chromatogram

Time

Figure 14.1 *(Continued)*

A chromatographic system typically involves a fluid-switching mechanism to change the fluid stream from the feed stream to the washing solution, and then to the eluent buffer. After elution and before it is ready to receive a fresh feed, a regeneration buffer has to be passed through the chromatographic column to restore it to its original state. At the point that liquid exits the column, a detector is present to measure the solute concentration. It directs the flow stream to either: (1) the product channel, for harvesting the solute; or (2) the waste channel, for disposal (Figure 14.2).

In this chapter, we will review a quantitative description of the adsorption process and elution chromatography.

14.2 Adsorbent

14.2.1 Types of Adsorbent

One of the most widely used classical adsorbents is activated carbon. It can be derived from different carbonaceous materials that are treated with high temperatures in the absence of oxygen. Activated carbon is used to remove small contaminating solutes from a feed stream. It also serves to remove color, odor, and other impurities (Panel 14.1).

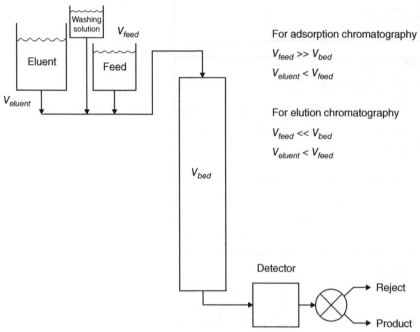

Figure 14.2 Components of a chromatographic system.

Panel 14.1

Examples of Adsorbents

- Carbon (active carbon)
  - Bone, coconut shell, or other carbonaceous material heated up to about 600 °C
  - Activation: selective oxidation of residual hydrocarbon by steam, air, or other oxidants
- Silica gel
- Ion exchange resins
  - Polyelectrolytes consist of cross-linked polymer matrices that are functionalized to provide ion exchange capacity
  - Example of matrices: styrene copolymerized with dinitrobenzene (Dowex), cross-linked Dextran, and polyacrylamide
  - Based on pK_A, three classes of ion exchange adsorbents:
    - o Strong acid cation
    - o Strong base anion
    - o Weak base anion
- Gel permeation chromatography
- Hydrophobic interaction chromatography
- Affinity chromatography

Ion exchange resins, on the other hand, are made of cross-linked polymeric matrices that are functionalized to be able to exchange ions. For instance, a cation exchange resin carries a negative charge so that it can immobilize positive counter-ions. They can then be "exchanged" with positive ions in a solution. Conversely, an anion exchange resin carries positively charged ligands on its matrices (Figure 14.3). At different pH levels, cation or anion exchange adsorbents will have different capacities for "exchanging" H^+ or OH^-, or other positive or negative ions.

The materials that make up an adsorbent, thus, often include a matrix and a functionalized group, or ligand. The matrix provides the mechanical support, the packing properties of the adsorbent that affect the fluid-flow characteristics in the column, and an internal pore structure that dictates the transport properties for the solute in the matrix. Many chromatographic matrices are made of hard materials (e.g., silica or polystyrene), which are not compressible. These materials retain their shape and porosity even under high flow rates and high-pressure conditions. These resins tend to be hydrophobic, and they are suitable for use with water and organic solvent mixtures, but not ideal for the adsorption of protein molecules in an aqueous solution.

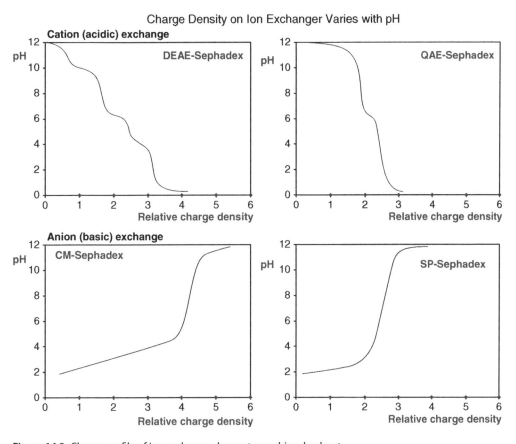

Figure 14.3 Charge profile of ion exchange chromatographic adsorbent.

To isolate proteins in an aqueous solution, other gel-like matrices are used. They are made of cross-linked polysaccharides (e.g., polyacrylamide, agarose, or a composite of those materials) and "swell" in an aqueous solution. These matrices are derivatized with different functional groups, such as diethylammoethyl (DEAE), phenyl, and butyl moieties, to give them specific binding characteristics. While these gel-type chromatographic matrices are suitable for the separation of protein molecules in an aqueous environment, they also become compressible and lose their structural integrity under high-pressure conditions. In general, the matrix material, the media structure (i.e., shape, porosity, pore size, and particle diameter), and the packing of the media give the chromatographic column its flow characteristics. In contrast, the functional group dictates its separation characteristics.

14.2.2 Ligand and Mechanism of Separation

An ion exchange adsorbent has one or more dissociable functional groups; its charge density is dependent on the pK_a of the functional group and the pH of the solution. At different pH or salt concentrations, the binding affinity between the ion exchange adsorbent and solute changes (Figure 14.3). This allows a solute to bind to the adsorbent under conditions that favor binding, and to dissociate from the adsorbent under conditions that promote dissociation.

Size exclusion chromatography separates molecules based on their size (Figure 14.4). Its media have internal pores with a restricted hydrodynamic size range. These pores

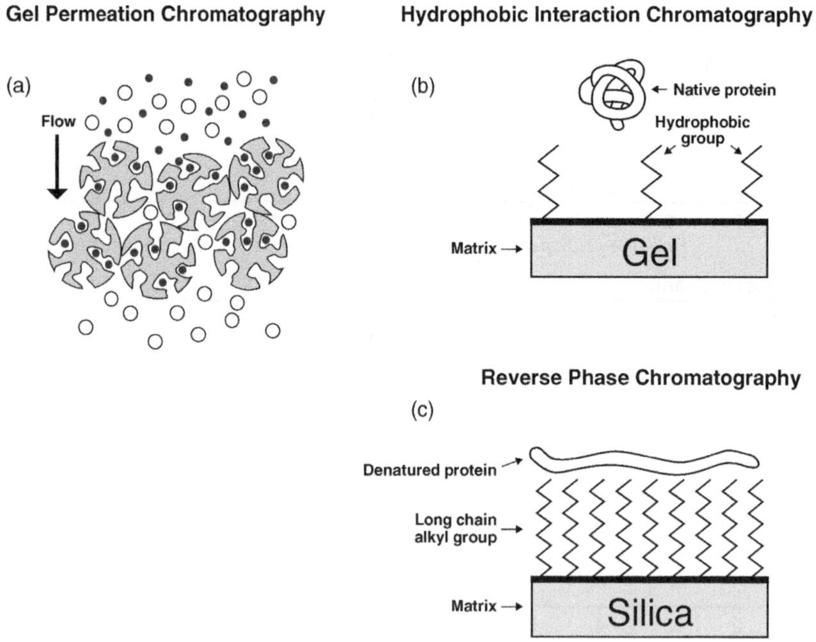

Gel Permeation Chromatography   **Hydrophobic Interaction Chromatography**

Figure 14.4 (a) Gel permeation (size exclusion) chromatography, (b) hydrophobic interaction chromatography, and (c) reverse phase chromatography.

only allow molecules that fall within the specified range to enter, and exclude all others. This effectively binds only molecules of a given size range. Unlike other adsorbents, size exclusion chromatography does not rely on a specific ligand or functional group to separate solutes. It is also sometimes referred to as "gel filtration chromatography" or "gel permeation chromatography." The latter is more often used in cases where an organic solvent is used.

Hydrophobic interaction chromatography (HIC) is frequently used in protein separation (Figure 14.4). Some amino acids in a protein molecule have a hydrophobic side chain that has a high affinity for hydrophobic ligands (such as butyl groups) in the HIC absorbent. HIC takes advantage of the variations in the amount of hydrophobic amino acid side chains between different protein molecules. To better separate proteins, HIC can also be performed under different salt conditions. Solvation is reduced in high salt conditions, and hydrophobic regions in the protein are better exposed.

The adsorbents described above are used in both adsorption and elution chromatography. Affinity chromatography is primarily used in adsorption, since it offers a very high selectivity and can achieve a high degree of purification. In affinity chromatography, the separation is accomplished by a molecular recognition between a specific solute and a ligand, such as the pairing of an antigen–antibody, enzyme–substrate, or enzyme–inhibitor.

One affinity separation process widely used in industry is Protein A chromatography. Thousands of kilograms of recombinant immunoglobulin G (IgG) produced from cell culture processes are purified annually using this method. It takes advantage of the specific binding of Protein A (in the cell wall of *Staphylococcus aureus*) to the *Fc* region of IgG molecules, and uses Protein A produced from recombinant DNA technology as the ligand of the adsorbent. The binding affinity is so high that an affinity chromatography step can achieve over 90% purity. Another high-affinity ligand is nickel (or copper) ion. It can capture protein molecules that are engineered to have an oligohistidine tag (usually four to six histidine residues). This allows a recombinant protein that has been engineered to have a histidine tag to be purified quickly for research use. Other ligand–solute pairs frequently used in affinity chromatography include antigen–antibody, enzyme–cofactor, and lectin–glycoprotein arrangements.

Because of its high selectivity, affinity chromatography is a highly desirable purification method. To find a high-affinity ligand for a given protein product, some researchers employ molecular design tools. Others use combinatorial libraries and resort to high-throughput screening. Importantly, the ideal ligand must not only have a very high affinity for binding, but also be able to later release the product under certain conditions, as will be discussed later in this chapter.

14.2.3 Types of Liquid Chromatography

Liquid chromatography is widely used for the recovery of biochemicals and proteins. It is also an important tool in chemical analysis for quickly separating and quantifying analytes. The power to separate different solutes that coexist in liquid chromatographic separation is referred to as "resolution." As will be discussed in this chapter, the resolution is greatly affected by the intraparticle mass transfer of the adsorbent. For instance, the use of smaller adsorbent beads would reduce the mass transfer limitation and increase a column's resolution. However, with a smaller particle size, the head pressure of the pump needs to be higher to maintain the same flow rate.

For faster and higher resolution operations, high-performance liquid chromatography (HPLC) is used. HPLC uses small, adsorbent particles based on silica or metals, which allows a high pressure to be applied to the column. For this reason, HPLC used to be called high-pressure liquid chromatography. Typical liquid chromatography uses adsorbents on the order of hundreds of micrometers in diameter. In comparison, the adsorbent used in HPLC has a diameter as small as 5 μL. Because the adsorbent is made of highly hydrophobic materials, HPLC is more suitable for use with less polarized buffer solutions. When even smaller particles are used (around 2 μL in diameter), the operation is often referred to as ultra-high-performance liquid chromatography (UPLC).

One specific type of HPLC is called reversed- (or reverse-) phase chromatography (Figure 14.4). Reverse-phase chromatography uses an adsorbent that is primarily silica with covalently attached, long alkyl chains, such as an 18-carbon tail (called C18). The binding affinity of solutes to these alkyl ligands is strongly affected by the polarity of the solvent. Reverse-phase chromatography differs from HIC in that it uses a different elution buffer. While HIC uses aqueous solutions to elute the bound proteins, reverse-phase chromatography employs a mixture of an aqueous solution and solvents (e.g., methanol and acetonitrile).

The elution of solutes from a chromatographic column can be performed with a single elution buffer. This is called "isocratic elution." Elution can also be performed with a series of buffers that have a gradient of solvents and chemical properties. For example, the fraction of acetonitrile or methanol in the elution buffer can be increased over time during elution, to decrease the column's polarity and modulate its binding affinity, allowing more polar solutes to be eluted first. This process can be used to increase the resolution of solute peaks. Gradient elution can also be carried out in increasing or decreasing pH or salt concentrations.

14.3 Adsorption Isotherm

14.3.1 Adsorption Equilibrium: Langmuir Isotherm

When a solution and an adsorbent are brought into contact, the solute molecule in the solution will begin to adsorb to the adsorbent. Given a sufficiently long time, equilibrium will be reached between the adsorbent and the solution. That equilibrium relationship is often called the "isotherm." How well a mixture of two solutes can be separated by liquid chromatography depends on the differences in their isotherms. In this chapter, we will study the adsorption behavior of a single solute. Keep in mind, however, that effective separation relies on the differences of the isotherms between solutes. The concentration of the solute in the solution (the liquid phase) will be given a symbol, y. That in the adsorbent (solid phase) will be q. Note that there are many different units for quantifying y and q (e.g., g solute/g adsorbent, mol/L adsorbent, mol/L solution, etc.).

The adsorption of molecules to a surface is described by the Langmuir isotherm (Panel 14.2). We assume that the total number of binding sites for the solute on the surface of a unit quantity of the adsorbent, q_m, is fixed. When in contact with the solution, a binding site is either occupied by a solute molecule (i.e., bound) or empty. The concentrations of these two types of sites are represented by q and q_0, respectively, while their sum is the total concentration of sites, q_m. At equilibrium, the relationship for the solution-adsorbent system can be written as shown in Eq. 14.1 using an equilibrium constant (K_{eq}) or its inverse (K_y). Through substitution and rearrangement, the Langmuir isotherm can then be obtained (Eq. 14.2).

Panel 14.2

Solute + vacant site $\rightleftharpoons$ occupied site

$$K_y = \frac{1}{K_{eq}} = \frac{[solute][vacant\ site]}{[occupied\ site]} = \frac{y \cdot q_0}{q} \tag{14.1}$$

$q_m = q + q_0 \qquad [total\ sites] = [vacant\ site] + [filled\ site]$

q_m represents the maximum available sites on the adsorbent:

$$K_y = \frac{q_0 \cdot y}{q} = \frac{(q_m - q) \cdot y}{q}$$

$$\therefore q = \frac{q_m y}{K_y + y} \tag{14.2}$$

The behavior of the Langmuir isotherm is similar to Michaelis–Menten kinetics for enzymatic reactions and Monod kinetics for cell growth. Note, however, that the Langmuir isotherm describes an equilibrium, whereas the Michaelis–Menten and Monod equations are related to the rate. Of note, the isotherm behaves linearly at low concentrations of solute. After a transition period, the adsorbed molecules become saturated at q_m. The adsorption of a solute to two different adsorbents, over a range of solute concentration, is shown in Figure 14.5. Though the two adsorbents have the same maximum adsorption capacity (q_m), they differ by their solute affinity. At a low solute concentration, more solute will be transferred from the solution phase to the adsorbent with a small K_y. Similar to Michaelis–Menten kinetics, one can obtain the value of the two parameters, K_y and q_m, from experimental data by plotting $1/y$ versus $1/q$.

14.3.2 Isotherm Dynamics in Adsorption and Desorption

In an adsorption process, the solute is first adsorbed and then desorbed. The transition from an adsorbed (bound) state to an unbound state is due to a change in the binding affinity of the solute to the adsorbent. As an example, we will use a strong acid-based cation exchange to illustrate the changing dynamics from adsorption to desorption or elution.

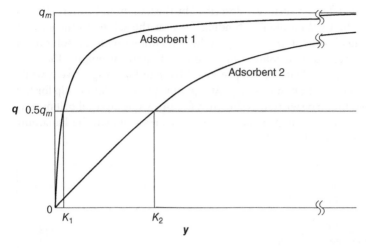

Figure 14.5 Langmuir adsorption isotherm.

The ion exchange between H^+ and another positively charged ion or molecule Y^+ is described in Eq. 14.3 (Panel 14.3) (for clarity, the concentration values in Panel 14.3 are enclosed in brackets, "[ ]"). With a strong acid as the functional group, a counter-ion (e.g., H^+ or Y^+) must be present to balance the charge. The equilibrium relationship is shown in Eq. 14.4. We again assume that the total number of binding sites per unit amount of adsorbent is fixed; also, that the number of unbound sites that have a net negative charge is negligible (Eq. 14.5). The total number of binding sites is thus the sum of those sites bound by H^+ and those sites bound by Y^+. We use the equilibrium relationship to express the concentration of solute (Y^+) adsorbed to the adsorbent (R^-Y^+) (Eq. 14.6), and to obtain the equation describing the solute concentrations in the liquid and adsorbent phases (Eq. 14.7).

Panel 14.3 Ion Exchange as an Example of Changing Adsorption Isotherm

Ion exchange reaction has stoichiometric correlation in reaction species. R^- is the ligand:

$$R^-H^+ + Y^+ \rightleftharpoons R^-Y^+ + H^+ \tag{14.3}$$

$$K_{eq} = \frac{[R^-Y^+][H^+]}{[R^-H^+][Y^+]} \tag{14.4}$$

We can assume that the total number of exchange sites is constant:

$$R_{total} = [R^-Y^+] + [R^-H^+] + [R^-] \tag{14.5}$$

Assume $[R^-]$ is negligible:

$$[R^-Y^+] = \frac{K_{eq}[R^-H^+][Y^+]}{[H^+]}$$

$$= \frac{K_{eq}[Y^+]}{[H^+]}\{[R]_{total} - [R^-Y^+]\} \tag{14.6}$$

$$[R^-Y^+] = \frac{K_{eq}[Y^+]\frac{[R]_{total}}{[H^+]}}{1 + \frac{K_{eq}[Y^+]}{[H^+]}} = \frac{K_{eq}[Y^+][R]_{total}}{[H^+] + K_{eq}[Y^+]} = \frac{[R]_{total}[Y^+]}{\frac{[H^+]}{K_{eq}} + [Y^+]}$$

$$q = \frac{q_m[Y^+]}{K_y + [Y^+]} \tag{14.7}$$

Note that the binding constant, K (equal to $[H^+]/K_{eq}$), is affected by pH. This allows one to use pH to control the binding (or adsorption) and dissociation of the solute. Figure 14.6 depicts a scenario in which pH is used to manipulate the adsorption and desorption of a solute. Let us consider a batch adsorption process in which a quantity of feed stream (Q_f) with a solute concentration y_f is in contact with a quantity (W) of

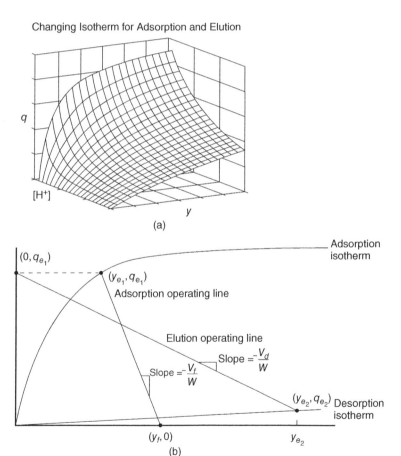

Figure 14.6 (a) Adsorption isotherm under different conditions (illustrated with pH) ranging from favoring adsorption to promoting desorption. (b) Material balances on adsorption and elution operations.

adsorbent that is free of solute, such that, after a short period of time, the solute in the two phases (y_{e1} and q_{e1}, respectively) is at equilibrium (Eq. 14.8 in Panel 14.4). From the balance on the solute, one can determine y_{e1} and q_{e1}. Ideally, the amount of solute transferred to the adsorbent phase should be large so that the recovery yield can be high.

Panel 14.4

Material Balance on Solute for Adsorption

$$V_f \cdot y_f + W \cdot q_0 = V_f \cdot y_{e_1} + W \cdot q_{e_1}$$

$$q_0 = 0$$

At equilibrium:

$$q_{e1} = \frac{V_f \cdot y_f - V_f \cdot y_{e_1}}{W} = \frac{-V_f}{W}(y_{e_1} - y_f) \tag{14.8}$$

Material Balance on Desorption

All adsorbent (W) from adsorption process (with q_{e_1}) will be desorbed with V_d of elution buffer.

$$W \cdot q_{e_1} = W \cdot q_{e_2} + V_d \cdot y_{e_2}$$

At equilibrium:

$$q_{e_2} - q_{e_1} = \frac{-V_d}{W} y_{e_2} \tag{14.9}$$

Since the pH used for adsorption favors the transfer of the solute to the adsorbent phase, q_{e1} is at a very high value. After adsorption, the adsorbent phase is "washed" to remove any unadsorbed solute and impurities. Then, an eluent at a pH that favors desorption is used to elute the solute. Typically, a small volume of eluent (Q_e) that contains no solute is used to desorb the solute from the W amount of adsorbent. The equilibrium relationship is now described by a new isotherm. From material balance between the liquid phase and adsorbent phase, one can calculate the new equilibrium concentrations (y_{e2} and q_{e2}) in both phases (Eq. 14.9). At the new equilibrium, q_{e2} should be very small. Note that none of the solute left in the liquid phase after adsorption or left in the adsorbent phase after desorption will be recovered. It is thus important that y_{e1} and q_{e2} are small, to achieve a high recovery yield.

To use chromatographic separation effectively, one manipulates the adsorption isotherm to enhance the difference of the binding affinities between the product solute and the contaminating solutes. When the selectivity is very high (i.e., the affinity of the product solute is drastically different from other undesired solutes), adsorption can be an effective tool to isolate a very pure product fraction.

14.4 Adsorption Chromatography

14.4.1 Discrete-Stage Analysis

Adsorption typically occurs in a packed bed, like a chromatographic column. To help understand the basic phenomenon seen in adsorption chromatography, we will examine an idealized process.

We assume that a chromatographic column can be divided into N number of invisible stages of equal volume. When the feed or the eluent moves along the column, it essentially moves stage by stage (Figure 14.7 and Panel 14.11). In an adsorption process, a feed stream is fed continuously. Note that we call this "stepfeeding," meaning that there is a sudden increase from no feed to a constant feed stream. We will idealize the process as

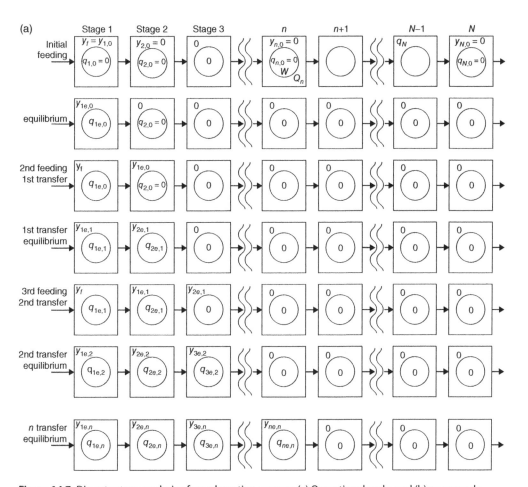

Figure 14.7 Discrete stage analysis of an adsorption process. (a) Operational cycle, and (b) an example of solute distribution in different stages over time.

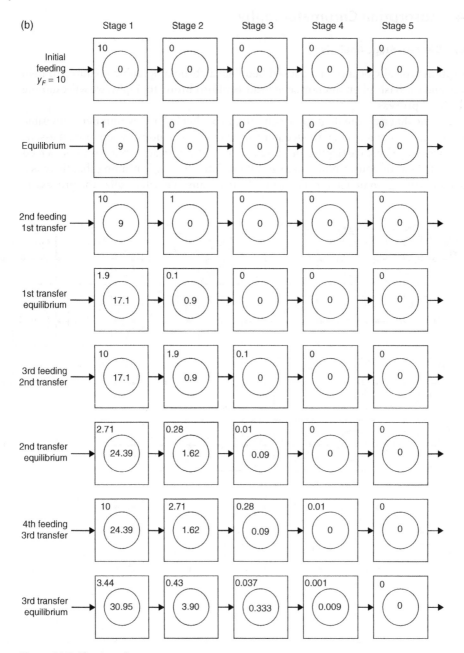

Figure 14.7 (*Continued*)

a stage-wise feeding wherein: a Q_n volume of feed stream enters the first stage (the stage number is denoted by the first subscript, i) to occupy the liquid phase that is in contact with a W_n amount of adsorbent. After the feed has been transferred into the first stage, the solute concentration in the fluid phase of stage 1 is $y_{1,0}$. The concentration in the adsorbent phase, $q_{1,0}$, remains 0. Of note, the second transcript here denotes the number of fluid transfers from one stage to the next stage. At this moment, except for the stage 1 fluid phase, the solute is not present anywhere else in the column. The adsorption process then reaches equilibrium. We assume a linear isotherm, so that $Q_n y_{1e,0} = KW_n q_{1e,0}$. All other stages remain without any solute.

Next, a fresh volume of the feed (Q_n) is transferred into stage 1 and begins to displace the liquid phase in every stage, pushing it to the next stage. However, the solute that was adsorbed in the adsorbent phase of each stage is not moving; it remains in its original stage. Thus, after the first transfer (denoted by the second subscript), the fluid phase concentration in stage 1 ($y_{1,1}$) becomes the same as that in the feed, while the concentration in stage 2 ($y_{2,1}$) is the same as that of stage 1 before the transfer ($y_{1e,0}$). We also assume that the fluid and adsorbent volumes in all stages (Q_n, W_n) are the same. New equilibrium is then reached in every stage to give rise to ($y_{1e,1}$, $q_{1e,1}$), ($y_{2e,1}$, $q_{2e,1}$), and so on. The process then goes on to the next transfer. The front of the feed moves further downstream after each transfer.

Figure 14.7b shows an example for a solute and adsorbent pair, for which the adsorption isotherm is linear ($q = 9y$). We assume the volume of the solution and adsorbent phases (i.e., Q_n and W_n) is the same for all stages. With this assumption, one can easily calculate the equilibrium concentrations in each phase by dividing the sum of the total amount of solute into ten parts; nine parts go to the adsorbent phase, and one part goes to the liquid phase. As can be seen from Figure 14.7b, as the front of the feed moves downstream, the solute profile in the liquid phase begins to lag behind, due to the transfer of the solute into the adsorbent phase. After many transfers, the solute concentration in upstream stages becomes very high.

This illustration uses a linear isotherm. Thus, as the concentration of solute in the adsorbent phase approaches 90, it will approach the equilibrium with the feed concentration. The transfer of solute from the liquid phase to the adsorbent phase will then diminish, and the adsorbent phase will become saturated. After that time, the fluid phase concentration will be the same as the feed in the upstream stages where it has been saturated, and then tapers off in the downstream stages. As more feed enters the column over time, the region where the solute concentration tapers off moves toward the downstream end, until it reaches the same concentration as in the feed.

14.4.2 Breakthrough Curve

Figure 14.8a illustrates an in-column view of the solute concentration in the liquid phase, as described in this chapter. As the "loading" of the column (or feeding) proceeds, the upstream region becomes "saturated." This is followed by a gradual decreasing of solute concentration toward the distal end of the column. Further downstream, the solute is

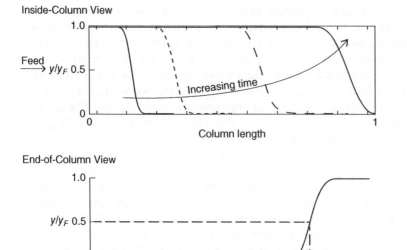

Inside-Column View

End-of-Column View

Figure 14.8 Solute profile in adsorption. In-column view and end-of-column view.

not detectable. With a strong adsorbent, practically all solute molecules in the feed are adsorbed in the early stages of the column, so one would not see any solute in the leading part of the feed stream further down. The lagging region that is saturated with solute is called the "equilibrium zone." The region where the solute concentration decreases is called the "adsorption zone." In the adsorption zone, the solute is transferred from the liquid to the adsorbent. The spread of the adsorption zone is affected by the affinity of the solute to the adsorbent: the stronger the affinity, the sharper the transition.

If one stands at the outlet of a column and monitors the solute concentration in the fluid phase over time, one will not see any solute for some time. Then, the feed solution exits from the column (Figure 14.8). When the solute emerges, its concentration profile is symmetrical to the solute profile of the in-column view at the point when the solute front reaches the end of the column. This concentration profile of the solute at the end of the column is called the "breakthrough curve."

One can calculate the time that the feed front and the solute will reach the end of the column. The column is packed with adsorbent beads that occupy a fraction of $1-\varepsilon$ of the entire column volume (V), while the liquid phase has a volume fraction of ε. Assuming that the feed flows through only in the void but not into the adsorbent, then the holding time (i.e., the time it takes the feed to reach the end of the column) θ is equal to $\varepsilon V/Q$ (Panel 14.5 in Eq. 14.10). When the entire adsorbent bed is saturated (i.e., it is in equilibrium with the feed stream), the concentration in the adsorbent phase is $q_f = Ky_f$. The total amount of solute that can then be adsorbed in the bed (i.e., the capacity of the column), $q_f(1-\varepsilon)V$, is thus $Ky_f(1-\varepsilon)V$ (Eq. 14.11).

Panel 14.5

$$\theta = \varepsilon \frac{V}{F} = \varepsilon \frac{l}{v} \tag{14.10}$$

When the column is saturated, $q_f = K y_f$.

- The total amount adsorbed is:

$$M = (1 - \varepsilon) V q_f = (1 - \varepsilon) V K y_f \tag{14.11}$$

- At the same time, the liquid phase has $\varepsilon V y_f$. The total solute in the column is:

$$(\varepsilon + (1 - \varepsilon) K) V y_f$$

Time to saturate the column with the solute is:

$$\theta_S = \left(\frac{(1 - \varepsilon) K}{\varepsilon} \right) \theta \tag{14.12}$$

Imagine that the solute front moves like a piston. Then, the time that the feed stream at a flow rate of Q can fill up the capacity of the column, θ_s, is also the time that the solute will emerge from the end of the column. When the column is filled to capacity with solute, the total amount of solute in the column is equal to the fraction that is in the adsorbent phase (i.e., $K y_f (1-\varepsilon) V$) and the fraction that is in the liquid phase ($\varepsilon y_f V$). θ_s, then, is equal to $(1+(1-\varepsilon)K/\varepsilon)\theta$ (Eq. 14.12 in Panel 14.5). In a typical adsorption process, the amount of solute in the liquid phase is very small compared to the amount that is in the adsorbent phase, or $\theta_s = ((1-\varepsilon)K/\varepsilon)\theta$. However, as discussed, the solute front does not move like a sharp band at the front of the feed stream. Rather, it develops into a breakthrough curve where θ_s is located at the midpoint of the curve.

14.4.3 An Empirical Two-Parameter Description of a Breakthrough Curve

A breakthrough curve is characterized by two properties: the breakthrough time, θ_s, and the spread of the curve. The spread of a solute concentration curve is affected by many factors. In the discussion thus far in this chapter, we assume that the solute reaches equilibrium between the liquid phase and the absorbent phase. In most cases, the diffusion of solute into the adsorbent is a rate-limiting step. Intraparticle mass transfer resistance causes further spread of the solute concentration profile. Another factor causing further spread is the dispersion of the fluid phase. That is, the fluid is not moving downstream like a plug flow.

One can characterize a breakthrough curve using its resemblance to a cumulative Gaussian distribution curve (Figure 14.9). The cumulative population density of a Gaussian distribution can be described using its mean and standard deviation (Eqs. 14.13 and 14.14 in Panel 14.6). Similarly, by using two parameters (θ_s along with a standard deviation, θ_σ), one can describe a breakthrough curve quantitatively using a Gaussian error function (Eq. 14.15). Experimentally obtained breakthrough data can be used to fit a cumulative Gaussian distribution function to obtain θ_s and θ_σ.

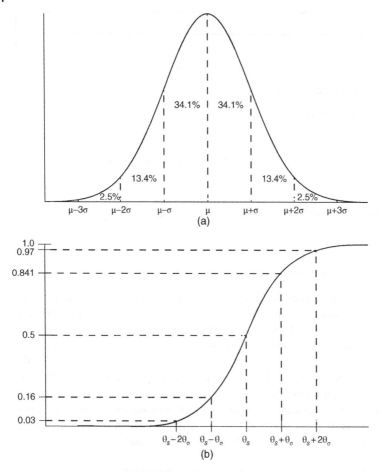

Figure 14.9 Population density of a Gaussian distribution and its use to describe a breakthrough curve.

Panel 14.6 Gaussian Distribution and Its Cumulative Population Density

$$f(x) = \frac{1}{\sigma\sqrt{2\pi}} Exp\left(\frac{-(x-\mu)^2}{(2\sigma^2)}\right) \tag{14.13}$$

The cumulative frequency is:

$$\int_0^x f(x)dx = \frac{1}{2}\left(1 - erf\left[\frac{x/\mu - 1}{\sqrt{2}\sigma}\right]\right) \tag{14.14}$$

Two-Parameter Model

Assuming a breakthrough curve can be fit with a cumulative frequency of a Gaussian distribution function:

$$\frac{y}{y_F} = \frac{1}{2}\left(1 - erf\left[\frac{\frac{t}{\theta_s} - 1}{\sqrt{2}\theta_\sigma}\right]\right) \tag{14.15}$$

14.4.4 One-Porosity Model for an Adsorption Process

Another way to model an adsorption process is to use a partial differential equation. This equation describes the flow of a fluid carrying a solute through a packed-bed reactor with the adsorption of the solute to the adsorbent beads. The schematic description of the reactor is shown in Panel 14.7. Here, the feed flow rate is Q. The length and cross-sectional area of the chromatographic column are L and A, respectively. The apparent flow velocity is thus u (equal to Q/A). However, because of the space occupied by the adsorbent, the cross-sectional area that is accessible to the fluid flow is only εA. So, the actual average velocity of fluid flow in the column is v (equal to u/ε).

Panel 14.7

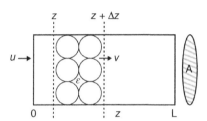

One-porosity model:

$$\varepsilon \frac{\partial y}{\partial t} = \left(-v\frac{\partial y}{\partial z} + D_z\frac{\partial^2 y}{\partial z^2} \right)\varepsilon - (1-\varepsilon)\frac{\partial q}{\partial t} \tag{14.16}$$

$$v = \frac{\mu}{\varepsilon}$$

Equilibrium is reached instantaneously.
Initial condition: $y = 0$, $q = 0$, for all z at $t = 0$.

Boundary conditions: $\begin{array}{ll} t \geq 0 & y|_{z=0} = y_F \\ y = 0, z \to \infty \end{array}$

and linear isotherm, $q = Ky$.

$\dfrac{\partial y}{\partial t} = K\dfrac{\partial q}{\partial t}$, substitute in :

$$\frac{(\varepsilon + (1-\varepsilon)K)}{E}\frac{\partial y}{\partial t} = -v\frac{\partial y}{\partial z} + D_z\frac{\partial^2 y}{\partial z^2} \tag{14.17}$$

Lapidus–Amundson Equation (1952)

$$\bar{y}(z, t) = \frac{y}{y_F} = \frac{1}{2}\left\{ 1 - erf\left[\frac{z - \dfrac{vt}{1 + \frac{1-\varepsilon}{\varepsilon}K}}{\left[\dfrac{4D_z t}{1 + \frac{1-\varepsilon}{\varepsilon}K} \right]^{1/2}} \right] \right\} \tag{14.18}$$

The equation can be used to simulate an "inside-column view":

$\bar{y} = \bar{y}(z)$ at different t

and "end-of-column view":

$\bar{y}(L)$ at different t

We can set up a material balance on the solute that exists in a thin section of the column, from z to $z+dz$ (Eq. 14.16 in Panel 14.7). This describes the change of solute concentration in the liquid phase. The contribution of fluid flow is considered in two respects, convection and liquid dispersion, as shown in the first two terms in the bracket on the right-hand side of the equation. These two terms describe the balance between the solute carried in and the solute carried out by the fluid phase. The amount of solute that is carried in by the fluid flow, but not carried out, is transferred to the adsorbent phase as described by the last term. Since the fluid phase and the adsorbent phase occupy only ε and the $1-\varepsilon$ fraction of the volume, their contribution terms are multiplied by ε and $1-\varepsilon$, respectively. With an assumption of instantaneous equilibrium, one can obtain a solute balance equation (Eq. 14.17).

We assume there is no solute in the column for $t < 0$ and, at $t = 0$, a step change to $y = y_f$ occurs at $z = 0$. One further assumes a boundary condition of $z = \infty$, $y = 0$.

The equation, with the initial and boundary conditions described here, was solved by Lapidus and Amundson at the University of Minnesota in 1952. The solution expressed as an error function is shown in Eq. 14.18 (Panel 14.7). This equation has two variables, z and t. One can use it to simulate the solute profile along the length of the column for a given time (Figure 14.10). By simulating the solute concentration at $z = L$ over time, one can obtain the breakthrough curve (end-of-column view). When using the equation to simulate the breakthrough curve, it can be seen that at $t = \theta_s$, equal to $\{\varepsilon + [(I - \varepsilon)/\varepsilon]K\}L/v$, the term inside the error function becomes zero and the normalized solute concentration is 0.5 (i.e., the concentration at the midpoint of the breakthrough curve, as we have discussed when the breakthrough curve was first presented in this chapter). From the model simulation of the inside-column view, we can see that the spread of the breakthrough increases as the solute front moves downstream.

Depending on the packing conditions of the adsorbent particles and velocity profile of the liquid, some degree of fluid back-mixing will occur. This will introduce "spreading" into the solute concentration profile. The effect of fluid dispersion is simulated in both inside-column and end-of-column views (Figure 14.11). The model considers only the porosity of the interparticle space, and assumes that the fluid phase is present only in the interparticle void space. In reality, the adsorbent has a large void space inside the adsorbent particle. Thus, the diffusion of solute into the intraparticle space is often the rate-limiting step. This issue is more prominent in elution chromatography and will be discussed later in this chapter.

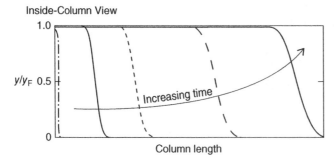

Inside-Column View

Figure 14.10 Solution profile in an adsorption column predicted by the Lapidus–Amundson equation.

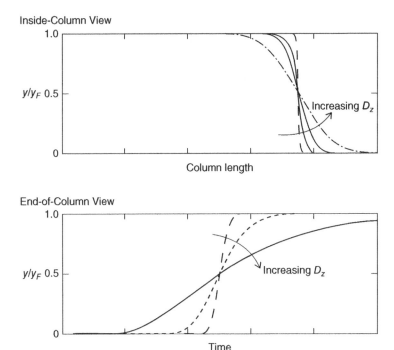

Figure 14.11 Effect of dispersion on the shape of a breakthrough curve.

14.4.5 Elution of Solutes from an Adsorption Column

In bioseparation, adsorption chromatography is used primarily to recover the product, as opposed to only removing contaminants. The elution of the product solute from the column after adsorption is thus very important. In most adsorption processes, the elution of the product solute is accomplished by manipulating the adsorption isotherm by switching from a buffer solution that favors adsorption to one that promotes desorption. Elution is frequently induced by changing the pH, salt concentration, or solvent composition.

A second elution mechanism is to add a displacing solute that has a much higher binding affinity to the ligand, which displaces the product solute. The product solute is then released to the fluid phase and eluted from the column. The displacement method is not frequently used. It is used only when the displacer is inexpensive or can be readily recycled.

To help understand the elution process, we go back to the idealized discrete-stage column as shown in Figure 14.7. Now, the adsorbent phase in every stage is loaded with the solute at an equal concentration (Figure 14.12). We start at the point after the residual solute in the liquid of the interstitial space has been removed. The solute concentration in the liquid phase is zero for all stages. The elution begins by feeding the elution buffer into Stage 1. After equilibrium is reached, the feeding of elution buffer into Stage 1 and the subsequent transfer of the liquid phase to the next stage are repeated. During the transfer, the liquid phase carries the solute into the next stage, while the solute in the adsorbent phase is stationary.

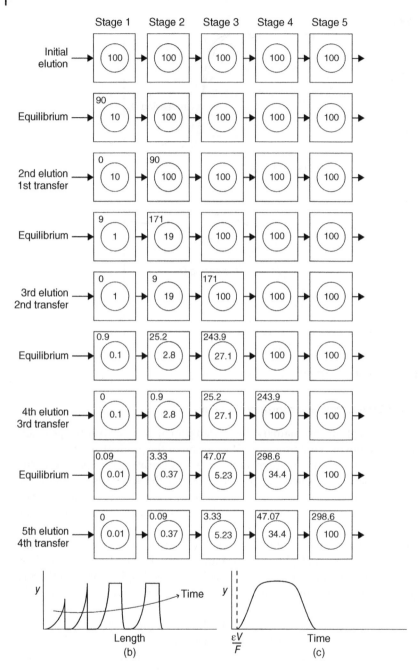

Figure 14.12 Discrete stage simulation of solute elution from an adsorption column; and the elution solute profile in an industrial adsorption chromatography.

The peak of the solute concentration moves along with the front of the elution buffer. The elution profile at the exit (this is the chromatogram seen in the chromatography) would show that the peak emerges at the time equivalent to $\theta_s = \varepsilon V/Q$.

As the front of the eluent and desorbed solute move to downstream stages, the solute concentration increases. At some point, it may reach equilibrium with the adsorbent phase. Afterward, the solute concentration in the moving eluent will remain constant as it moves downstream. During elution, the adsorption chromatographic column can thus be divided into three segments: (1) the pre-elution (the eluent has not yet reached) zone, (2) the equilibrium zone, and (3) the desorption zone. The idealized solute concentration profile in the elution chromatogram will thus show a sharp rise in solute concentration at θ_s. This is followed by a region of constant level caused by solute saturation, which then decreases rapidly after all the solute molecules are eluted. In practice, fluid dispersion will cause the dilution of the elution buffer at the frontal end of the elution stream, thus making the solute peak to rise less sharply at the front. As the solute concentration in the elution stream increases, the diffusion of solute from the adsorbent into the bulk fluid phase becomes limiting. Therefore, after an initial rapid rise in the solute concentration at the front of the elution stream, the concentration increase slows down. In industrial operations, the eluted solute may spread over a volume of the adsorption column (a bed volume) or more, in which the solute concentration may be flat over a wide region due to mass transfer resistance of solute diffusion from the adsorbent to the bulk liquid (Figure 14.12).

14.5 Elution Chromatography

14.5.1 Discrete-Stage Analysis

To analyze the behavior of elution chromatography, we return to discrete-stage analysis using a column that has N stages and initially contains no solute in all stages (Figure 14.13). A feed containing the solute is added to the first stage. After reaching the equilibrium between the adsorbent phase and the liquid phase, the elution buffer (containing no solute) is added to the first stage to push the liquid phase forward by one stage. The solute is allowed to reach equilibrium in each stage before elution buffer is again added to the first stage. The liquid phase in every stage is pushed to the next stage downstream. We assume the system reaches equilibrium very rapidly, using y_n and q_n to denote the concentrations in Stage n after reaching an equilibrium. The concentration of solute leaving Stage n is thus y_n. The fresh elution buffer entering Stage 1 is denoted as Stage 0 and has a solute concentration (y_0) of 0.

If we examine each stage separately, any stage (n) can be taken as a continuous well-mixed reactor. The reactor has a stage volume of V_n and a flow stream (at a flow rate of Q) that enters the reactor carrying the fluid phase from the preceding stage. The solute in the liquid phase and adsorbent then reaches equilibrium, and the exiting stream takes the liquid phase to the next stage. The material balance on the solute can be performed for Stage n, as shown in Eq. 14.19 (Panel 14.8). Since the adsorbent and liquid phases are in equilibrium (Eq. 14.20 in Panel 14.8), Eq. 14.19 can be simplified to an ordinary differential equation (Eq. 14.21). At $t = 0$, an M_F amount of solute is added to Stage 1 and is immediately equilibrated. This equilibrated solute concentration in the

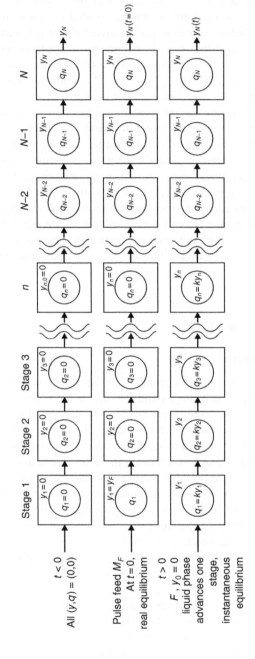

Figure 14.13 Discrete stage model of an elution chromatography.

liquid phase at Stage 1 is denoted as y_F, which can be obtained by balancing the solute on the first stage (Eq. 14.22). Thus, the initial condition for solving Eq. 14.21 is $y_1 = y_F$. Using the differential equation (Eq. 14.24) and the initial condition (Eq. 14.22), one can first solve for the concentration profile, with y_1 as a function of time (Eq. 14.25 in Panel 14.8).

Panel 14.8 Discrete Stage Analysis

Material Balance in Stage n:

$$\varepsilon \cdot V_n \frac{d}{dt} y_n + (1 - \varepsilon) \cdot V_n \frac{d}{dt} q_n = F(y_{n-1} - y_n) \tag{14.19}$$

Assuming linear isotherm:

$$(\varepsilon + (1 - \varepsilon) \cdot K) \cdot V_n \frac{d}{dt} y_n = F(y_{n-1} - y_n) \tag{14.20}$$

Substitution:

$$q_n = K \cdot y_n \qquad \frac{dq_n}{dt} = K \frac{d}{dt} y_n \tag{14.21}$$

Assume initially solute-free:

$t < 0, y_n = 0$ for all n

At $t = 0$, an amount M_F of a solute in a volume V_F is fed into Stage 1. At equilibrium, M_F gives to y_1 and q_1. Let the equilibrium value of y_1 immediately after the addition of the feed be y_F:

$$t = 0, y_1 = y_F \tag{14.22}$$

Assume that V_F is very small relative to V_n. Its addition does not change the volume of V_1.

$$y_F = \frac{M_F}{[(\varepsilon + (1 - \varepsilon) \cdot K) \cdot V_n)]} \tag{14.23}$$

We then proceed to solve the ordinary differential equation stage by stage. In elution chromatography, the feed is a pulse function. After M_F enters the bed, the elution solution does not contain solute, so y_0:

For Stage 1:

$$(\varepsilon + (1 - \varepsilon) \cdot K) \cdot V_n \frac{d}{dt} y_1 = F(y_0 - y_1)$$

Since $y_0 = 0$, therefore $(\varepsilon + (1 - \varepsilon) \cdot K) \cdot V_n \frac{d}{dt} y_1 = -F \cdot y_1 \tag{14.24}$

$$\frac{-(\varepsilon + (1 - \varepsilon) \cdot K) \cdot V_n}{F \cdot y_1} \cdot dy_1 = dt$$

The solution for Stage 1 is:

$$y_1 = y_F \cdot \exp\left[\frac{-F}{(\varepsilon + (1 - \varepsilon) \cdot K) \cdot V_n} \cdot t \right] \tag{14.25}$$

(Continued)

Panel 14.8 (Continued)

For $n = 2$, plug in the solution of y_1:

$$(\varepsilon + (1 - \varepsilon) \cdot K) \cdot V_n \frac{d}{dt} y_2 = F[y_1 - y_2]$$

$$= F \left[y_F \cdot \exp \left[\frac{F}{(\varepsilon + (1 - \varepsilon) \cdot K) \cdot V_n} \cdot t \right] - y_2 \right]$$

Solve the equation progressively for each stage for a large number of stages. The solution for any stage y_n:

$$y_n = y_F \left[\frac{\tau^{(n-1)} \cdot \exp(-\tau)}{(n-1)!} \right] \tag{14.26}$$

where $\tau = \dfrac{F \cdot t}{(\varepsilon + (1 - \varepsilon) \cdot K) \cdot V_n}$ (14.27)

τ: dimensionless time.

Substituting $V_n = V_B / N$:

$$\tau = \frac{F \cdot t \cdot N}{(\varepsilon + (1 - \varepsilon) \cdot K) \cdot V_B} = \frac{t \cdot N}{\left[\frac{(\varepsilon + (1 - \varepsilon) \cdot K) \cdot V_B}{F} \right]} = \frac{t}{t_0} \cdot N \tag{14.28}$$

where $t_0 = \dfrac{(\varepsilon + (1 - \varepsilon) \cdot K) \cdot V_B}{F}$ (14.29)

Next, we examine Stage 2. The equation for Stage 2 contains y_1, for which we have already obtained a solution. This solution can then be used to substitute y_1 to solve for the concentration profile of y_2. The process can be repeated for each stage downstream.

As the stage number increases, a general solution describing the concentration of y_n in Stage n over time can be obtained (Eq. 14.26 in Panel 14.8). For convenience, the equation is expressed in terms of dimensionless time (τ) that is the elution time (t) relative to the solute holding time in each stage, $\{Q/[\varepsilon + (I - \varepsilon) K] V_n\}$ (Eq. 14.27). Recall that we have hypothesized that the column has N stages. Instead of using a stage volume (V_n), we replace it with the total column bed volume (V_B) by multiplying the stage volume by N (Eq. 14.28). Note that the term inside the bracket in Eq. 14.28 is the time at which the solute peak reaches stage N, or the solute peak elution time (denoted as t_N). Using t_N as the reference time, τ can also be viewed as the relative position of the solute peak in the column with respect to stage N.

Using Eq. 14.28, one can simulate the time profile of solute concentration in any Stage n when the total number of stages is N. Figure 14.14a shows such a time-profile plot at four different time points in different stages of a column. One can see that the concentration stays at early stages over a period of time, and then moves downstream over time. As it moves downstream, the peak concentration decreases and the spread of the peak widens. The equation can also be used to simulate the concentration at different stages for a given time (Figure 14.14b).

A closer look at Eq. 14.26 will reveal that it has the same form of a Poisson distribution (Eq. 14.30) that describes the probability of x occurrences, $f(x)$, of random events, for which the average of all events is μ (Panel 14.9). Poisson distribution is a one-parameter

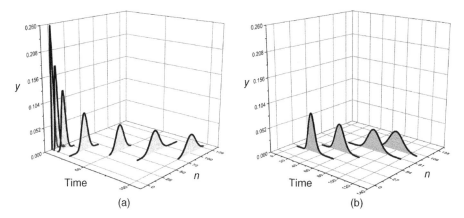

Figure 14.14 In-column view of solute profile in an elution chromatography distribution (a) in the column at four different times, and (b) in four different positions in the column (four different stages) over time.

distribution. The value of $f(x)$ is determined for a given μ. A Poisson distribution function can be approximated by a Gaussian distribution when μ is large (>20) (Eq. 14.31). In the case that a Poisson distribution is approximated by a Gaussian distribution, the standard deviation is the square root of the mean.

Panel 14.9 Poisson distribution:

$$f(x) = \frac{\mu^x \cdot \exp(-\mu)}{x!} \tag{14.30}$$

When μ is large (>20), a Poisson distribution can be approximated by a Gaussian distribution and $\sigma = \sqrt{\mu}$:

$$f(x) = \frac{1}{\sigma\sqrt{2 \cdot \pi}} \cdot \exp\left[\frac{-(x-\mu)^2}{2\sigma^2}\right]$$

$$= \frac{1}{\sqrt{2\pi\mu}} \cdot \exp\left[\frac{-(x-\mu)^2}{2\mu}\right] \tag{14.31}$$

The equation derived from discrete-stage analysis can thus be described using a Gaussian distribution function. It can be presented in terms of the dimensionless number, τ (Eq. 14.32), or in terms of time (Eq. 14.33 in Panel 14.10). We define the reference time (t_N) as the solute peak elution time at stage N. Setting $t = t_N$ (i.e., $\tau = \tau_N$) in Eq. 14.32 (which makes the exponential function on the right-hand side of the equation zero) gives the highest (peak) concentration of the solute in terms of y_F (Eq. 14.34 in Panel 14.10). y_F can also be expressed in terms of the amount of solute fed to the column, M_F (Eq. 14.35). The equation for stage N (Eqs. 14.32 and 14.33) can be rewritten to obtain the solute concentration profile in terms of the peak concentration, y_p, that is readily measurable (Eqs. 14.36 and 14.37 in Panel 14.10). The time profile at stage N (i.e., at the outlet of the column) is the chromatogram from elution chromatography. Note that the stage number is imaginary, since the column is continuous in space. Experimentally, one uses the chromatogram to determine a value for N, and uses N to characterize the power of separation of the column and its adsorbent.

Panel 14.10

By analogy, the concentration profile y for given stages at different τ (i.e., chromatogram) can be evaluated by a normal distribution with τ_o as the mean.

Since $\sigma = \sqrt{\tau_O}$ and $\tau_O = N$

$$y_N(\tau) = y_F \frac{1}{\sqrt{2\pi\tau_O}} \cdot \exp\left[\frac{-(\tau - \tau_O)^2}{2\tau_O}\right]$$

$$= y_F \frac{1}{\sqrt{2\pi N}} \cdot \exp\left[\frac{-(\tau - N)^2}{2N}\right] \tag{14.32}$$

$$y_N(t) = y_F \frac{1}{\sqrt{2\pi N}} \cdot \exp\left[\frac{-\left(\frac{t}{t_0} \cdot N - N\right)^2}{2N}\right]$$

$$= y_F \frac{1}{\sqrt{2\pi N}} \cdot \exp\left[\frac{-\left(\frac{t}{t_0} - 1\right)^2 \cdot N}{2}\right] \tag{14.33}$$

The highest peak concentration is when $\tau = \tau_o$. The peak concentration y_p is:

$$y_p = y_F \frac{1}{\sqrt{2\pi\tau_o}} \cdot \exp\left[\frac{-(\tau_o - \tau_o)^2}{2\sigma^2}\right]$$

$$= y_F \frac{1}{\sqrt{2\pi\tau_o}}$$

$$= y_F \frac{1}{\sqrt{2\pi N}} \tag{14.34}$$

Recall Eq. 14.4:

$$y_F = \frac{M_F}{[(\varepsilon + (1 - \varepsilon) \cdot K) \cdot V_n)]}$$

$$y_p = \frac{M_F}{[(\varepsilon + (1 - \varepsilon) \cdot K) \cdot V_n)]} \frac{1}{\sqrt{2\pi N}} \tag{14.35}$$

The equation for chromatogram (Eqs. 14.32 and 14.33) can be rewritten as:

$$y_N(\tau) = y_p \exp\left[\frac{(\tau - N)^2}{2 \cdot N}\right] \tag{14.36}$$

$$y_N(\tau) = y_p \exp\left[\frac{-\left(\frac{t}{t_0} - 1\right)^2 \cdot N}{2}\right] \tag{14.37}$$

With the discrete-stage model, we have one equation for each stage in the column. We can also employ the equations to simulate at a given time point (Eq. 14.38 in Panel 14.10). The plot generated will be the same as those shown in Figure 14.14.

14.5.2 Determination of Stage Number

When plotted over different times, Eq. 14.37 yields the chromatogram that can be observed experimentally. Note that when the solute peak reaches the end of the column, which has a length of L, the time is denoted as t_N and the column is defined as having Stage N. The solute concentration profile is described by a Gaussian distribution function, in which the mean is t_N or N and the standard deviation is the square root of the mean.

With an experimentally obtained chromatogram, we would not know the value of N. To determine this, one can measure the time and the corresponding solute concentration at the peak (t_N, y_p) and at another point along the shoulder. As can be seen in Panel 14.11, N can be calculated from Eq. 14.38. The value of constant C for a number of positions at the shoulder is tabulated in Panel 14.11.

Panel 14.11

Using Eq. 14.37, take the ratio of a time point t of a peak to the highest point of the peak:

$$\frac{y_n}{y_p} = \exp\left[\frac{-\left(\frac{t}{t_0} - 1\right)^2 \cdot N}{2}\right]$$

N can be calculated from solute concentrations:

$$\ln\frac{y_n}{y_p} = \left[\frac{-\left(\frac{t}{t_0} - 1\right)^2 \cdot N}{2}\right]$$

$$N = \frac{\ln\frac{y_n}{y_p}}{-((t/t_0) - 1)^2} = -\ln\left(\frac{y_n}{y_p}\right) \cdot \left(\frac{t_0}{t - t_0}\right)^2$$

$$= C \cdot \left(\frac{t_0}{(t - t_0) \cdot 2}\right)^2$$

$$= C\left(\frac{\text{peak elution time}}{\text{peak width span}}\right)^2 \tag{14.38}$$

∴ The stage number N is determined by Eq. 14.39.
C can be calculated for different Gaussian peaks:

(Continued)

Panel 14.11 (Continued)

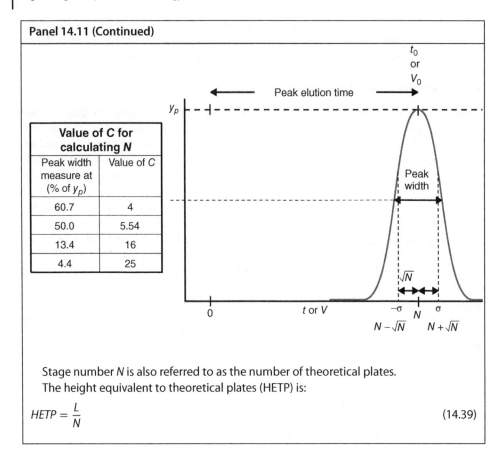

Value of C for calculating N	
Peak width measure at (% of y_p)	Value of C
60.7	4
50.0	5.54
13.4	16
4.4	25

Stage number N is also referred to as the number of theoretical plates. The height equivalent to theoretical plates (HETP) is:

$$HETP = \frac{L}{N} \qquad (14.39)$$

The number of stages, also called the number of theoretical plates (NTP), is a commonly used quantity to compare the efficiency of separation in a liquid chromatographic operation. The physical length of a column, L, is imaginarily divided into N stages, based on the calculation of the peak of the solute as described in this chapter (Eq. 14.39 in Panel 14.11). The length of the column divided by the stage number is called the stage height, or height equivalent to a theoretical plate (HETP). With a given column length, a smaller HETP will give a better separation of the solute.

14.5.3 Effect of Stage Number and Number of Theoretical Plates

Figures 14.14 illustrates that as the solute peak moves downstream, the peak concentration (y_p) decreases and the spread of the solute increases. This effect of increasing stage number on decreasing peak concentration (y_p) is seen in Eqs. 14.35 and 14.37. Increasing N increases the standard deviation (and thus the spread of the peak) and decreases the peak concentration in a proportionality of $N^{1/2}$.

To separate two solutes by elution chromatography, their two peaks must be distinctly separate by the time the faster moving one (with a lower affinity to the adsorbent) reaches the end of the column. The overlapping region of the shoulder of the two peaks must be kept at a low level to maintain the contaminant below a prescribed range. The spread of a solute peak can be assessed by the standard deviation time, t_σ. In a Gaussian distribution, 95% of the solute falls within the bounds of $t_N - 2t_\sigma$ and $t_N + 2t_\sigma$. To achieve a good resolution, one would like to give $t_{N1} + 2t_{\sigma1}$ of solute 1 (or the tail of the first peak) and $t_{N2} - 2t_{\sigma2}$ of solute 2 (the front of the second peak) enough distance. The spread of a peak (i.e., the standard deviation) is proportional to $N^{1/2}$, while the elution time, t_N, is proportional to column length, or N. The general relationship (Eq. 14.40 in Panel 14.12) shows that the resolution between two peaks increases with increasing N and increasing difference of K between the two solutes. However, as N (or the column length) increases, the elution time increases, as does the volume of eluent that must be collected to yield the desired recovery. Using a longer column in elution chromatography yields increased separation, but is panelized by a more dilute product and longer operating time.

Panel 14.12

$$t_{01} = \frac{(\varepsilon + (1 - \varepsilon) \cdot K_1) \cdot V_B}{F}$$

From Eq. 14.39:

$$N = C \cdot \left(\frac{t_{01}}{t_{1,spread}} \right)^2 = C \cdot \left(\frac{t_{01}}{\Delta t_1} \right)^2$$

$$\frac{t_{01}}{\Delta t_1} = \sqrt{\frac{N}{C}} \quad (C : \text{ constant})$$

$$\Delta t_1 = t_{01} \sqrt{\frac{C}{N}}$$

$$(t_{01} + \Delta t_1) = t_{01} \left(1 + \sqrt{\frac{C}{N}} \right)$$

$$(t_{02} - \Delta t_2) = t_{02} \left(1 - \sqrt{\frac{C}{N}} \right)$$

typically $N \gg C \therefore$.

$$(t_{01} + \Delta t_1) - (t_{02} - \Delta t_2) \approx (t_{01} - t_{02})$$

$$= \frac{V_B}{F} ((\varepsilon + (1 - \varepsilon) \cdot K_1) - (\varepsilon + (1 - \varepsilon) \cdot K_2))$$

$$= \frac{V_B}{F} (1 - \varepsilon) \cdot (K_1 - K_2)$$

$$= \frac{N \cdot V_n}{F} (1 - \varepsilon) \cdot (K_1 - K_2) \tag{14.40}$$

It is important to note that the discussion so far has been based on a number of important assumptions: (1) linear isotherm, (2) instantaneous equilibrium (no mass transfer limitation), (3) uniform packing and uniform properties in the packed adsorbent, and (4) the bed comprises a liquid phase and an adsorbent phase. With this model, the number of stages calculated using concentration profiles of different solutes in the same chromatogram should be the same. They arrive at the end stage at different time points due to their difference in K. In practice, because the assumptions listed in this section are not true, the number of stages is usually determined using the product solute, as different solutes will likely give different N's.

In laboratory or industrial operations, it is a good practice to keep the operating conditions close to ideal. It is very important to pack the column uniformly. Adsorbents are rarely monodispersed, but are usually heterogeneous in size. If a column is not packed carefully, larger particles may settle faster, thus creating a particle-size gradient. Uneven packing may also lead to fluid channeling. When adding the feed stream, it is important not to perturb the chromatographic bed and to lay the feed in as small of a volume as possible. Ideally, a stage volume is the bed volume divided by stage number (V_B/N). If the feeding process causes a large perturbation in the bed at the top of the column, the initial stage volume will become large and lead to changes in the final separation outcome. Although the discussion in this chapter deals with idealized chromatographic separation, the conclusions on the behavior of separating two solutes are generally applicable in practice.

14.5.4 Two-Porosity Model, Mass Transfer Limitation

In the discussion in this chapter, the chromatographic column is considered to have two phases: a liquid phase occupying an ε fraction of the bed volume, with the remainder being an adsorbent phase. The model is a one-porosity model because it considers only the porosity of the interstitial fluid outside the adsorbent phase. In reality, the adsorbent phase consists of a small fraction of solid adsorbent material and a larger volume fraction of intraparticle fluid. Some adsorbent used for protein separation has only 10% of the adsorbent volume as a solid.

A more realistic description of chromatography employs a two-porosity model that considers an intraparticle (intra-adsorbent) liquid phase (denoted by subscript I) that occupies an ε' fraction of the adsorbent volume, as shown in Panel 14.13 for a general case that solute, j, is to be purified from other contaminating solutes. The balance on solute y_j thus entails an additional term to account for the intraparticle liquid phase. A more detailed model may even consider the "accessibility" of the intraparticle liquid phase to different solutes by diffusion. For example, a large protein molecule would not have the same access to the intraparticle liquid space as smaller molecules. A dispersion term may be introduced to deal with molecular diffusion, due to the random motion of the solute molecules, and the eddy diffusion caused by fluid flow (Eq. 14.41 in Panel 14.13). The balance of the solute j in the liquid phase is written as a partial differentiation equation that accounts for the advection in the liquid phase and both the eddy and molecular diffusions, the transfer in the intraparticle liquid phase, as well as adsorption to the adsorbent phase.

Panel 14.13

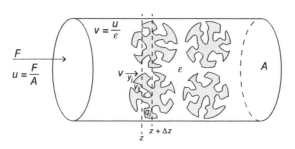

Two-Porosity Model

$$\varepsilon\frac{\partial y_j}{\partial t} + \delta_j(1-\varepsilon)\varepsilon'\frac{\partial y_{lj}}{\partial t} + (1-\varepsilon)(1-\varepsilon')\frac{\partial q_j}{\partial t} + \varepsilon\frac{\partial(vy_j)}{\partial z} - \varepsilon(E_{D_j} + D_{M_j})\frac{\partial^2 y_j}{\partial z^2} = 0 \qquad (14.41)$$

Key Assumptions:

Packing is homogenous.
Radial gradient of y is negligible.
No phase change.
No chemical reaction.

Describe mass transfer of y to the adsorbent phase using an overall mass transfer expression:

$$(1-\varepsilon)(1-\varepsilon')\frac{\partial q_j}{\partial t} + \delta_j\varepsilon'(1-\varepsilon)\frac{\partial y_{lj}}{\partial t} = -K_T a(y^*{}_{lj} - y_{lj}) \qquad (14.42)$$

Substituting into Eq. 14.41:

$$\varepsilon\frac{\partial y_j}{\partial t} + -K_T a(y^*{}_{lj} - y_{lj}) + \varepsilon\frac{\partial(vy_j)}{\partial t} - \varepsilon(E_{D_j} + D_{M_j})\frac{\partial^2 y_j}{\partial z^2} = 0 \qquad (14.43)$$

To deal with mass transfer limitation, the terms on the intraparticle fluid phase and the adsorbent phase are lumped together. They are described as being dependent on an overall mass transfer coefficient and an overall driving force of solute concentration difference. The overall driving force is conceptualized as being the difference between an apparent equilibrium intraparticle concentration (y_I^*) and an actual intraparticle concentration (y_I) (Eq. 14.42 in Panel 14.13). y_I^* is the solute concentration in the adsorbent phase if it were at an equilibrium with the bulk liquid phase. In reality, y_I^* and y_I are very difficult to determine experimentally. Using such a mass transfer control–based expression, we can obtain an equation (Eq. 14.43) that contains only one dependent variable, y.

In addition to mass transfer resistance, the dispersion in the liquid phase is another major reason that elution chromatography deviates from the idealized discrete stage analysis. Using a one-porosity model as an illustration (Eq 14.44 in Panel 14.14), the partial differential equation for elution chromatography is the same as the one for adsorption (Eq. 14.16 in Panel 14.7), except that the effect of dispersion is written as being bolstered by eddy diffusion in the bulk fluid flow and molecular diffusion. These two contributions are often combined to form a dimensionless number, called the Peclet number (Pe_z). The Peclet number thus reflects the relative contribution of advection and diffusion on mass transfer (Eq. 14.45 in Panel 14.14).

Panel 14.14 Effect of Dispersion

$$\varepsilon \frac{\partial y}{\partial t} + (1 - \varepsilon)\frac{\partial q}{\partial t} = (E_D + D_M)\frac{\partial^2 y}{\partial z^2} - v\frac{\partial y}{\partial z} \tag{14.44}$$

The solution is expressed with a dimensionless parameter, the Peclet number:

$$Pe_z = \frac{Lv}{(E_D + D_M)}$$

With a pulse feed and assuming instantaneous equilibrium, the solution is:

$$\left(\frac{y}{y_F}\right) = \frac{1}{2}\frac{V}{V_\ell}\left(\frac{Pe_z}{\pi}\right)^{1/2} Exp\left[\frac{-Pe_z(V - V_\ell)^2}{4VV_\ell}\right]$$

$$V_\ell = (1 - \varepsilon)KV_B \tag{14.45}$$

In the case of mass transfer control, the term for the transfer to the adsorbent phase can be written as:

$$\frac{\partial q}{\partial t} = K_T a(y^* - y_I) = K_T a(Ky - k'y) = K_T a(K - k')y \tag{14.46}$$

The solution also behaves like a Gaussian distribution.

We consider two cases, instantaneous equilibrium and mass transfer control, assuming a linear isotherm (and $q = Ky$) in both cases. In the case of instantaneous equilibrium, we use the same initial conditions and boundary conditions as those used for solving the adsorption problem (except that the feed is a delta function or a pulse addition of solute). The solution thus gives an exponential function that, when plotted, also gives a curve like a Gaussian curve. In the case of mass transfer control, we describe the diffusion using a mass transfer coefficient and a driving force ($y_I - y^*$) (Eq. 14.46). With a linear isotherm assumption, the mass transfer rate is simplified to be linearly dependent on y, just as in the case of instantaneous equilibrium. Thus, the shape of the peak is the same as in the two cases. Importantly, different solutes that have different mass transfer characteristics will behave differently. This is observed experimentally and is the reason that different solute peaks in the same chromatogram may not give rise to the same stage number if mass transfer is limiting.

14.6 Scale-Up and Continuous Operation

14.6.1 Mass Transfer Limitation and the van Deemter Equation

As discussed in this chapter, HETP is a descriptor of the effectiveness of chromato-graphic separation. Mass transfer limitation imparts a negative effect on solute sepa-ration. That effect is described by the van Deemter equation (Eq. 14.47 in Panel 14.15). It states that HETP is affected by fluid dispersion and mass transfer resistance from the bulk to the adsorbent. The dispersion effect is inversely proportional to fluid velocity. In contrast, HETP is proportional to the mass transfer resistance and the fluid velocity. At a faster flow rate, there is less time for the solute to diffuse to the adsorption site in the adsorbent phase. Mass transfer resistance can be written as the inverse of the mass transfer conductance ($1/K_T a$). The dispersion is further divided into molecular diffu-sion, which is independent of the flow rate, and eddy diffusion, which is proportional to liquid velocity.

Panel 14.15 van Deemter Equation

$$HETP = \frac{2D_z}{v} + \left[2\left(\frac{(1-\varepsilon)K}{\varepsilon + (1-\varepsilon)K} \right)^2 \frac{\varepsilon}{K_T a_p} \right] v \tag{14.47}$$

$$D_z = E_D + \gamma D_M$$

$E_D \propto v d_p$ D_M is independent of v.

D_z divides into two terms: one velocity dependent, the other independent.

$$HETP = A + \frac{B}{v} + Cv \tag{14.48}$$

The last term *(Cv)* dominates.

$$\therefore HETP \propto \frac{1}{K_T a} \tag{14.49}$$

Overall mass transfer resistance is the sum of external liquid film resistance and internal pore diffusion resistance.

$$\frac{1}{K_T a} = \frac{1}{k_M a_p} + \frac{1}{k_i k_s a_p} \tag{14.50}$$

$k_M \propto \dfrac{1}{d_p}$ $a \propto \dfrac{1}{d_p}$; $k_s \propto \dfrac{1}{d_p}$

$k_M a_p \propto \dfrac{1}{d_p^{\,2}}$

$k_s a_p \propto \dfrac{1}{d_p^{\,2}}$

Overall : $HETP \propto d_p^{\,2}$ $\tag{14.51}$

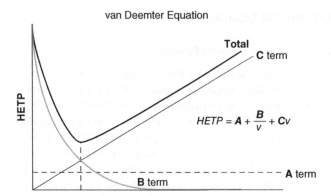

Figure 14.15 van Deemter equation depicts the effect of fluid velocity on HETP.

Eq. 14.47 can then be written as having three components, as shown in Eq. 14.48. A is the contribution of eddy diffusion (as a component of D_z in Eq. 14.47). Its flow rate dependence is cancelled out after dividing by v. B is the contribution of the molecular diffusion term, and C is the mass transfer term.

The dependence of HETP on fluid velocity is shown in Figure 14.15. The contribution of eddy diffusion (the A term) is constant at all velocities. The molecular diffusion (the B term) decreases with increasing velocity. Mass transfer (the C term) increases linearly with velocity. When the contributions from all three components are combined, HETP first decreases with increasing velocity. After reaching a minimal value, it increases linearly with velocity. At the velocity where HETP is minimum, the efficiency of separation is highest. Unfortunately, the velocity at which HETP is minimum is so slow that it is of no practical use. Therefore, in practice, elution chromatography is operated at a velocity that falls under the mass transfer control region. In short, the dominating factor affecting HETP is the mass transfer limitation.

The relationship between HETP and flow rate revealed by the van Deemter equation is generally consistent with experimental observation. With very small molecules and little mass transfer resistance from the fluid phase into adsorbent particles, the effect of the flow rate is relatively minor; however, for medium- and large-sized proteins, the HETP is much higher at a higher flow rate.

14.6.2 Scale-Up of Chromatography

In the discussion of mass transfer, we refer to an overall mass transfer that can be divided into the separate components that contribute to mass transfer. For example, overall oxygen transfer in the air–liquid interface has two components: liquid-phase transfer and gas-phase transfer. Recall that the inverse of the transfer coefficient can be taken as resistance and that the overall mass transfer resistance is the sum of its component resistances. The mass transfer resistance contribution to HETP can be described as overall resistance, $1/K_T a$ (Eq. 14.49 in Panel 14.15), which consists of two components: (1) external mass transfer resistance, or the resistance of transfer from the bulk liquid to the adsorbent's external surface; and (2) internal mass transfer resistance, or the resistance of solute diffusion into the pore in the interior of the adsorbent (Eq. 14.50).

The magnitude of the external mass transfer resistance depends on the external mass transfer coefficient (k_M) and the external surface area. Both the external mass transfer coefficient and the specific external surface decrease with increasing particle diameter. Thus, the external resistance increases with increasing particle diameter, to the power of two.

Internal to the particle, the solute molecules first diffuse along the tortuous pore before adsorbing to the surface. For internal resistance, the diffusion distance also increases with increasing diameter, thus making the mass transfer coefficient decrease with diameter. The change of the specific intrapore surface due to particle size is affected by changes in pore size distribution. Overall, the internal mass resistance increases with the particle diameter by a power greater than one. It is sometimes approximated by a power of two.

Taken together, the mass transfer resistance $(1/K_T a)$, and thus HETP, increases with the particle diameter to the power of two (Eq. 14.51). In other words, the size of the adsorbent has a profound effect on the separation power of elution chromatography; reducing the diameter by half reduces the HETP by fourfold. Given the same column length, the effective N increases fourfold. The effect of particle size on chromatographic separation is observed experimentally. However, using a smaller adsorbent also imposes constraints. Most fluid flow in a packed column is in the laminar-flow regime. To maintain the same flow rate, the changes in the pressure drop across the column are inversely proportional to the square of the diameter of the adsorbent.

In scaling up, a larger chromatographic column is used to take up a larger feed volume and to accommodate the increased volume of the process stream. The need for an increased chromatographic capacity cannot be met by merely increasing the diameter of the column. In scaling up a chromatographic process, the column diameter can only be increased to a certain limit. For example, in protein separation using gel-based adsorbent, the largest column we can use is about 1.5 to 2 m in diameter. Beyond that, it becomes difficult to evenly distribute the fluid across the chromatographic bed. For this reason, scaling up typically relies on increasing the column length.

To maintain the same degree of separation, one wants to maintain the same number of stages, N, in both large- and small-scale columns. For that, one needs to maintain the same L/HETP (Eq. 14.39 in Panel 14.11). To keep N constant, one can increase L and HETP proportionally while scaling up. In that way, the stage volume (L/N times the cross-sectional area of the column, or the total bed volume divided by N) increases, as does the feed volume. Since HETP is proportional to the diameter to the second power, one can use an adsorbent with a larger diameter that keeps L/d_p^2 constant (Eq. 14.52 in Panel 14.16). The flow rate can be kept constant, so that the pressure drop across the bed also remains unchanged because of the larger particle diameter (Eq. 14.53). The peak elution time in the large-scale column will be longer, and will be proportional to the column length. Thus, the cycling time will be longer in the large scale. To maintain the same recovery yield, one has to recover the product solute over the same span of peak spread (e.g., the same $t_N \pm 2t_\sigma$). As a result, the harvest time will also be longer.

Panel 14.16 Scale-Up of Elution Chromatography

- Mass transfer resistance is the dominant factor in determining HETP.
- In scaling up:
  - The feed volume increases proportionally to the stage volume, which is proportional to HETP.
  - N is kept constant to achieve the same degree of separation in different scales.
  - The length of the column has to increase in proportion to the stage volume or HETP to keep N constant.

$$L = N \cdot HETP \tag{14.40}$$

From Eq. 14.51: $HETP \propto d_p^2$:

$$N = \frac{L}{HETP} \propto \frac{L}{d_p^2} \tag{14.52}$$

  - By increasing particle diameter in proportion to $L^{1/2}$, N can be maintained constant in scaling up.
- From Kozeny equation (Eq. 13.2 in Panel 13.1):

$$v = K \frac{\Delta p \cdot d_p^2}{L} \tag{14.53}$$

If we keep L/d_p^2 constant, then columns of different scales can be operated at the same ΔP and v while maintaining N constant:
By collecting the same spread [i.e., the same number (n) of standard deviation] of the eluted solute, the recovery yield is also kept unchanged.

$$\frac{4\sqrt{N}}{N}\text{(relative zone spreading)} = \text{constant}$$

14.6.3   Continuous Adsorption and Continuous Elution Chromatography

Adsorption and elution chromatography are both intrinsically periodic batch operations. At a given time, the adsorption of solute to the adsorbent, and the separation of product solute from other contaminating solutes, takes place only in a section of the chromatographic column. In an adsorption process, the adsorption zone (the region where the solute is being transferred to the adsorbent) is small. In the other regions in the column, the adsorbent either is saturated already or is still waiting for the solute to arrive. An elution chromatography column is like a racetrack separating the fast runner and the slow runner. Increasing the length of the track increases the distance of separation between the fast runner and the slow runner. However, the segment of the track that is actually used by the two runners at a given time is only a small fraction of the entire track. Thus, in both adsorption and elution chromatography, only a small portion of the column is actually used at a given moment – the rest is merely in a holding mode. Just like a batch reactor, the efficiency of a column is also hampered by the equipment turnaround time. Thus, adapting the operation to a continuous mode can possibly increase the productivity.

Continuous operation of chromatography can be done through the use of multiple columns. Consider an adsorption process that uses a chromatographic column of volume V to support a feed flow of F. The feed is introduced for a period of θ_f. This is followed by flow-throughs of washing buffer and elution buffer, each applied for a time period of θ_w and θ_e, respectively. Even though the feed flow rate is θ_f, the process throughput is only $F\,\theta_f/(\theta_f + \theta_w + \theta_e)$.

To simplify our discussion, let us assume that θ_f, θ_w, and θ_e are equal. We now divide the column into three smaller columns, each with one-third of the volume (Figure 14.16). Under the same operating conditions, with the length of each smaller column reduced to one-third, it takes only one-third of the time to load, wash, and elute. At the beginning, the feed is introduced into the first column. After one-third of θ_f, the feed is directed to the second column while the first column is now switched to washing. After another time period equivalent to one-third of θ_f, the feed is switched to the third column, while the second column is undergoing washing and the first column is being eluted. With a cyclic operation, which rotates the feed, washing, and elution solutions through the three columns, the process becomes a continuous operation. The feed stream is fed into one of the three columns at the original feed flow rate. The throughput thus increases, using the same volume of adsorbent.

To convert elution chromatography to a continuous process, one may take lessons from multistage countercurrent separation processes, like distillation and liquid–liquid extraction. In an idealized countercurrent operation, two streams (like the aqueous-phase and organic-phase streams in liquid–liquid extraction) are in contact and allowed to reach equilibrium in each of the many stages. As one phase is transferred to the next stage upstream, the other phase is transferred to the next stage downstream.

There are major differences between a countercurrent, stage-wise, liquid–liquid extraction and stage-wise elution chromatography. For example, each stage of a liquid–liquid system resembles a mixing tank that allows its two-phase content to reach equilibrium. In contrast, each stage or subcolumn of a chromatographic column is intrinsically a packed-bed system. While the liquid phase exiting a stage can be continuously directed to the next stage, the solid adsorbent from one stage cannot be easily transferred to the next stage in the opposite direction.

Two adaptations have been made to simulate a countercurrent operation. First, instead of moving the solid adsorbent phase continuously, the content of a physical stage can be "transferred" periodically, after the liquid phase has been flown to the next stage for a period of time. Second, instead of moving the adsorbent to the next stage downstream, the connection of feed, eluent, and product streams can be "re-wired" to distribute the flow, to resemble moving the content of every stage by one stage in the opposite direction of the liquid stream.

There are a variety of different schemes for accomplishing a continuous operation of elution chromatography. A simple design may have as few as three subcolumns with a feed, an eluent, and two product streams. Others have a large number of stages and may allow for multiple solutes to be purified. The product stream that contains the low-affinity solute (eluted first) is often referred to as "raffinate." The product stream with the high-affinity solute (eluted last) is called an "extract."

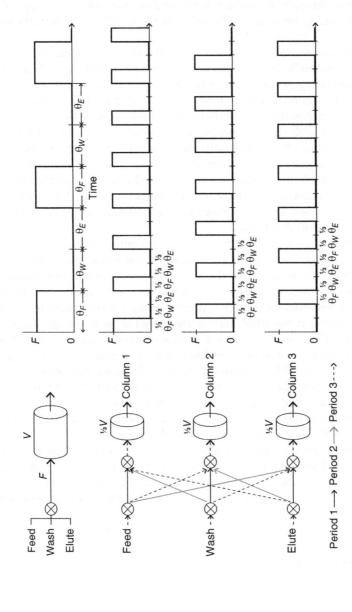

Figure 14.16 A sequential-multicolumn system for continuous adsorption chromatography.

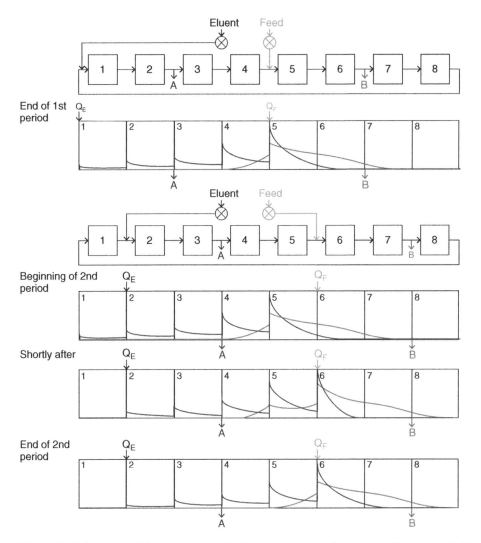

Figure 14.17 Continuous (simulated moving bed) chromatography. Solutes A and B are separated by an eight-subcolumn elution. Each subcolumn goes through a periodic operation. A fluid flow switching system directs the addition points of the feed, elution buffer, and the exit points of the raffinate (containing the faster moving solute) and extract (containing the slower moving solute) to different subcolumns. After each period, the addition and exit points are all advanced by one substage. The solute profile for each subcolumn, at the beginning and at the end of each period of operation, is shown.

An example of a simulated moving-bed chromatography with eight subcolumns is shown in Figure 14.17. Each subcolumn plays the role of receiving the feed, separating the product from contaminants, eluting the product, and then eluting the contaminants at different timepoints, depending on the arrangement of the fluid flow distribution. Each subcolumn rotates through those roles in a periodic fashion.

A simulated moving-bed chromatography can potentially achieve a high degree of separation, but it requires a more complex operation than a batch process. It also has a large number of operating parameters, especially if there are a large number of stages that need to be optimized. Such a system can also benefit from on-line monitoring sensors to increase the process robustness. Nevertheless, for the separation of a complex mixture of difficult-to-purify compounds, the benefits of having higher throughput and better purification certainly outweigh the effort of developing the process.

14.7 Concluding Remarks

Liquid chromatography is a very important unit operation for the recovery of bioproducts. Its separation power is due to the difference in the isotherm between the product solute and the contaminating solutes. Adsorption chromatography gives a high capacity and produces a more concentrated product stream, but it requires a highly selective adsorbent. Elution chromatography has high-resolution power, but also a relatively poor capacity. Selecting a good adsorbent and optimizing the operating conditions to maximize the resolution power are key to achieving a highly effective product recovery.

In recent years, there have been increasing efforts to develop continuous adsorption and elution chromatography. These efforts are complemented by work on the design of adsorbent ligands. Together, they will facilitate the integration of product recovery and continuous culture for continuous bioproduct manufacturing.

Further Reading

1 Jungbauer, A 2013, 'Continuous downstream processing of biopharmaceuticals', *Trends in Biotechnology*, **31**, 479–492.
2 Steinebach, F, Müller-Späth, T & Morbidelli, M 2016, 'Continuous counter-current chromatography for capture and polishing steps in biopharmaceutical production', *Biotechnology Journal*. Available from: http://onlinelibrary.wiley.com/doi/10.1002/biot .201500354/abstract. [19 July 2016].
3 Wankat, PC 1986, 'Physical picture and simple theories for adsorption and chromatography', in *Large scale adsorption and chromatography*, pp. 7–54. CRC Press, Boca Raton, FL.
4 Wankat, PC 1990, 'Linear theory of sorption and chromatography', in *Rate-controlled separations*, pp. 288–364. Elsevier Applied Science, London.

Nomenclature

A	cross-sectional area of total	L^2
a	apparent particle surface area	L^2
a_p	external surface of particle	L^2
A, B, C	molecular diffusion, eddy diffusion, and mass transfer term in van Deemter equation	
D_M	molecular diffusion	L^2/t
D_z	effective axial diffusivity	L^2/t
d_p	particle diameter	L
E_D	eddy diffusivity	L^2/t
F	volumetric flow rate	L^3/t

HETP	height equivalent to theoretical plates	L
subscript *I*	internal of adsorbent	
subscript *j*	solute species *j*	
K	partition coefficient in liquid extraction, linear isotherm equilibrium constant	(mole/L^3)/ (mole/L^3), (M/L^3)/ (M/L^3)
K_y	half-saturation constant in Langmuir isotherm	(mole/L^3)/ (mole/L^3), (M/L^3)/ (M/L^3)
K_{eq}	equilibrium constant	Dependent on the number of species involved
K_T	overall mass transfer coefficient	L/t
k_i	weighting factor	
$k_M a_p$	external liquid resistance	1/t
$k_s a_p$	internal resistance	1/t
L	height of the column	L
M_F	amount of solute fed to the column	m
N	number of plates or stages	
Pe_z	Peclet number	
q	solute concentration in the adsorbent	(mole/L^3)/ (mole/L^3), (M/L^3)/ (M/L^3)
q_m	maximal concentration of solute adsorbed to adsorbent	mole/L^3, M/L^3
R	concentration of solute adsorption (ligand) site	
R_{total}	total concentration of solute adsorption	
Δt_σ	peak width or peak spread at a given position from *tN*	t
t	time	t
t_N	time at peak maximum	t
u	bulk linear velocity of fluid in the column	L/t
V	volume	L^3
V_N	peak evolution volume in chromatogram	L^3
V_B	bed volume of column	L^3
V_F	volume of the feed	L^3
V_d	volume of the eluent	L^3
v	effective linear velocity (adjusted for voidage) in the column	L/t
W	volume of adsorbent	L^3
y	solute concentration in bulk liquid	mole/L^3, M/L^3
y^*	solute concentration in liquid phase at equilibrium with adsorbent phase	mole/L^3, M/L^3
y_f	solute concentration in the feed	mole/L^3, M/L^3
y_F	solute concentration in the solution phase after reaching equilibrium in the first stage upon the pulse feed to the first stage	mole/L^3, M/L^3

y_p	solute concentration at peak maximum	mole/L^3, M/L^3
z	length coordinate in a column	L
δ_j	fraction of pores available to solute	
σ	standard deviation	
μ	mean in a distribution function	
θ_σ	standard deviation in a breakthrough curve	t
θ_s	breakthrough time	t
ε	interparticle porosity	
ε'	intraparticle porosity	
τ	dimensionless time	
τ_N	dimensionless time at peak maximum	

Problems

A. Adsorption

A1 The adsorption isotherm $\left(\frac{\text{mole of oxygen bound}}{\text{hemoglobulin}} \text{ vs } \frac{\text{mole of oxygen solution}}{\text{volume of solution}} \right)$ for hemoglobulin is a sigmoid curve, but not a linear isotherm. Why does this help narrow the difference of dissolved oxygen levels between blood upstream and downstream?

A2 Use the Lapidus–Amundson equation to simulate the concentration profile of the solute in the liquid phase in the bed at different t. Also, simulate breakthrough curves. Try different k values. Discuss the effect of K on breakthrough.

A3 You are selecting the adsorbent for affinity purification of IgG from a cell culture fluid with an IgG concentration of 5 g/L. The q_m and K for the adsorbents A and B are (0.7 g/L-adsorbent, 0.001 g/L-fluid) and (0.4 g/L–adsorbent, 0.0001 g/L–fluid), respectively. For both adsorbents, the bound IgG can be eluted at pH 3.5 with a desorption isotherm of ($q_m = 0.005$ g/L-adsorbent, k = 0.005 g/L-fluid) and ($g_m = $ g/L adsorbent, K = 0.05 g/L-fluid) for A and B. 500 L of cell culture medium will be processed in each batch of isolation. Which adsorbent will you use? Justify your selection.

A4 Many lymphocytes (a kind of blood cells) are morphologically very similar but have very different surface antigens. An immunoaffinity column, using an antibody specific to a T-cell type, can be used to purify the T cell from the bloodstream. Two immunoaffinity columns were evaluated for their ability to remove T cells and enrich natural killer (NK) cells in the blood.
Figure P.14.1 shows the breakthrough curves for the columns tested.
a) Estimate the number of T cells bound to column A and column B.
b) Which column will you choose?

A5 Show that the midpoint of a breakthrough curve will appear at $t_m = \frac{l}{v}(1 + ((1 - \varepsilon)/\varepsilon)K)$ using the Lapidus–Amundson solution. What is its effect on the "capacity" (i.e., the amount of solute adsorbed on the column)?

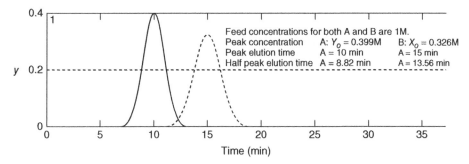

Figure P.14.1

a) Use the Lapidus–Amundson equation to simulate an inside-column view of a feed with $y=10$, $q=Ky$, and $K=1$, 10, 100.
b) Use the Lapidus–Amundson equation to simulate a breakthrough curve of $y=10$, $K=100$.

A6 A student investigated the suitability of a variety of materials for use in an implantation device. A criterion used in the evaluation is the adhesion of platelet cells to nonporous beads coated with these materials. He decided to load each type of beads into a glass column and then pass a suspension of platelets at 10^6 cells/mL through the column continuously. He also included a dye that is not adsorbed to the sample as a marker. The dimensions of the packed bed (porosity $\varepsilon = 0.28$) were 2 cm in diameter and 20 cm in height. Beads were 1 mm in diameter. The platelet suspension was fed at 10 mL/min. Shown in Table P.14.1 are the concentration profiles of the dye (which is the same in every experiment with the same flow rate) and the platelets in the testing of two materials.
a) Which material will you choose to use?
b) The student then repeated the experiment for Material 1 but with a flow rate of 3 mL/min. Predict when (in terms of time) the sample will have a dye concentration of 5 mg/L.

A7 A gear pump is used to deliver fluid to a chromatographic column 10.0 cm in diameter and 10.0 cm in height. The chromatographic medium used is solid, incompressible beads 100 μm (0.01 cm) in diameter. The separation of the product is deemed unsatisfactory, and the chief engineer decided to use beads of 5.0 cm instead. The same pump will need to be used to deliver the same pump head. To improve the separation, it is also decided that a column 5 cm in diameter will be used to allow for an increase in column height. How high should the new column be?

A8 An adsorption column was used for the separation of lysozyme. A macroporous adsorbent based on a styrene–divinylbenzene copolymer is loaded in a chromatographic column with an internal diameter of 0.3 cm and length of 10 cm. The breakthrough curve of lysozyme has a θ_s of 50 min and θ_σ of 1.5 min at a flow rate of 0.5 mL/min. The feed concentration is $C_F = 1$ mg/mL. Calculate the capacity of the column if all the adsorbent in the column is "saturated" when the loading is stopped. Consider the second case that the feed continues until the solute concentration in the effluent stream is half of the feed concentration. How much

Table P.14.1 Concentrations of the dye, and the platelets in the testing of two materials.

Sample	Dye (mg/L)	Material 1[a]	Material 2[a]
1	0	0	0
2	0	0	0
3	0	0	0
4	0	0	0
5	0	0	0
6	0	0	0
7	0.2	0	0
8	0.7	0	0
9	1.6	0.2	0
10	3.1	1.6	0
11	5	5	0
12	6.9	8.4	0
13	8.4	9.8	0
14	9.3	10	0
15	9.8	10	0
16	10	10	0.2
17	10	10	0.8
18	10	10	2.4
19	10	10	5
20	10	10	7.6
21	10	10	9.2
22	10	10	9.8
23	10	10	10
24	10	10	10
25	10	10	10

a) Platelets ($\times 10^5$/mL).

solute is "loaded" on the adsorbent in the entire column? Compare it to the first case and discuss the difference between the two types of solute loading by considering the relative value of the solute and the adsorbent, as well as the cost of the adsorbent.

A9 A 15 cm × 2 cm (height × diameter) antibody affinity column is used to purify urokinase. The interparticle voidage is 0.22; the intraparticle void volume is neglected. The flow rate of the feed, containing 1000 units/mL of urokinase, is 2 mL/min. The breakthrough curve is characterized with a two-parameter model as having a mean elution time of 220 min and a standard deviation of 10 min. Because the adsorbent is fairly expensive, the flow of feed is continued until the whole column is "saturated." Part of the feed stream after coming out of the column is re-fed into the column in the next cycle of the adsorption process.

a) Estimate the total capacity of the column.After washing, the adsorbed urokinase is eluted with acetate buffer at 2 mL/min. The eluted urokinase forms a Gaussian peak having a mean elution time and standard deviation of 24 min and 2 min, respectively.
b) Estimate the peak concentration of urokinase after elution.
c) You are to achieve 98% recovery. When should you collect the eluent?

A10 An affinity chromatographic column is used to isolate a peptide antibiotic from a feed solution containing this antibiotic at a concentration of 5 g/L. The adsorbent has a maximum capacity of 10 mg antibiotic/mL adsorbent. The void volume in the interior of the adsorbent can be neglected. A column of 2 cm inner diameter is packed with this adsorbent to 15 cm in height. The interparticle void fraction in the column is 0.2. The adsorption isotherm is strongly favorable, and equilibrium is reached relatively rapidly. The breakthrough curve can be described very well with a two-parameter model. In this particular case, the standard deviation is the square root of θ_s. Draw the breakthrough curve.

A11 A two-parameter model is used to characterize the loading of EPO to an antibody affinity chromatographic column. The solvent front is eluted at $t = 2$ min. The breakthrough curve (i.e., unbound EPO) has a $\theta_s = 50$ min and a standard deviation of 2 min. The antibody concentration in the feed is 0.5 g/L. After loading, the column is washed and then EPO is eluted by a low-pH buffer. Sketch the concentration profile of EPO in the elution (use the back of this sheet if necessary). State your assumptions.

A12 The breakthrough curve of an adsorption process is shown in Table P.14.2. Determine the mean and the standard deviation of the breakthrough curve.

Table P.14.2 Breakthrough curve of an adsorption process.

t(min)	y
<30	0
30	0.009
31	0.052
32	0.181
33	0.411
33.3	0.500
34	0.671
35	0.862
36	0.957
37	0.99
38	0.998
39	1
40	1
>41	1

A13 Crocodiles can submerge in water for a long time. Their hemoglobulin has a different oxygen adsorption isotherm that is very responsive to the CO_2 concentration. As the CO_2 concentration increases (after they submerge in water for a long time), more O_2 is released than that in humans. Sketch the oxygen isotherm of crocodile hemoglobulin at low and high CO_2 concentrations.

B. Elution Chromatography

B1 Gel permeation chromatography is used to separate molecules that are very different in molecular weight. Using an adsorbent with very small pores, large proteins are excluded and flow through the column (i.e., $K=0$), while the salts (like NaCl) get retained much longer (K is very large). This process is referred to as desalting. It used to be used after protein was eluted from an adsorption column to remove salts used for elution. However, this process has practically disappeared from industrial processing. Why? What is the replacement of gel permeation chromatography for desalting?

B2 Determine the number of stages for peak A of the following chromatogram (Figure P.14.2) from gel permeation chromatography.
  a) A column three times longer is to be used. How far apart will peaks A and B be separated? Answer in a number of minutes. What will the peak concentration of A and B be? State all assumptions.

B3 In elution chromatography, what happens to the height of the peak of the eluted solute when the length of the column is made twice as long while keeping all the other operating variables constant?

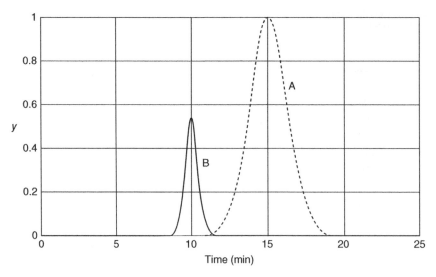

Figure P.14.2

B4 Two solutes are eluted at 15 and 17 min, respectively, in gel permeation chromatography. If the column is made twice as long while the operating conditions remain unchanged, when will they be eluted?

B5 In an elution chromatographic run, the peak of solute is eluted from a 40 cm long column at 200 min. The solute concentration profile on the chromatogram is perfectly symmetrical, and the half heights (i.e., the position at which the concentration is 50% of the maximum) occur at 190 and 210 min. The same solute is now eluted in a 1 m long column. The linear flow rate remains unchanged.
a) What is the total number of theoretical plates in the first column?
b) When will the peak and half height appear in the second column?

B6 A gear pump is used to deliver fluid to a chromatographic column (10.0 cm in diameter and 10.0 cm in height). The chromatographic medium used is solid, incompressible beads (100 μm in diameter). The separation of the product is deemed unsatisfactory, and the chief engineer decided to use beads of 50 μm instead. To improve the separation, it is also decided that a column 5 cm in diameter will be used to allow for an increase in column height. The same volumetric flow rate will be used. How much of an increase in stage number can be attained?

B7 1 mL of a solution containing 2 g/L of polysaccharide A and 1 g/L of polysaccharide B is fed into a gel permeation chromatography column that is 2 cm in diameter and 50 cm in height. It is eluted with a buffer at a flow rate of 5 mL/min. The front of the elution buffer, which is marked by a nonadsorbing dye, is eluted at 6 min after the beginning of elution. The center of peak (maximum) A appears at 60 min, while the points at which the concentration is half of the maximum peak height appear at 59 and 61 min, respectively. When A is eluted, the maximum of peak B in the column has a concentration of 0.0002 g/L. The adsorption isotherm is linear. Assume that the concentration profile of solutes can be adequately described by the discrete stage model.
a) What is the concentration of A at its maximum in the eluted chromatogram?
b) What is the number of plates (or stages) for A?
c) How many minutes are the maxima of peak A and peak B separated from each other in the chromatogram?
d) You are asked to reduce the separation of the maxima of the two peaks to 50% of its current value by changing the column height. While using the same flow rate, how can you do it?

B8 An adsorbent has a relationship between HETP, flow velocity (v), and the diameter (d_p) of adsorbent, as shown hee:
$$\text{HETP} \propto v\sqrt{d_p}$$
Can you use the same column (i.e., the same column length and diameter) to decrease the peak elution time by half while maintaining the same feed?

B9 In gel permeation chromatography, the product peak is eluted at 100 min, and the standard deviation of the peak as a Gaussian profile is 1 min. The target recovery yield is 98%. In what time interval should the product be collected from the eluent?

B10 A solute is purified using a 1.0 cm × 30 cm (diameter × height) gel permeation chromatographic column packed with 90 μm diameter absorbent. It is eluted at a linear velocity of 2 cm/min. The solute peak shows at 150 min on the chromatogram. The number of theoretical plates is calculated to be 1000. The process is to be scaled up to handle 90 times of the loading (feed) volume by using a 4.5 cm diameter column. Furthermore, it was determined that the number of theoretical plates can be reduced to 600 and still give a satisfactory resolution with other solutes.

Assume that the same linear velocity is to be used. You can also assume that discrete stage analysis can be used to calculate the number of theoretical plates and to characterize the shape of peak. How would you scale up the column? Specify the height of the scale-up column, and sketch the chromatogram of the scaled-up column.

B11 Elution chromatography is used to separate two proteins. The elution profile can be described by the discrete stage model. The means of the two peaks, A and B, are 96 min and 112 min, respectively. The standard deviation of peak A is 4.8 min. The shoulders of the two peaks overlap significantly.

The edge of mean ± one standard deviation of the two peaks are to be kept 10 min apart from each other. You are to keep the operative conditions the same except for using a longer column to accomplish that.

a) Calculate the number of theoretical equivalents of plates in the original column.

b) How much longer should the column be, as compared to the first column?

c) What will be the new number of theoretical plates and peak elution time for A?

Note: The operating conditions (elution flow rate, sample loading volume, and column diameter) are to be kept the same. So the number of theoretical plates is proportional to the length of the column.

B12 Gel permeation chromatography is used to separate two solutes, A and B. Using a 4 × 30 cm column and a superficial velocity of 1.0 cm/min, the peak of A is eluted at 400 min. The half peaks (at which the concentration of A is half of its peak) are at 390 and 410 minutes, respectively.

a) Estimate the number of theoretical plates and HETP.

b) Under these operating conditions, the resolution between A and B is less than satisfactory. It is proposed to decrease the elution linear velocity to 0.4 cm/min. The relationship between HETP and linear velocity is: $\frac{HETP_1}{HETP_2} = \frac{0.1+v_1^{0.5}}{0.1+v_2^{0.5}}$.

Estimate the time at which the peak and half-peak heights appear on the chromatogram.

B13   Solutes A and B are satisfactorily separated in a laboratory-scale column of dimensions 2 cm in diameter and 30 cm in height. The diameter of the adsorbent is 30 μm. In scaling up, the linear velocity of the liquid flow rate is to be kept the same, while the diameter of the column is to be increased to 20 cm and the adsorbent diameter increased to 120 μm. The HETP is approximately proportional to the second power of the diameter of the adsorbent:

$$\text{HEPT } \alpha d_p{}^2$$

What is the new column height needed for achieving the same separation as the small-scale column? How many fold increase in the volume of the feed is permissible?

B14   A separation column of 250 mm long, 4.6 mm bore, packed with an adsorbent with a diameter of 7 μm, is used to separate benzene, naphthalene, and diphenyl. The separation conditions are as follows:

$$\text{Mobile phase}: \text{ CH}_3\text{OH/H}_2\text{O, 8:2, v/v, 1 mL/min, 170 bar,}$$

$$\text{room temperature}$$

$$\text{Sample}: \quad 3 \text{ μL in CH}_3\text{OH}$$

$$1.\ 6 \times 10^{-5}\text{g benzene}$$

$$2.\ 6 \times 10^{-6}\text{g naphthalene}$$

$$3.\ 6 \times 10^{-6}\text{g diphenyl}$$

The total retention time (t_o; the time at which the maximum of the peak leaves the column) and the peak width at half height ($W_{0.5}$) are shown in seconds in Table 14.3.
a) Calculate the theoretical plate number and theoretical plate height.
b) Fill in x and y in Table 14.3.

B15   Reverse-phase chromatography was used to separate two peptides. The elution chromatogram along with its peak elution times and other data are shown in Figure P.14.3. To separate A from B further, it is decided to make the column twice as long while keeping other operating conditions (linear flow rate, adsorbent) constant.
a) Predict the new peak elution time and standard deviation for A and B.
b) What will be the new peak concentrations of A and B in the eluted product?
c) 90% of B eluted will be collected and combined. How much more diluted is it in the product from the longer column than from the shorter column?

Table P.14.3 Total retention time and the peak width at half height.

Compound	1	2	3
t_o	316	530	y
$W_{0.5}$	8.5	x	19.1

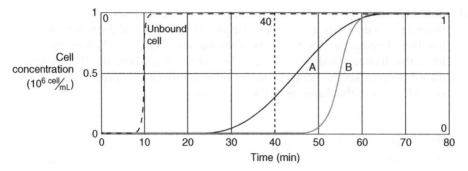

Figure P.14.3

B16 In an elution chromatographic process, the size of adsorbent varies with the elution solution. In an original process, a product A is satisfactorily purified from the impurities. The size of the adsorbent beads is 100 μm. In order to stabilize the product, more salts are added to the elution solution, which changes the adsorbent bead size to 70 μm. You can assume that the porosity of the bed remains unchanged at 0.2. Of course, this means that the height of the column bed will be shorter under the new elution conditions. The addition of salts does not change the adsorption isotherm but changes the mass transfer characteristics because of size change of the adsorbent. Under these operating conditions, its effect is described as:

$$HETP \propto vd^{1.2}$$

where d is the diameter of adsorbent; and v is the linear velocity (flow rate divided by bed cross-sectional area) of elution.
a) If the same pressure drop across the bed, ΔP, is used, how will the linear velocity and total number of stages (or plates) (N) be affected?
b) Propose a set of operating conditions that keep N the same as the original operating conditions.

B17 You are given a fixed volume of adsorbent to perform a separation process and are asked to evaluate two alternatives.
a) Use a shorter and wider column, with a for column height l and cross-sectional A area and $HETP1$.
b) A longer but narrower column of $4\,l \times \frac{1}{4}\,A$.
   1. Compare the number of stages you can accomplish with each column if you keep the operating linear flow rate the same. What happens to the pressure drop($\Delta P/\Delta L$)? Assume that your sample loading volume is very small and has no effect on the separation performance.
   2. Sketch the elution profile of one solute in both cases (plot concentration vs. time), and mark the ratio of the time quantitatively.
   3. There are two ways to decide how to collect the eluent from the column: collect a fraction of eluent at a fixed time interval (from t_a to t_b), or collect

after a fixed column is eluted (from v_a to v_b) (basically, you turn on the collection device at t_a or v_a):

3.1 If you choose to use t, should t_a-t_b be the same in the two cases?

3.2 How about if you choose to use volume?

B18 Monodispersed agarose beads with a diameter of 50 μm are packed into a 5 cm diameter column to a bed height of 15 cm to purify a lipoprotein. The voidage of the bed is 0.23, and intraparticle porosity can be neglected. After adding a 10 μL feed, the column is eluted with elution buffer at a flow rate of 3 mL/min. The peak of lipoprotein at a concentration of 0.02 mg/mL appears at 90 min. At 95 and 85 min, the concentration of lipoprotein in the chromatogram is half of the peak.

a) What is the number of theoretical stages in this column?

b) In a refined process, 30 μm diameter beads will be used. In general, HETP or the height of each stage can be assumed to be proportional to the diameter to the second power of chromatographic beads used. Assume the adsorption isotherm and voidage are not affected by the diameter of the chromatographic beads. Predict the peak concentration and the times that half-peak concentration will appear if the same volumetric flow rate is used.

c) It turns out that only a single-speed pump is available. This column with 30 μm beads will thus be eluted at the same pressure drop across the bed as in (a). At this range of elution solution velocity, the height of each stage is proportional to the liquid velocity. You can assume that the flow rate is the laminar-flow regime. What is the new predicted elution time and peak concentration?

C. Advanced Problems

C1 The breakthrough curve of a 2 cm × 15 cm $(d \times l)$ affinity (adsorption) chromatography column is characterized with a two-parameter model, and it has a θ_s of 100 min and a standard deviation of 5 min. The linear velocity used is 2.0 cm/min. After the column is fully loaded and washed, the solute is eluted with a low-pH buffer at the same linear velocity. A total of 40 g solute is eluted with a θ_s of 5 min and a standard deviation of 0.5 min.

Now, in scaling up, a column of 10 cm × 45 cm will be used. The capacity of the column should be used to its maximum. Sketch the new breakthrough curve and the elution solute profile as quantitatively as you can. State all the assumptions. The operating linear velocity remains unchanged.

Note: There are many different ways to estimate the standard deviation in scaling up. From the Amundson equation. you can see the important parameters and see which one will affect standard deviation when scaling up.

C2 You are asked to scale up a chromatographic process. In the lab scale, adsorbent beads of 50 μm diameter are used. The breakthrough curve has a mean time of 80 min and a standard deviation of 3 min (as evaluated with a two-parameter model). In the production scale, the bed height will be increased fivefold.

However, due to the limitation of the pump used, the pressure drop across the bed can be increased only twofold. Nevertheless, the linear velocity of the liquid flow through the bed must be maintained. It is proposed to increase the diameter of the adsorbent beads while keeping the voidage of the bed constant.
a) What should the new adsorbent bead diameter be?
b) What will the new mean time of the breakthrough curve be?

C3 Cross-linked dextran beads are often used in bioseparations as an adsorbent. The beads are usually packed in a column, and the feed is added from the top of the bed. In one such separation, beads of 80 μm diameter are loaded in a column with a diameter of 6.0 cm to a bed height of 30 cm. The feed containing erythropoietin (EPO) in 0.05 M NaCl is passed through the column at a rate of 10 mL/min. The beads in the column can be considered to be closest-packed identical spheres with a porosity of 0.27. After the adsorption stage, the column is washed with 0.05 M NaCl, followed by elution with 1.0 M NaCl. However, upon exposure to 1.0 M NaCl, the diameter of the beads shrinks to 40 μm. Due to the fluid flow, the beads rearrange to pack the column, and they do so in a fashion that can be considered to be closest-packed. The viscosity of 0.05 M and 1.0 M NaCl are virtually the same. What will be the new flow rate of the 1.0 M NaCl solution after the beads are repacked and settled in a new "stable" height?
Note: The pressure drop across the bed does not change.

C4 One way to make the adsorption a continuous process is to employ a moving bed adsorber. In such a system, the adsorbent beads and the feed solution are moved countercurrently from stage to stage. A moving bed adsorber is used to remove pentachlorobenzene (PCB) from a waste stream containing 1000 g/m³ of PCB. The polluting PCB must be reduced to 5 g/m³ before discharge. The isotherm of PCB adsorption with silica is linear and can be described as:

$$q[\text{g}/1000 \text{ kg adsorbent}] = 4[\text{m}^3/1000 \text{ kg adsorbent}] \, y \, [\text{g/m}^3]$$

The flow rate of the stream to be treated is 10 m³/h, and that of the adsorbent is 10,000 kg/h. The bulk density of the adsorbent is 2000 kg/m³, and the intraparticle porosity is 0.7. After the adsorption process, the adsorbent is regenerated to be free of PCB. After each stage, the fluid and adsorbent are separated and transferred to the respective next stage.
a) If the fluid and adsorbent can be completely separated after reaching the equilibrium in every stage, how many stages are needed? You can assume that completely dried, regenerated, adsorbent beads are used.
b) However, in practice, only bulk fluid can be completely separated; the fluid inside the adsorbent filling the void space cannot be removed and is transferred with adsorbent to the next stage. You can assume that equilibrium is achieved in every stage. To simplify the analysis, you can assume that the intraparticle void volume in the regenerated beads is filled with a solution similar to the feed but without PCB. How many stages are needed?
c) Compare the answers to (a) and (b). Which requires more stages? Why?

Table P.14.4 Concentration profile.

Time (min)	Ca^{+2} concentration (ppm)	Cd^{+2} concentration (ppm)
0	<0.1	<0.1
5–30	<0.1	<0.1
35	0.15	<0.1
40	0.62	<0.1
45	4.65	<0.1
50	15	<0.1
55	25.4	<0.1
60	29.3	<0.1
65	29.9	<0.1
70	30±0.2	<0.1
75–300	30±0.2	0.14
315	30±0.2	0.53
330	30±0.2	4.37
345	30±0.2	14.9
360	30±0.2	29.4
375	30±0.2	29.8
390	30±0.2	30±0.3
405–500 (termination)	30±0.2	30±0.3

C5 A recombinant microorganism has been developed to adsorb heavy metal ion Cd^{+2} reversibly. Under oxygenated conditions, these microorganisms adsorb Cd^{+2} from a very dilute stream (as dilute as 0.01 ppm) and deposit the metal as a protein–ion complex in periplasmic space. Under oxygen-depleted conditions, Cd^{+2} is released. Both adsorption and desorption are very rapid. The organism is now to be immobilized on the external surface of solid spherical beads and used as adsorbent in a packed-bed reactor to treat polluted water. A packed-bed reactor (or column) in the pilot plant is 30 cm in diameter and 200 cm in height. The interparticle void fraction is 0.4. The water to be treated contains 30 ppm (mg/L) of Cd^{+2} and is fed from the top of the column. It also contains 30 ppm of Ca^{+2}, which is not adsorbed. As soon as water is fed into the column, samples are taken periodically from the effluent stream for Ca^{+2} and Cd^{+2} measurement. The flow rate is 20 cm/min. The concentration profile is shown in Table P.14.4.

a) What is the capacity of the column? (*Note*: You should not allow the effluent stream to have a Cd^{+2} concentration higher than 0.5 ppm.)

b) The adsorbed Cd^{+2} is then to be released by elution with an oxygen-depleted water stream (which also contains a minute amount of sodium sulfite to ensure oxygen depletion). The properties and the flow rate of the elution stream can be considered to be identical to the feeding stream. Assuming the solubility of Cd^{+2} in the elution stream is very high, draw a concentration profile of Cd^{+2} in the elution process. State your assumptions.

C6 Gelatin–sepharose (GS) is used for affinity chromatography to purify fibronectin (FN) from plasma. The plasma contains 400 mg/L of fibronectin. The adsorption isotherm is:

$$q(mg\ FN/mL\ GS) = \frac{1\ (mg\ FN/mL\ GS)\ y}{0.01 + y}$$

where y is the FN concentration in solution (mg/L). An adsorption column is 20 cm long and 3 cm in diameter. The interstitial (interparticle) voidage is 0.2 in the packed adsorption column, and the (internal) voidage in the GS beads is 0.90. Plasma is relatively inexpensive compared to FN and GS. We would like to use up the maximum capacity of GS in each run. The GS is very well made, so that you can assume the mass transfer resistance in GS beads or externally is negligible

a) How much plasma will you pass through the column?

b) What is the percentage of the FN from plasma that is retained in the column after washing with buffer? Assume that during washing, there is no loss of FN from the adsorbent.

c) The bound FN will be eluted with urea solution. The isotherm in urea solution is:

$$q = \frac{0.1e^{0.15x}y}{10e^{0.15x} + y}$$

where X is the urea concentration in M (moles/L). FN will be eluted with 4 M urea at a flow rate of 20 mL/min. Sketch the FN concentration on the chromatogram. Describe your reasoning. Be as quantitative as you can.

d) After elution, the column will be washed with a higher concentration of urea to ensure that 99% of binding sites are regenerated and available for the next cycle of purification. What is the urea concentration for the regeneration of the adsorption column?

C7 Gel permeation chromatography is used to separate a mixture of two proteins. Using a 15 cm × 2 cm (height × diameter) column, the peak maxima of A and B appear at 60 and 65 min, respectively. The half peak (50% of peak maximum) of A appears at 58 and 62 min. To improve the resolution, all the adsorbent is taken from the original column and loaded into another column with a diameter of 1.0 cm. The elution is then repeated using the same pressure drop across the column height (i.e., ΔP is kept constant). The sample loading volume can be assumed to be very small and well loaded, so it does not affect discrete stage calculation.

a) Applying the discrete stage analysis, what is the half-peak elution time for protein B in the original column?

b) Predict the elution times for peak A and B, as well as for the half peaks, in the new column.

c) Calculate the peak elution times for A and B, if the pressure drop across the bed in the two columns is proportional to the column height. What will be the half-peak time? Use discrete stage analysis.

d) In reality, the process is mass transfer controlled (i.e., the intra-adsorbent-bead diffusion is limiting). Will it make any difference in separating A and B whether one keeps ΔP or $\Delta P/l$ (where l is adsorbent bed height) constant?

C8 Gel chromatography is used to separate the surface antigen of a virus; most larger contaminating molecules flow through with the front of the elution solution. The column used in the laboratory is 2 cm in diameter and 10 cm in height. Assuming the adsorbent has a uniform diameter of 15 μm, the linear velocity during the elution is 10 cm/min (defined as liquid flow rate/cross-sectional area). The process has to be scaled up 1000-fold with a single column. However, there are two constraints: (1) the diameter of the column cannot be greater than 20 cm, and (2) the pressure drop across the bed cannot be greater than 3 times the Δp in the lab column. How would you achieve this? Draw a chromatogram (with an elution time t_1 and standard deviation σ_1) for the laboratory-scale column; then draw your predicted chromatogram according to your scale-up specifications. State all of your assumptions and reasoning.

Note: **There is no unique solution to this problem. There may not even be a best solution to this problem. It is probably a judgment call.**

C9 A process analogous to displacement chromatography is used for an enzyme reaction requiring a cofactor. The reactions are:

$$E_1 + A \rightarrow E_1 A$$
$$E_1 A + NADH \rightarrow E_1 + B + NAD$$
$$E_2 + C \rightarrow E_2 C$$
$$E_2 C + NAD \rightarrow E_2 + D + NADH$$

In these equations, E_1 and E_2 are the two enzymes co-immobilized on the adsorbent. A is the reactant; B is the product. The conversion of A to B requires the cofactor NADH, which is expensive and has to be regenerated for reuse. C is used to convert NAD back to NADH through the reaction catalyzed by E_2; a byproduct of this is D, which can be easily separated from B.

Traditional methods use a single enzyme packed-bed reactor in which only E_1 is immobilized and the feed contains both A and NADH. NAD is then recovered from the eluent and fed along with C into the second column, in which E_2 is immobilized for regeneration.

a) Explain how this new method works.
b) Sketch qualitatively a "inside-column view" and a breakthrough curve for A, B, C, D, NADH, and NAD. Mark also the front of the feed stream if the flow is plug flow.

Index

Page numbers in *italics* or **bold** refer to figures and tables respectively.

Engineering Principles in Biotechnology, First Edition. Wei-Shou Hu.
© 2018 John Wiley & Sons Ltd. Published 2018 by John Wiley & Sons Ltd.
Companion Website: www.wiley.com/go/hu/engineering_fundamentals_of_biotechnology

Printed and bound by CPI Group (UK) Ltd, Croydon, CR0 4YY

16/04/2025

14658558-0004